International Conference on Neutron Scattering 2009

AA008882

Journal of Physics: Conference Series Volume 251

Knoxville, Tennessee, USA
3-7 May 2009

ISBN: 978-1-61782-243-8
ISSN: 1742-6588

Printed from e-media with permission by:

Curran Associates, Inc.
57 Morehouse Lane
Red Hook, NY 12571

CURRAN ASSOCIATES INC. proceedings.com

Printed by Curran Associates, Inc. (2011)

Institute of Physics
Dirac House, Temple Back
Bristol BS1 6BE UK

Phone: 44 1 17 929 7481
Fax: 44 1 17 920 0979

techtracking@iop.org

Additional copies of this publication are available from:

Curran Associates, Inc.
57 Morehouse Lane
Red Hook, NY 12571 USA
Phone: 845-758-0400
Fax: 845-758-2634
Email: curran@proceedings.com
Web: www.proceedings.com

TABLE OF CONTENTS

CONDENSED MATTER SCIENCE

SOFT MATERIALS

ENGINEERING MATERIALS

NEUTRON FACILITIES

INSTRUMENT DESIGNS AND CONCEPTS

INSTRUMENTAL TECHNIQUES

INSTRUMENTAL DEVICES

SAMPLE ENVIRONMENT

SOFTWARE

Suppression of hidden order in URu_2Si_2 under pressure and restoration in magnetic field

E Hassinger, D Aoki, F Bourdarot, G Knebel, V Taufour, S Raymond, A Villaume and J Flouquet

CEA Grenoble, 17 rue de Martyrs, 38000 Grenoble, France

E-mail: elena.hassinger@cea.fr

Abstract. We describe here recent inelastic neutron scattering experiments on the heavy fermion compound URu_2Si_2 realized in order to clarify the nature of the hidden order (HO) phase which occurs below $T_0 = 17.5$ K at ambient pressure. The choice was to measure at a given pressure P where the system will go, by lowering the temperature, successively from paramagnetic (PM) to HO and then to antiferromagnetic phase (AF). Furthermore, in order to verify the selection of the pressure, a macroscopic detection of the phase transitions was also achieved in situ via its thermal expansion response detected by a strain gauge glued on the crystal. Just above $P_x = 0.5$ GPa, where the ground state switches from HO to AF, the $Q_0 = (1,0,0)$ excitation disappears while the excitation at the incommensurate wavevector $Q_1 = (1.4,0,0)$ remains. Thus, the $Q_0 = (1,0,0)$ excitation is intrinsic only in the HO phase. This result is reinforced by studies where now pressure and magnetic field H can be used as tuning variable. Above P_x, the AF phase at low temperature is destroyed by a magnetic field larger than H_{AF} (collapse of the AF $Q_0 = (1,0,0)$ Bragg reflection). The field reentrance of the HO phase is demonstrated by the reappearance of its characteristic $Q_0 = (1,0,0)$ excitation. The recovery of a PM phase will only be achieved far above H_{AF} at $H_M \approx 35$ T. To determine the P-H-T phase diagram of URu_2Si_2, macroscopic measurements of the thermal expansion were realized with a strain gauge. The reentrant magnetic field increases strongly with pressure. Finally, to investigate the interplay between superconductivity (SC) and spin dynamics, new inelastic neutron scattering experiments are reported down to 0.4 K, far below the superconducting critical temperature $T_{SC} \approx 1.3$ K as measured on our crystal by diamagnetic shielding.

1. Introduction

URu_2Si_2 is a heavy fermion compound with a body centered tetragonal crystal structure in its paramagnetic phase. At ambient pressure, on lowering the temperature below $T_0 = 17.5$ K a transition occurs into a hidden order phase which cannot be a conventional magnetic phase. Over the last decade, the picture was that of small moment antiferromagnetism (as all neutron scattering experiments point out a tiny ordered moment $M_0 \approx 0.02\mu_B$ with the propagation vector $Q_{AF} = (0,0,1)$ [1]) associated with an electronic instability corresponding to a loss of electronic carriers at T_0 [2–4]. The departure from the conventional AF picture is clearly demonstrated by the fact that the entropy loss at T_0, $\Delta S \approx 0.2R\ln(2)$, is too high to be accounted for exclusively by the formation of localized tiny ordered moments. NMR and μSR spectroscopy rule out an intrinsic origin of the tiny sublattice magnetization [5; 6]; the small moment seems generated by lattice defaults. This high sensitivity to lattice imperfections is clearly related with the observation that at a rather weak pressure $P_x \approx 0.5\,\mathrm{GPa}$ [7; 8] the ground state switches from HO to AF ground state with a substantional sublattice magnetization $M_0 = 0.4\mu_B$ at $T \to 0$ with the same Q_{AF}. Here we tune from HO to AF under pressure to clarify the main properties of each phase and in particular noting their inelastic neutron scattering spectra.

At the opposite to the elastic AF neutron scattering response at ambient pressure, the inelastic neutron scattering signals are very robust [1]: excellent agreement exists between results. Below T_0, inelastic neutron scattering experiments indicate two sharp and intense excitations, one with a symmetric shape in the energy spectrum at the incommensurate position $Q_1 = (1.4, 0, 0)$ with a gap $E_1 \approx 4.8$ meV and another one with an asymmetric shape at the wavevector $Q_0 = (1, 0, 0)$ with a gap $E_0 \approx 2\,\mathrm{meV}$. Note that the wavevector $Q_0 = (1, 0, 0)$ is in another Brillouin zone equivalent to the ordering wavevector $Q_{AF} = (0, 0, 1)$. On warming above T_0, both excitations respond differently. The signal at Q_0 is quasielastic and strongly damped, whereas at Q_1 it remains inelastic but so strongly damped that the spectrum is not fully gapped any more. The T-evolution of the excitation at Q_1 and equivalent positions can explain the specific heat anomaly at T_0 [9]. It is worthwhile to recognize that if no phase transition occured at T_0, URu_2Si_2 would be a classical intermediate valence compound; it is the feedback between spin dynamics and band structure reconstruction which leads to the fact that below T_0 sharp excitations at Q_0 and Q_1 emerge [10].

2. Results and discussion

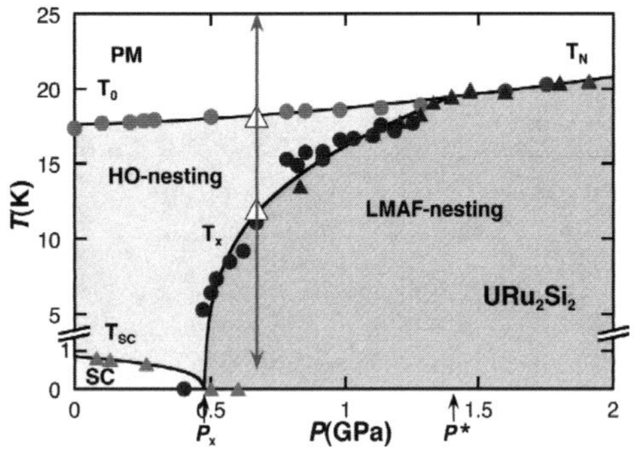

Figure 1. Pressure-temperature phase diagram of URu_2Si_2 determined by specific heat and resistivity measurements [11]. The green vertical line indicates the pressure where the excitations were measured.

Figure 1 shows the (T,P) phase diagram of URu_2Si_2 as recently determined by resistivity and microcalorimetry experiments [11]. Special attention was given in the determination of the three lines: $T_0(P)$ of the transition from PM to HO, $T_x(P)$ from HO to AF and $T_N(P)$ from PM to AF. In agreement with thermal expansion and elastic neutron scattering experiments [7; 8; 13], $P_x \approx 0.5\,\mathrm{GPa}$ at $T \to 0$, while the three lines seem to meet at the critical pressure $P_c \approx 1.3\,\mathrm{GPa}$ with $T_c \approx 18.5\,\mathrm{K}$. The P-evolution of the two excitations were reported nine years ago [14]; four different pressures were measured on crystals where P_x was reported near 1.5 GPa. Both excitations exist at 0.87 GPa but disappear simultaneously at 1.86 GPa. Taking into account progress in the determination of the (P,T) phase diagram and also our recent ability to measure simultaneously neutron scattering and thermal expansion via a strain gauge in the same pressure cell, we decided to study at a selected pressure P between P_x and P_c (vertical green line in figure 1) the evolution of the inelastic neutron scattering spectrum in the three phases PM, HO and AF for both Q_0 and Q_1 using the high performance of the respective triple axis spectrometers IN12 and IN22 at the Institut Laue Langevin (ILL) [12]. The figure 2 shows the results for the two wavevectors. Clearly, the collective excitation at $Q_0 = (1, 0, 0)$ is characteristic of the HO phase as it disappears below $T_x \approx 13\,\mathrm{K}$ while the signal at $Q_1 = (1.4, 0, 0)$ remains but shifts in energy.

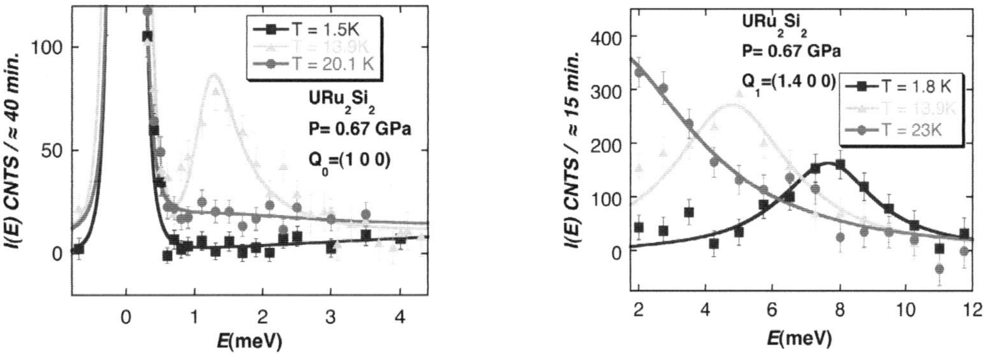

Figure 2. On the left panel are shown the energy spectra at the wave vector $Q_0 = (1, 0, 0)$. The excitation present at 13.9 K in the HO state (green triangles) disappears completely at low temperature in the AF phase (blue squares). On the right panel the excitation at $Q_1 = (1.4, 0, 0)$. The lines are guides to the eye.[12]

Considering the fact that the $Q_0 = (1, 0, 0)$ excitation is characteristic of strong longitudinal fluctuations at the wavevector Q_{AF} and that no drastic difference is detected in the carrier number between both sides of P_x [11], the proposal is that the Fermi surface reconstruction at T_0 and T_N are identical. This statement is supported by our recent study of Shubnikov-de Haas oscillations through P_x [Hassinger *et al.* to be published], where it was observed that the three observed frequencies (α, β, γ) are P invariant. In the AF phase the two uranium sites are inequivalent and the symmetry changes. In excellent agreement with band structure calculations [15–17], a drop of carrier density is associated with this change. In the HO phase, our measurements strongly suggest that Q_{AF} will also be the wavevector of the HO state. Recently it was stressed out from band structure calculations, that the strong longitudinal fluctuations at Q_{AF} in the HO phase preserve a Fermi surface reconstruction similar to the AF state [15].

A recent complementary experiment was to continue the inelastic neutron scattering studies with the addition of a magnetic field [18]. The experiments were performed on the IN14 spectrometer at the ILL and restricted to the field evolution of the Q_0 excitations. The reduction

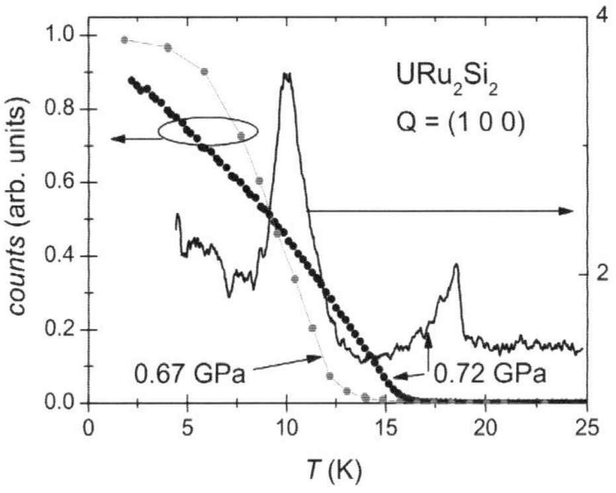

Figure 3. Scattering intensity at the magnetic Bragg peak position as a function of temperature at a fixed pressure of 0.72 GPa in zero field (black dots, left scale) compared to the same signal in reference [12](red dots + line) at 0.67 GPa. Additionally the temperature derivative of the strain gauge resistance (black line, right scale) is shown, which is proportional to the thermal expansion coefficient α_c and our indicator for phase transitions. [18]

of signal by the combined use of a pressure cell in a cryomagnet device (high magnetic field up to 14.5 T) led us to prepare a large crystal of 6 mm diameter and 11 mm length oriented with

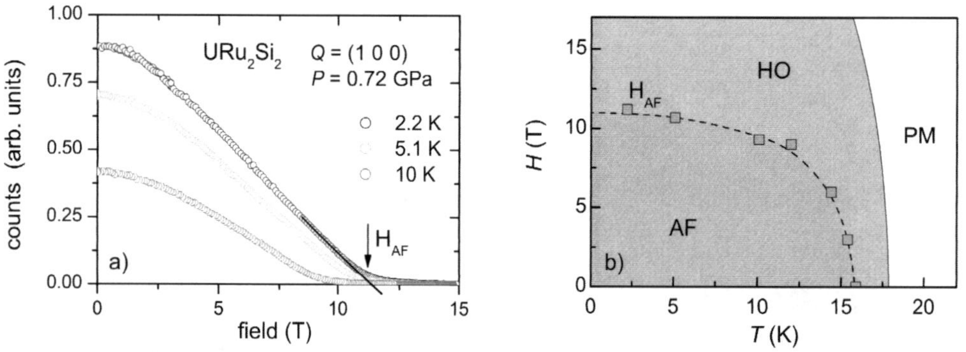

Figure 4. a) Field dependence of the intensity of the magnetic Bragg peak at $Q_0 = (1, 0, 0)$ at three temperatures at $P = 0.72$ GPa. Above H_{AF}, the AF phase is completely destroyed. b) The phase diagram obtained from the curves on the left panel and from temperature sweeps at constant fields. The transition line between HO and PM follows the behavior measured by thermal expansion at 0.75 GPa interpolated with high field data at 0.8 GPa [12; 18; 19].

the easy c magnetization axis along the magnetic field H; the selected transmitted medium was a mixture of fluorinert FC 84/FC 87. As before, the selected pressure $P \approx 0.72$ GPa was verified via the detection of the phase transitions PM-HO-AF via the thermal expansion measured by strain gauge. Figure 3 shows the temperature dependence of the magnetic Bragg peak intensity (measurement of the intensity at the $Q_0 = (1, 0, 0)$ position) at zero field at 0.72 GPa (black dots). By comparison, the signal obtained with the previous pressure experiments is drawn as well as the relative variation of the thermal expansion coefficient. It is obvious that pressure inhomogeneities exist in the present set up and that furthermore the deviation from hydrostatics is larger than the previous set up. From the H and T evolution of the Bragg reflection, we can estimate that 10% of "parasitic HO" phase remains even at $T = 0$ and $P = 0.72$ GPa while ideal conditions will lead to the complete disappearence of the HO component as in the previous measurement (see figure 2, left panel). However, following the field evolution of the intensity of the Bragg reflection at different temperatures (figure 4a), a complete disappearence of AF occurs above $H_{AF}(T = 0) = 11$ T. The figure 4b represents the temperature evolution of the field $H_{AF}(T)$ where AF disappears.

Inelastic neutron scattering experiments lead to the interesting result that a nice excitation reemerges above H_{AF} for $Q_0 = (1, 0, 0)$ (shown in figure 5), the weak contribution detected at $H = 0$ being a consequence of the 10 % parasitic HO survivance. Taking into account our previous P study, our proposal is that the disappearance of AF is associated with the field reentrance of the HO phase. Increasing further the magnetic field, the HO will be destroyed above $H_M \approx 35$ T (the field to suppress the HO at zero pressure is $H_M \approx 35$ T and increases with pressure [19]). In this high field domain, the original higher carrier number of the PM phase will be recovered and different phases occur[20].

Neutron scattering experiments are crucial to point out the nature of the (P, H) induced phases, however it is very arduous to realize a systematic pressure study. This was achieved by detecting the (P,H) evolution of the transition temperatures by thermal expansion via a strain gauge as represented in figure 6.

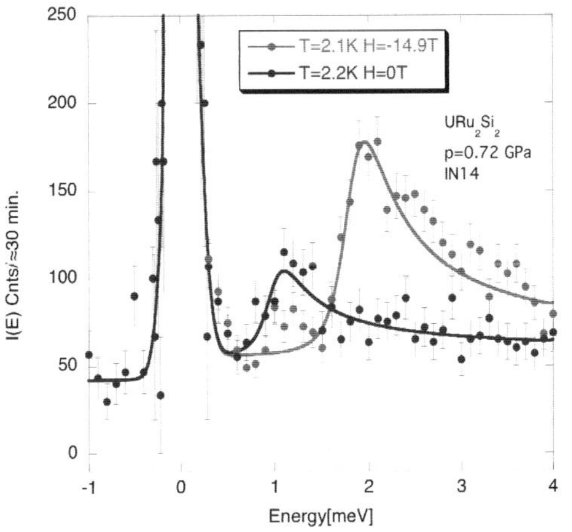

Figure 5. Energy spectra at the wave vector $Q_0 = (1, 0, 0)$ without magnetic field (blue dots) and in 14.9 T (red dots) at 0.72 GPa. The lines are guides to the eye.[18]

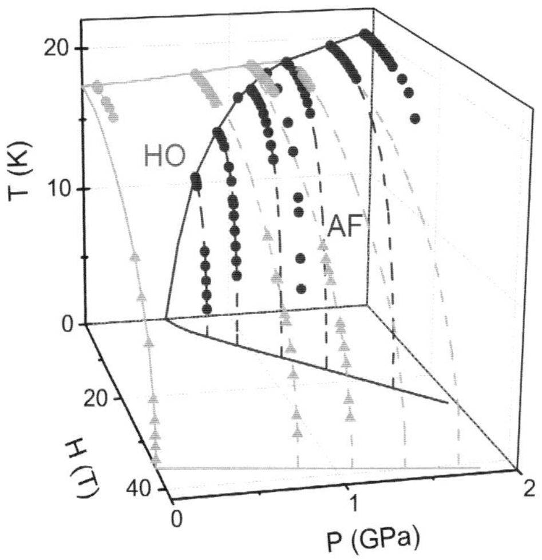

Figure 6. Three dimensional phase diagram from thermal expansion measurements [18]. The high field data (green triangles) is from reference [19]. The lines are guides to the eye.

At P_x bulk superconductivity collapses as well as the excitation at $Q_0 = (1, 0, 0)$. Surprisingly, no inelastic neutron scattering experiment at this wavevector has been reported in the SC phase. That led us to conduct an accurate inelastic scattering experiment with a high quality crystal ($T_{SC} \approx 1.2$ K from the measurement of its diamagnetic shielding) down to 0.4 K on IN12 spectrometer at ILL [21]. The temperature dependence of the inelastic spectrum for both Q_0 and Q_1 excitations were recorded. Figure 7 shows the intensity of the inelastic response at Q_0 for $T = 2$ K ($> T_{SC}$) and $T = 0.4$ K ($< T_{SC}$). At 2 K, no quasielastic contribution can be detected at low energy; the same lack of quasielastic component was observed for the energy response at

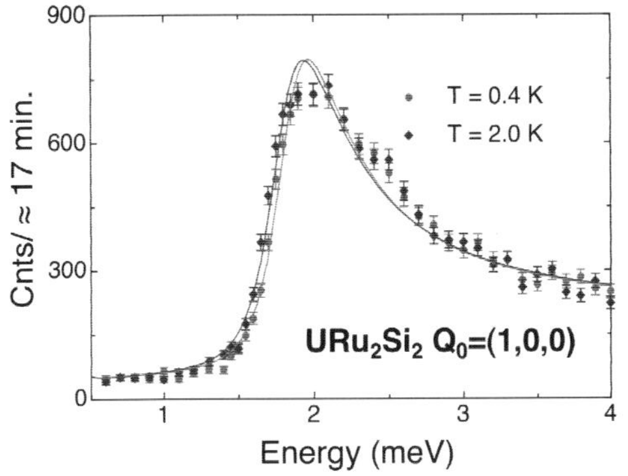

Figure 7. Energy spectra at the position $Q_0 = (1, 0, 0)$ above the superconducting transition at 2 K and below it at 0.4 K at ambient pressure. The lines are guides to the eye.

Q_1. At a first glance, no drastic change occurs below T_{SC} at $T = 0.4$ K for both wavevectors Q_0 and Q_1: no emergence of a low energy resonance characteristic of the superconducting pairing as observed for the heavy fermion superconductors UPd_2Al_3, $CeCu_2Si_2$ and $CeCoIn_5$ [22–25]. Looking more closely, the excitation at Q_0 shifts by 40 μeV, while the excitation at Q_1 remains unchanged. The shift in energy of the excitation at Q_0 through T_{SC} as well as the collapse of the inelastic response above P_x may suggest that the origin of the SC pairing can be magnetic excitations as proposed for UPd_2Al_3.

The absence of detectable, low energy quasielastic ($T > T_{SC}$) or resonant ($T < T_{SC}$) contributions at Q_0 by comparison to UPd_2Al_3, $CeCu_2Si_2$ and $CeCoIn_5$ can be explained partly by the weakness of its Sommerfeld coefficient γ ($\gamma = 70\,\mathrm{mJmol^{-1}K^{-2}}$ in URu_2Si_2, $140\,\mathrm{mJmol^{-1}K^{-2}}$ in UPd_2Al_3, about $1000\,\mathrm{mJmol^{-1}K^{-2}}$ in $CeCu_2Si_2$ and $CeCoIn_5$[26]). Indeed a weak γ corresponds to a large relaxation rate for the normal state fluctuations ($T > T_{SC}$) and therefore the excitation spectrum is spread out in energy space. It is also possible that this magnetic response in URu_2Si_2 is not strongly structured in momentum space as well contrarily to the three previous cases where the quasielastic response is concentrated along the antiferromagnetic hot spot. The case of URu_2Si_2 appears rather similar to that of the skutterudite system $PrOs_4Sb_{12}$ where despite a rather large γ-term ($\approx 300\,\mathrm{mJmol^{-1}K^{-2}}$) no quasielastic contribution has been detected around any wavevector[27; 28]. For the case of $PrOs_4Sb_{12}$, it was recently stressed that the mass enhancement may originate from the aspherical Coulomb scattering of conductions electrons from the very low energy singlet-triplet crystalline electric field excitation [29]. This mechanism is related to the dominant quadrupolar degrees of freedom at play in this compound. Multipolar degreees of freedom may be involved in the formation of complex hidden order phase in URu_2Si_2 .

On entering under pressure above P_x in the AF phase, it was observed from resistivity measurements that the gap energy jumps to a higher value at P_x[18]. According to P studies realized on 1-1-5 heavy fermion systems such as $CeRhIn_5$ on the duality between AF and SC, this jump can explain the collapse of SC in the AF phase[30]. Thus the proof that the $Q_0 = (1, 0, 0)$ excitation may be the source of the SC pairing deserves more theoretical treatment.

3. Conclusion

New neutron scattering experiments show clearly that the excitation at $Q_0 = (1, 0, 0)$ is characteristic of the hidden order state. It disappears under pressure at P_x where the ground state switches from HO to AF and is restored above H_{AF} where the groundstate switches from AF to HO. There are strong indications from spin dynamics, but also from transport measurements that the wavevector of the HO phase is $Q_{AF} = (0, 0, 1)$. It is also demonstrated that the Q_0 excitation is directly coupled with superconductivity (P and T variation). Experimentally, URu_2Si_2 is a beautiful case where the main effect at the first order transition at P_x occurs on the sublattice magnetization, on the inelastic neutron scattering response and on the thermal expansion, while other probes like the bulk magnetization, the resistivity, the Shubnikov-de Haas oscillations have only minor changes [31]. For neutron scattering, the next targets are to determine the relation between the sharp excitations at Q_0 and Q_1 and the Fermi surface reconstruction. Theoretical work on the absence of quasielastic response, on the source of SC pairing and on the (P-H) instability of HO and AF phases is strongly needed.

Acknowledgments

Financial support has been given by the French ANR within the programs ECCE and NEMSICOM.

References

[1] Broholm C, Kjems J K, Buyers W J L, Matthews P, Palstra T T M, Menovsky A A and Mydosh J A 1987 *Phys. Rev. Lett.* **58** 1467–1470

[2] Maple M B, Chen J W, Dalichaouch Y, Kohara T, Rossel C, Torikachvili M S, McElfresh M W and Thompson J D 1986 *Phys. Rev. Lett.* **56** 185–188

[3] Schoenes J, Schoeneberger C, Franse J J M and Menovsky A A 1987 *Phys. Rev. B* **35** 5375–5378

[4] Behnia K *et al.* 2005 *Phys. Rev. Lett.* **94** 156405

[5] Matsuda K, Kohori Y, Kohara T, Amitsuka H, Kuwahara K and Matsumoto T 2003 *J. Phys.: Condens. Matter* **15** 2363–2373

[6] Amato A, Graf M J, de Visser A, Amitsuka H, Andreica D and Schenck A 2004 *J. Phys.: Condens. Matter* **16** S4403–S4420

[7] Motoyama G, Nishioka T and Sato N K 2003 *Phys. Rev. Lett.* **90** 166402

[8] Bourdarot F, Fåk B, Mineev V P, Zhitomirsky M E, Kernavanois N, Raymond S, Lapierre F, Lejay P and Flouquet J 2004 *Physica B* **350** e179–e181

[9] Wiebe C R *et al.* 2007 *Nature Physics* **3** 96–100

[10] Hassinger E, Derr J, Levallois J, Aoki D, Behnia K, Bourdarot F, Knebel G, Proust C and Flouquet J 2008 *J. Phys. Soc. Jpn.* **77** Suppl. A, 172–179

[11] Hassinger E, Knebel G, Izawa K, Lejay P, Salce B and Flouquet J 2008 *Phys. Rev. B* **77** 115117

[12] Villaume A, Bourdarot F, Hassinger E, Raymond S, Taufour V, Aoki D and Flouquet J 2008 *Phys. Rev. B* **78** 012504

[13] Amitsuka H, Matsuda K, Kawasaki I, Tenya K and Yokoyama M 2007 *J. Magn. Magn. Mater.* **310** 214–220

[14] Amitsuka H *et al.* 2000 *J. Phys. Soc. Jpn.* **69** Suppl. A, 12

[15] Elgazzar S, Rusz J, Amft M, Oppeneer P M and Mydosh J A 2009 *Nat. Mater.* **8** 337–341

[16] Ohkuni H *et al.* 1999 *Philos. Mag. B* **79** 1045–1077

[17] Yamagami H and Hamada N 2000 *Physica B* **284** 1295–1296

[18] Aoki D, Bourdarot F, Hassinger E, Knebel G, Miyake A, Raymond S, Taufour V and Flouquet J 2009 *J. Phys. Soc. Jpn.* **78** 053701

[19] Jo Y J, Balicas L, Capan C, Behnia K, Lejay P, Flouquet J, Mydosh J A and Schlottmann P 2008 *Physica B* **403** 749–751

[20] Kim K H, Harrison N, Jaime M, Boebinger G S and Mydosh J A 2003 *Phys. Rev. Lett.* **91** 269902

[21] Bourdarot F, to be published

[22] Sato N K *et al.* 2001 *Nature* **410** 340–343

[23] Hiess A *et al.* 2006 *J. Phys.: Condens. Matter* **18** R437–R451

[24] Stock C, Broholm C, Hudis J, Kang H J and Petrovic C 2008 *Phys. Rev. Lett.* **100** 087001

[25] Stockert O, Arndt J, Schneidewind A, Schnelder H, Jeevan H S, Geibel C, Steglich F and Loewenhaupt M 2008 *Physica B* **403** 973–976

[26] Flouquet J 2005 in *Progress in Low Temperature Physics* ed. W Halperin (Amsterdam: Elsevier) p 139

[27] Kuwahara K *et al.* 2005 *Phys. Rev. Lett.* **95** 107003

[28] Raymond S *et al.* 2009 *J. Phys.: Condens. Matter* **21** 215702

[29] Zwicknagl G, Thalmeier P and Fulde P 2009 *Phys. Rev. B* **79** 115132

[30] Knebel G, Aoki D, Brison J P and Flouquet J 2008 *J. Phys. Soc. Jpn.* **77** 114704

[31] Nakashima M, Ikeda S, Okubo T, Inada Y, Settai R, Ohkuni H, Yamamoto E, Haga Y and Onuki Y 2003 *Physica B* **329** 566–567

Crystal fields in UO_2 - revisited

H Nakotte[1,2], R Rajaram[1,3], S Kern[4], R J McQueeney[5,6], G H Lander[7] and R A Robinson[8]

[1] Department of Physics, New Mexico State University, Las Cruces, NM 88003, USA
[2] Los Alamos Neutron Science Center, Los Alamos National Laboratory, Los Alamos, NM 87545, USA
[3] Department of Physics and Astronomy, University of North Carolina at Chapel Hill, Chapel Hill, NC 27599, USA
[4] Department of Physics, Colorado State University, Fort Collins, CO 80524, USA
[5] Ames Laboratory, Ames, IA 50011, USA
[6] Department of Physics and Astronomy, Iowa State University, Ames, IA 50011, USA
[7] European Commission, JRC, Institute for Transuranium Elements, Postfach 2340, D-76125 Karlsruhe, Germany
[8] Bragg Institute, ANSTO, Menai, NSW 2234, Australia

E-mail: hnakotte@nmsu.edu

Abstract. We performed inelastic neutron scattering (INS) in order to re-investigate the crystal-field ground state and the level splitting in UO_2. Previous INS studies on UO_2 by *Amoretti et al.* [Physical Review B **15**, 1989, 1856] uncovered four excitations at low temperatures in the 150-180 meV range. Considering the dipole-allowed transitions, only three of these transitions could be explained by the published crystal-field model. Our INS results on a different UO_2 sample revealed that the unaccounted peak at about 180 meV is a spurious one, and thus not intrinsic to UO_2. In good agreement with Amoretti's results, we corroborated that the ground-state of UO_2 is the Γ_5 triplet, and we computed that the fourth- and six-order crystal field parameters are V_4 = -116 meV and V_6 = 26 meV, respectively. We also studied the INS response of the non-magnetic $U_{0.4}Th_{0.6}O_2$. The splitting for this thorium-doped compound is similar to the one of UO_2, which orders antiferromagnetically at low temperatures. Therefore, we can conclude that magnetic interactions only weakly perturb the energy level splitting, which is dominated by strong crystal fields.

1. Introduction

Uranium dioxide (UO_2) crystallizes in the cubic CaF_2 structure (space group: $Fm\bar{3}m$) and it is an important nuclear fuel material [1]. In basic sciences, UO_2 attracted much attention because it is one of very few actinide compounds, for which some properties are consistent with localized *5f* electrons [2]. The cubic crystalline environment of the U^{4+} ions and application of Hund's rules result in a nine-fold degenerate $^3H_4$ ground state for the $5f^2$ electrons in UO_2 [3]. Strong crystal fields lift the degeneracy into a Γ_1 singlet, a Γ_3 doublet, and two triplets, Γ_4 and Γ_5. Here, Γ_n represents the different states of the irreducible representation of the cubic point group.

The cubic crystal field Hamiltonian in the Russell-Saunders or the intermediate coupling (IC) schemes is given by [4]

$$H_{CF} = V_4\beta[\hat{O}_4^0(J) + 5\hat{O}_4^4(J)] + V_6\gamma[\hat{O}_6^0(J) - 21\hat{O}_6^4(J)], \quad (1)$$

where $\hat{O}_n^m$ are the Stevens operator equivalents, β and γ are the reduced matrix elements of the J manifold, and the parameters V_n are the coefficients of the crystal field potential.

Using a point charge model, *Rahman and Runciman* [5] successfully diagonalized the complete Hamiltonian for paramagnetic UO_2 taking into account J mixing due to crystal field effects. Combining the matrices of the cubic crystal field potential for the f^2 configuration with the electrostatic and spin-orbit matrices, they predicted that Γ_5 is the ground state for UO_2 and they computed V_4 = -409 meV and V_6/V_4 = 0.06. It should be noted, however, that the point charge model tends to underestimate the sixth-degree term, as it does not take into account overlap and covalency effects.

2. Previous neutron scattering data and analysis

With the advent of spallation sources and chopper spectrometers, UO_2 was one of the first materials to be examined for high-energy excitations. *Kern et al.(1985)* [6] reported excitations near 160 meV. Later, with the better resolution of HET at ISIS, a more detailed study was reported [7].

The experimental results corroborated the theoretically predicted Γ_5 ground state; however, the values for V_4 = -123 meV and V_6/V_4 = 0.22 determined from experiments were very different from the theoretical predictions above. Moreover, for temperatures below 40 K, the experimental data did provide evidence for four excitations with energy transfers of 149, 157, 169 and 179 meV. Cubic symmetry does not permit four transitions.

On the other hand, it is well established that the oxygen sublattice in UO_2 exhibits a dynamical Jahn-Teller distortion, when the compound orders antiferromagnetically at T_N =30.8 K [8, 9]. Both effects reduce the point symmetry at the U site. Nevertheless, even a sophisticated lattice-sum point charge model (PCM) assuming a triple-**k** monoclinic distortion and intermediate coupling (IC) predicts only three transitions at 147, 154 and 165 meV [7], while the fourth excitation around 180 meV is still unaccounted for.

3. Motivation and experimental approach

It is possible to avoid the complication due to magnetic ordering and the accompanied structural distortions by diluting the uranium sublattice so that magnetic long-range order is suppressed. In a study of UO_2-ThO_2 solid solutions, it was found that the Néel temperature decreases linearly, and magnetic order vanishes when about half of the uranium is replaced by thorium [10].

The Institute for Transuranium Elements (Karlsruhe, Germany) provided two high-quality samples, pure UO_2 and $U_{0.4}Th_{0.6}O_2$. The latter composition was chosen as a non-magnetic counterpart to the magnetic UO_2.

Inelastic neutron scattering experiments were performed using the time-of-flight spectrometer PHAROS at the Los Alamos Neutron Science Center. For the studies, we used powdered samples of UO_2 with total mass of about 190 g and $U_{0.4}Th_{0.6}O_2$ with mass of about 170 g. Data for UO_2 were collected at 30 and 50 K and data for $U_{0.4}Th_{0.6}O_2$ were collected at 17 and 50 K. Data were collected for 12-24 hours at each temperature, and we used incident beam energies of 243 meV. Correction for background, incident beam intensity, detector efficiencies and absorption was applied.

4. Results and discussion

All of the PHAROS data exhibit three distinct sets of features: a strong elastic peak around zero-energy transfer, three phonon peaks between 30 and 75 meV energy transfers, and multiple crystal-field excitations in the 145-175 meV range. For this contribution, we concentrate on the features arising from the crystal fields in both compounds.

Figure 1 shows the INS data for $U_{0.4}Th_{0.6}O_2$ at 17 K (left) and pure UO_2 at 50 K (right). It should be noted that we observed only very little difference in the response at 17 and 50 K for the non-magnetic $U_{0.4}Th_{0.6}O_2$, as expected. As can be seen in figure 1, there are two clear excitations in the INS response of $U_{0.4}Th_{0.6}O_2$ at energy transfers of ~149 and 169 meV. Comparing the results of the thorium-doped sample with the INS data on pure UO_2 at 50 K (paramagnetic region), we find that the positions of the two excitations are essentially unchanged and that the integrated intensities scale roughly with the uranium content, although the peaks are slightly broader in UO_2 (not resolution-limited). Nevertheless, our comparison provides evidence that the energy level splitting in UO_2 is only marginally affected (if at all) by short-range magnetic correlations at temperatures just above T_N.

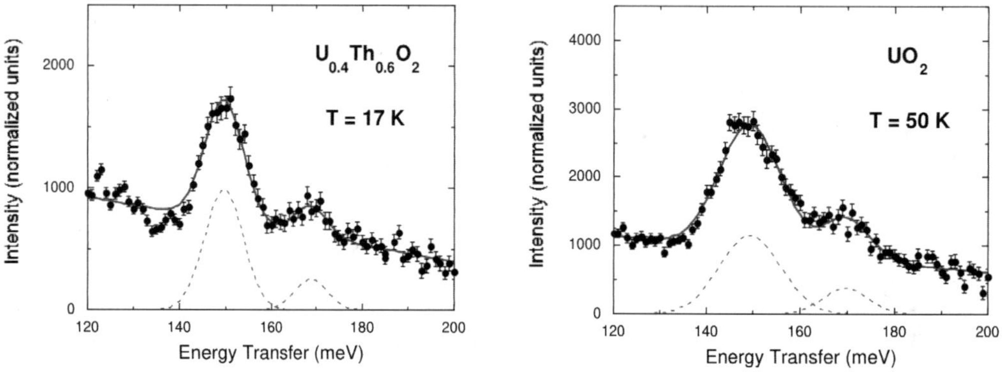

Figure 1. Normalized inelastic neutron scattering spectra for $U_{0.4}Th_{0.6}O_2$ at 17 K (left) for pure UO_2 at 50 K (right), as taken on PHAROS. The experimental data were fitted using the two Gaussians shown. Note the different scales for the two samples.

Figure 2 shows the INS data on pure UO_2 taken at 30 K (just below T_N). Unlike the 50-K UO_2 data, fitting the observed spectrum at that temperature requires at least three Gaussians, and we were able to achieve a good fit assuming Gaussians centered at energy transfers of ~150, 158 and 170 meV. The peak positions of the three excitations and the overall shape of our INS curve are in excellent agreement with the previously published data taken at comparable temperatures [7] with one important exception: namely, the peak around 180 meV is absent. Therefore, we conclude that the previously observed excitation at ~180 meV is not intrinsic to UO_2.

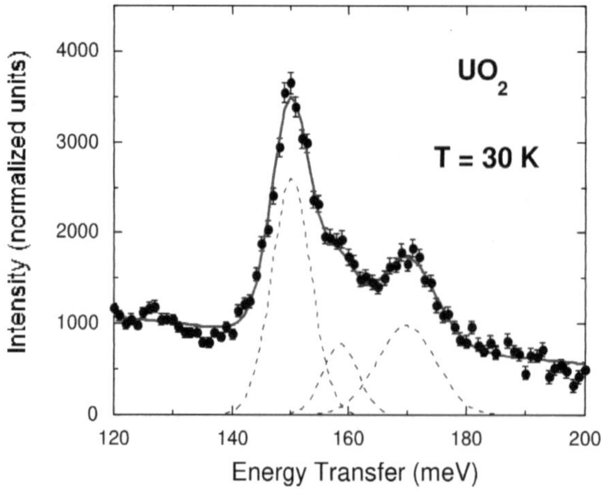

Figure 2. INS spectrum of UO_2 taken at 30 K (below T_N). The experimental data were fitted using the three Gaussians shown.

We used our data then to compute the crystal-field coefficients for cubic symmetry given in equation (1). The crystal-field energy matrices were calculated using the Stevens' operator equivalent method. We used a computer code by *Crosswhite and Crosswhite* [11] to determine the crystal field parameters for UO_2. Results for V_4 and V_6 in UO_2 for this and previous studies are listed in table 1.

Table 1. Cubic fourth- and sixth-order crystal field coefficients for UO_2. Column 1 lists our experimental results, column 2 lists previously published experimental results [7], column 3 lists the theoretical predictions from *Rahman and Runciman* [5], and column 4 lists the predictions from a lattice-sum point-charge model using an intermediate coupling scheme [7].

	experimental data (this result)	experimental Data [7]	PCM with J mixing [5]	lattice-sum PCM and IC [7]
V_4 (meV)	-116.2	-123.0	-409.0	-80.9
V_6 (meV)	25.8	26.5	24.8	8.7
V_6/V_4	-0.22	-0.22	-0.06	-0.11

Our crystal-field coefficients are in very good agreement with values determined from previous experiments, but neither theoretical model provides a satisfactory agreement with the experimental data. Both theoretical models assumed that the ground state is the Γ_5 triplet, followed by the Γ_3 doublet, the Γ_4 triplet and the Γ_1 singlet. In that sequence, one would label ΔE_1 ~150 meV as due to transitions between Γ_5 and Γ_3, ΔE_2 ~158 meV as due to transitions between Γ_5 and the lower two energy levels of Γ_4, and ΔE_3 ~170 meV as due to Γ_5 and the third level of the Γ_4 triplet.

Considering the transition strength between energy levels in UO_2, we calculated theoretical ratios of $\Delta E_2/\Delta E_1 = 0.57$ and $\Delta E_3/\Delta E_1 = 0.29$, while our experimental values (determined using the integrated intensities) are essentially reversed, i.e. $\Delta E_2/\Delta E_1 = 0.26$ and $\Delta E_3/\Delta E_1 = 0.55$.

In general, for the light actinide dioxides, the transition energies are consistent with the crystal-field scheme [12]. However, the individual matrix elements, especially those involving the ground states, still present a problem. Despite their simple crystal structure and many investigations, a full understanding requires consideration of higher-order terms [13].

References

[1] Belle J 1961 *Uranium Dioxide: Properties and Nuclear Applications* (Washington DC: U.S. Atomic Energy Commission)
[2] Joyce J J, Durakiewicz T, Olson C B, Guziewicz E, Graham K S, Moore DP and Morales L A 2007 *Materials Research Society Symposium Proceedings* **986** 35
[3] Cracknell A P and Daniel MR 1967 *Phys. Soc. Proc.* **92** 705
[4] Lea K R, Leask M J M and Wolf W P 1962 *J. Phys. Chem. Solids* **23** 1381
[5] Rahman H U and Runciman W A 1966 *J. Phys. Chem. Solids* **27** 1833
[6] Kern S, Loong C-K and Lander G H 1985 *Phys. Rev.* B **32**, 3051
[7] Amoretti G, Blaise A, Caciuffo R, Fournier J M, Hutchings M T, Osborn R and Taylor A D 1989 *Phys Rev.* B **40** 1856
[8] Faber J jr, Lander G H and Cooper B R 1975 *Phys Rev. Lett.* **35** 1770
[9] Ippolito D, Martinellli L and Bevilacqua G 2005 *Phys Rev.* B **71** 64419
[10] Hinatsu Y and Fujino T 1985 *J. Solid State Chem.* **60** 195
[11] Crosswhite H and Crosswhite H M 1984 *J. Opt. Soc. Am.* B **1** 246
[12] Magnani N, Santini P, Amoretti G and Caciuffo R 2005 *Phys. Rev.* B **71** 054405
[13] Santini P, Carretta S, Amoretti G, Caciuffo R, Magnani N and Lander G H 2009 *Rev. Mod. Phys.* **81**, 807

Low Lying Spin Excitation in the Spin Ice, $Ho_2Ti_2O_7$

G. Ehlers and E. Mamontov
SNS, Oak Ridge National Laboratory, Oak Ridge, TN 37831-6475, USA.
E-mail: ehlersg@ornl.gov

M. Zamponi
SNS, Oak Ridge National Laboratory, Oak Ridge, TN 37831-6473, USA and
Jülich Center for Neutron Science, FZ Jülich, 52425 Jülich, Germany.

J. S. Gardner
Indiana University, Bloomington, IN 47408, USA and
NCNR, NIST, Gaithersburg, MD 20899-6102, USA.
E-mail: jsg@nist.gov

Abstract. The high flux and low background of the new backscattering spectrometer at the SNS combine to produce an excellent signal to noise ratio, allowing us to investigate a low lying weak excitation never seen before in the spin ice, $Ho_2Ti_2O_7$. This non-dispersive excitation has been observed at $E = 26.3$ μeV below 100 K but is resolution limited only below $\sim$ 65 K. It is indifferent to magnetic fields below $\mu_0 H = 4.5$ T, at 1.6 K. These characteristics help us to identify the excitation as due to the nuclear spin system.

1. Introduction

When studying low temperature magnetism, especially in systems with delicately balanced and competing interactions, like for example in spin ice [1, 2], Cs_2CoCl_4 [3] or $LiHoF_4$ [4], one has to not only consider the electronic spins, but also the nuclear spins, particularly in compounds where there is a large nuclear spin in high abundance (e.g. Ho, Mn, Nd, Co etc.)

Geometrically frustrated magnets are systems where magnetic moments are placed on a lattice that is incompatible with the dominant spin-spin interactions [5]. In the rare-earth titanate pyrochlores [6], large rare-earth moments reside on a sublattice that makes up a network of corner sharing tetrahedra, and all the other elements are non magnetic. $Ho_2Ti_2O_7$ [1, 2] and $Tb_2Ti_2O_7$ [7, 8] have exchange and dipole interactions of approximately the same strength, and the single ion anisotropy gives both systems an Ising like character. However, as shown by Melko and Gingras [9], a slight change in interaction strength, albeit in sign too, can drive such a system between different ground states (spin liquid in the case of $Tb_2Ti_2O_7$ and spin ice in $Ho_2Ti_2O_7$). Another important difference between these two systems is the energy gap to the first excited state. In $Tb_2Ti_2O_7$ the gap is small ($\sim$ 20 K) and reduces as the temperature is lowered. In $Ho_2Ti_2O_7$ the gap is on the order of $\sim$ 300 K. These differences may or may not be the complete answer as to why the ground states are so different, but when interpreting the slow spin dynamics at low temperature of these and other magnets, one must not forget the nuclear spin system that co-exists with the electronic spins.

The spin ice, $Ho_2Ti_2O_7$ was recently shown to be an ideal candidate for studying the nuclear spin system [10]. The slow, low temperature spin dynamics have been studied [1, 11, 12, 13, 14, 15], and it was speculated that residual fluctuations seen below 2 K are a consequence of the nuclear spins perturbing the electronic spins off their $\langle 111 \rangle$ equilibrium axes causing a *wobble*. In both spin ices, $Ho_2Ti_2O_7$ and $Dy_2Ti_2O_7$ a high temperature, thermally activated process dominates the magnetic spectrum [13, 14, 15, 16, 17]. As the temperature is lowered below 30 K, the systems enter a quantum-like regime where temperature plays almost no role in the spin relaxation processes [12, 13]. This phase, which exists in $Ho_2Ti_2O_7$ between 2 K and 30 K, can be described as a slowly relaxing ($\approx$ 20 ns) [14], concentrated spin ensemble. In $Dy_2Ti_2O_7$ the spins are significantly slower ($\approx$ 5 ms) [13]. Recently this region has been associated with the propagation of magnetic monopoles through the system [18, 19]. Below 2 K, a unique reentrant behaviour from a quantum to a thermally activated regime has been observed in both spin ice compounds [13, 14]. The residual spin dynamics eludes the best models that describe spin ice but one possible explanation involves a coupling of the nuclear spin system to the electronic spins [12, 21].

In this paper we extend to a lower temperature the observation of a single non-dispersive excitation at $E_0 = 26.3\ \mu$eV in $Ho_2Ti_2O_7$. This has been reported below 100 K in $Ho_2Ti_2O_7$ and is absent in isotopic $^{162}Dy_2Ti_2O_7$.

2. Experimental Details

Neutron experiments were performed at the new backscattering spectrometer BASIS on beam line 2 at the Spallation Neutron Source (SNS) in Oak Ridge [20]. The energy resolution of this instrument is $\sim 3.3\ \mu$eV full width at half maximum (FWHM) for our particular sample size. In one neutron experiment, a cryomagnet was used with a maximum field of $\mu_0 H = 5$ T and a base temperature of $T = 1.6$ K. In others, a ^{3}He-refrigerator was used to extend the temperature range in zero field. The powders were very tightly packed in their containers to avoid reorientation of the grains in a magnetic field and to provide good thermal contact.

Natural dysprosium has two isotopes with large absorption cross sections for cold neutrons. Therefore an isotopically enriched $^{162}Dy_2O_3$ oxide was used in the preparation of the neutron sample. Thus our $Dy_2Ti_2O_7$ sample had only a very small fraction of Dy isotopes containing a nuclear spin ($\leq 0.6\%$).

3. Results and Discussion

Our experimental findings can be summarized as follows.

The excitation is observed in $Ho_2Ti_2O_7$ in zero field and applied field below 100 K, but is absent in isotopically enriched $^{162}Dy_2Ti_2O_7$. It is resolution limited below 65 K. The energy of the excitation is nearly independent of temperature, with a possible small softening at the highest temperatures. The intensity is constant below 50 K and then drops to disappear at around 100 K.

The lack of a dependence on magnetic field clearly indicates that the excitation does not originate from within the electronic spins. In zero field the sample is in the spin ice state, a disordered state with short ranged magnetic correlations. In a field, on the other hand, the system is long range ordered, which is directly confirmed in our data by the presence of magnetic Bragg peaks, and the short ranged dynamic correlations are largely suppressed.

The interpretation as a transition between split nuclear states of Ho leads to an estimate of the hyperfine field of $B_{\mathrm{hf}} = 7/2 \cdot E_0/\mu_{\mathrm{Ho}} = (702 \pm 1)$ T, a value that is close to a previous estimate, $B_{\mathrm{hf}} \sim 720$ T [22], and is comparable to pure Ho metal (770 T) [23]. At this magnitude of the hyperfine field, an external applied field of 5 T is negligible.

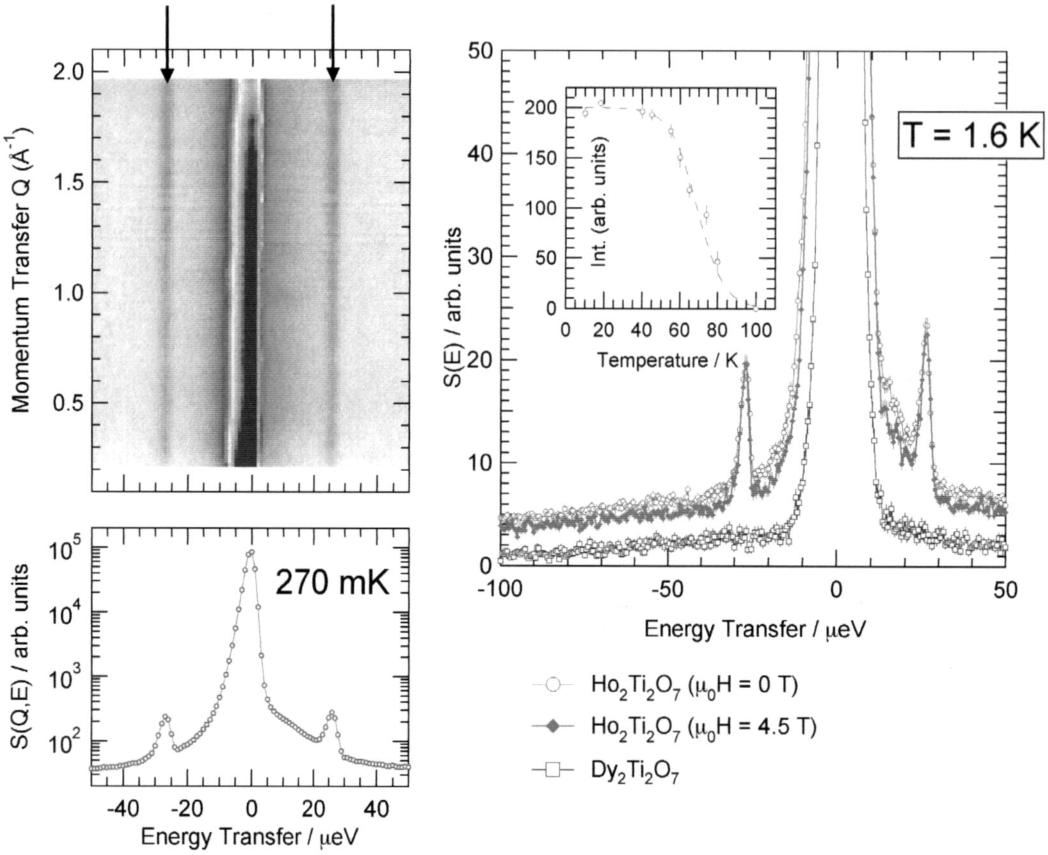

Figure 1. Top left: False colour contour plot of the inelastic spectra from $Ho_2Ti_2O_7$ at $T = 270$ mK. It is clear that the mode at $E_0 = 26.3$ μeV is independent of Q. Right: The energy spectra from the two spin ice compounds at $T = 1.6$ K. In $Ho_2Ti_2O_7$ the excitation can clearly be seen at 26.3 μeV with and without magnetic field, while in $Dy_2Ti_2O_7$ it is entirely absent. The temperature evolution of the mode intensity in zero field is shown in the inset for $Ho_2Ti_2O_7$. Error bars represent $\pm 1\sigma$. Bottom left: Spectra averaged over all Q from the powder sample. While the nominal reading of the cold head was at 270 mK, the intensity ratio in the data between neutron energy gain and loss sides is about ~ 1.5, indicating that the temperature of the powder sample (by detailed balance) may be as high as ~ 750 mK.

Clearly, the hyperfine field has to be *static* on the time scale of the electronic spin fluctuations for this scenario to be valid. In other words, the Larmor frequency of the Ho nuclear moment in the hyperfine field (estimated at $\nu_L \sim 6.36$ GHz, with the gyromagnetic ratio of the Ho nucleus equal to $\gamma/2\pi = 9.06$ MHz/T) has to be slower than the electronic spin fluctuations. In the relevant temperature range, between 50 K and 100 K, the electronic spin fluctuation rate follows an Arrhenius law, $\nu = \nu_0 \cdot \exp(-\Delta/T)$, with $\nu_0 = 1.1 \cdot 10^{11}$ Hz and $\Delta = 293$ K [14], and thus equals ν_L at $T_L \sim 103$ K. Hence one expects nuclear level transition to be observable at $T \ll T_L$, and to disappear around $T \sim T_L$, in accordance with the experimental result.

There are two more arguments that make the interpretation as a nuclear excitation very plausible. (i) The published low temperature specific heat of $Ho_2Ti_2O_7$ shows a Schottky anomaly for nuclear Ho spin ($I = 7/2$) of $0.9R$, and the observed level splitting of 0.3 K equals ~ 26 μeV [2]. (ii) In absolute terms, the observed scattering intensity of the excitation is very small, ~ 500 times smaller than the quasielastic magnetic intensity (see Fig. 1). Again, this is expected, since the incoherent scattering cross section of Ho (which is purely spin incoherent since there is only one isotope) is 0.36 bn, about ~ 1000 times smaller than the scattering length of an electronic magnetic moment of ~ 10 μ_B.

To conclude, we have observed a low energy excitation in $Ho_2Ti_2O_7$ spin ice which is ascribed to the Ho nuclear moment. It may have a significant impact on the low temperature spin dynamics of the system for which residual traces have been found in the ice phase, and may account for the differences between the two spin ices in this regard.

Acknowledgments

We would like to thank R. A. Mills for support during our experiments. ORNL/SNS is managed by UT-Battelle, LLC, for the U.S. Department of Energy under contract DE-AC05-00OR22725.

References

[1] Bramwell S. T. and Gingras M. J. P., Science **294**, 1495 (2001).
[2] Bramwell S. T., Harris M. J., den Hertog B. C., Gingras M. J. P., Gardner J. S., McMorrow D. F., Wildes A. R., Cornelius A. L., Champion J. D. M., Melko R. G. and Fennell T., *Phys. Rev. Lett.* **87** 047205 (2001).
[3] Kenzelmann M., Coldea R., Tennant D. A., Visser D., Hofmann M., Smeibidl P. and Tylczynski Z., *Phys. Rev.* B **65** 144432 (2002).
[4] Bitko D., Rosenbaum T. and Aeppli G., *Phys. Rev. Lett.* **77** 2035 (1992).
[5] Gardner J. S., Gingras M. J. P. and Greedan J. E., accepted by *Rev. of Mod. Phys.* (2009).
[6] Greedan J. E., J. Alloys and Compounds, **408**, 444 (2006).
[7] Gardner J. S., Dunsiger S. R., Gaulin B. D., Gingras M. J. P., Greedan J. E., Kiefl R. F., Lumsden M. D., MacFarlane W. A., Raju N. P., Sonier J. E., Swainson I. and Tun Z., *Phys. Rev. Lett.* **82** 1012 (1999).
[8] Gardner J. S., Gaulin B. D., Berlinsky A. J., Waldron P., Dunsiger S. R., Raju N. P. and Greedan J. E., *Phys. Rev.* B **64** 224416 (2001).
[9] Melko R. G. and Gingras M. J. P., *J. Phys.: Condens. Matter* **16** R1277 (2004).
[10] Ehlers G., Mamontov E., Zamponi M., Kam K. C. , and Gardner J. S., *Phys. Rev. Lett.* **102** 016405 (2009).
[11] Snyder J., Slusky J. S., Cava R. J. and Schiffer P., Nature (London) **413**, 48 (2001).
[12] Ehlers G., Cornelius A. L., Orendáč M. , Kajňaková M., Fennell T., Bramwell S. T. and Gardner J. S., *J. Phys.: Condens. Matter* **15** L9 (2003).
[13] Snyder J., Ueland B. G., Slusky J. S., Karunadasa H., Cava R. J., Mizel A. and Schiffer P., *Phys. Rev. Lett.* **91** 107201 (2003).
[14] Ehlers G., Cornelius A. L., Fennell T., Koza M., Bramwell S. T. and Gardner J. S., *J. Phys.: Condens. Matter* **16** S635 (2004).
[15] Orendáč M., Hanko J., Čižmár E., Orendáčová A., Shirai M., and Bramwell S. T., *Phys. Rev.* B **75** 104425 (2007).
[16] Ehlers G., Gardner J. S., Booth C. H., Daniel M., Kam K. C., Cheetham A. K., Antonio D., Brooks H. E., Cornelius A. L., Bramwell S. T., Lago J., Häussler W. and Rosov N., *Phys. Rev.* B **73** 174429 (2006).
[17] Ehlers G., Mamontov E., Zamponi M., Faraone A., Qiu Y., Cornelius A. L., Booth C. H., Kam K. C., Toquin R. L., Cheetham A. K. and Gardner J. S., *J. Phys.: Condens. Matter* **20** 235206 (2008).
[18] Castelnovo C., Moessner R. and Sondhi S. L., Nature (London) **451**, 42 (2008).
[19] Jaubert L. D. C. and Holdsworth P. C. W, Nature Physics **5**, 258 (2009).
[20] Mason T. E., Abernathy D., Anderson I., Ankner J., Egami T., Ehlers G., Ekkebus A., Granroth G., Hagen M., Herwig K., Hodges J., Hoffmann C., Horak C., Horton L., Klose F., Larese J., Mesecar A., Myles D., Neuefeind J., Ohl M., Tulk C., Wang X.-L. and Zhao J., Physica B **385-386**, 955 (2006).
[21] Lago J., Blundell S. J. and Baines C., *J. Phys.: Condens. Matter* **19** 326210 (2007).
[22] Jana Y. M. and Ghosh, D., *Phys. Rev.* B **61** 9657 (2000).
[23] Ofer S., Nowik, I., and Cohen, S. G., in "Chemical Applications of Mössbauer Spectroscopy", editd by V. I. Goldanskii and R. H. Herber, Academic Press, New York (1968).

International Conference on Neutron Scattering 2009 IOP Publishing
Journal of Physics: Conference Series **251** (2010) 012004 doi:10.1088/1742-6596/251/1/012004

Magnetic excitations of Er_2PdSi_3 studied by inelastic neutron scattering in fields up to 12 T

Fei Tang[1], Peter Link[2], Matthias Frontzek[1], Astrid Schneidewind[1,2], Wolfgang Löser[3] and Michael Loewenhaupt[1]

[1] Institut für Festkörperphysik, TU Dresden, Dresden, Germany
[2] Forschungsneutronenquelle Heinz-Maier-Leibnitz, TU München, Garching, Germany
[3] Leibniz-Institut für Festkörper- und Werkstoffforschung, Dresden, Germany

E-mail: [1] `ftang@ifp.phy.tu-dresden.de`

Abstract. The series R_2PdSi_3 (R = heavy rare earth) has been found to exhibit rich magnetic phase diagrams with a large variety of magnetic states. This complex magnetic behavior results from the interplay between RKKY interaction, magneto-crystalline anisotropy based on crystalline-electric field (CEF) effects and geometric frustration due to the AlB_2 derived hexagonal crystal structure. Within the series the Er_2PdSi_3 compound has been found to exhibit in zero field a simple CEF excitation spectrum with just one strong transition at 3.5 meV from the ground state to the first excited state. Neutron spectroscopy in magnetic fields up to 12 T has been performed on the cold-triple axis spectrometer PANDA at FRM II. In this study two transitions within the 3.5 meV excitation could be resolved showing a linear dependency in applied magnetic fields with an almost identical slope of $\mu_B g_L$ for fields above 2 T. The results can be explained in a picture of two slightly different sets of CEF parameters for two inequivalent Er^{3+} sites due the crystallographic superstructure observed in all single crystals of type R_2PdSi_3.

1. Introduction

The series of R_2PdSi_3 (R = rare earth) compounds has been a focus of interest for over 15 years[1]. The compounds crystallize in an AlB_2-derived hexagonal structure (P6/mmm) with a latent geometric frustration. The magnetic rare-earth ions occupy the Al positions of the AlB_2 structure while the non-magnetic Pd and Si atoms are assumed to be statistically distributed on the B positions[2]. However, recent studies on single crystalline samples in the series revealed that Pd and Si atoms are forming some kind of ordering on the B sites instead of being randomly distributed[3]. This crystallographic superstructure can be characterized by the extra reflections at Q positions (1/2, 1/2, n/8) with n being integer. Consequently, the unit cell is doubled in the basal plane and octupled along the ***c*** axis resulting in multiple inequivalent CEF environments for the rare earth ions.

It was previously assumed that the magnetic properties of the compounds are not influenced by the crystallographic superstructure. However, preliminary studies indicate that the crystallographic superstructure is closely related to a generic magnetic structure observed in applied magnetic field for most of the compounds in the series. Thus it is very intriguing to ask the question: how does the crystallographic superstructure influence the magnetic properties of the system?

2. Experimental details

The single crystal preparation and characterization are described elsewhere[4, 5]. The inelastic neutron scattering experiments were carried out on the cold triple axis spectrometer PANDA at the FRM II in Garching. The Er_2PdSi_3 (m $\simeq$ 1300 mg) single crystal used in this study was of high quality. It was mounted in a 15 T vertical magnet with the [001] axis parallel to the field direction making the ***ab*** plane the scattering plane. All measurements have been performed at 1.5 K. The spectrometer was operated in its fully focused mode at constant k_f = 1.55 $Å^{-1}$. A liquid Nitrogen cooled Be filter was mounted before the analyzer to suppress the $\lambda/2$ contamination.

3. Crystal Electrical Field Effect in Er_2PdSi_3

The Er^{3+} ion has a total angular momentum of 15/2 with the Landé factor $g_L = 1.2$. The compound orders at $T_N = 7.0$ K with the magnetic moment parallel to the [001] direction and the propagation vector $\boldsymbol{\tau} = (1/9,\ 1/9,\ 0)$ in the basal plane[6]. At a relative low critical field (1.2 T along [001] direction at 2 K) the compound is already in the induced ferromagnetic state with the fully saturated magnetic moment on the Er^{3+} ion[7]. In this state the RKKY exchange interaction is negligible compared to the external field and the CEF.

For a hexagonal crystal structure, there are four non-zero B^n_m CEF parameters in the CEF Hamiltonian: $H_{CEF} = B^0_2O^0_2 + B^0_4O^0_4 + B^0_6O^0_6 + B^6_6O^6_6$ with the Stevens operators O^n_m [8]. Consequently, for Er_2PdSi_3 there are 8 doublets with wave functions of the form $|\psi> = a|M_J + 6> + b|M_J> + c|M_J - 6>$ where $|M_J>$ is the eigenfunction of the angular momentum operator J_z. The magnetization curve of Er_2PdSi_3 suggests that the Er^{+3} ion reaches its saturation moment (9 μ_B/ion) at 1.2 T when an external magnetic field is applied along the [001] direction[6]. This implies that the CEF ground state wave function must be of the form: $|\psi_0> = a|15/2> + b|3/2> + c|-9/2>$. In magnetic fields $\mu_0 H > 1.2$ T there should be only one CEF transition for neutron scattering from the ground state $|\psi_0>$ to the excited state $|\psi_1> = a_1|13/2> + b_1|1/2> + c_1|-11/2>$ with an additional energy transfer proportional to the Zeeman energy of $\mu_B \cdot g_L \cdot \mu_0 H$.

4. Results

4.1. Field dependence of CEF excitations

In the measurements for magnetic fields above 1.2 T we could resolve two peaks for the CEF transition around the energy transfer at 3.5 meV. Figure 1 shows the energy spectra obtained at $\mathbf{Q}$ =(0.0, -1.0, 0.0) in the reciprocal space at T = 1.5 K with $\mu_0 H$ = 2 T and 12 T. The two transitions have an energy difference of around 0.3 meV. In figure 2 we present the field dependence of the positions of the two transitions for fields between 0 T to 12 T. In the field range above 2 T the energy transfer of the two transitions increases linearly with field with almost the identical rate of $\sim$ 0.07 meV/T which has to be compared to the theoretical value $\mu_B \cdot g_L = 0.0695$ meV/T. Thus both transitions are connecting the ground state $|\psi_0>$ with the first excited state $|\psi_1>$ as given in the previous section.

This suggests that the reason for the two transitions must be coming from two different Er^{3+} sites with slightly different local environments. This agrees with the findings about the crystallographic superstructures in the series: the superstructure with an enlarged unit cell of the size (2x2x8) times the original unit cell could produce more than one local environment for the rare earth site. Furthermore, preliminary results for the superstructure suggest one hexagonal and three monoclinic symmetric sites. This ratio roughly agrees with the ratio between the intensities of the two transitions suggesting that the 3.6 meV transition at 2 T in figure 1 originates from the hexagonal sites and the 3.3 meV transition from the monoclinic sites.

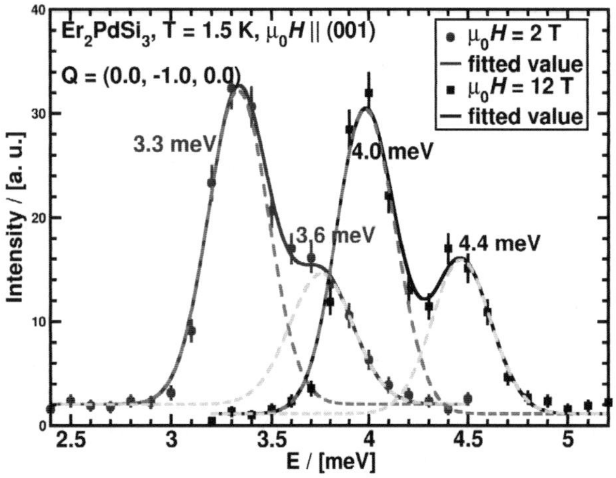

Figure 1. Energy spectra for Er_2PdSi_3 obtained at $\mathbf{Q}$ = (0.0, -1.0, 0.0) at T = 1.5 K: blue ● and black ■ symbols are the experimental data points for 2 T and 12 T, respectively. Full lines are total fits for both transitions, dashed lines indicate the individual contributions.

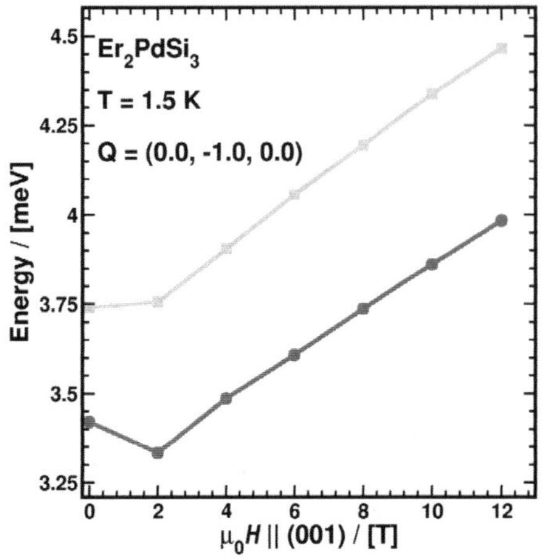

Figure 2. Field dependence of the energy transfers of the two CEF transitions in Er_2PdSi_3 as shown in figure 1 for 2 T and 12 T ; the lines are guides to the eyes.

4.2. Directional dependence of CEF excitations

To further elucidate the origin of the two CEF transitions, measurements have been made on its $\mathbf{Q}$ dependence in the reciprocal $\boldsymbol{a}^*\boldsymbol{b}^*$ plane. The intensity of the two transitions shows a strong directional dependence between the (*H*, 0, 0) and (*H*, *H*, 0) directions. Shown in figure 3 are the energy spectra taken at the $\mathbf{Q}$ positions (0, -1, 0) and (-0.5, -0.5, 0) at 1.5 K and 2 T with the field direction parallel to the [001] direction. The intensity of the second transition at higher energy transfer is almost directional independent while the intensity of the first transition at lower energy transfer almost doubled going from (-0.5, -0.5, 0) to (0, -1, 0).

The neutron scattering cross section of a CEF transition is dependent on the transition matrix element of angular momentum operator J_α where α = x, y, z[9]. For hexagonal symmetry with CEF wave functions of the form $|\psi>=\sum_{|M_J+6i|\leq J} a_i|M_J+6i>$ the transition matrix elements of J_x and J_y are equal so that the scattering cross sections for $\mathbf{Q}$ vectors in the $\boldsymbol{a}^*\boldsymbol{b}^*$ plane are directional independent. The strong directional dependence of the intensity of the first transition at lower energy transfers highly suggests that the local environment is no longer hexagonal, but most probably monoclinic with non-zero CEF parameter B_2^2. With non-zero B_2^2 the CEF wave functions will take the form $|\psi>=\sum_{|M_J+2i|\leq J} a_i|M_J+2i>$ with which transition matrix elements of J_x and J_y are not the same.

5. Conclusions

Detailed inelastic neutron scattering measurements have been performed on Er_2PdSi_3 to examine the magnetic excitations in the compound. In the induced ferromagnetic phase two CEF transitions have been observed while only one is expected in a simple AlB_2 hexagonal structure. The energy transfers of the two transitions increase linearly with the rate 0.07 meV/T for applied magnetic fields along the [001] direction when the field is larger than 2 T. This is in excellent agreement with a calculated slope of $\mu_B \cdot g_L$. Taking into account the crystallographic superstructure which has been observed in the series R_2PdSi_3, the two CEF transitions can be

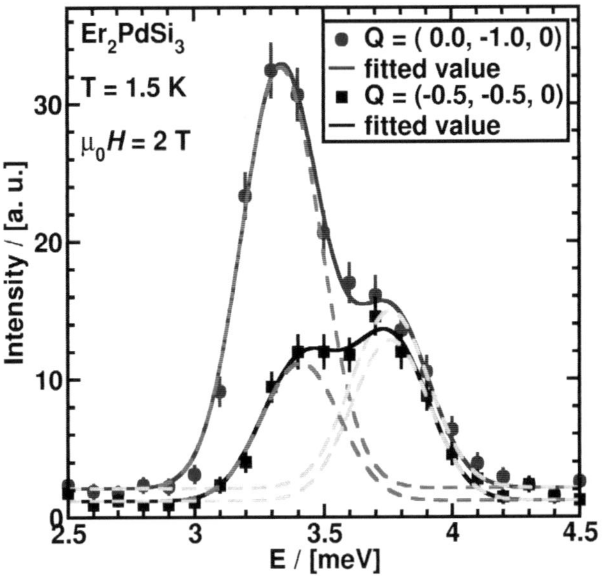

Figure 3. Directional dependence of the intensity of two CEF transitions in Er_2PdSi_3: spectra taken at 1.5 K and 2 T with the magnetic field parallel to [001] direction. The blue ● and black ■ symbols are the experimental data for **Q** = (0.0, -1.0, 0.0) and (-0.5, -0.5, 0), respectively. Full lines are total fits for both transitions, dashed lines indicate the individual contributions.

attributed to two inequivalent rare earth sites surrounded by different distributions of Pd and Si atoms on the B positions.

Moreover, the intensity of the first transition at lower energy transfers shows a strong directional dependence with a factor around two between (0.0, -1.0, 0.0) and (-0.5, -0.5, 0.0) positions while the second transition is almost directional independent. This highly suggests that the local environment for the first transition is monoclinic while the local environment for the second transition has still hexagonal symmetry.

The results demonstrate that although the crystallographic superstructure is a relative weak modification of the major crystal structure, its influence through the modification of the CEF on the magnetic properties of the R_2PdSi_3 compounds is not negligible. More experiments are planned to examine the magnetic excitations of other members of the R_2PdSi_3 series to further understand the effects of the crystallographic superstructure on the magnetic properties of these compounds.

References

[1] Kotsanidis P A and Yakinthos J K 1990 *J. Magn. Magn. Mater.* **87** 199
[2] Szytula A, Hofmann M, Penc B, Slaski M, Majumdar S, Sampathkumaran E V and Zygmuntand A 1999 *J. Magn. Magn. Mater.* **202** 365
[3] Dshemuchadse J 2008 diploma thesis TU Dresden, to be published
[4] Graw G, Bitterlich H, Löser W, Behr G, Fink J and Schultz L 2000 *J. Alloys and Compounds* **308** 193
[5] Mazilu I, Frontzek M, Löser W, Behr G, Teresiak A and Schultz L 2005 *J. Cryst. Growth* **275** e103
[6] Frontzek M, Kreyssig A, Doerr M, Rotter M, Behr G, Lser W, Mazilu I and Loewenhaupt M 2006 *J. Magn. Magn. Mater.* **301** 398
[7] Frontzek M, Kreyssig A, Doerr M, Schneidewind A, Hoffmann J U and Loewenhaupt M 2007 *J. Phys. : Condens. Matter* **19** 145276
[8] Jensen J and Mackintosh A R 1991 *Rare Earth Magnetism* (Oxford: Clarendon Press) p 42
[9] Chatterji T 2006 *Neutron Scattering from Magnetic Materials* (Amsterdam: Elsevier B. V.) p 22

Neutron Scattering Studies of $LiCoPO_4$ & $LiMnPO_4$

Wei Tian[1], Jiying Li[2,3], Haifeng Li[1], Jeffrey W. Lynn[2], Jerel L. Zarestky[1], David Vaknin[1]

[1] Ames Laboratory and Department of Physics and Astronomy, Iowa State University, Ames, Iowa 50011, USA

[2] NCNR, National Institute of Standards and Technology, Gaithersburg, MD 20899, USA

[3] Department of Materials Science and Engineering, University of Maryland, College Park, MD 20742, USA

Abstract. $LiCoPO_4$ ($T_N \approx 21.8$ K) & $LiMnPO_4$ ($T_N \approx 34$ K) are antiferromagnetic insulators exhibiting large magnetoelectric effects. We performed inelastic neutron scattering (INS) experiments to investigate the spin dynamics of these systems and analyzed the measured magnetic spectra by linear spin-wave theory, taking into account intra- and inter-plane nearest, next nearest neighbor magnetic exchange interactions and single ion anisotropy. The INS results indicate that the single ion anisotropy in $LiCoPO_4$ is comparable to the nearest-neighbor magnetic exchange interaction rendering Ising-type behavior of $LiCoPO_4$. Neutron diffraction studies of $LiMnPO_4$ in applied magnetic fields reveal a spin-flop transition at ~ 3.5 Tesla with characteristics of a second order phase transition.

1. Introduction

$LiCoPO_4$ and $LiMnPO_4$ are antiferromagnetic (AFM) insulators belonging to the olivine family of lithium orthophosphates LiMPO$_4$ ($M = Mn^{2+}, Fe^{2+}, Co^{2+}, Ni^{2+}$) [1, 2]. These materials are known for their exceptionally large magnetoelectric (ME) effect [3, 4] and continue to attract much attention. The recent observations of weak ferromagnetism [5], ME "butterfly loop" anomaly [6], and in particularly the ferrotoroidic domain structure in $LiCoPO_4$ [7] have ignited renewed interest in these materials [8, 9, 10, 11, 12].

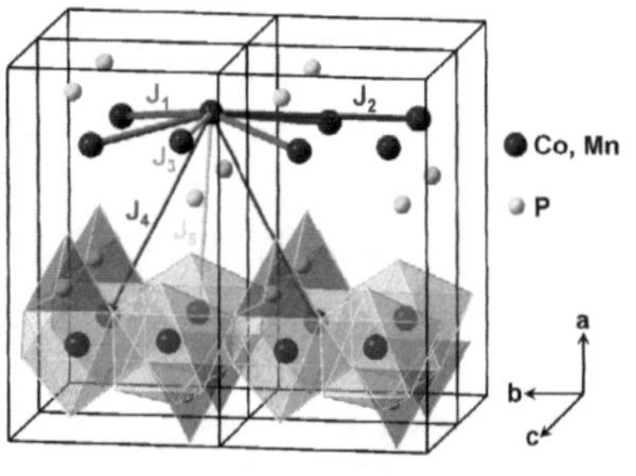

Figure 1. (Color online) Schematic crystal structure of LiMPO$_4$ (M = Co, Mn). The P and M^{2+} magnetic ions are shown, and only one layer of the MO_6 octahedrons and PO_4 tetrahedrons is illustrated for clarity. The in-plane nearest-neighbor (J_1), next-nearest-neighbor (J_2, J_3), and inter-plane nearest, next-nearest-neighbor (J_4, J_5) magnetic exchange interactions included in the spin wave theory are labeled.

$LiCoPO_4$ and $LiMnPO_4$ crystallize in the orthorhombic crystal structure, space group *Pnma* (no. 62) at room temperature. As illustrated in Fig. 1 where only the P and M^{2+} (M = Co, Mn) ions are shown for clarity, the magnetic M^{2+} ($S = 3/2$ for Co^{2+} and $S = 5/2$ for Mn^{2+}) ions are at the center of a slightly distorted MO_6 octahedron that share an oxygen anion with the PO_4 tetrahedra forming buckled MO layers stacked along the a-axis. $LiCoPO_4$ (T_N $\approx$ 21.8 K) and $LiMnPO_4$ ($T_N \approx$ 34 K) undergo long range AFM transitions with a collinear AFM ground state. They adopt the same magnetic structure ($Pnm'a$) differing only in spin orientation. The spins are oriented along the b-axis for $LiCoPO_4$ and along the a-axis for $LiMnPO_4$ [2, 8]. As depicted in Fig. 1, the magnetic coupling in the layer is through the M-O-M super-exchange interactions, and the coupling between adjacent layers is mediated through the PO_4 phosphate group, rendering the magnetic system quasi-two-dimensional. We have recently reported inelastic neutron scattering (INS) studies and determined the microscopic magnetic interactions in these two systems (the detailed results have been published in Ref.13 and Ref.14). In this manuscript, we briefly present the main INS results of $LiCoPO_4$ and preliminary neutron diffraction measurements under applied magnetic field (H $\parallel$ a-axis) of $LiMnPO_4$ that reveal a spin-flop transition at $\sim$ 3.5 Tesla.

2. Experimental Techniques, Results and Discussions

Neutron scattering experiments were carried out using single crystal samples grown by the standard LiCl flux method similar to that reported in Ref.11,15. The crystals were characterized by X-ray diffraction measurements and no impurity phases were detected. Inelastic neutron scattering measurements of $LiCoPO_4$ were performed on the HB1A triple-axis spectrometer (TAS) at HFIR, and the BT7, BT9 and SPINS TAS spectrometers at NIST. The neutron diffraction measurements of $LiMnPO_4$ in applied magnetic fields were performed on the BT7 TAS employing a 11 Tesla vertical magnet with the crystal oriented in the (0 K L) scattering plane and the magnetic field applied along the a-axis, $H \parallel a$. All data have been normalized to beam monitor counts.

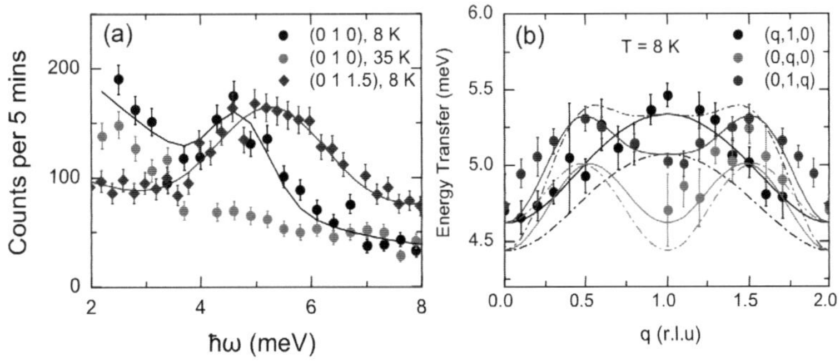

Figure 2. (Color online) (a) Constant wave-vector scans of $LiCoPO_4$ measured at (0 1 0) (T = 8 K and 35 K) and at (0 1 1.5)(T = 8 K). (b) $LiCoPO_4$ spin-wave dispersion curves along three reciprocal lattice directions. Data points are obtained based upon a Gaussian peak approximation. The solid and dashed lines are calculations based upon a global fit to the linear spin wave approximation theory as described in the text.

Figure. 2 (a) depicts representative constant wave-vector scans of $LiCoPO_4$ (error bars in this paper are statistical in origin and represent one standard deviation). It shows that a single excitation of $\hbar\omega \approx$ 4.7 meV is detected at (0 1 0) at T = 8 K ($T < T_N$). At T = 35 K ($T > T_N$) the peak intensity is significantly reduced demonstrating the excitation is magnetic in origin. Comparing T = 8 K constant wave-vector scans measured at (0 1 0) and (0 1 1.5), which typically correspond to the minimum and maximum spin wave excitations along the L direction, the $\sim$ 4.7 meV excitation at (0 1 0) shifts to higher energy transfer, $\sim$ 5.3 meV at (0 1 1.5) indicating weak dispersion along the L direction. Measurements along the (H 1 0) and (0 K 0) directions indicate the $\sim$ 4.7 meV excitation propagating weakly along both the

H and K directions as well. The spin wave dispersion curves of $LiCoPO_4$ determined from a series of constant wave-vector energy-scans below T_N are shown in Fig. 2 (b). Weak dispersion was observed in $LiCoPO_4$ with a band width less than 1 meV exhibiting Ising-type behavior. The large uncertainties in the $LiCoPO_4$ experimental data are associated with instrumental energy resolution which for the TAS used is $\triangle$ E $\sim$ 1 meV at the elastic position. The large experimental uncertainty has a significant effect on the theoretical modeling of $LiCoPO_4$.

The measured spin wave dispersion curves are analyzed in the linear spin wave theory framework. The proposed spin Hamiltonian [16] taking into account different magnetic exchange interactions (J_1 to J_5 in Fig. 1) can be written by the following equation:

$$\mathcal{H} = \sum_{i,j} J_{ij}\mathbf{S}_i \cdot \mathbf{S}_j + \sum_{i,\alpha} D_\alpha (S_i^\alpha)^2, \tag{1}$$

where D_α ($\alpha = x, y, z$) represents the single-ion anisotropy along the x , y, and z-directions. Within the linear spin wave approximation, the derived spin wave dispersion from Eq. (1) is given by:

$$\hbar\omega = \sqrt{A^2 - (B \pm C)^2}. \tag{2}$$

The definition of parameters A, B and C in Eq. (2) is described in Ref.13. The $(B \pm C)$ in Eq. (2) indicates that there are two non-degenerate spin wave branches due to the different values of D_x and D_y.

The magnetic spectra of $LiCoPO_4$ can be adequately described by the proposed spin-wave model. The non-linear least squares fit of the obtained $LiCoPO_4$ magnetic spectra using Eq. (2) ("B-C" branch) yields: $J_1 = 0.771 \pm 0.144$ meV, $J_2 = 0.129 \pm 0.113$ meV, $J_3 = 0.208 \pm 0.102$ meV, $J_4 = -0.167 \pm 0.067$ meV, $J_5 = -0.193 \pm 0.102$ meV, $D_x = 0.73 \pm 0.15$ meV, $D_y = 0.808 \pm 0.159$ meV. The solid and dashed lines in Fig. 2 (b) are the calculations of the two spin wave branches using the obtained parameters. It is remarkable that large single-ion anisotropy was obtained that is comparable to the strongest nearest-neighbor magnetic interaction J_1, $D_x \sim D_y \sim J_1$. Such relatively strong anisotropy may split the $S = 3/2$ quartet of the Co^{2+} ion into two doublets effectively resulting in the suggested Ising-type character of $LiCoPO_4$ [9]. Our studies indicate that single ion anisotropy plays an important role in the spin dynamics of $LiCoPO_4$. As shown in Fig. 2 (b), the calculated spin wave dispersion predicts a maximum separation of $\sim$ 0.3 meV at (0 1 0). Both thermal neutron TAS data (with a resolution of $\sim$ 1 meV) and cold neutron SPINS high resolution data (with a resolution of $\sim$ 0.28 meV) show only one excitation around $\sim$ 4.7 meV. In our measurements, we could not resolve the two branches. It is possible that the second excitation is very weak in intensity since the model does not predict intensities. Another possibility is that the intrinsic linewidth of the observed excitations are broader than the resolution ($\sim$ 1 meV) suggesting contributions from both branches overlap and cause the broadening. Detailed studies together with the observation of an anomalous low energy magnetic excitation are reported in Ref.13.

Neutron diffraction measurements of $LiMnPO_4$ in magnetic field reveal a spin-flop transition at $\sim$ 3.5 Tesla. Fig. 3 shows typical scans monitoring the (0 1 0) and (0 0 1) strong magnetic reflections as a function of applied magnetic field. The magnetic field was applied along the moment direction, H $\parallel$ a-axis. The data of (0 0 1) (Fig. 3 (a)) at 2 K shows that the peak intensity of (0 0 1) disappears above $H_c \sim$ 3.5 Tesla indicating a spin-flop transition, we refer to H_c as the critical field associated with the spin-flop transition. On the other hand, as shown in Fig. 3 (b), the peak intensity of (0 1 0) remains nearly the same at fields both above and below H_c. For magnetic scattering, only those spin components which are perpendicular to the scattering vector have a non-vanishing cross-section. Hence the disappearance of the (0 0 1) peak, and the nearly unchanged (0 1 0) peak indicate that the flopped spins are aligned nearly along the c-axis above H_c. As depicted in the inset in Fig. 3 (a), no hysterisis is observed upon

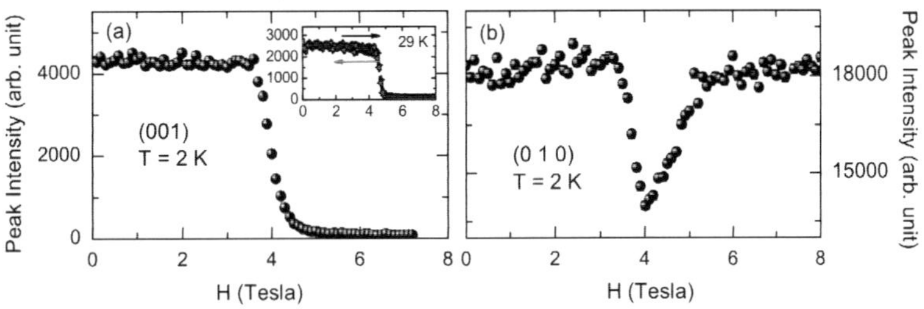

Figure 3. (Color online) Magnetic field induced spin-flop transition in $LiMnPO_4$, the magnetic field was applied along the *a*-axis, H $\parallel$ *a*. (a) (0 0 1) peak intensity vs. H at 2 K. Inset: (0 0 1) peak intensity measured with increasing and decreasing H at 29 K. (b) (0 1 0) peak intensity vs. H at 2 K.

increasing and decreasing the applied magnetic field through the transition, indicating this may be a second order phase transition. An anomalous dip is observed in the (0 1 0) field scan (Fig. 3 (b)) suggesting the existence of an intermediate phase with the critical field H_c extending over a range of $\sim$ 1.6 Tesla at 2 K. We derived the magnetic field versus temperature phase diagram based on field and temperature dependent neutron diffraction and magnetization measurements. The phase diagram and detailed data analysis of this spin-flop transition will be reported elsewhere [17]. Our preliminary results of the characteristics of the spin-flop transition suggest the existence of anisotropic magnetic exchange that competes with the single ion anisotropy.

Acknowledgments
Ames Laboratory is supported by the U. S. Department of Energy Office of Basic Energy Science under Contract No. DE-AC02-07CH11358. HFIR is funded by the United States Department of Energy, Office of Basic Energy Sciences, Materials Science, under Contract No. DE-AC05-00OR22725 with UT-Battelle, LLC. The work has benefited from the use of the NIST Center of Neutron Research at the National Institute of Standards Technology, where SPINS is supported in part by the National Science Foundation under Agreement No. DMR-0454672.

References

[1] H. D. Megaw, Crystal Structure - A working Approach (Saunders Philadelphia, 1973) p. 249.
[2] R. P. Santoro, R. E. Newnham, and S. Nomura, J. Phys. Chem. Solids **27**, 655 (1966); P. Santoro, D. J. Segal, and R. E. Newnham, J. Phys. Chem. Solids **27**, 1192 (1966); R. P. Santoro and R.E. Newnham, Acta Crystallogr. **22**, 344 (1967).
[3] M. Mercier, J. Gareyte, and E. F. Bertaut, C. R. Acad. Sci. Paris B **264**, 979 (1967); M. Mercier, Rev. Gen. Electr. **80**, 143 (1971).
[4] J.-P. Rivera, Ferroelectrics **161**, 147 (1994).
[5] Yu. Kharchenko, N. Kharchenko, M. Baran, and R. Szymczak, cond-mat: 0310156.
[6] I. Kornev, M. Bichurin, J.-P. Rivera *et al*, Phys. Rev. B **62**, 12247(2000).
[7] B. B. van Aken, J.-P. Rivera, H. Schmid, and M. Fiebig, Nature **449**, 702 (2007).
[8] D. Vaknin, J. L. Zarestky, J. E. Ostenson *et al*, Phys. Rev. B **60**, 1100 (1999).
[9] D. Vaknin, J. L. Zarestky, L. L. Miller *et al*, Phys. Rev. B **64** 224414 (2002).
[10] D. Vaknin, J. L. Zarestky, J.-P. Rivera, and H. Schmid, Phys. Rev. Lett. **92**, 207201 (2004).
[11] J. Li, V. O. Garlea, J. L. Zarestky, and D. Vaknin, Phys. Rev. B **73**, 024410 (2006).
[12] T. B. S. Jensen, N. B. Christensen, M. Kenzelmann, H. M. Rønnow *et al*, Phys. Rev. B **79**, 092412 (2009).
[13] Wei Tian, Jiying Li, Jeffrey W. Lynn *et al*, Phys. Rev. B. **78**, 184429 (2008).
[14] Jiying Li, Wei Tian, Ying Chen *et al*, Phys. Rev. B **79**, 144410 (2009).
[15] V. I. Fomin, V. P. Genezdilov, V. S. Kurnosov *et al*, Low Temp. Phys. **28**, 203 (2002).
[16] T. B. S. Jensen, N. B. Christensen,M. Kenzelmann, H. M. Rønnow,*et al*, Phys. Rev. B, **79**, 092413 (2009).
[17] Haifeng Li *et al*, in preparation.

International Conference on Neutron Scattering 2009 IOP Publishing
Journal of Physics: Conference Series **251** (2010) 012006 doi:10.1088/1742-6596/251/1/012006

Very High Energy Neutron Scattering from Hydrogen

R. A. Cowley[1], C. Stock[2], S. M. Bennington[2], J. Taylor[2] and N. I. Gidopoulos[2]

1. Department of Physics, University of Oxford, Parks Road, Oxford OX1 3PU, UK
2. ISIS Facility, Rutherford Appleton Laboratory, Chilton, Didcot, Oxfordshire, OX11 0QX, UK

r.cowley@physics.ox.ac.uk

Abstract

The neutron scattering from hydrogen in polythene has been measured with the direct time-of flight spectrometer, MARI, at the ISIS facility of the Rutherford Appleton Laboratory with incident neutron energies between 0.5 eV and 600 eV. The results of experiments using the spectrometer, VESUVIO, have given intensities from hydrogen containing materials that were about 60% of the intensity expected from hydrogen. Since VESUVIO is the only instrument in the world that routinely operates with incident neutron energies in the eV range we have chosen to measure the scattering from hydrogen at high incident neutron energies with a different type of instrument. The MARI, direct time-of-flight, instrument was chosen for the experiment and we have studied the scattering for several different incident neutron energies. We have learnt how to subtract the gamma ray background, how to calibrate the incident energy and how to convert the spectra to an energy plot . The intensity of the hydrogen scattering was independent of the scattering angle for scattering angles from about 5 degrees up to 70 degrees for at least 3 different incident neutron energies between 20 eV and 100 eV. When the data was put on an absolute scale, by measuring the scattering from 5 metal foils with known thicknesses under the same conditions we found that the absolute intensity of the scattering from the hydrogen was in agreement with that expected to an accuracy of ± 5.0% over a wide range of wave-vector transfers between 1 and 250 Å^{-1}. These measurements show that it is possible to measure the neutron scattering with incident neutron energies up to at least 100 eV with a direct geometry time-of-flight spectrometer and that the results are in agreement with conventional scattering theory.

...

Ever since the experiments of Chatzidimitriou-Dreismann et al [1] in 1997 found that the intensity of the hydrogen scattering from partially deuterated water was approximately only 60% of that expected from conventional theory there has been considerable interest in the reason for this decrease. Later measurements showed that the decrease occurred for most other hydrogen containing systems. On the theoretical side it was proposed that the decrease in intensity arose from quantum entanglement of the hydrogen but it is now generally agreed that this was an incorrect explanation [2-3]. Alternative explanations have been put forward but these also have not yet been shown to explain the observed decrease in the intensity [4-5]. The experiments at high energy transfers have been conducted using VESUVIO, which is an indirect geometry time-of-flight instrument at the ISIS facility of the Rutherford Appleton Laboratory. The instrument and the analysis methods have been steadily refined but no aspect has been identified that is clearly responsible for the observed decrease in the intensity of the scattering from hydrogen.

This paper describes a different approach. We have used a conventional direct time-of-flight instrument, MARI, at the ISIS pulsed source. Before this experiment the instrument has been used with incident neutron energies of 1ev but this is too small to enable the scattering to be measured at the same energies as those employed on VESUVIO. We have therefore investigated if MARI could be used with higher incident neutron energies. After some difficulties we have successfully operated the spectrometer at a range of energies between 0.5 and 600 eV. This necessitated removal of the nimonic chopper that normally reduces the background on MARI and so developing a way of measuring and subtracting this background reliably from the data. We also had to rewrite the spectrometer soft-ware so that it correctly enabled us to operate at higher incident neutron energies.

Our most recent set of experiments used samples held at room temperature and supported so that there was no other material in the neutron beam at the sample position. We studied three samples of polythene, low density at two thicknesses and high density at one thickness and to obtain an accurate absolute normalisation we used five metal foils of Pb, Ni, Cu, Nb and V. The samples were all chosen to be of a thickness such that less than 5% of the incident neutron beam was scattered and the counting times were typical a few hours for each sample. We shall describe the results in detail for three different incident neutron energies of 20, 40 and 100eV.

The scattering expected in the impulse approximation from hydrogen from a direct time-of-flight spectrometer shows a peak in the energy transfer on the energy loss side equal to the kinetic energy of the recoiling hydrogen atom. As the angle increases the recoil energy increases until it becomes close to the incident neutron energy for a scattering angle of 90^0. The scattering then cuts off abruptly due to the kinematics of the scattering. In the recoil approximation the width of the scattering is expected to increase linearly with the energy transfer while the intensity is expected to be independent of scattering angle. The Waller- Froman J factor [6] is unity for hydrogen recoil.

The scattering was binned into 5 degree segments, the background subtracted, the changing efficiency of the detectors with energy allowed for, and the time-of-flight spectrum converted to an energy spectrum. The results are shown in fig 1 for several different scattering angles. At high scattered energies, low angles, the background subtraction has worked well and the result is flat on the high energy side of the peak. At small scattered energies the statistics become less satisfactory because of the increasing effect of the time to energy conversion. Because of this data with a scattering angle greater than 70^0 was not used in the analysis. The data for each spectrum was fitted to two Gaussians. One arises from the hydrogen recoil and the other from the carbon recoil. The parameters for the carbon peak were obtained from the data taken at high scattering angles 100^0 to 130^0 and extrapolated to lower angles to give the starting parameters (this data is not easily visible in the plots in fig. 1). The intensities obtained from some of these analyses are shown in fig. 2. It is clear that the intensity for each set of measurements is constant within error over the accessible range.

The absolute normalisation of the data was performed using the scattering from the metal foils, The data is similar to that for polythene except that the recoil energies are much smaller. The data was analysed in the same way and the average results are shown in fig. 3. For each incident neutron energy the intensity of the scattering was calculated from the known cross-section and thickness of the foil and compared with the measured scattering. The scatter about each straight line gives a measure of the accuracy of the absolute intensity measurements. These results were then used to calculate the absolute intensity of the hydrogen scattering from the polythene and the two are compared in fig.4 as a function of the intensity of the scattering.

Journal of Physics: Conference Series **251** (2010) 012006 doi:10.1088/1742-6596/251/1/012006

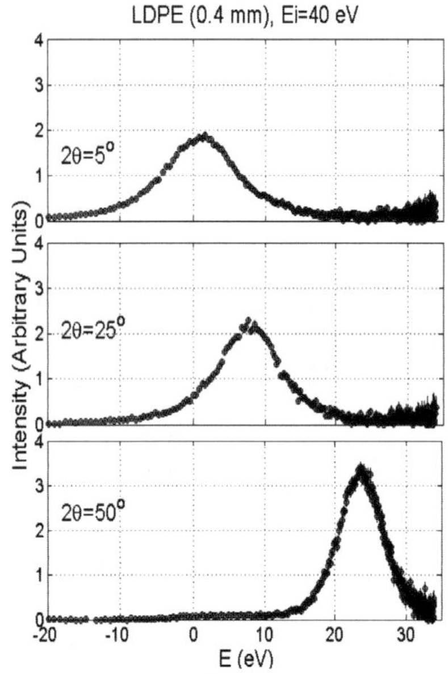

Figure 1 The scattering observed from polythene for one sample and at three angles as an energy scan,

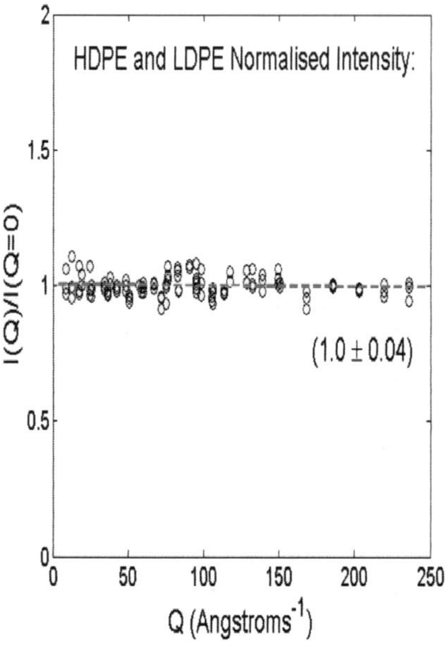

Figure 2. The intensity of the scattering observed for each three samples and incident energies of 20, 40, 100 eV.

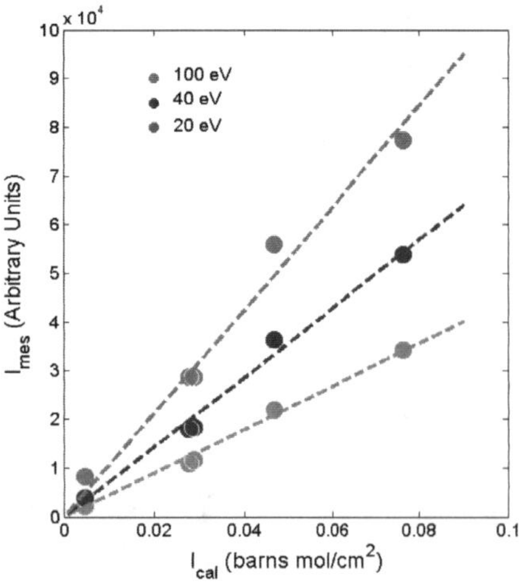

Figure 3 The scattering from the metal foils for each of the 3 incident energies plotted against the calculated scattering.

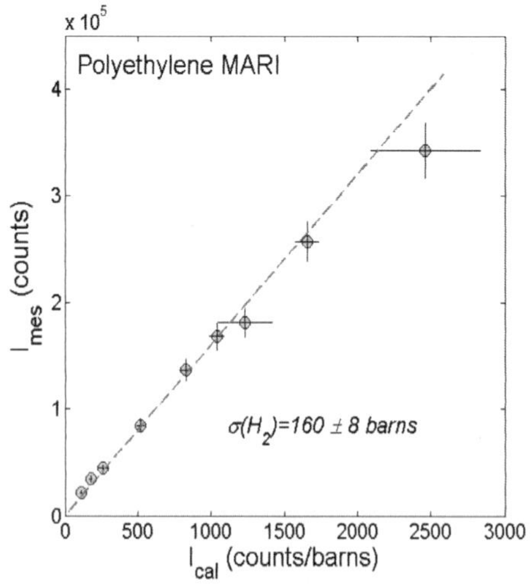

Figure 4 A summary of all the data for the hydrogen scattering plotted as absolute intensities against the calculated scattering. The straight line gives the scattering for each hydrogen atom as 80±4 barns.

The results shown in fig.2 were normalised to the scattering when the wave-vector became small. The results clearly show for all the results that within error the scattering is independent of the wave-vector transfer. This is different from the results found for polythene with VESUVIO because in that case the polythene scattering decreased in intensity as the wave-vector transfer increased.. Our second result is illustrated in fig.4 where the results for the different samples of polythene are shown plotted against the calculated cross-section so that the slope of the line gives the scattering cross-section for two hydrogen atoms. This result of averaging to obtain the best estimate of the cross-section for hydrogen gives a result that is within 2% of the nominal value and we estimate the error to be ± 4%. These results clearly show that the scattering from hydrogen does not vary with the wave-vector transfer and that it has a value that is agreement with other measurements of that cross-section. A large part of the error in these measurements, particularly from the metal foils, arises from uncertainty in the scattering cross-sections particularly in the eV energy range. The measurements in this part of the energy range often differ by a few percent points and this makes a substantial contribution to the error.

Nevertheless we have shown with a conventional direct geometry chopper that quantitative experiments can be performed with 100 eV incident neutrons. The energy resolution is increasing rapidly at these energies and better measurements would require a redesign of the choppers available for experiments using MARI. This and also possibly improvements to shorten the time bins measuring the time-of-flight would make a larger improvement to the resolution and experiments would still be possible because the neutron flux was very large. Some experiments using the MARI spectrometer to measure electronic transitions are in progress to exploit this new range of possible experiments.

The conclusion about our results is that experiments using MARI have some considerable advantages over the earlier measurements on VESUVIO. This is largely because it is technically possible to perform measurements with different incident energies so as to have the same wave-vector transfer at differing scattering angles. Because VESUVIO has very few absorption foils the wave-vector transfer can usually only be obtained with one set of instrumental parameters. The detector coverage is also larger on MARI. Nevertheless these features do not provide an explanation of the discrepancy between the VESUVIO and MARI results. We were surprised in our experiments how strong the gamma ray flux was at high energies. Possibly this has not been taken account of properly in the VESUVIO experiments at least in part because it was not possible to obtain the same energy transfer with different scattering energies. Needless to say we consider that since we obtain the expected theoretical results our conclusion is that the MARI experiment is to be preferred. Since completing these measurements we have learnt of the measurements using an indirect time-of-flight instrument similar to VESUVIO but with an electron linac used to produce its neutrons. This experiment showed that the recoil from hydrogen in mixtures of deuterated water agrees with the conventional theory [7]. Taken together the experiments on the electon linac and ours suggest conclusively that the scattering from hydrogen at high energies agrees with conventional impulse-approximation theory.

We are grateful for the help and support that we have had from the staff at RAL and for financial support from the Leverhulme Trust and from the STFC.

References

1. Chatzidimitriou-Dreismann C A, Abdul-Redah T, Streffer R M F and Mayers J 1997 *Phys. Rev. Lett.* **79**, 2839
2. Karlsson E B and Lovesey S W 2000 *Phys. Rev.* A **61** 062714
3. Cowley R A 2003 *J. Phys: Condens Matt.* **15** 4143 .
4. Gidopoulos N 2005 *Phys. Rev. B* **71** 054106
5. Reiter G F and Platzman P M *Phys Rev. B* **71** 054107
6. Waller I and Fröman P O 1952 *Ark. Fys.* **4** 183
7. Blostein J J, Rodriguez Palomino L A and Dawidowski J 2009 *Phys. Rev. Lett.* **102** 097401

Negative thermal expansion in the Prussian Blue analog $Zn_3[Fe(CN)_6]_2$: X-ray diffraction and neutron vibrational studies

S Adak[1,2], L L Daemen[2] and H Nakotte[1,2]

[1]Department of Physics, New Mexico State University, Las Cruces, NM 88003, USA
[2]Los Alamos Neutron Science Center, Los Alamos National Laboratory, Los Alamos, NM 87545, USA

E-mail: sadak@lanl.gov

Abstract. The cubic Prussian Blue (PB) analog, Zn_3 $[Fe(CN)_6]_2$, has been studied by X-ray powder diffraction and inelastic neutron scattering (INS). X-ray data collected at 300 and 84 K revealed negative thermal expansion (NTE) behaviour for this material. The NTE coefficient was found to be -31.1 x 10^{-6} K^{-1}. The neutron vibrational spectrum for $Zn_3[Fe(CN)_6]_2.xH_2O$, was studied in detail. The INS spectrum showed well-defined, well-separated bands corresponding to the stretching of and deformation modes of the Fe and Zn octahedra, all below 800 cm^{-1}.

1. Introduction

Negative thermal expansion (NTE), expansion on cooling or contraction upon heating, is an interesting phenomenon from both fundamental and applied physics viewpoints. It occurs in different classes of materials. In many cases, the physical mechanism behind this phenomenon is often not completely understood.

It seems to be "common knowledge" that many PB compounds display NTE, although quantitative information on thermal expansion coefficients is often missing. Furthermore, not all compounds in this family exhibit NTE. Therefore we started a comprehensive study of these compounds. The possibility of varying ion size and charge in this material offers an interesting playground for studying NTE systematically, and to investigate possible correlations with electronic and crystal structures. The linkage between the octahedral units in PB analogs with a linear cyanide ligand introduces more degrees of freedom in these cubic structures.

We studied the PB analog $Zn_3[Fe(CN)_6]_2$. xH_2O in some detail. X-ray powder diffraction showed that this compound exhibits strong NTE behavior. We performed in depth studies of the inelastic neutron scattering (INS) spectrum of $Zn_3[Fe(CN)_6]_2.xD_2O$, and compared the experimental results with simple semi-empirical cluster calculations.

2. Experiment and Computation

A polycrystalline sample of $Zn_3[Fe(CN)_6]_2$. xH_2O was prepared via standard chemical precipitation. All reagents were ACS quality and were used as-received without further purification. Inelastic neutron spectroscopic measurements were done using a deuterated sample, $Zn_3[Fe(CN)_6]_2.\ xD_2O$, prepared via the same synthesis route using D_2O instead of H_2O.

X-ray diffraction (XRD) patterns at room temperature (T_2 = 300K) and low temperature (T_1 = 84K) were collected under vacuum for the powder sample on a Rigaku Ultima III X-ray diffractometer, operated in Bragg-Brentano geometry with monochromatic CuK_α (λ = 1.54 Å) radiation. The lattice parameters were obtained from Rietveld refinements of the diffraction patterns using the General Structure Analysis System (GSAS) software [1].

The inelastic neutron scattering (INS) spectrum was collected using the Filter Difference Spectrometer (FDS) at Los Alamos Neutron Scattering Center (LANSCE). The interpretation of the INS spectrum was done with the help of simple cluster calculations (semi empirical, PM3 [2]) performed using HyperChem computational chemistry package [3], a-climax program [4], and the use of a few reference spectra [5].

3. Results & Discussions

$Zn_3[Fe(CN)_6]_2.xH_2O$ crystallizes into a cubic structure with space group $Fm\bar{3}m$. The structure consists of two types of octahedra ($Zn^{II}(NC)_6$ and $Fe^{III}(CN)_6$) arranged on a cubic lattice and linked by cyanide ligands. Water molecules, which contribute to the existence of structural disorder, appear in the space between the octahedrons. The structure is shown in figure 1.

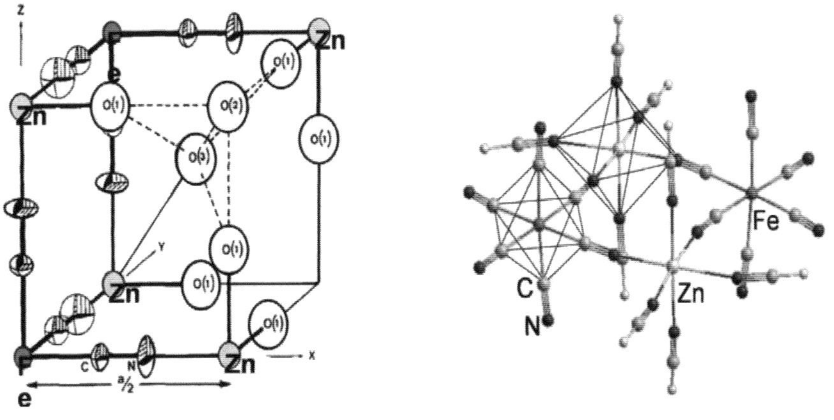

(a) (b)

Figure 1. The structure of $Zn_3[Fe(CN)_6]_2$ showing (a) the possible sites for water in $Zn_3[Fe(CN)_6]_2.14H_2O$ [6], and (b) a simple cluster model used to calculate the INS spectrum.

To determine if this PB analog shows NTE behavior, we studied the lattice parameters for the compound at two temperatures. We determined an average coefficient of thermal expansion (CTE) from the optimized lattice parameters obtained from the Rietveld refinement of X-ray diffraction (XRD) data collected at 300 and 84 K. For the refinements, crystallographic information available in the ICSD database [6] was used as a starting point. The lattice parameters, thermal parameters, atom positions, peak profiles, and background parameters were refined. Our refinements produced accurate values of the lattice parameters, even if atomic positions and thermal parameters are less reliable. From the lattice parameters at 300 and 84 K, we determined that the linear thermal expansion coefficient is -31.1 x 10^{-6} K^{-1}.

To gain more insight in the mechanism behind the NTE behavior in this compound, we studied the neutron vibrational spectrum. The modes of vibration of an octahedral molecule, $X(YZ)_6$, are easily obtained. The 33 modes of vibration of $X(YZ)_6$ where Y and Z are on 4-fold axes are:

$$\Gamma = 2.A_{1g} + 2.E_g + 1.F_{1g} + 4.F_{1u} + 2.F_{2g} + 2F_{2u}$$

There are now 13 distinct types of modes: 4 stretches and 9 deformations.

The INS results are shown in figures 2 and 3. The INS spectra were interpreted with the help of simple cluster calculations. Figure 1(b) shows the cluster model used for our calculation. The linkages in this compound comprise 2 atoms and have internal dynamics, which makes the situation more complex than in silicates. Therefore progress requires a reasonably complete interpretation of the vibrational spectrum of the PB analog. A structurally optimized cluster using the HyperChem package [3] was used to obtain harmonic frequencies. These results were then used with a-climax [4] to calculate the neutron vibrational spectrum of the material. Spectral interpretation was done from model compounds, group theory, group frequencies, and comparison with reference IR and Raman spectra.

Figure 2 shows the INS spectrum for $Zn_3[Fe(CN)_6]_2.xD_2O$ below 700 cm^{-1}. Analysis of the INS data shows well-defined, well-separated bands corresponding to the stretching and deformation modes of the Fe and Zn octahedra, all below 800 cm^{-1}.

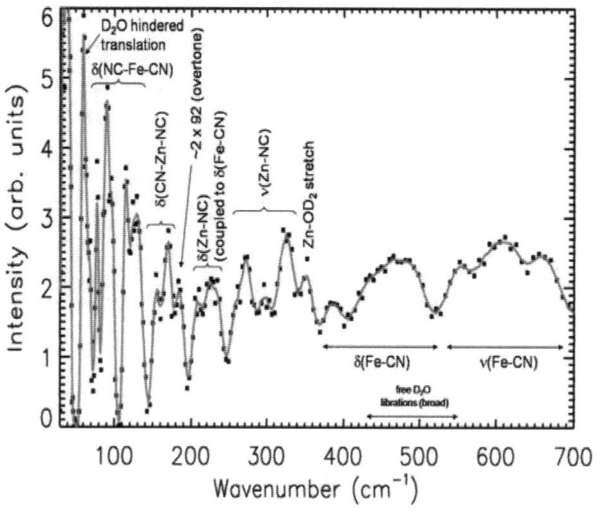

Figure 2. The inelastic neutron scattering spectrum (INS) of $Zn_3[Fe(CN)_6]_2.xD_2O$ below 700 cm^{-1} showing the different bands of stretching and deformation modes of Fe and Zn octahedra.

The INS spectrum in the 500-2500 cm^{-1} range is shown in figure 3. We observed weak CN stretches, which is not unexpected given the small scattering cross section of C and N and the weak amplitude of their motion. In $Zn_3[Fe(CN)_6]_2.14H_2O$ [6], Zn occupies the Wyckoff position 4b with unit occupancy. By contrast, Fe, C, and N occupy Wyckoff sites 4a, 24e, and 24e, respectively –all with occupancy 2/3. Fe, C, or N sites may be occupied by oxygen (from water). If so, $Fe(CN)_6$ octahedra should be intact, but isocyanide ligands may be missing around Zn. Instead, Zn-O(H) may be present.

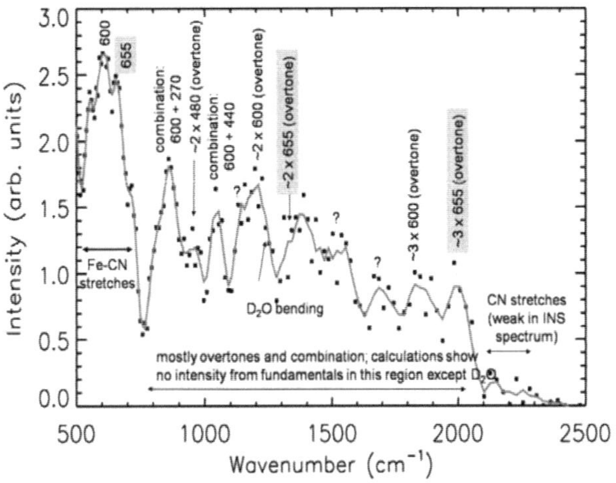

Figure 3. The inelastic neutron scattering spectrum (INS) of $Zn_3[Fe(CN)_6]_2.xD_2O$ within 500-2500 cm^{-1} energy range showing combination, overtones and other vibrational modes.

In addition to the water at well-defined sites in the lattice, lattice water may be present in the cavities between the octahedrons. Relatively strong Zn-OD_2 stretching mode at 380 cm^{-1} in the INS spectrum confirms that some water occupies interstitial positions in the lattice.

4. Conclusion

To our knowledge this is the first time the INS spectrum of a PB analog has been calculated and interpreted in greater detail. More detailed calculations of the vibrational modes are in progress to finalize the interpretation of the vibrational spectrum and help with future measurements of mode Grüneisen parameters. Indeed, this will open up the possibility of correlating NTE behavior with the temperature dependence of the mode Grüneisen parameter.

Acknowledgement

The Basic Energy Sciences program of the US Department of Energy supported the work at Los Alamos Neutron Science Center (LANSCE).

References

[1] Larson A C and von Dreele R B 2000 "General structure analysis system (GSAS)" *Los Alamos National Laboratory Report* **No. LAUR** 86-748
[2] Stewart J P P 1989 *J. Comput. Chem.* **10** 209-220
[3] HyperChem(TM) Professional 7.51 Hypercube Inc. 1115 NW 4th Street Gainesville Florida 32601 USA
[4] Ramirez-Cuesta A J 2004 *Comput. Phys. Commun.* **157** 226
[5] Nakamoto K 1997 *Infrared and Raman Spectra of Inorganic and Coordination Compounds Part: B Applications in Coordination, Organometallic, and Bioinorganic Chemistry* (New York: Wiley–Interscience)
[6] Gravereau P and Garnier E 1984 *Acta Cryst. C* **40** 1306

Anomalous phonons in $CaFe_2As_2$ explored by inelastic neutron scattering

R. Mittal[1,2], L. Pintschovius[3], D. Lamago[3,4], R. Heid[3], K-P. Bohnen[3], D. Reznik[3], S. L. Chaplot[2], Y. Su[1], N. Kumar[5], S. K. Dhar[5], A. Thamizhavel[5], and Th. Brueckel[1,6]

[1]Juelich Centre for Neutron Science, IFF, Forschungszentrum Juelich, Outstation at FRM II, Lichtenbergstr. 1, D-85747 Garching, Germany
[2]Solid State Physics Division, Bhabha Atomic Research Centre, Trombay, Mumbai 400 085, India
[3]Forschungszentrum Karlsruhe, Institut für Festkörperphysik, P.O.B. 3640, D-76021 Karlsruhe, Germany
[4]Laboratoire Leon Brillouin, CEA-Saclay, F-91191 Gif sur Yvette Cedex, France
[5]Department of Condensed Matter Physics and Material Sciences, Tata Institute of Fundamental Research, Homi Bhabha Road, Colaba, Mumbai 400 005, India
[6]Institut fuer Festkoerperforschung, Forschungszentrum Juelich, D-52425 Juelich, Germany

E-mail: rmittal70@yahoo.co.in

Abstract. Extensive inelastic neutron scattering measurements of phonons on a single crystal of $CaFe_2As_2$ allowed us to establish a fairly complete picture of phonon dispersions in the main symmetry directions. The phonon spectra were also calculated by density functional theory (DFT) in the local density approximation (LDA). There are serious discrepancies between calculations done for the optimized structure and experiment, because the optimised structure is not the ambient pressure structure but is very close to the "collapsed" structure reached at p = 3.5 kbar. However, if the experimental crystal structure is used the calculation gives correct frequencies of most phonons. The most important new result is that linewidths/frequencies of certain modes are larger/softer than predicted by DFT-LDA. We also observed strong temperature dependence of some phonons near the structural phase transition near 172 K. This behavior may indicate anomalously strong electron phonon coupling and/or anharmonicity, which may be important to the mechanism of superconductivity.

1. Introduction

The discovery of superconductivity at temperatures exceeding 50 K in iron arsenide compounds with general compositions RFeAsO (R = rare earth) and MFe_2As_2 (M = alkaline earth metal) and MFeAsF has attracted great interest [1-15] in these materials. At present, it is hotly debated whether these compounds are unconventional metals similar to the cuprate superconductors or can be understood within the same theoretical framework as conventional intermetallic compounds like the borocarbides or MgB_2. Superconductivity in these compounds appears either at a critical doping level of the parent compound, or by application of pressure above a critical value. The role of the phonons for the mechanism of superconductivity is not known at present. DFT calculations predict weak electron-phonon coupling [9] with a negligible contribution to the superconductivity mechanism. Inelastic x-ray scattering investigation [10] of the phonon density of states in $LaFeAsO_{1-x}F_x$ and NdFeAsO, as well as measurements of a few phonon branches [10(a),11] on single crystals of $BaFe_2As_2$ and $PrFeAsO_{1-y}$ showed that DFT is only moderately [10(a),12] successful in predicting phonon frequencies in these compounds. The phonon density of states in $BaFe_2As_2$, $Sr_{0.6}K_{0.4}Fe_2As_2$ and $Ca_{0.6}Na_{0.4}Fe_2As_2$ was investigated on polycrystalline samples using inelastic neutron scattering [13,14]. Empirical models used to analyze the data again had limited success.

2. Experiment and phonon calculations

Single crystals of $CaFe_2As_2$ (15 mm × 10 mm × 0.4 mm) were grown from a high temperature solution using Sn as flux [15]. The details of crystal characterization are given in Ref. [15]. The neutron measurements were performed on the 1T1 triple-axis spectrometer at the Laboratoire Léon Brillouin, Saclay. Measurements were done with pyrolytic graphite (PG002) as a monochromator and analyzer. Most measurements were carried out at 300K with open collimations and double focusing on both the analyzer and the monochromator. Selected phonons were studied as a function of temperature down to T = 100 K, which is well below the magnetic/structural phase transition at 172 K.

The calculations were carried out within the framework of the LDA and GGA using a mixed basis pseudopotential method [16]. A density functional perturbation approach was used for calculating the phonon frequencies and phonon eigenvectors [17]. We employed norm-conserving pseudopotentials and a plane-wave cutoff of 22 Ryd, augmented by local functions at the Ca and Fe sites. Brillouin zone (BZ) summations were done with a Gaussian broadening technique using a broadening of 0.2 eV and 40 wavevector points in the irreducible part of the BZ.

3. Results and discussion

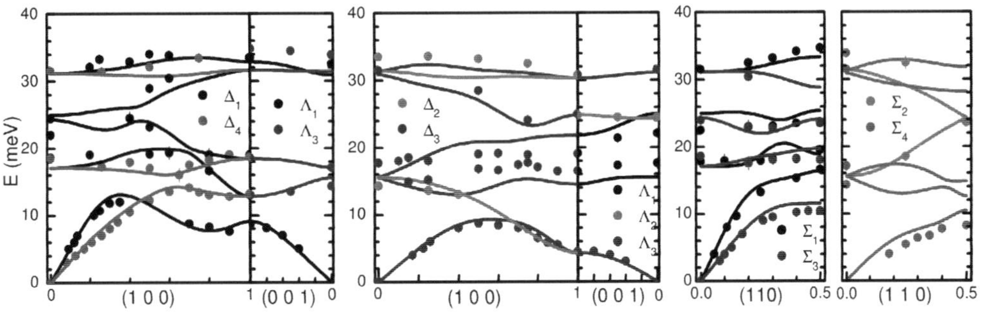

Fig. 1 Comparison of experimentally determined [21] phonon frequencies (solid circles) in the (100), (001) and (110) directions at T = 300 K with results of density functional theory (solid lines). The calculations were based on the experimental crystal structure. The 15 phonon modes along the Δ(100), Λ(001) and Σ(110) directions can be classified as $\Delta: 5\Delta_1 + 2\Delta_2 + 5\Delta_3 + 3\Delta_4$; $\Lambda: 4\Lambda_1 + \Lambda_2 + 5\Lambda_3$; $\Sigma: 4\Sigma_1 + 2\Sigma_2 + 4\Sigma_3 + 5\Sigma_4$.

The optimized structure was initially used for the phonon dispersion calculation. However, as reported previously [18,19], we soon realized that the optimized structure is relatively far away from the experimental one. Figure 1 shows that DFT in the LDA is quite successful in predicting the phonon frequencies in $CaFe_2As_2$ if, instead of the relaxed structure, the experimental one is used. It is important to emphasize here that one must impose a nearly perfect tetrahedral environment of the Fe atoms as observed in experiment in order to obtain best agreement between calculated and experimental phonon frequencies. However, even in this case it is worse than in many other compounds, including conventional superconductors with high T_c like MgB_2 [20]. This finding is similar to the previous observations on the Ba122 compounds by inelastic x-ray scattering [10(a)].

Although the calculations based on the experimental structure appear to be more accurate, some important differences with experiment remain. The main one is for the phonons of Δ_3 symmetry between q=(0.5 0 0) and (1 0 0). Some appear around 19 meV but predicted at 22 meV and others are observed around 16 meV but predicted 2 meV lower. One must also keep in mind that the good agreement for other phonons may be somewhat misleading in the sense that the predicted eigenvectors may differ from the experimental ones even where the phonon frequencies agree. In experiment, phonon eigenvectors determine observed phonon intensities. When phonons are nearly degenerate as in $CaFe_2As_2$ near 20 meV, different phonon branches may hybridize. In this case small differences between calculated and experimental frequencies result in large differences in the eigenvectors and the comparison between predicted and calculated phonon intensities is not very meaningful. However, in the case of the Σ_3-frequencies observed at the zone boundary, the situation is clear-cut: because of the high symmetry of the zone boundary point, the Σ_3-phonons decompose into sub-groups. The mode observed at E = 18 meV with very high intensity at (2.5,1.5,0) is single and hence, its eigenvector is completely determined by symmetry. It therefore can be unambiguously assigned to a mode calculated at 23 meV by theory. On these grounds, the disagreement between calculated and observed frequencies of Σ_3-symmetry at the **q**=(0.5, 0.5, 0) zone boundary is stronger than one might guess from inspection of Fig. 1 when considering the eigenvectors, the data points shown at 18 meV and 23 meV correspond to calculated frequencies at 23 meV and 20 meV, respectively, i.e. the phonon frequencies are "flipped".

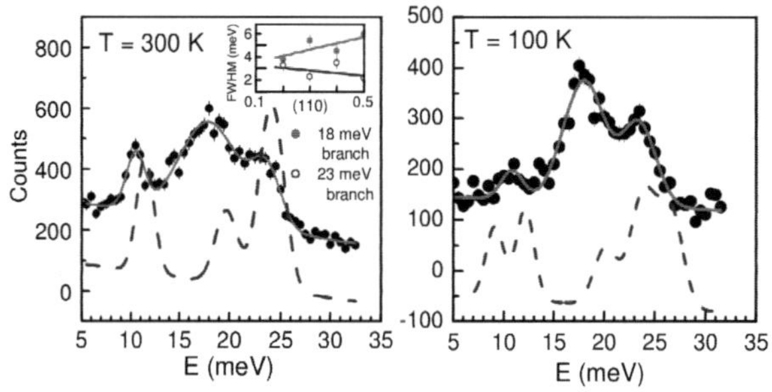

Fig. 2 Energy scans [21] taken at Q = (2.5,1.5,0) at room temperature and at a temperature far below the structural phase transformation. The calculated phonon structure factors for non-magnetic and spin-polarized are shown in left and right panels respectively. For better visibility the calculated profiles (dashed lines) in the lower and upper panels are shifted down by 200 counts. The insert in the left panel shows the q-dependence of the phonon line widths of the branch around 18 meV (red dots) and around 24 meV (blue dots).

We also found substantial line broadenings for a number of phonons. For instance, an energy scan at a wavevector Q = (2.5,1.5,0) and T = 300 K shows a pronounced broadening for a mode at 18 meV (Fig. 2, left), which, based on its intensity, can be unambiguously assigned to Fe vibrations. The line broadening of this branch is maximum at the zone boundary, which becomes a reciprocal lattice point in the low temperature orthorhombic phase. As already mentioned above, its frequency is considerably lower than 23 meV calculated by DFT. There is very little change of this mode on cooling from 300 K to 190 K but its linewidth shrinks considerably below the tetragonal-to-orthorhombic phase transition at 172 K (Fig. 2, right). These observations indicate a close relationship between the line broadening and the structural instability. However, there is no direct relationship between the elongation patterns of the 18 meV mode and the displacements during the phase transition.

We have also carried out spin-polarized DFT-GGA calculations in the orthorhombic phase of $CaFe_2As_2$. The calculated phonon spectra for non-magnetic/spin-polarized structures are shown as dashed lines Fig. 2 left/right respectively. It appears that the calculated line widths of phonon modes are larger in the orthorhombic phase because the orthorhombic distortion leads to a splitting of modes. The agreement between our experimental results and the calculation is poor, which further suggest anomalous phonons in $CaFe_2As_2$. Unlike the case of $BaFe_2As_2$ [11], including magnetism does not improve agreement with experiment (Fig. 2 right).

Simple anharmonicity is unlikely to account for the observed line widths. Strong coupling of phonons to electron-hole excitations is another possibility. However, we calculated the electron-phonon coupling induced phonon line widths using density functional perturbation theory and found that the calculated line widths are much smaller than observed. Also, no strong broadenings appear in $BaFe_2As_2$ [3], which has similar electronic and crystal structure, and also orders magnetically below 170 K. Since the main difference between $BaFe_2As_2$ and $CaFe_2As_2$ that the former is not close to the "collapsed" high pressure phase, this points to the proximity to the "collapsed" high pressure phase as the most probable explanation of the phonon anomalies in $CaFe_2As_2$. A mechanism for this behavior, which accounts for the difference between $CaFe_2As_2$ and $BaFe_2As_2$, has been proposed in [18]. On the other hand, our observation that the broadening of the 18 meV mode becomes much smaller in the magnetic phase leads to the conclusion that the proximity to the magnetic phase is important.

4. Conclusions

More work is necessary to understand the effects we report here. In any case, our findings indicate that the interplay between magnetism and the lattice vibrations is in some way responsible for the anomalous phonons in $CaFe_2As_2$. That is to say, the coupling of the vibrational and the electronic degrees of freedom is stronger than calculated by DFT, and hence phonons might play an important role in superconductivity in the doped compounds.

[1] Y. Kamihara et al, J. Am. Chem. Soc. **130**, 3296 (2008).
[2] G. Wu et al, J. Phys: Condens. Matter **21**, 142203 (2009).
[3] Zhi-An Ren et al, Chin. Phys. Lett. **25**, 2215 (2008).
[4] M. Rotter et al, Phys. Rev. Lett. **101**, 107006 (2008).
[5] M. S. Torikachvili et al, Phys. Rev. Lett. **101**, 057006 (2008).
[6] A. Kreyssig et al, Phys. Rev. B **78**, 184517 (2008).
[7] A. I. Goldman et al, Phys. Rev. B **78**, 100506(R) (2008).
[8] D. J. Singh, Phys. Rev. B **78**, 094511 (2008).
[9] L. Boeri et al, Phys. Rev. Lett **101**, 026403 (2008).
[10] (a)T. Fukuda et al, J. Phys. Soc. Japan. **77**, 103715 (2008). (b) M. Le Tacon, et al, Phys. Rev. B **78**, 140505 (2008).
[11] D. Reznik et al, arXiv: 0810.4941.
[12] A. D. Christianson et al, Phys. Rev. Lett. **101**, 157004 (2008).
[13] R. Mittal et al, Phys. Rev. B **78**, 104514 (2008).
[14] R. Mittal et al, Phys. Rev. B **78**, 224518 (2008).
[15] N. Kumar et al, Phys. Rev.B **79**, 12504 (2009).
[16] B. Meyer, C. Elsässer, M. Lechermann, and F. Fähnle, FORTRAN90, a program for mixed-basis pseudopotential calculations for crystals, MPI für Metallforschung, Stuttgart.
[17] R. Heid and K.-P. Bohnen, Phs. Rev. B **60**, R3709 (1999).
[18] T. Yildirim, Phys. Rev. Lett. **102**, 037003 (2009).
[19] D. Kasinathan et al., New J. Phys. **11**, 025023 (2009).
[20] A. Q. R. Baron et al. Phys. Rev. Lett. **92**, 197004 (2004).
[21] R. Mittal et al, Phys. Rev. Lett. **102**, 217001 (2009).

The crystal structure of superconducting $FeSe_{1-x}Te_x$ by pulsed neutron diffraction

M C Lehman[1], A Llobet[2], K Horigane[3] and D Louca[1]
[1]University of Virginia, Department of Physics, VA 22904, USA
[2]Los Alamos National Laboratory, MS H805, Los Alamos, NM 87545, USA
[3]WPI Advanced Institute for Materials Research, Tohoku University, Katahira 2-1-1, Aoba, Sendai 980-8577, JAPAN

Email: mcl4v@virginia.edu

Abstract. A transition to a superconducting state was recently observed in the binary alloy of $FeSe_{1-x}Te_x$ system where T_C rises with increasing x. The substitution of the larger Te for Se ion results in no additional charges but increases the internal chemical pressure. Earlier studies suggested that the crystal structure maintains the tetragonal *P4/nmm* symmetry with the substitution of Te where the average bond angle, α, decreases considerably from ~ 104° in FeSe to 100.5° in the mixed phase of $FeSe_{0.5}Te_{0.5}$. With the use of pulsed neutron power diffraction and the Rietveld analysis, the crystal structure refinement for $FeSe_{0.5}Te_{0.5}$ yielded very large thermal factors in the superconducting phase indicative of the presence of structural distortions that may be significant in understanding the electronic and magnetic properties of this system.

1. Introduction

The recent discovery of superconductivity in $FeSe_{1-x}Te_x$ has resulted in a simple binary system with tetrahedrally coordinated Fe layers, similar to the original FeAs oxy-pnictide based superconducting compounds [1]. While density functional calculations indicate this system is not an electron-phonon type superconductor [2-4], recent NMR measurements suggested that antiferromagnetic (AFM) fluctuations may even increase with T_C in FeSe [5]. Additionally, the substitution of Se^{2-} with Te^{2-} implies that charge doping is not critical to raise T_C in this class of superconductors, unlike in the case of high T_C cuprates where hole or electron doping in the AFM parent lattice is needed to make the system superconducting. Although the mechanism of superconductivity is not well understood, and perhaps the role of magnetism is not solely important to the pairing mechanism, this is a new class of superconductors with distinctly unique properties from the cuprates and possibly with a new mechanism. At the same time, the application of external pressure can enhance T_C directly implicating that the pairing mechanism is in some way influenced by the lattice structure. In particular, an interesting structural property relevant to the oxy-pnictide superconductors is that T_C becomes a maximum when the Fe tetrahedra become most regular [6]. It thus appears that the crystal structure is a key element to the understanding of the ground state properties of the ferrous compounds. With the use of high resolution pulsed neutron powder diffraction, we investigated the crystal structures of FeSe and $FeSe_{0.5}Te_{0.5}$ above and below T_C. We find that nominally FeSe can be described using the tetragonal, *P4/nmm* symmetry above T_C and the orthorhombic symmetry *Cmma* below. For $FeSe_{0.5}Te_{0.5}$, no clear structural transition is observed but the thermal factors increase significantly below T_C indicative of non-periodic structural distortions. Details of the refinement will be discussed.

2. Methods

The samples were prepared by solid-state reaction from Fe (99.9 %), Se (99.9 %), and Te (99.9 %). Stoichiometric amounts of the starting materials were mixed, and the pressured powders were sealed in an evacuated quartz tube and heated at 700°C for 24 hours. The details of sample preparation can be found in Ref. [7]. The pulsed neutron diffraction measurements were performed using the high intensity powder diffractometer (HIPD) of Los Alamos National Laboratory on polycrystalline samples of FeSe and $FeSe_{0.5}Te_{0.5}$ at temperatures above and below the superconducting transition. The Rietveld crystallographic method of analysis was used to determine the crystal structure.

3. Results and Discussion

Stoichiometric α-FeSe is superconducting with a $T_C \sim 7$ K. Consistent with earlier findings, the crystal symmetry above T_C is tetragonal and the refined parameters are summarized in Table 1(a). This

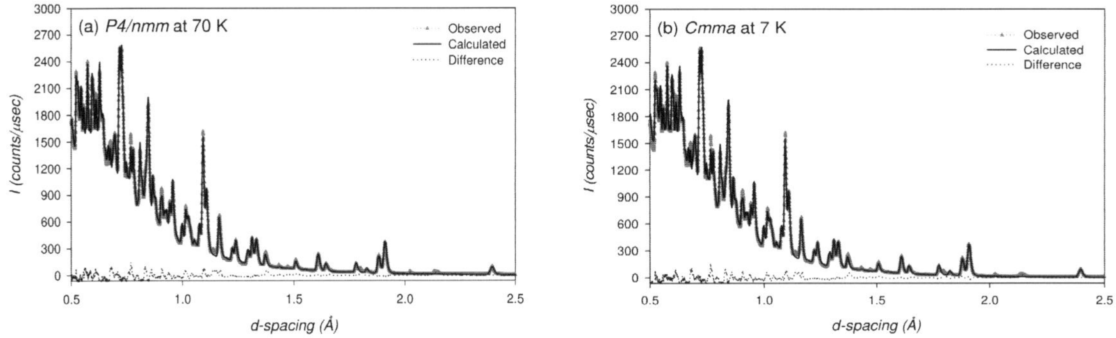

Figure 1 The diffractograms at 70 K (left) and 7 K (right) of FeSe with a T_C of ~ 7 K, are compared to a tetragonal and orthorhombic crystal model, respectively. The difference is shown as well.

Table 1: (a) The refined parameters for FeSe at 70 K using the *P4/nmm* space group. The lattice constants are given by a = b = 3.75993(3) Å and c = 5.48220(6) Å. Rp = 3.31 % and wRp = 5.74 %.

Atom	Site	x	y	z	U11	U22	U33 (Å^2)	Frac
Fe	2a	0.75	0.25	0	0.0017(3)	0.0017(3)	0.0028(4)	1.0
Se	2c	0.25	0.25	0.26628	0.0008(3)	0.0008(3)	0.0033(6)	1.0

(b) The refined parameters for FeSe at 7 K using the *Cmma* space group. The lattice constants are given by a = 5.32491(5) Å, b = 5.30890(1)(0) Å and c = 5.47905(1) Å. Rp = 3.36 % and wRp = 5.76 %.

Atom	Site	x	y	z	U11	U22	U33 (Å^2)	Frac
Fe	4a	0.25	0	0	0.0007(8)	0.0003(1)	0.0013(4)	1.0
Se	4g	0	0.25	0.26598	0.005(1)	0.005(1)	0.0033(5)	1.0

particular sample contained about 16 % of the Fe_7Se_8 phase [see Ref. [7] for details of the refinement of the impurity phase]. Anisotropic thermal factors were used in the refinement that showed a tendency for larger $\langle U \rangle$ factors along the c-axis. The diffraction data are compared to a model calculated using the *P4/nmm* symmetry as shown in Fig. 1a. With cooling, a subtle structural change is observed. Recently it was reported that this change is due to a structural transition to an orthorhombic symmetry, *Cmma* [8]. The result of the comparison between the 7 K data to this crystal symmetry is shown in Fig. 1(b) and the parameters are summarized in Table 1(b). Note that the thermal factors increase below T_C.

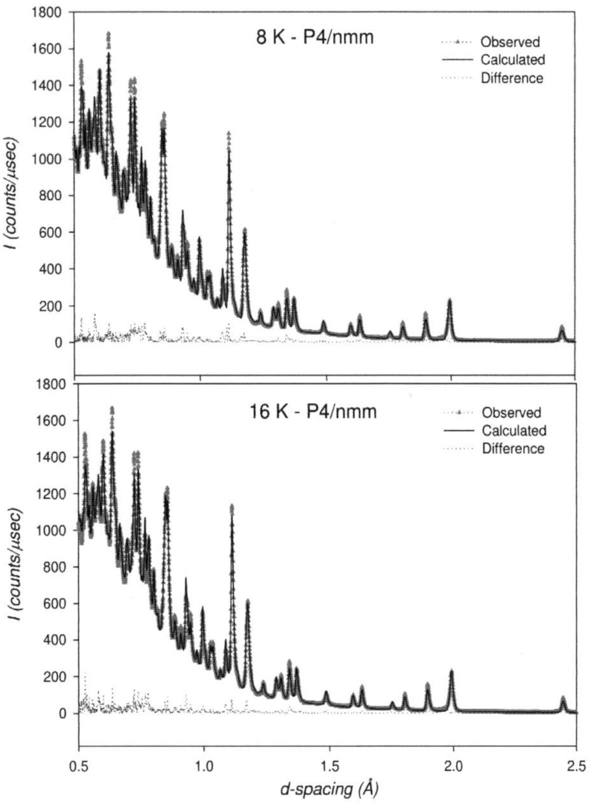

Figure 2 The diffraction data at 8 K (top) and 16 K (bottom) are compared to a crystal model for the *P4/nmm* symmetry. The difference is shown as well in black.

With the substitution of Te for Se, it is presumed that no structural transition is observed below T_C and that the symmetry remains in the tetragonal phase of *P4/nmm*. The refinement results are summarized in Table 2(a) and (b). This is different from the FeSe discussed above. At the same time, the absence of a structural transition is analogous to the $LaO_{1-x}F_xFeAs$ that also showed no structural transition when the system entered the superconducting regime [9]. This effect was in turn correlated with the absence of long-range antiferromagnetic order in the superconducting state and the onset of a spin-density wave (SDW). Shown in Fig. 2 are the diffraction patters at 8 and 16 K compared to the tetragonal crystal symmetry. The fits are reasonable. However, with cooling, we observed that the thermal factors, <U> $Å^2$, increase in an uncharacteristic way, indicating that atomic distortions are enhanced in the superconducting phase. In most solids upon cooling, the thermal factors are reduced because of the decrease in thermal vibrations. The increase in thermal factors is probably in response to local lattice distortions that may be key to the superconducting mechanism. This calls for a study of the local atomic structure that can directly probe that local lattice without the assumption of crystal periodicity.

4. Summary

In the Fe-Se-Te system, T_c is enhanced by changing the ionic size from Se to Te with no additional charges added to the system. This is uniquely different from the cuprate superconductors where the onset of superconductivity is coupled with either hole or electron doping. The negative pressure induced by substituting the larger Te for Se is what probably drives the tetrahedra to become more regular. This also changes the bond angle between the Se-Fe-Se ions that is in turn linked to the higher T_C in the doped system. At the same time, the large thermal factors observed in the refined structural parameters below T_C suggest that local distortions are present that are not manifested in the long-range symmetry.

Table 2: (a) The refined parameters for $FeSe_{0.5}Te_{0.5}$ using space group *P4/nmm* at 16 K. a = 3.791266 Å and c = 5.939199 Å; R_p = 4.17 %, wR_p = 5.85 %;

Atom	Site	x	y	z	U (Å^2)	Frac
Fe(1)	2a	0.75	0.25	0	0.0015(2)	1.0
Se/Te	2c	0.25	0.25	0.266902	0.0029(1)	1.0
Fe(2)	2c	0.25	0.25	0.808282	0.0081(1)	0.03

(b) The refined parameters for $FeSe_{0.5}Te_{0.5}$ at 8 K using the *P4/nmm* space group. The lattice constants are given by a = 3.79151(1) Å and c = 5.93898(0) Å. Rp = 4.12 % and wRp = 5.89 %.

Atom	Site	x	y	z	U (Å^2)	Frac
Fe(1)	2a	0.75	0.25	0	0.0109(4)	1.0
Se/Te	2c	0.25	0.25	0.267573	0.0113(8)	1.0
Fe(2)	2c	0.25	0.25	0.808282	0.0081(1)	0.03

5. Acknowledgments

The authors would like to acknowledge valuable discussions with K. Yamada. This work is supported by the U. S. Department of Energy, Office of Basic Energy Sciences under contract DE-FG02-01ER45927. This work has benefited from the use of HIPD at the Lujan Center at Los Alamos Neutron Science Center, funded by DOE Office of Basic Energy Sciences. Los Alamos National Laboratory is operated by Los Alamos National Security LLC under DOE Contract DE-AC52-06NA25396.

6. References

[1]. Hsu F C *et al.* 2008 *Proc. Nat. Acad. Sci.* **105** 14262.
[2]. Subedi A, Zhang L, Singh D J, Du M H 2008 *Phys. Rev. B* **78** 134514.
[3]. Singh D H and Du M-H 2008 *Phys. Rev. Lett.* **100** 237003.
[4]. Yildirim T 2009 *arXiv*:0902.3462v1.
[5]. Imai T, Ahilan K, Ning, F L, McQueen, T M and Cava, R J 2009 *arXiv:*0902.3832.
[6]. Lee C-H *et al.* 2008 *J. Phys. Soc. Jpn* **77** 083704.
[7]. Horigane K, Takeshita N, Lee C H, Hiraka H, Yamada K 2009 *J. Phys. Soc. Jpn.* **78** 063705.
[8]. Margadonna S *et al.* 2008 *Chem. Comm.* 14262.
[9]. De la Cruz C *et al.* 2008 *Nature* **453** 899.

Single-crystal neutron diffraction studies on Ni-based metal-pnictide superconductor $BaNi_2As_2$

K Kothapalli[1], F Ronning[2], E D Bauer[2], A J Schultz[1] and H Nakotte[1,3]

[1] *Department of Physics, New Mexico State University, Las Cruces, NM 88003, USA*
[2] *MPA-10, Los Alamos National Laboratory, Los Alamos, NM 87545, USA*
[3] *LANSCE, Los Alamos National Laboratory, Los Alamos, NM 87545, USA*

E-mail: karunakark@lanl.gov

Abstract. We report the results of single-crystal neutron diffraction studies of the superconductor $BaNi_2As_2$. The experiments were performed on a tiny crystal of mass 0.8 mg at several temperatures between 20 and 200 K using the Single Crystal Diffractometer, SCD, at the Los Alamos Neutron Science Center. Above 130 K, $BaNi_2As_2$ crystallizes in the tetragonal $ThCr_2Si_2$ structure. Our neutron diffraction data corroborate a first-order structural transition around 130 K with a relatively large hysteresis of about 10 K, in agreement with observations from bulk studies. The anisotropic thermal displacement coefficients are enhanced along *c*-axis approaching the transition, and a splitting is observed for in-plane type reflections below the transition, which is evidence for a change in crystal structure.

1. Introduction

The discovery of superconductivity in the metal-pnictide AT_2Pn_2 (A = Ba, Sr, Ca, Eu, T = Fe, Ni, Pn = P, As) compounds has furthered the effort to understand the relation between structure, superconductivity and magnetism. Fe-based pnictides have been studied extensively and the Ni-based pnictide compounds have gained attention because of their close structural relationship to their Fe-counterparts and because of the possibilities they offer due to the tunability of the T_2Pn_2 layer [1, 2]. Here, we report the initial single-crystal neutron diffraction studies on $BaNi_2As_2$ performed on the newly installed Single Crystal Diffractometer (SCD) at Los Alamos Neutron Science Center. $BaNi_2As_2$ has been reported to be a fully gapped BCS superconductor with superconducting transition at 0.7 K [3].

2. Experimental

The experiments were performed on a very small plate-like single crystal of $BaNi_2As_2$, grown by Pb flux, with approximate dimensions of 1.3 x 2.0 x 0.1 mm^3 (volume ~1/4 mm^3) and a total weight of only ~0.8 mg. The experiments were performed on the Single-Crystal Diffractometer at the Los Alamos Neutron Science Center (LANSCE), Los Alamos National Laboratory. SCD had been in operation at the Intense Pulsed Neutron Source (IPNS), Argonne National Laboratory [4], and it was reinstalled at LANSCE following the recent closure of the IPNS. At LANSCE, SCD benefits from significantly higher neutron beam currents compared to IPNS, and it is therefore possible to obtain sufficient statistics at reasonable measuring times (3-4 hours per histogram) on samples that are much smaller (~0.1 mm^3) than samples typically used for neutron diffraction experiments. The capability of measuring small samples was particularly important for the present study, where we had only a small single crystal available. We collected data for seven different angular settings at 5 different temperatures (200 K, 150 K, 120 K, 90 K and 60 K). At 200 K, statistically significant intensities ($>3\sigma$) were found for more than 300 unique *(hkl)* reflections. We had similar numbers of observations for the other temperatures, for which we collected 7 histograms. Finally, we collected additional data for one particular angular setting containing the *(0kl)*-type reflection at another 13 temperatures between 21 and 165 K. For each histogram, we collected data for at least 3 hours. Integrated intensities of the diffraction spots were computed using subroutines in the Integrated Spectral Analysis Workbench (ISAW) software. Next, we determined the structural parameters of our $BaNi_2As_2$ single crystal by fitting integrated intensities to the structural models using the Generalized Structural Analysis Software (GSAS) [5].

3. Results and discussion

The refined structural parameters and the R-factors at 200 K are listed in Table 1 and our data agree well with previously determined results [2]. Using GSAS we refined the structural parameters and the anisotropic thermal factors. For the present study, absorption and extinction were found to be negligible.

Table1. Structural parameters and atom positions of $BaNi_2As_2$ at 200 K

space group: I4/*mmm* (no. 139) - tetragonal
lattice parameters: a = 4.136(6) Å; c = 11.66(4) Å
χ^2 = 1.065, R(F) = 6.09 %, $R_w(F^2)$ = 15.04%

Atom	*Position*	*x*	*y*	*z*
Ba	*2a*	0	0	0
Ni	*4d*	0	½	¼
As	*4e*	0	0	0.3469(3)

A full structural refinement was performed also for the 150 K data and the parameters differ mostly in the thermal parameters. The thermal ellipsoids for each atom at two representative temperatures are shown in figure 1. The vertical direction is the *c*-axis direction. Upon cooling, the atomic displacement parameters of all atoms are increased in the *c*-axis direction. This is a clear indication that the nearby transition involves atom displacements along this direction.

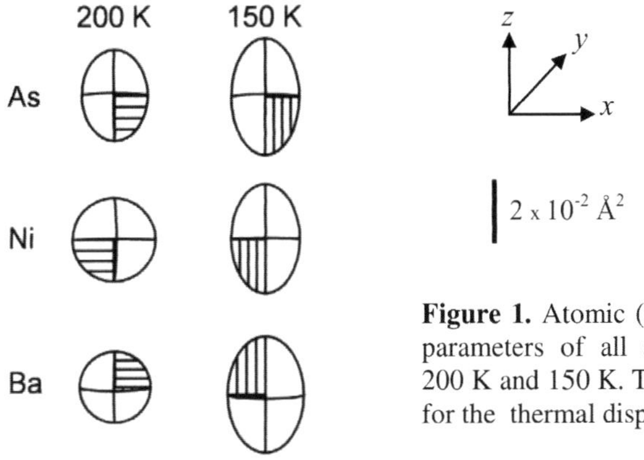

Figure 1. Atomic (thermal) displacements parameters of all atoms in $BaNi_2As_2$ at 200 K and 150 K. The line shows the scale for the thermal displacements.

Next we cooled the sample below the transition temperature at ~130 K, as indicated from bulk measurements [2]. There is clear evidence for a structural phase transition in $BaNi_2As_2$ at that temperature. This is shown by the observation that some of the 'high-temperature' tetragonal *(hkl)* peaks broaden or split, as shown in figures 2 and 3.

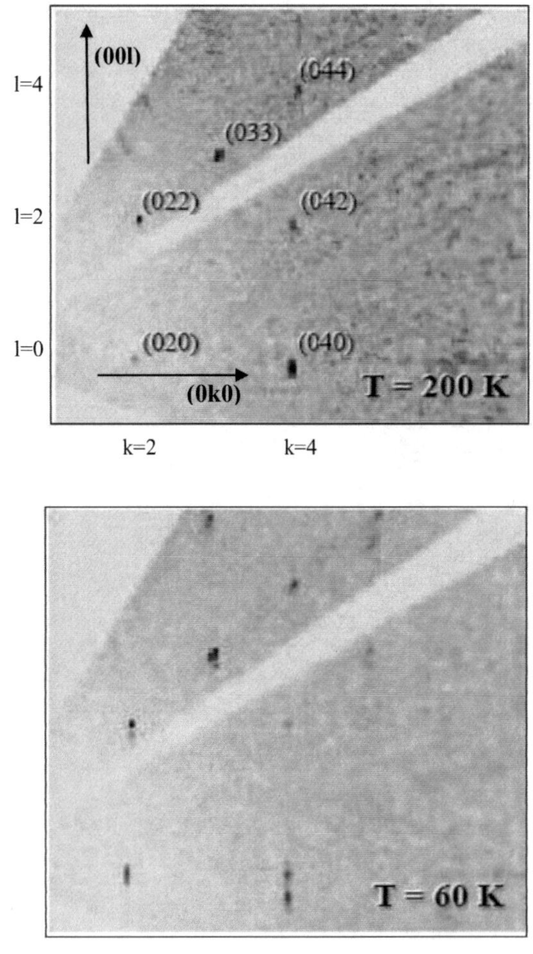

Figure 2. Exemplary (0kl) slice of the data at 200 K and 60 K. Tetragonal indices for the peaks at 200 K are provided. Note that the reflections seen at 200 K are found to be broadened or split at 60 K.

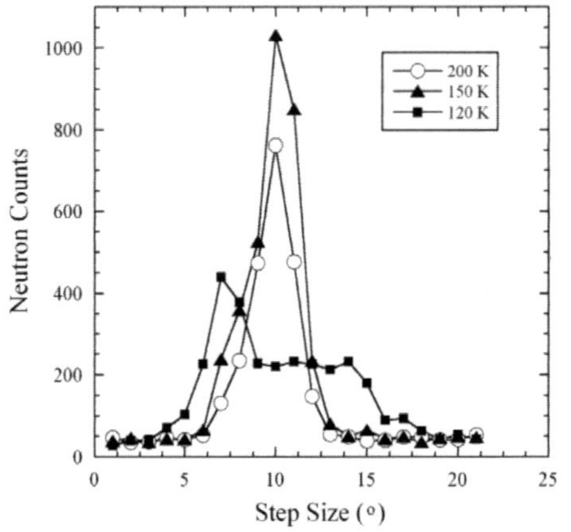

Figure 3. Peak shape of the (040) reflection of at 200,150 and 120K.

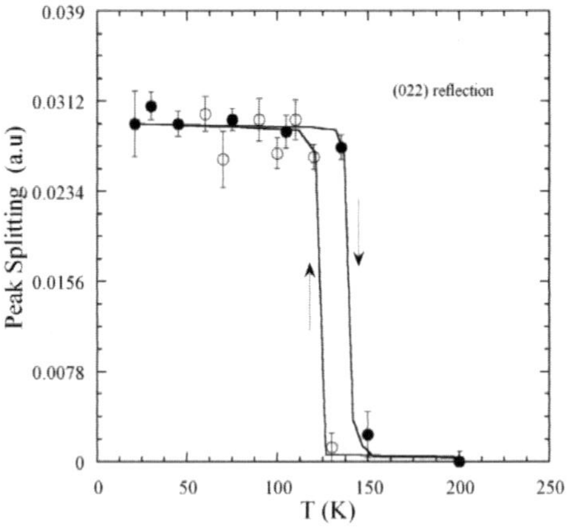

Figure 4. Peak splitting of the (022) reflection as a function of temperature. Closed and open symbols are measured upon heating and cooling respectively. The lines are guides to the eye.

Figure 2 shows an exemplary *(0kl)* plot for $BaNi_2As_2$ taken at 200 and 60 K. At 200 K, the pattern in reciprocal space exhibits clear Laue diffraction spots, all of which are nicely indexed in the tetragonal structure reported in table 1. At 60 K, however, we observe a clear splitting for some of the peaks, which is evidence for a structural phase transition. Figure 3 shows the shape of the *(040)* peak at 200, 150 and 120 K. The *(040)* peak can be fit by a single Gaussian curve at 200 K, there is a slight broadening at 150 K and a clear split into two peaks at 120 K. We can produce similar graphs for the

(022) reflection at all 18 different temperatures, some measured during cooling down and some measured during heating. The data were then fit with a function assuming two Gaussian curves. We constrained the width of the two Gaussians, and then determined the difference in positions as a measure of the splitting that occurs at low temperatures. Not surprisingly, we find that the fit collapses into one Gaussian curve for temperatures above ~135 K. The temperature dependence of the peak splitting is shown in figure 4. The heating and cooling curves exhibit some hysteresis indicating a first-order transition, which is consistent with previously reported bulk results [3]. The splitting of peaks provides immediate evidence for the formation of two domains, similar to the structural transition to an orthorhombic phase as reported in $BaFe_2As_2$ [7, 8]. It should be noted however, that a tetragonal to triclinic transition was reported recently for a different $BaNi_2As_2$ crystal grown from NiAs flux [6]. Our data are insufficient to determine the exact nature of the structural transition in our sample.

Our $BaNi_2As_2$ data provide no evidence for the long-range magnetism in the temperature range 21 – 300 K. None of the histograms had extra peaks with respect to others and there is no **Q**-dependence of the observed peaks. On the other hand, possible magnetic contributions to the peaks in $BaNi_2As_2$, if any, are likely much smaller than the nuclear ones, and our statistics on the small crystal may not be good enough to extract those. With the present data, we estimated that, for collinear moment configurations, we may not be able to resolve long-range magnetic order with moments less than 1μ_B per Ni atom.

Acknowledgements
We gratefully acknowledge the support of the Basic Energy Sciences program of the DoE to the facilities at Los Alamos National Laboratory.

References

[1] Ronning F, Bauer E D, Park T, Kurita N, Klimczuk T, Movshovich R, Sefat A S, Mandrus D and Thompson J D 2009 *Physica* C **469** 396

[2] Ronning F, Kurita N, Bauer E D, Scott B L, Park T, Klimczuk T, Movshovich R and Thomson J D 2008 *Journal of Physics: Condens. Matter* **20** 342203

[3] Kurita N, Tokiwa Y, Bauer E D, Subedi A, Singh D J, Thompson and Movshovich 2009 *Phys. Rev. Lett.* **102** 1047004

[4] Schultz A J, De Lurgio P M, Hammonds J P, Mikkelson D J, Miller M E, Naday I, Peterson P F, Porter R R and Worlton T G 2006 *Physica* B **385-386** 1059-1061

[5] Larson A C and Von Dreele R B 1986, Training Manual, General Structure Analysis System (GSAS) LANL report LA-UR-86-748 pp-1-179

[6] Sefat A S, McGuire M A, Rongying J, Sales B C, Mandrus D, Ronning F, Bauer E D and Mozharivskyj Y 2009 *Phys. Rev.* B **79** 094508

[7] Huang Q, Qiu Y, Bao W, Lynn J W, Green M A, Chen Y, Wu T, Wu G and Chen X H 2008 *reprint at http:// arxiv.org/abs/0806.2776*

[8] Ni N, Budko S L, Kreyssig A, Nandi S, Rustan G E, Goldman A I, Gupta S, Corbett J D, Kracher A and Canfielf PC 2008 *Phys. Rev.* B **78** 0145079

Study of diffuse scattering under hydrostatic pressure in $PbMg_{1/3}Nb_{2/3}O_3$

G-M Rotaru[1], B Padmanabhan[1], S N Gvasaliya[1], B Roessli[1], Th Strässle[1], R A Cowley[2], S G Lushnikov[3], S. Klotz[4]

[1] Laboratory for Neutron Scattering, ETH Zurich and Paul Scherrer Institut, CH-5232 Villigen, Switzerland
[2] Clarendon Laboratory, Oxford University, Parks Road, Oxford OX1 3PU, UK
[3] Ioffe Physical Technical Institute, 26 Politekhnicheskaya, 194021 St Petersburg, Russia
[4] Physique des Milieux Denses, IMPMC, CNRS UMR 7590, Université P. M. Curie, 4 Place Jussieu, 75252 Paris, France

E-mail: gelu.rotaru@psi.ch

Abstract. We report measurements of the evolution of the diffuse scattering in a single crystal $PbMg_{1/3}Nb_{2/3}O_3$ as a function of hydrostatic pressure. Upon applying pressure the diffuse scattering intensity decreases and is suppressed at about 3 GPa, while no change in the line shape is observed. Correlations between Pb displacements, diffuse scattering and relaxor properties are discussed.

1. Introduction

Relaxors are of scientific and commercial interest due to their dielectric properties. The giant piezoelectric and electrostriction effects present in these materials are exploited, for example, in sensors and actuators. $PbMg_{1/3}Nb_{2/3}O_3$ (PMN) is a well-known complex perovskite that exhibits relaxor ferroelectric properties. The average structure stays centrosymmetric with the space group Pm3m down to low temperatures although the dielectric permittivity has a broad maximum around 270 K when measured at 1 kHz [1]. It has been shown that the lowest transverse optic mode does not condense at this temperature [2], [3] as happens in the classical ferroelectrics at the structural phase transition. On the other hand a quasielastic mode was found whose susceptibility has a similar behavior with the temperature dependence of the dielectric permittivity [4]. The current interpretation is that the dielectric properties in PMN find their origins in nanoscale correlated regions, the so-called polar nano-regions (PNR). These polar nanoregions appear at the Burns temperature of about 620 K and grow when the temperature is lowered. Evidence for the formation of PNR in relaxor ferroelectrics comes from the presence of temperature dependent anisotropic diffuse scattering close to the Bragg reflections in X-rays and neutron scattering experiments. It is generally accepted that these are correlated displacements of Pb-ions that are mainly responsible for the formation of PNR. Systematic studies of diffuse scattering as a function of temperature have been reported, e.g., in Refs [4], [5], [6].

Another parameter that changes the dielectric properties is the hydrostatic pressure [7]. Pressure reduces the peak in the dielectric permittivity and shifts its maximum towards lower temperatures [1]. Suppression of the intensity of the X-ray diffuse scattering [8] and structural phase transitions are reported in PMN at about 4.5 GPa [8], [9], [10] and in $PbSc_{1/2}Ta_{1/2}O_3$ at 1.9 GPa [11]. Neutron powder diffraction experiments [12] show that the atomic displacements of Pb diminish and anisotropy of oxygen temperature factor increases with pressure. The aim of the present paper is to study the neutron diffuse scattering as a function of pressure in a single crystal of PMN.

2. Experimental details

The measurements were carried out with the cold neutron three-axis spectrometer TASP [13] at SINQ. For this experiment a 3x3x2 mm^3 single crystal of PMN was oriented with the [1,0,0] and [0,1,0] crystallographic directions in the scattering plane. The use of the spectrometer was justified by the need to reduce the background and hence to improve the signal-to-noise ratio. The spectrometer was configured in elastic mode with $k_f = 1.97$ Å^{-1}, collimation open-$-80' - 80' - 80'$.

To apply pressure, the crystal was embedded in a lead matrix placed into a Cu-Be gasket. Here, lead acts as pressure transmitting medium. Then the crystal - gasket assembly was loaded in a Paris-Edinburgh cell [14]. The entire pressure cell is mounted in a cryostat. The lead matrix in which the sample is embedded is also used as a pressure calibrant, the pressure being determined from the change in volume of the Pb unit cell using the third-order Birch-Murnaghan equation of state [15]. As the diffuse scattering in PMN is most intense at lower temperatures [4] the sample was cooled down to 80K. The spectra were collected in the (1,0,0) Brillouin zone.

3. Results

We have measured the diffuse scattering distribution around the (1,0,0) Bragg reflection at 0.5 GPa and 2.2 GPa. Systematic measurements of the diffuse scattering as a function of pressure were performed along the $[1, -1, 0]$ direction in the (1,0,0) Brillouin zone. We remind here that the diffuse scattering has in the (1,0,0)-zone the butterfly shape [6] with the wings elongated along $< 1, 1, 0 >$.

In Fig. 1 the distribution of diffuse scattering around (1,0,0) is shown at 0.5 GPa-(a) and 2.2 GPa (b). As one can see the butterfly shape is not observed. This is due to the high background resulting from the sample environment that makes the diffuse scattering indistinguishable at $|q|$ bigger than 0.1 rlu. A better understanding of the evolution of the

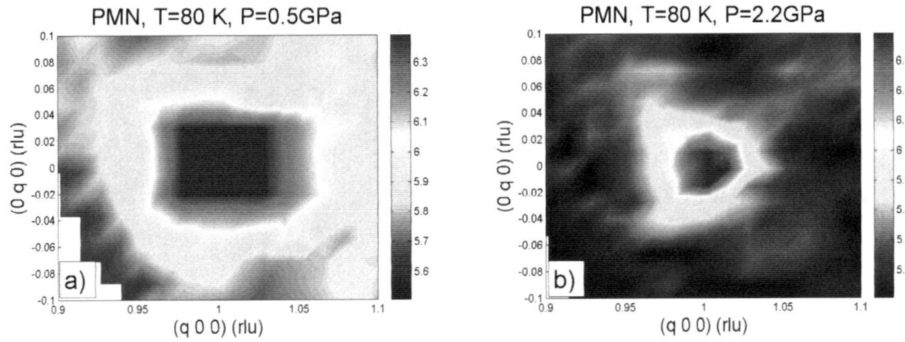

Figure 1: Normalized diffuse scattering distribution in the (1,0,0) Brillouin zone measured at 80 K under 0.5 GPa-(a) and 2.2 GPa-(b). The diffuse scattering intensity is given in logarithmic scale.

diffuse scattering can be obtained by comparing the Q-scans at different pressures. In Fig. 2, for clarity only two scans at 0.5, and 3.2 GPa measured at 80 K are shown. The spectra contains the Bragg peaks reflecting the long range order present at both pressures. The diffuse scattering decreases with pressure such that at 3.2 GPa only the resolution limited peak is left. We have fitted the data assigning a Gaussian function for the Bragg peak and a Lorentzian function for the diffuse scattering.

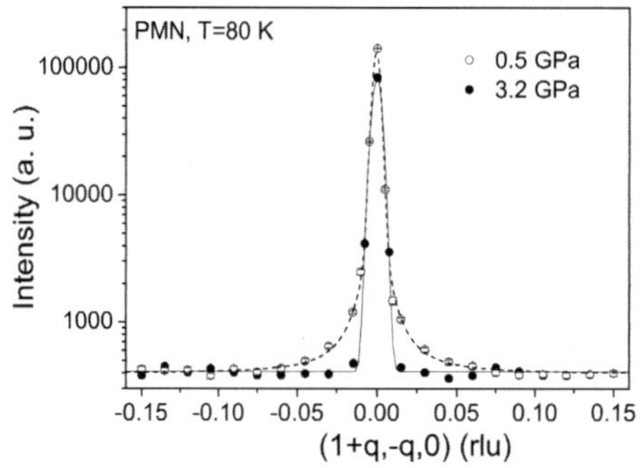

Figure 2: Constant Q-scans around the (1,0,0) Bragg peak in the [1,-1,0] direction measured at 0.5 GPa and 3.2 GPa respectively. The dashed and solid lines correspond to the fit (see text). Notice the pressure induced suppression of diffuse scattering.

A summary of the evolution of the diffuse scattering intensity as a function of pressure is shown in Fig. 3. Although we cannot resolve the pressure dependence of the correlation length of the short-range correlations, we do observe a decrease in the diffuse scattering. Hydrostatic pressure finally suppresses the polar short-range order around 3 GPa.

In conclusion, our results indicate that there exists a strong correlation between atomic displacements, that give rise to the PNR, diffuse scattering and the behavior of the dielectric

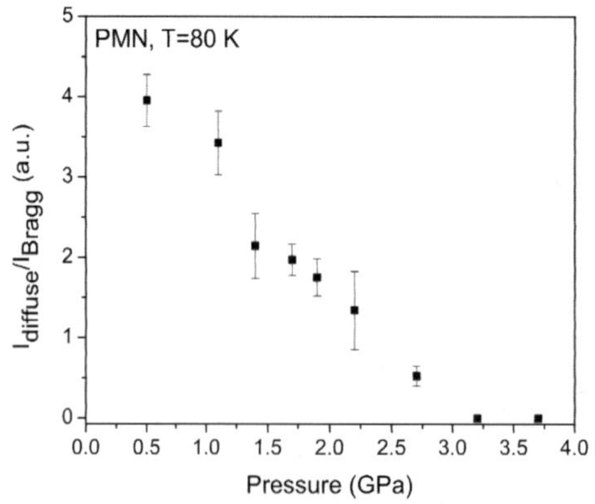

Figure 3: Diffuse scattering intensity measured along the [1,-1,0] direction in the (1,0,0) Brillouin zone, normalized to the Bragg intensity of the (1,0,0) reflection taken in two-axis mode is plotted as a function of pressure.

permittivity in PMN. These results combined with the X-rays, neutron and dielectric studies provide further evidence that the local structure plays an important role in the dielectric properties of PMN.

Acknowledgments
This work was performed at the spallation neutron source SINQ, Paul Scherrer Institut, Villigen (Switzerland) and was partially supported by the Swiss National Foundation (Project No. 20002-111545).

4. References

[1] Nawrocik W, Czarnecki P, Hilczer A and Blinc R 2003 *Phase Transitions* **76** 693
[2] Cao H, Stock C, Xu G, Gehring P M, Li J and Viehland D 2008 *Phys. Rev. B* **78** 104103
[3] Hlinka J and Kempa M 2008 *Phase Transitions* **81** 491
[4] Gvasaliya S N, Roessli B, Cowley R A, Hubert P and Lushnikov S G 2005 *J. Phys.: Condens. Matter* **17** 4343
[5] Matsuura M, Hirota K, Gehring P, Chen W, Ye Z and Shirane G, 2006 *Physica B: Condens. Matter* **385-386** 123
[6] Xu G, Shirane G, Copley J R D and Gehring P M 2004 *Phys. Rev. B* 69 064112
[7] Samara G A 2000 *AIP-Conf.-Proc.* vol **535** (New York: AIP) p 344
[8] Chaabane B, Kreisel J, Dkhil B, Bouvier P and Mezouar M 2003 *Phys. Rev. Lett.* **90** 257601
[9] Kreisel J, Dkhil B, Bouvier P, and Kiat J -M 2002 *Phys. Rev. B* **65** 172101
[10] Muhtar A, Maddury S, Zuo-Guang Y, Cohen R E, Ho-kwang M and Russell J H 2009 *Phys. Rev. B* **79** 132103
[11] Mihailova B, Angel R J, Welsch A M, Zhao J, Engel J, Paulmann C, Gospodinov M, Ahsbahs H, Stosch R, Guttler B and Bismayer U 2008 *Phys. Rev. Lett.* **101** 017602
[12] Rotaru G-M, Gvasaliya S N, Pomjakushin V, Roessli B, Strässle Th, Lushnikov S G, Shaplygina T A and Günter P 2008 *J. Phys.: Condens. Matter* **20** 104235
[13] Semadeni F, Roessli B, and Böni P 2001 *Physica B* **297** 152
[14] Klotz S, Strässle Th, Rousse G, Hamel G and Pomjakushin V 2005 *Appl. Phys. Lett.* **86**, 031917
[15] Birch F J 1947 *Phys. Rev.* **71** 809; Murnaghan F D 1937 *Am. J. Math.* **49** 235

Journal of Physics: Conference Series **251** (2010) 012012 doi:10.1088/1742-6596/251/1/012012

Nanocrystalline Pr_2Fe_{17} studied by neutron powder diffraction

Pedro Gorria[1], Pablo Álvarez[1], Jorge Sánchez Marcos[2], José L Sánchez Llamazares[1] and Jesús A. Blanco[1]

[1] Departamento de Física, Universidad de Oviedo, Av. Calvo Sotelo, s/n, 33007 Oviedo, Spain.
[2] Instituto de Ciencia de Materiales de Madrid, CSIC, Cantoblanco, 28049 Madrid, Spain

E-mail: pgorria@uniovi.es

Abstract. We have used neutron powder diffraction to make a comprehensive study of the crystal structure in Pr_2Fe_{17} nanocrystalline compounds. These nanostructured materials have been obtained by high-energy ball milling from an arc-melted bulk polycrystalline alloy. The Th_2Zn_{17}-type rhombohedral crystal structure of the starting bulk alloy is maintained after milling for 10 and 20 hours. The studied alloys are ferromagnetic below $T_C = 286\pm1$ K (bulk) and $T_C \approx 300\pm20$ K (milled). These materials exhibit strong magneto-volume anomalies below T_C, such as a negative coefficient of thermal expansion ($\alpha_V \approx -30 \times 10^{-6}$ K^{-1}).

Mechanical processing by means of high-energy ball milling technique (HEBM) is commonly used to obtain a huge number of alloys far from the thermodynamic equilibrium [1]. In particular, different Fe-TM binary compounds, where TM is a transition metal, can be synthesized with distinct degrees of disorder, i.e. metastable FeCu solid solutions [2], nanocrystalline FeNi powders [3] or FeZr amorphous alloys [4], among others. Furthermore, drastic microstructural changes in a bulk alloy, such as nanostructure formation, can be induced using HEBM, giving rise to modifications of the physical-chemical properties respect to those exhibited by the starting bulk alloy [5,6]. We report here a neutron powder diffraction study of the induced changes in the microstructure and in the magneto-volume anomalies when a Pr_2Fe_{17} bulk intermetallic compound is milled for 10 and 20 hours.

Fe-rich R_2Fe_{17} alloys crystallize in the Th_2Zn_{17}-type rhombohedral crystal structure ($R\bar{3}m$) when R is one of the light rare earths, and exhibit ferromagnetic order with Curie temperatures, T_C, around room temperature in the case of Pr or Nd [7,8]. The are four inequivalent crystallographic Fe sites in Pr_2Fe_{17} [9] (in Wyckoff notation, 6c, 9d, 18f and 18h respectively) and the local environments of these Fe atoms determine the ferromagnetic character of these alloys, because the sign and the magnitude of the magnetic interactions depends on the number of Fe next neighbours and especially on the Fe-Fe interatomic distances [7,10]. In particular, the distance between 6c-6c Fe atoms (located along the c-axis), the so-called "dumbbell site", is lower than the critical distance of 2.45 Å [11,12], which induces a negative local magnetic interaction and therefore, antiferromagnetic coupling is favoured. The rest of the iron magnetic moments are parallel due to Fe-Fe positive interactions, and the magnetic energy is stored around the shortest interatomic distances [Fe(6c)-Fe(6c)] where negative interactions exist. The observed magneto-volume effects below the ordering temperature such as the large and negative thermal expansion along the c-axis are direct effects of the compensation between magnetic and elastic energies. Moreover, ferromagnetic R_2Fe_{17} compounds have attracted interest for their moderate magneto-caloric effect around room temperature (being the magnetic entropy change for a magnetic field variation from 0 to 50 kOe $|\Delta S_M| \approx 6$ J kg^{-1} K^{-1} in the case of R = Pr or Nd) [13].

In Figure 1 we show the room temperature neutron powder diffraction patterns collected on the D2B two-axis high-resolution diffractometer (ILL, Grenoble, France). The peaks observed in the patterns correspond to the Bragg reflections of the rhombohedral Pr_2Fe_{17} phase with a $R\bar{3}m$ crystal structure, thus indicating that the milling procedure does not give rise to any structural transformation and/or the appearance of new crystalline phases. However, a large decrease of the peak intensity together with a line-broadening of the Bragg reflections is clearly observed in both BM-10h and BM-20h milled samples. From the peak-profile analysis [15,16], we obtain a mean crystalline size of 20±2 nm (in good agreement with that estimated from transmission electron microscopy [14]), while the induced microstrain is almost vanishing in contrast with that reported for Fe-TM milled alloys [3,16]. The latter suggest the milling process mainly produces a progressive grain fracturing due to the brittleness of these compounds. Moreover, an increase of the background intensity due to the induced structural disorder at the grain boundaries is clearly observed. In Table I we present the values for some characteristic structural parameters estimated from the Rietveld analysis of the diffraction patterns shown in Figure 1. It is worth noting that only subtle changes are observed in the values of the unit cell parameters and in the atomic positions. Only those corresponding to the Fe1(6c) and Pr(6c) show a tendency to increase and decrease with the milling time, respectively.

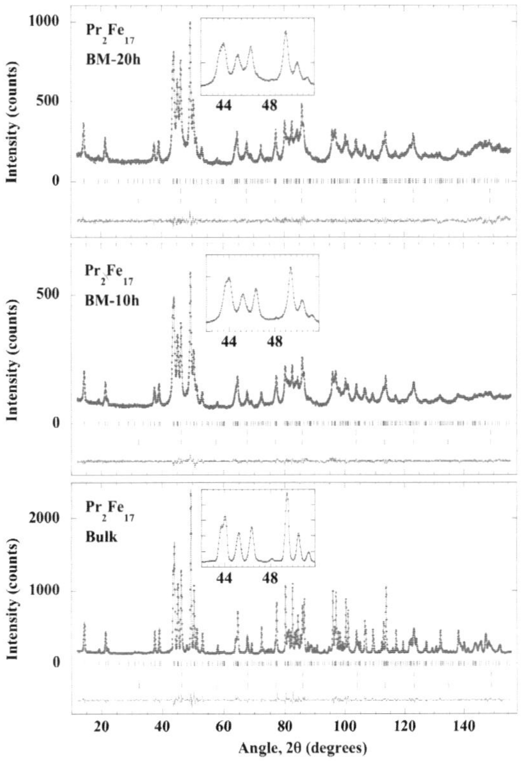

Figure 1. Observed (dots) and calculated (solid lines) neutron diffraction patterns. The position of the Bragg reflections is represented by vertical bars: first row corresponds to the Pr_2Fe_{17} phase while the second one is associated with α-Fe impurity. The observed-calculated difference is depicted at the bottom of each pattern. The insets show the peak broadening of the most intense reflections.

As previously mentioned, the magnetism of these compounds is strongly sensitive to the distance between Fe atoms at the "dumbbell sites", any increase of such Fe-Fe distance favors ferromagnetism [10-12]. Therefore, magneto-volume coupling acts as a driven force for the selective slight distortions induced by the milling process, thus affecting mainly to the 6c sites. In addition, it is observed a substantially increase in the value of the Debye-Waller temperature factor, B, for all the atomic positions in the milled samples. The significant increase of atoms at the grain surfaces and/or grain boundaries due to the drastic reduction in the average grain size during the milling procedure gives rise to structural disorder [16]; and can be the responsible for the higher values of B.

TABLE I. Structural data of Bulk and BM compounds at T = 300 K obtained from the neutron powder diffraction patterns shown in Fig. 1. Data for single crystal [9] is also reported for comparison

Compounds	Single crystal	Bulk	BM-10h	BM-20h
a (Å)	8.585(5)	8.5849(1)	8.5880(2)	8.5870(2)
c (Å)	12.464(8)	12.4659(1)	12.4699(2)	12.4679(3)
Pr(6c) z	0.3436(1)	0.3443(3)	0.3425(7)	0.3417(9)
Fe1(6c) z	0.0960(4)	0.0967(1)	0.0970(3)	0.0981(4)
Fe3(18f) x	0.2868(3)	0.2877(1)	0.2874(2)	0.2878(3)
Fe4(18h) x	0.1696(2)	0.1690(1)	0.1691(2)	0.1689(2)
y	0.8304(2)	0.8310(1)	0.8309(2)	0.8311(2)
z	0.4897(2)	0.4896(1)	0.4891(2)	0.4893(3)
B ($Å^2$) Pr(6c)	0.54(12)	0.65(6)	0.91(9)	1.01(12)
Fe1(6c)	0.55(15)	0.39(3)	1.43(7)	1.52(8)
Fe2(9d)	0.47(12)	0.29(2)	0.91(5)	0.98(7)
Fe3(18f)	0.66(12)	0.35(2)	1.09(4)	1.23(6)
Fe4(18h)	0.68(13)	0.35(2)	0.97(4)	1.02(6)
R_B(%)	3.9	3.3	2.5	2.4
χ^2		7.0	1.5	1.4

In order to check for the persistence of the magneto-volume anomalies after the strees-induced microstructural modification of the alloys, we have performed neutron powder thermo-diffraction experiments using the D1B two-axis diffractometer (ILL, Grenoble, France) in the temperature range between 5 and 300 K (see [4,16] for details). The temperature dependence of the cell volume for the three samples is depicted in figure 2.

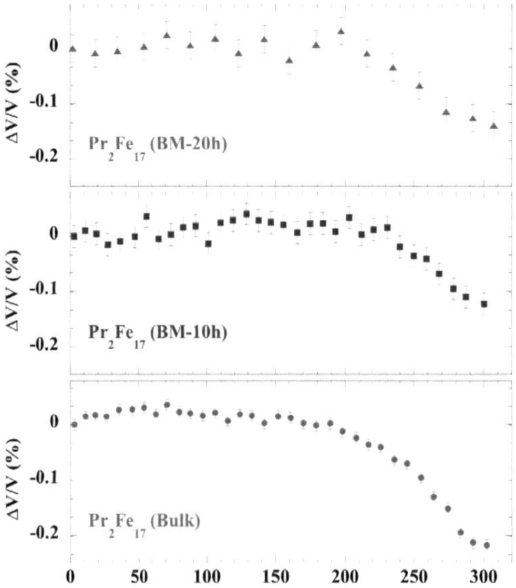

Figure 2. Temperature dependence of the unit cell volume, V, as deduced from neutron thermodiffraction experiments.

The three samples exhibit an almost constant value for the cell volume below 200 K, thus indicating a zero thermal expansion coefficient and the invar character in this temperature range. Taking into account that the Curie temperature of the Bulk alloy is $T_C \approx 286\pm1$ K [7,14], a minimum of the V(T) curve is expected around room temperature [10-12], as it can be observed in figure 2. For the milled samples this minimum is broader and shifted to higher temperatures, similar to that observed for the magnetization vs temperature curves, M(T) [13,14], where the decrease of M for T > 250 K is slower, making the estimation of an unique value for the Curie temperature ($T_C \approx 300\pm20$ K) extremely difficult in these disordered nanocrystalline alloys. Therefore, the temperature dependence of the $\Delta V/V$ is quite reminiscent of that associated with the magnetization M(T) [14]. This feature indicates the strong magneto-volume coupling in these materials, which is responsible for the moderate magneto-caloric effect found [13]. The slight increase of the estimated magnetic moments of the BM materials obtained from magnetization measurements is difficult to be observed from low temperature neutron diffraction patterns, owing to the significant line-broadening that mask the analysis.

In conclusion, high-energy ball milling allows obtaining nanocrystalline Pr_2Fe_{17} alloys with a mean crystalline size around 20 nm. The severe mechanical processing does not modify the crystal structure and the magneto-volume anomalies, including invar behaviour below 200 K and negative coefficient of thermal expansion around room temperature. The slight induced increase in the Fe-Fe interatomic distances at the "dumbbell" 6c sites causes the shift to higher temperatures for both the minimum of the temperature dependence of the cell volume as well as the T_C value. However, the decrease of the magnetization as well as that of the cell volume above 250 K is slower in the milled samples, probably due to the large amount of disorder generated at the grain boundaries. The latter suggests that a distribution of Curie temperatures must be considered as the most realistic picture for these disordered nanostructured materials, similarly to that occurring in other nanocrystalline systems [17,18].

Acknowledgments

Research supported from FEDER and the Spanish MICINN (NAN2004-09203-C04-03 and MAT2008-06542-C04). We acknowledge ILL and Spanish CRG-D1B for allocating neutron beam time. PA and JLSL thank FICyT for PhD grant and research contract (COF07-013) respectively. JSM thanks MICINN for "Juan de la Cierva" research contract.

References

[1] Suryanarayana C 2001 *Prog. Mater. Sci.* **46** 1.
[2] Yavari A R *et al.* 1992 *Phys. Rev. Lett.* **68** 2235.
[3] Gorria P *et al.* 2005 *J. Magn. Magn. Mater.* **294** 159.
[4] Gorria P *et al.* 2009 *Phys. Status Solidi (RRL)* **3** 28.
[5] Gorria P *et al.* 2005 *Phys. Rev. B* **72** 014401.
[6] Gorria P *et al.* 2009 *Phys. Status Solidi (RRL)* **3** 115.
[7] Buschow K H J 1977 *Rep. Prog. Phys.* **40** 1179.
[8] Álvarez P *et al.* 2008 *J. Non-Cryst. Solids* **354** 5172.
[9] Johnson Q *et al.* 1968 *Acta Cryst.* **B24** 274.
[10] Givord G and Lemaire R 1974 *IEEE Trans. Mag.* **MAG-10** 109.
[11] Long G J *et al.* 1993 *J. Appl. Phys.* **74** 504.
[12] Yelon W B 1995 *IEEE Trans. Mag.* **31** 3689.
[13] Gorria P *et al.* 2008 *J. Phys. D: Appl. Phys.* **41** 192003.
[14] Gorria P *et al.* 2009 *Acta Mater.* **57** 1724.
[15] Rodríguez Carvajal J 1993 *Physica B* **192** 55.
[16] Martinez-Blanco D *et al.* 2008 *J. Phys: Condens. Matter* **20** 335213.
[17] Hernando A *et al.* 1995 *Phys. Rev. B* **51** 3281.
[18] Garitaonandia J S *et al.* 2000 *Phys. Rev. B* **61** 6150.

Structure study of chalcogenide glasses from high Q-range neutron diffraction experiment and RMC modelling

M Fábián[1], E Sváb[1], S Vogel[2], V Pamukchieva[3] and A Szekeres[3]
[1]*Research Institute for Solid State Physics and Optics, H-1525 Budapest, P.O.B. 49, Hungary*
[2]*Los Alamos Natl. Lab., Lujan Neutron Scattering Ctr, Los Alamos, NM 87545, USA*
[3]*Institute of Solid State Physics, Bulgarian Academy of Sciences, Tzarigradsko Chaussee 72, 1784 Sofia, Bulgaria*

E-mail: fabian@szfki.hu

Abstract. A neutron diffraction structure study has been performed on newly synthesized $Ge_{20}As_{20}Se_{50}Te_{10}$ and $Ge_{27}As_{13}Se_{50}Te_{10}$ chalcogenide glasses. Oscillations in the structure factor, S(Q) have been measured with good sign-to-noise up to 35 $Å^{-1}$. The reverse Monte Carlo simulation was used to model the 3-dimensional atomic configuration. The partial atomic correlation functions and structure factors have been revealed. Several first and second neighbour distances, and coordination numbers have been calculated. We have established that several first neighbour atomic distances are overlapping at two characteristic distances, namely the Ge-Ge, Ge-Se and As-Se are centred at 2.42 Å, while the Ge-Te, As-Te and Se-Te are centred at 2.60 Å.

1. Introduction

Chalcogenide glasses are transparent in the infrared region and exhibit high third order non-linear optical properties [1,2]. Thus, they are very attractive as high-speed optical elements for application such as data processing devices, electronic switches, and optical memories.

Among such materials is the quaternary Ge-As-Se-Te telluride system, which is in the focus of our interest. Here we present our neutron diffraction experiments on $Ge_{20}As_{20}Se_{50}Te_{10}$ and $Ge_{27}As_{13}Se_{50}Te_{10}$ compositions, and preliminary results revealed by reverse Monte Carlo modelling characterising the short range structure.

2. Experimental details

2.1. Sample preparation

The glassy samples $Ge_{20}Sb_{20}S_{50}Te_{10}$ and $Ge_{27}Sb_{13}S_{50}Te_{10}$ were synthesized by the conventional melt-quenching method from 5N purity elements. The syntheses were performed in evacuated quartz ampoules (10^{-3} Pa) using a rotary furnace. The specimens were heated up to 950°C, kept at this temperature for 24 h meanwhile rotating the furnace for homogeneous melting. The ampoules were quenched in air.

2.2. Neutron diffraction experiments

Neutron diffraction measurements have been performed in a relatively broad momentum transfer range, combining the data measured by the 2-axis 'PSD' monochromatic neutron diffractometer (λ_0=1.068 Å; Q=0.45-9.8 $Å^{-1}$) [3] at the 10 MW Budapest research reactor and by the time-of-flight 'HIPPO' instrument at the LANSCE pulsed neutron source Q=0.9-50 $Å^{-1}$ [4]. The ND patterns have shown that the specimens are fully amorphous. The structure factor, S(Q) was obtained with a good signal-to-noise ratio up to Q_{max}=35 $Å^{-1}$ (see figure 1), which made possible high-resolution r-space analyses. The S(Q)'s for the two compositions are fairly similar to each other, however, several fine details are somewhat different, i.e. the position of the first peak at around 1 Å^{-1} (called also pre-peak), or the oscillations in the 6-8 Å^{-1} Q-range.

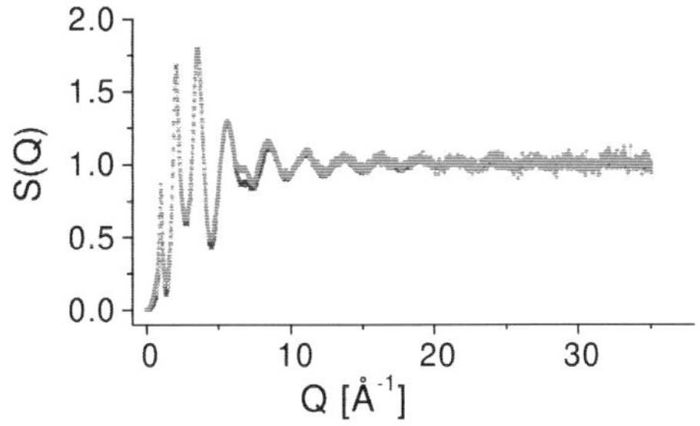

Figure 1. Experimental structure factors for the $Ge_{20}As_{20}Se_{50}Te_{10}$ (blue cross) and $Ge_{27}As_{13}Se_{50}Te_{10}$ (orange circle) glasses.

3. Data evaluation – Reverse Monte Carlo modelling

The S(Q) data displayed in figure 1 have been simulated by the RMC method [5]. For the RMC starting model a disordered atomic configuration was built up with a simulation box containing 10 000 atoms, ρ_0=0.033/0.036 atom density and half-box lengths of 33.5/32.6 Å for $Ge_{20}As_{20}Se_{50}Te_{10}$/ $Ge_{27}As_{13}Se_{50}Te_{10}$ specimens, respectively. The RMC modelling was performed in two main steps. As a first step, the MCGR method was applied [6] with the aim to lighten the atomic motions in the RMC calculations. MCGR is a 1-dimensional version of RMC to produce a total atomic pair correlation function, g(r), by allowing atomic movements to minimize the difference between calculated and experimental structure factors and, contains some fine corrections for the background as well. The MCGR data were used as input experimental data for the RMC modelling. In the RMC simulation procedure several runs have been performed with different interatomic cut-off distances used as constraints and, the corresponding results have been carefully analysed. Because of the several overlapping first neighbour distributions, we decreased the number of $g_{ij}(r)$ by applying further constraints, supposing that no Se-Se, Ge-As, As-As and Te-Te bonds are present. In a first approach this is feasible, because of the small probability of being first neighbours. We have revealed that with the following cut-off atomic neighbour distances the results proved to be highly stable: Ge-Ge: 2.2/2.2Å, Ge-Se:2.18/2.17Å, Ge-Te:2.5/2.55Å, As-Se:2.17/2.19Å, As-Te:2.58/2.56Å and Se-Te:2.5/2.56Å, for $Ge_{20}As_{20}Se_{50}Te_{10}$/ $Ge_{27}As_{13}Se_{50}Te_{10}$, respectively.

The final RMC run matched very well the experimental one, as it is illustrated in figure 2.

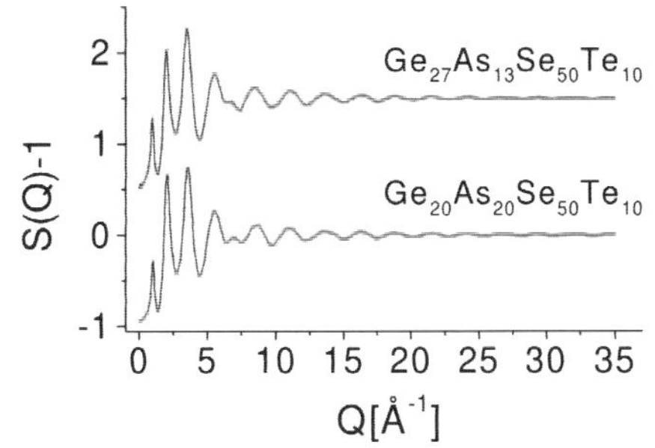

Figure 2. Structure factors: experimental MCGR data (magenta cross) and RMC modelling (black line). The curves are shifted vertically for clarity.

Several partial atomic correlation functions, $g_{ij}(r)$ and corresponding partial structure factors, $S_{ij}(Q)$ have been revealed from the RMC simulation with a fairly good stability. They are presented in figure 3 and 4 for both compositions. The first neighbour coordination number distributions are shown in figure 5. The 1st and 2nd neighbour interatomic distances and first neighbour coordination numbers are collected in Table 1. The Ge-Ge, Ge-Se and As-Te distribution are fairly similar to each other, both the 1st and 2nd neighbour distances are at the same distance within limit of error, at 2.42±0.02 Å and at 3.78±0.05 Å, respectively. Similarly, the distributions for the Ge-Te, As-Te and Se-Te are fairly similar to each other, the 1st neighbour distance is at 2.60±0.02 Å and the 2nd neighbour distance is at 3.72±0.05 Å.

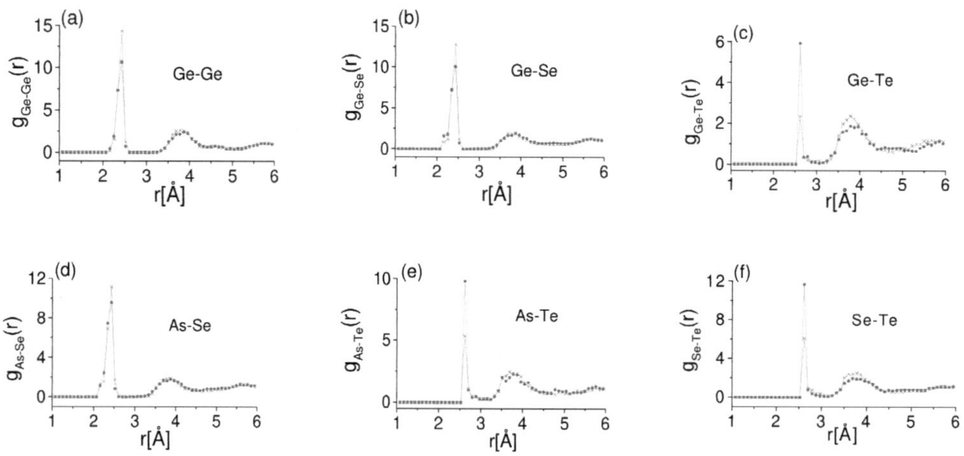

Figure 3. The partial atomic correlation functions obtained by RMC simulation, for the $Ge_{20}As_{20}Se_{50}Te_{10}$ (olive cross) and $Ge_{27}As_{13}Se_{50}Te_{10}$ (red circle) glasses.

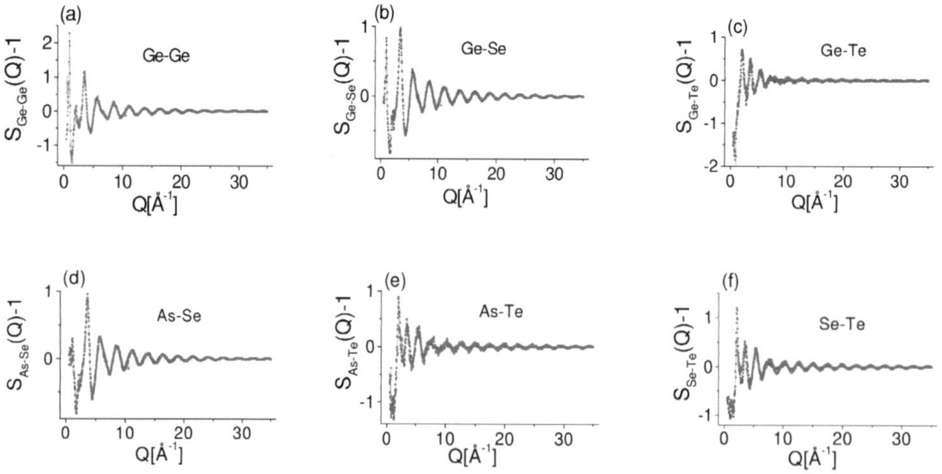

Figure 4. The partial structure factors obtained by RMC simulation, for the $Ge_{20}As_{20}Se_{50}Te_{10}$ (olive cross) and $Ge_{27}As_{13}Se_{50}Te_{10}$ (red circle) glasses.

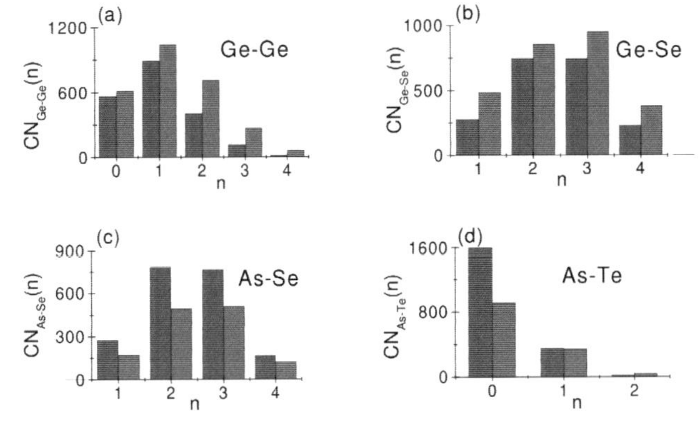

Figure 5. Coordination number distributions for the $Ge_{20}As_{20}Se_{50}Te_{10}$ (olive) and $Ge_{27}As_{13}Se_{50}Te_{10}$ (red) glasses.

Table 1. Interatomic distances, r_{ij}(Å) and coordination numbers, CN_{ij} obtained from RMC simulation. The corresponding intervals are indicated in brackets.

Atom pairs	$Ge_{20}As_{20}Se_{50}Te_{10}$			$Ge_{27}As_{13}Se_{50}Te_{10}$		
	Interatomic distances, $\mathbf{r_{ij}}$(Å)		Coordination number, CN_{ij} (±10%)	Interatomic distances, $\mathbf{r_{ij}}$(Å)		Coordination number CN_{ij} (±10%)
	1st (±0.02)	**2nd** (±0.05)		**1st** (±0.02)	**2nd** (±0.05)	
Ge-Ge	2.42	3.78	1.08	2.43	3.87	1.31
Ge-Se	2.43	3.78	2.46	2.42	3.86	2.45
Ge-Te	2.60	3.70	0.15	2.61	3.77	0.19
As-Se	2.42	3.80	2.41	2.42	3.88	2.40
As-Te	2.60	3.70	0.20	2.60	3.80	0.31
Se-Te	2.61	3.75	0.21	2.61	3.73	0.36

4. Summary

We have performed a neutron diffraction structural study on newly synthesized 4-component chalcogenide specimens with composition $Ge_{20}As_{20}Se_{50}Te_{10}$ and $Ge_{27}As_{13}Se_{50}Te_{10}$. The structure factors have been measured over a broad momentum transfer range, up to 35 $Å^{-1}$, which made available high r-space resolution. RMC simulation was successfully applied to calculate most of the partial atomic pair correlation functions, the partial structure factors and coordination number distributions. Several first and second neighbour distances and coordination numbers have been revealed. We have established that the Ge-Ge, Ge-Se and As-Se are centred at 2.42 Å, and the Ge-Te, As-Te and Se-Te distances are centred at 2.60 Å. Further analyses of the data are in progress.

Acknowledgement - This study has benefited from the use of PSD at Budapest funded by EU-F7 N226507-NMI3 and from the use of HIPPO at the LANSCE, by US-DOE Contract W-7405-ENG-36.

5. References

[1] Kapany N S and Simms R J 1965 *Infrared Phys.* **5** 69
[2] Schardt C R, Simmons J H, Lucas P, Neindre L L and Lucas J 2000 *J. Non-Cryst. Solids* **274** 23
[3] Sváb E, Deák F and Mészáros Gy 1996 *Mater. Sci. Forum.* **228** 247, www.bnc.hu
[4] Wenk H R, Lutterotti L ans Vogel S 2003 *Nucl. Instr. Methods A* **515** 575
[5] McGreevy R L and Pusztai L 1988 *Molec. Simul.* **1** 359
[6] Zetterström P and McGreevy R L 2000 Physica B 276-278 178

International Conference on Neutron Scattering 2009 IOP Publishing
Journal of Physics: Conference Series **251** (2010) 012014 doi:10.1088/1742-6596/251/1/012014

Magnetic structure and local lattice distortion in giant negative thermal expansion material $Mn_3Cu_{1-x}Ge_xN$

S. Iikubo[1], K. Kodama[2], K. Takenaka[3,4], H. Takagi[4,5] and S. Shamoto[2]

[1]Kyushu Institute of Technology, Kitakyushu, Fukuoka 808-0196, Japan
[2]Quantum Beam Science Directorate, Japan Atomic Energy Agency, Tokai, Ibaraki, 319-1195, Japan
[3]Department of Crystalline Materials Science, Nagoya University, Nagoya, Aichi, 464-8603, Japan
[4]RIKEN (The Institute of Physical and Chemical Research), Wako, Saitama, 351-0198, Japan
[5]CREST, Japan Science and Technology Agency, Kawaguchi, Saitama, 332-0012, Japan

E-mail: `iikubo@life.kyutech.ac.jp`

Abstract. Magnetic and local structures in an antiperovskite system, $Mn_3Cu_{1-x}Ge_xN$, with a giant negative thermal expansion have been studied by neutron powder diffraction measurement. We discuss (1) an importance of an averaged cubic crystal structure and a Γ^{5g} antiferromagnetic spin structure for the large magneto-volume effect (MVE) in this itinerant electron system, (2) an unique role of a local lattice distortion well described by the low temperature tetragonal structure of Mn_3GeN for the broadening of MVE.

1. Introduction

Negative thermal expansion (NTE) materials have already been used in a wide area of technical applications in which it is desperately needed to control the thermal expansion[1, 2]. The NTE occurs as a result of the gradual volume expansion accompanied by magnetic ordering, the so-called magneto-volume effect (MVE). The MVE of itinerant electron systems has been investigated since the discovery of extraordinarily small thermal expansion in Invar alloys [3]. Antiperovskite manganese nitrides Mn_3AN, where A is a metal or a semiconducting element, are well known for their large MVE [4]. However, this system has not been considered as a practical NTE material to date, because all the MVEs reported in Mn_3AN members have been associated with first-order phase transitions. Recently, Takenaka and Takagi reported that the MVE is broadened against temperature T in $Mn_3Cu_{1-x}Ge_xN$ and leads to a giant negative thermal expansion coefficient [5, 6, 7]. At $x{\sim}0.5$, linear thermal expansion $\Delta L/L$ is almost linear to T in the temperature range of $270 \leq T \leq 350$ K. The large negative coefficient of linear thermal expansion α is about -2×10^{-5}/K, the largest value among all NTE materials. The clarification of the microscopic mechanism of the MVE may provide us a useful guideline for designing NTE materials with better performance. Especially, to clarify the mechanism for the broadening of MVE (the Invar problem) has been a challenge in solid state physics over a century. We have studied the magnetic and local structures in $Mn_3Cu_{1-x}Ge_xN$ using neutron

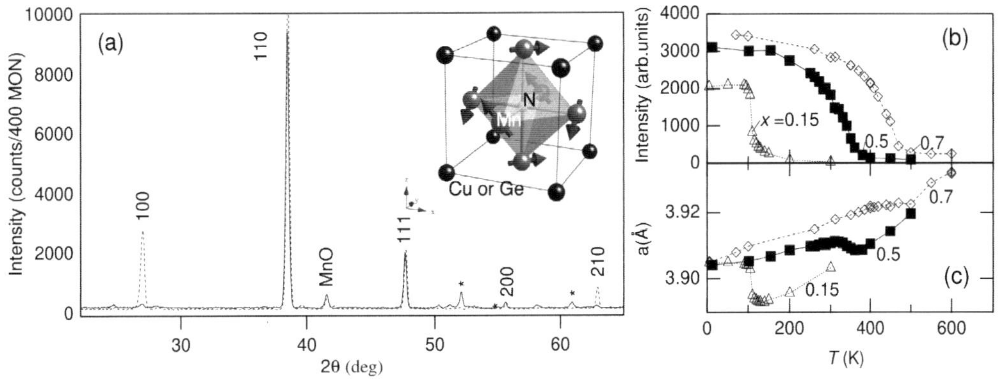

Figure 1. (a) Neutron powder diffraction patterns of $Mn_3Cu_{0.5}Ge_{0.5}N$ above and below magnetic ordering temperatures, which are at T=400 (solid line) and 15 K (dotted line), respectively. The diffraction pattern contains weak reflections from MnO impurity and Al_2O_3, marked by stars, in the furnace. The inset shows the Γ^{5g} Antiferromagnetic structure of $Mn_3Cu_{1-x}Ge_xN$. (b) T dependence of (100) magnetic reflection intensity and (c) T-dependence of lattice constants of $Mn_3Cu_{1-x}Ge_xN$ for x=0.15, 0.5 and 0.7.

powder diffraction[8, 9]. In this paper, we review the present understanding of the Ge-doping effect on the magnetic properties and MVE in $Mn_3Cu_{1-x}Ge_xN$.

2. Magnetic structure

Figure 1 (a) shows neutron diffraction patterns of $Mn_3Cu_{0.5}Ge_{0.5}N$. Solid and dotted lines show the data collected above and below magnetic ordering temperatures. Miller indices are given in reciprocal lattice units of primitive cubic perovskite. We can see magnetic reflections corresponding to magnetic ordering vector $\boldsymbol{q}$=(0 0 0), where the (100) and the (210) magnetic reflection are strong. Figures 1 (b) and (c) show the T dependence of the (100) magnetic reflections and T-dependence of lattice constants of $Mn_3Cu_{1-x}Ge_xN$ for x= 0.15, 0.5, and 0.7. The T-dependence of lattice constants are consistent with the linear thermal expansions previously reported by Takenaka and Takagi [5]. Both the 1 0 0 magnetic reflection and the lattice constant exhibit sharp increases at x=0.15 with decreasing T. For x=0.5, they gradually increase with decreasing T in the temperature range from 360 to 320 K. The magnetic reflection intensity grows in a progressively wider T range with increasing Ge content. The width of the magnetic peaks is nearly resolution-limited, suggesting long-range Neel ordering. Based on the neutron and the NMR results[9], we conclude that the systems with a gradual volume change exhibit a gradual change in the magnitude of the ordered moments.

Ge-doped samples have a cubic structure (space group: $Pm\bar{3}m$) and magnetic ordering vector $\boldsymbol{q}$=(0 0 0). On the basis of these conditions, three possible models have been proposed by Fruchart and Bertaut[4]. Among three magnetic structures, so-called Γ^{5g} antiferromagnetic (AF) structure shown in inset of Fig. 1 has large intensity of (100) reflection. Calculated intensities of all reflections assuming Γ^{5g} AF structure can reproduce the observed intensities.

Let us now look at other antiperovskite materials. Mn_3GaN and Mn_3ZnN are well known for their large MVEs[10, 11]. Their volumes show a sudden and pronounced increase with decreasing temperature at the first-order transition, and exhibit the Γ^{5g} antiferromagnetic structure in the cubic crystal structure below the phase transition temperature[4]. The MVE of an itinerant electron system has been discussed in terms of the amplitude of magnetic moment. However,

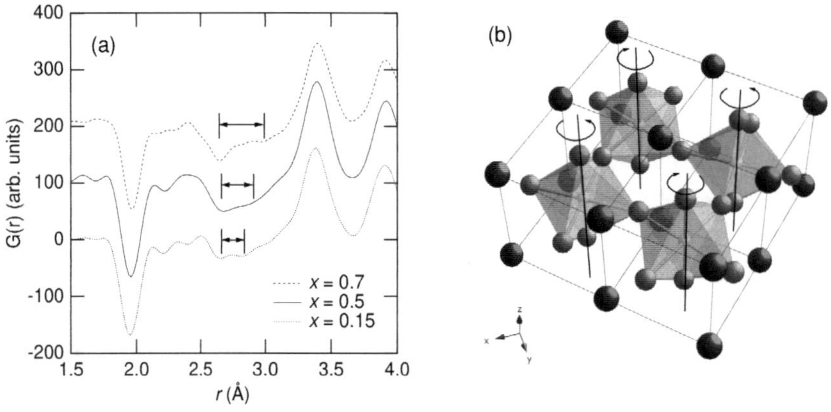

Figure 2. (a) PDFs of $Mn_3Cu_{1-x}Ge_xN$ for $x = 0.15$, 0.5 and 0.7 at 300 K. (b) Local octahedral rotation of Mn_6N in $Mn_3Cu_{1-x}Ge_xN$. Alternate rotation of Mn_6N octahedra occurs locally.

the intimate relationship between the Γ^{5g} antiferromagnetic cubic structure and large MVE in $Mn_3Cu_{1-x}Ge_xN$ indicates the necessity of a new theoretical framework for MVE, in which the ordered magnetic structure is taken into account.

3. Local structure

The technological essence of NTE in Mn_3AN is the discovery of Ge and Sn dopants that broaden the volume change. The strong preference of Ge and Sn for broadening the MVE in Mn_3AN revealed in previous studies [5, 6, 7] indicates importance of a local distortion caused by the atomic and/or chemical characteristic of dopants. Therefore we have investigated the local structure by the atomic pair distribution function (PDF) analysis [12, 13]. Although the overall crystal structure remains cubic for $0.15 \leq x \leq 0.7$ in the whole T region, interestingly, the PDF shows considerable deviation of the local structure in short length from the average cubic structure.

Figure 2 (a) displays the experimental PDF obtained for x=0.15, 0.5 and 0.7 at 300 K. Negative peaks at ~1.9 Å and ~2.6 - 3.0 Å correspond to atomic pair bond distances related Mn atoms, because Mn nucleus has negative neutron scattering length. The first negative peak, which comes from the Mn-N correlation in Mn_6N octahedra. On the other hand, the second negative peak comes mainly from negative Mn-Cu (or Ge) contribution. The second negative peak around 2.8 Å is much wider than the first peak and has a double-peak structure shown in the figure, for all samples. In the ideal cubic structure, the second peak should be as sharp as the first peak at ~1.9 Å. The observed PDF provides clear evidence for the local distortion on the Mn-Cu (or Ge) correlation against relatively rigid Mn-N bonds. The observed local distortion is related to the low-temperature tetragonal structure of Mn_3GeN. Mn_3GeN shows a transition from the high T cubic to the low T tetragonal T_4 ($I4/mcm$) phase at T_t ~540 K [14, 15]. The transition primarily involves alternate rotation of the Mn_6N octahedra as shown in Fig. 2(b). By this rotation, short and long Mn-Cu(Ge) bonds are generated, resulting in the splitting of the second atomic pair correlation peak.

From the structure refinement with the T_4 model, we found that the rotation angle θ of Mn_6N octahedra is a good indicator of the broadening of the MVE. At 300 K, the θs are in order of 2.3(3)(x =0.15) < 4.1(2)(x=0.5) < 4.6(3)(x=0.7). In $Mn_3Cu_{1-x}Ge_xN$, θ systematically increases with increasing Ge doping level x, corresponding to the increase of the splitting width of the second negative peak shown in Fig. 2(a). The local octahedral rotation angle clearly correlates with the T dependence of the ordered magnetic moment. Observed structural instability seems to lead to the instability of the amplitude of magnetic moment. For the Invar alloy system, the Invar effect or an instability of the amplitude of magnetic moment also appears near the phase boundary between the fcc and bcc phases [16]. So far, the variation of magnetic phase transition has been discussed on the basis of the spatially averaged information. The present study provides the first experimental result emphasizing the importance of the local structure to the related Invar problem.

4. Summary

The system with the Γ^{5g} antiferromagnetic cubic structure exhibits a large MVE in $Mn_3Cu_{1-x}Ge_xN$. The present results establish a new MVE paradigm that will require a new theoretical framework that takes into account the ordered magnetic structure. A local T_4 structure in average cubic phase appears as Ge substitution proceeds. This strongly suggests that the gradual growth of magnetic moment leading to the broadening of MVE is related to the structural instability between cubic and tetragonal induced by Ge substitution.

5. Acknowledgments

We are grateful to J. Matsuno, K. Fukamichi, Y. Kakehashi, A. Fujita and T. Egami for helpful discussions. The authors thank N. Igawa, M. Matsuda, and Th. Proffen for their help in the neutron diffraction measurements. This work was performed under the NIMS-RIKEN-JAEA Cooperative Research Program on Quantum Beam Science and Technology. This work has benefited from the use of NPDF at the Lujan Center at Los Alamos Neutron Science Center. Los Alamos National Laboratory is funded by DOE under Contact No. W-7405-ENG-36. This work was partly supported by NEDO, Japan.

6. References

[1] A. W. Sleight, Inorg. Chem. **37**, 2854 (1998).
[2] A. W. Sleight, Current Opinion in Solid State and Materials Science, **3**, 128 (1998).
[3] C. E. Guillaume, Recherches sur les aciers au nickel. Dilatations aux temperatures elevees; resistance electrique. CR Acad. Sci. **125**, 235 (1897).
[4] D. Fruchart and E. F. Bertaut, J. Phys. Soc. Jpn. **44**, 781 (1978).
[5] K. Takenaka and H. Takagi, Appl. Phys. Lett. **87**, 261902 (2005).
[6] K. Takenaka and H. Takagi, Mater. Trans. **47**, 471 (2006).
[7] K. Takenaka *et al.*, Appl. Phys. Lett. **92**, 011927 (2008).
[8] S. Iikubo, K. Kodama, K. Takenaka, H. Takagi and S. Shamoto, Phys. Rev. B **77**, 020409(R) (2008).
[9] S. Iikubo, K. Kodama, K. Takenaka, H. Takagi and S. Shamoto, Phys. Rev. Lett. **101**, 205901 (2008).
[10] Ph. l'Heritier, D. Boursier, R. Fruchart, and D. Fruchart, Mat. Res. Bull. **14**, 1203 (1979).
[11] R. Fruchart, R. Madar, M. Barberon, E. Fruchart, and M. G. Lorthioir, J. Phys. (Paris) **32** C1, 982 (1971).
[12] T. Egami and S. J. L. Billinge, *Underneath the Bragg Peaks: Structural Analysis of Complex Materials* (Pergamon, Elsevier, 2003).
[13] Th. Proffen and S. J. L. Billinge, J. Appl. Crstallogr. **32**, 572, (1999).
[14] M. Barberon *et al.*, Mat. Res. Bull. **5**, 1 (1970).
[15] Although the appearance of the successive phase transition in Mn_3GeN depends on sample preparation condition, the structure at the lowest temperature (T<510 K) is the identical T_4 structure. See Ref. [14].
[16] E. F. Wasserman: *Ferromagnetic Materials Vol. 5*, edited by K. H. J. Bushow and E. P. Wohlfarth (Elsevier Science Publishers, 1990) p.237.

Journal of Physics: Conference Series **251** (2010) 012015 doi:10.1088/1742-6596/251/1/012015

A complex magnetic structure in the magnetoresistive $La_{0.82}Ba_{0.18}CoO_3$

P Tong[1], Q Huang[2], M Kofu[1*], M C Lehman[1], J Yu[1] and D Louca[1]
[1]University of Virginia, Department of Physics, VA 22904, USA
[2]NIST Center for Neutron Research, Gaithersburg, MD 20899, USA

Email: tongpeng@virginia.edu

Abstract. The magnetic structure of the perovskite cobaltite $La_{0.82}Ba_{0.18}CoO_3$ was investigated using powder neutron diffraction. Although macroscopically this composition is thought to be in the spin-glass phase from bulk magnetic measurements, the presence of Bragg-like magnetic features in the neutron diffractograms suggests otherwise. At 4 K, two long-range magnetic orders are observed with two propagation wave vectors, k_1 = (0,0,0) and k_2 = (0,-0.5,0.5), corresponding to ferromagnetic and antiferromagnetic orders, respectively. However, above 100 K, the k_2 component almost disappears, while k_1 persists.

1. Introduction

The intricate interplay of spin, charge, orbital and lattice degrees of freedom in complex transition metal oxides often leads to emergent behaviours with unusual magnetic and electronic characteristics as in the case of the cobalt perovskite $La_{1-x}A_xCoO_3$ (A = divalent ion, Ca, Sr or Ba) [1]. The parent compound, $LaCoO_3$, has a non-magnetic insulating ground state, with the Co^{3+} ion in the $t_{2g}^6e_g^0$ (low spin, LS) electronic configuration [2]. However, as the temperature rises, an electronic excitation occurs to a higher spin state, with possibly one of two configurations, $t_{2g}^5e_g^1$ (intermediate spin, IS) or $t_{2g}^4e_g^2$ (high spin, HS). This transition is coupled with a 0.6 meV excitation due to a single-ion effect of the Co^{3+} [3]. From the elastic neutron scattering on single crystals, it was previously shown that dynamic ferromagnetic (FM) and antiferromagnetic (AFM) correlations between the Co^{3+} ions are present, consistent with the thermally excited transition from the LS to the IS state [3]. Upon doping $La_{1-x}A_xCoO_3$ by divalent ions such as Ba, a long-range FM state is presumed to develop by $x \sim 0.2$. This is followed by an insulator-to-metal transition at $x \sim 0.25$, brought upon the spin-to-charge coupling via the Double Exchange (DE) mechanism [4]. However, this seemingly simple picture fails to describe the true magnetic state when the Ba concentration reaches 18%. In this paper, we report on the magnetic neutron diffraction analysis of a polycrystalline sample of $La_{0.82}Ba_{0.18}CoO_3$ under magnetic field. Two commensurate magnetic structure with propagation vectors k_1 = (0,0,0) and k_2 = (0,-0.5,0.5) corresponding to the FM and AFM ordering, respectively, were observed [5]. The diffraction patterns were refined and a possible configuration of the magnetic moments is presented and discussed.

* Present address: Institute for Solid State Physics, University of Tokyo, Kashiwa 277-8581, Japan.

2. Methods

The sample was prepared by standard solid state reaction. The powder neutron diffraction experiments were performed using the BT1 diffractometer at the NIST Center for Neutron Research using a 7 Tesla vertical magnet. Very weak $\lambda/3$ reflections are observed in the diffraction patterns. The measurements were carried out at two temperatures, 4 and 100 K, with and without the magnetic field turned on. At 4 K, data were also collected in zero-field cooling (ZFC) and field cooling (FC) modes. Each data set was collected for a total of about 18 hours. The programs Fullprof and SARAh [6] were employed for the data analysis. The crystal structure does not change with the magnetic transition within the resolution of our experiment and remains in the rhombohedral phase with the $R\bar{3}c$ space group. Based on the nuclear symmetry and the observed magnetic propagation vectors, the basis vectors (BVs) for each wave vector were calculated using SARAh following the irreducible representation (IR) theory. The calculated results are listed in table 1 for k_1 and k_2. In the $R\bar{3}c$ symmetry, there are two distinct Co ion sites, with coordinates at (0, 0, 0) for Co1 and (0, 0, 0.5) for Co2. The two characteristic wave vectors may originate either from the same domain or from two different domains. It is possible to combine the two k-vectors if originating from a single magnetic domain with a double-k structure. On the other hand, if the two k-vectors are independent of each other because they originate from different magnetic domains, then forming single-k structure is more appropriate. The former scenario was assumed in the present case.

3. Results and Discussion

At 4 K, two sets of magnetic peaks are observed, and indexed using two propagation vectors, k_1=(0, 0, 0) and k_2=(0, -0.5, 0.5), as shown in figure 1(a), corresponding to static FM and AFM magnetic structures. The two propagation vectors are commensurate with the crystal unit cell. When the magnetic field is turned on to H = 7 T under ZFC mode [figure 1(b)], the peaks of k_1 are significantly enhanced, while those of k_2 are somewhat suppressed but not absent. Under the FC mode [figure 1(c)], however, the background changes and the k_1 peaks become even stronger while the intensity of the k_2 peaks is reduced further. By increasing the temperature to 100 K as in figure 1(d), both sets of magnetic diffraction peaks decrease in intensity. Particularly, the diffraction pattern stemming from the k_2 structure nearly disappears.

By constraining the population (i.e., phase fraction) of the k_1 and k_2 components to be equal to that of the nuclear phase, several possible moment configurations have been tested on the basis of symmetry-allowed BVs from table 1. For k_1, there are three IRs allowed by the crystal symmetry. Among the six BVs, only ψ_2 of Γ_3 can reproduce the observed patterns alone, which means a FM structure with all moments along c axis as shown in figure 2(a). As for k_2, there is only one nonzero IR as shown in table 1. We find that ψ_2 or ψ_5 alone can reproduce the diffraction patterns of k_2. The involvement of any other BVs does not lead to an improvement of the refinement. The magnetic structures resulting from ψ_2 or ψ_5 are non-collinear AFM with moments along [010] and ±[110] for Co1(0, 0, 0) and Co2 (0, 0, 0.5) sites, respectively. The case of ψ_5 is shown in figure 2 (b). The AFM unit cell is doubled along b and c axes, while the FM unit cell is as same as the nuclear one. The refined magnetic moments $\mu(k_1)$ and $\mu(k_2)$ are listed in figure 1 for k_1 and k_2 phases, respectively. It can be seen that the magnetic field enhanced the Co moment of k_1, while it suppressed that of k_2. Compared to the 4 K data, at 100 K both the moments of k_1 and k_2 are suppressed. In comparison with the k_1 component, the weaker k_2 magnetic intensities yield a higher value of Rietveld factor R_{mag}, as shown in figure 1.

If the two k-vectors are coupled with each other, the combination of them may give rise to a canted magnetic structure (i.e., a single domain structure). As discussed above, the possible moment of k_1 is orthogonal to those of k_2, which ensures each Co site has the same magnitude of total moment (μ_{total}) after combination. The values of μ_{total} are shown in figure 1. Alternatively, if the single-k structure is more applicable, the volume fraction for each k-vector can be estimated (not shown here) by supposing that the moment magnitudes are equal in both k_1 and k_2 domains, and that the whole sample volume is magnetically ordered. At present, however, it is not easy to distinguish between the two models in our case because whether and how the two *k*-vectors modulate with each other are still unclear. In addition, the powder diffraction performed can not distinguish the moment orientation within the *ab* plane of such a rhombohedral crystal unit cell. Further studies on other chemical compositions near 18 % and polarized neutron diffraction with the aid of magnetic field on single crystals may shed light on these issues.

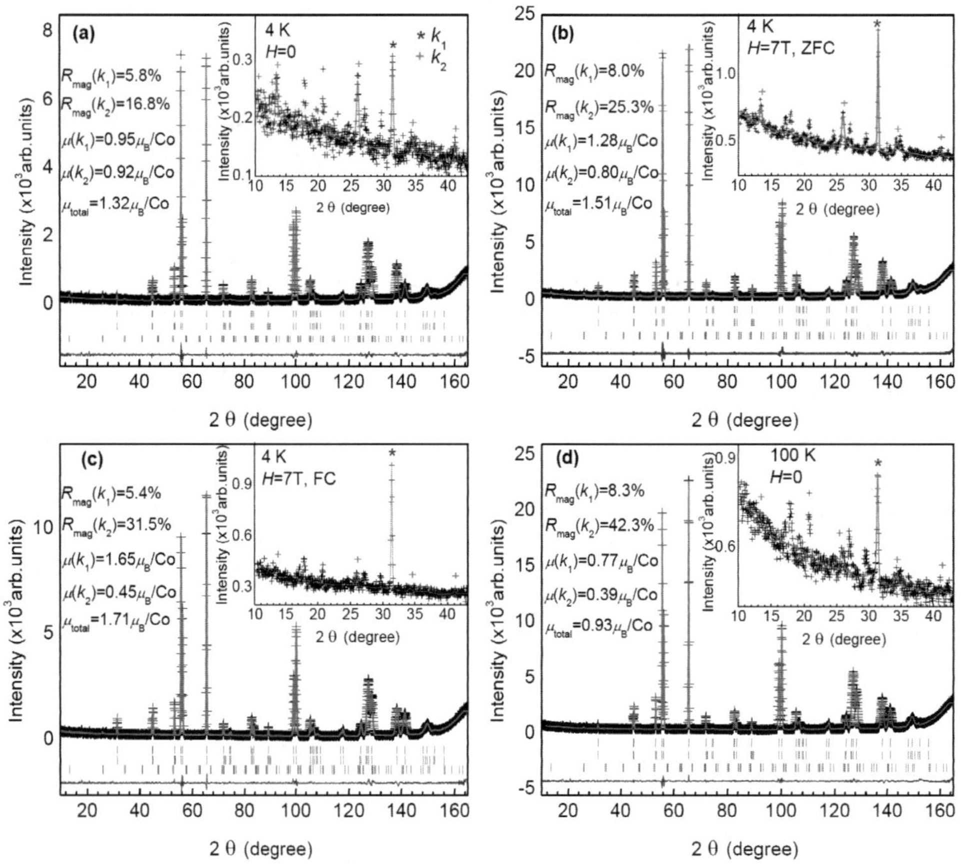

Figure 1. (Color online) Rietveld refinements of neutron diffraction patterns. The observed intensities are shown by cross symbols and calculated ones by a solid line. The vertical bars indicate the Bragg peak positions of nuclear phase, k_1 and k_2 from top to bottom. The solid line at the bottom of each panel indicates the difference between the observed and calculated intensities. R_{mag} is the agreement parameter for magnetic phase. The magnetic moments $\mu(k_1)$ and $\mu(k_2)$ are obtained by supposing the single domain structure with two *k*-vectors. μ_{total} is the total moment after combination of $\mu(k_1)$ and $\mu(k_2)$. An enlarged view of the profiles at low angles is shown in the inset of every panel. The magnetic diffractions from propagation vectors k_1=(0, 0, 0) (blue asterisk) and k_2=(0, -0.5, 0.5) (pink plus) are indicated. Some weak diffraction peaks unmarked are due to the $\lambda/3$ contamination in the BT1 instrument. Relative intensity of the peak at $2\theta \sim 27$ degree is nearly invariable with temperature and magnetic field, indicating its non-magnetic origin. It may come from a secondary phase, thus was excluded in the indexing and refinement.

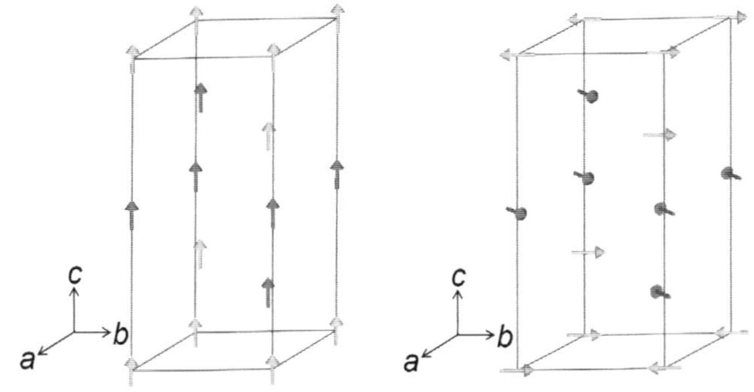

(a) (b)

Figure 2. (Color online) Possible magnetic structure for k_1 (a) and k_2 (b) plotted within one nuclear unit cell. The moments at Co (0,0,0) sites are plotted in green, and at Co (0,0,0.5) sites in pink.

Table 1. The calculated irreducible representation (Γ) and related basis vectors (BVs) at the two Co sites (Co1 and Co2) for k_1 and k_2 with respect to the space group $R\bar{3}c$. The components of BV, M_a, M_b and M_c are given with respect to the crystallographic axes a, b and c respectively.

		k_1=(0, 0, 0)								**k_2=(0, -0.5, 0.5)**					
		Co(1)			**Co(2)**					**Co(1)**			**Co(2)**		
IR	**BV**	M_a	M_b	M_c	M_a	M_b	M_c	**IR**	**BV**	M_a	M_b	M_c	M_a	M_b	M_c
Γ_1	Ψ_1	0	0	1	0	0	-1	Γ_1	Ψ_1	1	0	0	1	0	0
Γ_3	Ψ_2	0	0	1	0	0	1		Ψ_2	0	1	0	-1	-1	0
Γ_6	Ψ_3	1	0	0	0	0	0		Ψ_3	0	0	1	0	0	-1
	Ψ_4	0	0	0	0	1	0		Ψ_4	1	0	0	-1	0	0
	Ψ_5	0	0	0	2	1	0		Ψ_5	0	1	0	1	1	0
	Ψ_6	-1	-2	0	0	0	0		Ψ_6	0	0	1	0	0	1

4. Summary

To summarize, we report on the powder neutron diffraction on $La_{0.82}Ba_{0.18}CoO_3$. Two magnetic propagation vectors k_1 = (0,0,0) and k_2 = (0,-0.5,0.5) were observed corresponding to the FM and AFM structures. Further studies are necessary to understand the details of the coupling between the magnetic orders and the orientations of the magnetic moments.

Acknowledgments

The authors would like to acknowledge valuable discussions with K. Yamada. This work is supported by the U. S. Department of Energy under contract DE-FG02-01ER45927 and the U. S. DOC through NIST-70NANBH1152.

References

[1] Phelan D *et al.* 2006 *Phys. Rev. Lett.* **97** 235501.
[2] Raccah P M *et al.* 1967 *Phys. Rev.* **155** 932.
[3] Phelan D *et al.* 2006 *Phys. Rev. Lett.* **96** 027201.
[4] Phelan D *et al.* 2007 *Phys. Rev.* B **77** 104111.
[5] Yu J *et al.* submitted.
[6] Wills A S 2000 *Physica* B **276-278** 680-681.

Investigation of the magnetic structure in $NaFeGe_2O_6$ using neutron powder diffraction

T Drokina[1], G Petrakovskii[1], L. Keller[2] and J. Schefer[2]

[1]L.V. Kirensky Institute of Physics, Siberian Branch of Russian Academy of Science, Krasnoyarsk 660036, Russia
[2]Laboratory for Neutron Scattering, ETH Zürich and Paul Scherrer Institut, CH-5232 Villigen PSI, Switzerland

E-mail: tvd@iph.krasn.ru

Abstract. $NaFeGe_2O_6$ is a one of the representatives of the pyroxene family with the general formula $Me^{1+}Me^{3+}Si(Ge)_2O_6$ (Me^{1+}- monovalent metal cation, Me^{3+}-3-d transition metal cation), which regained strong interest in spintronics as a new class of multiferroic materials and in solid-state physics as a quasi-low-dimensional insulating system. The sodium iron germanate $NaFeGe_2O_6$ powder was synthesized at 800°, 900° C for 4 days from the stoichiometric mixture of Na_2CO_3, Fe_2O_3, GeO_2 by solid state reaction after several cycles of regrinding. Using neutron powder diffraction the magnetic order below $T_N \approx 13$ K was found and the magnetic structure has been determined. It is incommensurate and consists of antiferromagnetically coupled Fe^{3+} pairs with helical modulation within the *a-c* plane of the crystal lattice.

1. Introduction

The sodium iron germanate $NaFeGe_2O_6$ is a one of the representatives of the pyroxenes family with the general formula $Me^{1+}Me^{3+}Si(Ge)_2O_6$ (Me^{1+} - monovalent metal cation, Me^{3+} - 3-d transition metal cation). These systems as a new class of multiferroic materials, which are simultaneously magnetic, ferroelectric and ferroelastic, have interesting physical properties and promise important applications [1].

The room temperature crystallographic $NaFeGe_2O_6$ structure was first determined by L. Solovyova et al. (1967) [2]. This compound belongs to the monoclinic type of pyroxene structures that crystallize in the space group C 2/c. The structure of this dielectric material contains GeO_4 tetrahedra that are connected into infinite chains extended along the c axis. The oxygen polyhedra around the Fe atoms are octahedra. They are bound through shared edges into continuous zigzag ribbons running along the c axis. The two types of ribbons alternate along the b axis.

The structure of the magnetic subsystem in pyroxenes is quasi-one-dimensional [3, 4]. The theoretical and experimental peculiarities of low-dimensional insulating systems with a low magnetic transition temperature were discussed that emphasized the importance of the investigation of the magnetic ordering in such systems [3]. The general topology of exchange interactions is such that the magnetic

[1] tvd@iph.krasn.ru

system may be frustrated [3]. It may lead to nontrivial magnetic structures and properties of pyroxenes, which depend both on intrachain and interchain magnetic exchange interactions.
The SQUID magnetic measurement of Fe-based pyroxenes was carried out in a magnetic field of 100 Oe on a sample cooled to 4.2 K in zero magnetic field [4]. This magnetic measurement showed that the inverse susceptibility above T = 100 K in $NaFeGe_2O_6$ can be approximated by a Curie-Weiss law.

The negative asymptotic temperature (Θp = -135 K) shows the prevailing presence of the antiferromagnetic exchange interactions in $NaFeGe_2O_6$. Below 100 K the susceptibility deviates from the Curie–Weiss behavior and exhibits a maximum at the temperature T_m=25 K. The magnetic transition temperature is defined as the inflection point at T_N=15 K. The value of the effective magnetic moment (molar value) μ_{eff} = 5.89 μ_B (μ_B is the Bohr magneton) close to the theoretically expected value of the Fe^{3+} (s=5/2) magnetic moments 5.91 μ_B.
The $NaFeGe_2O_6$ Mössbauer studies were carried out at room temperature on powder samples 5–10 mg/cm^2 in thickness in the natural iron content using a ^{57}Co(Cr) source [4]. In paramagnetic phase the ^{57}Fe Mössbauer spectrum (300 K) is a doublet with the isomer shift Θ = 0.40 mm/s relatively to α - Fe, the quadrupole splitting 0.34 mm/s. This spectrum is attributed to Fe^{3+} ions with oxygen octahedral surroundings.
The aim of this study was to investigate the magnetic structure in the $NaFe^{3+}Ge_2O_6$.

2. Sample preparation
The $NaFeGe_2O_6$ powder (2 cm^3) was synthesized at 800°- 900° C in air for 4 days using reagents: 16% Na_2CO_3, 23% Fe_2O_3, 61% GeO_2 by the solid state reaction after several cycles of regrinding. The product contained some value of $Na_4Ge_9O_{20}$ as an impurity phase. The unit cell parameters of our specimen, which were found by X-ray powder diffraction method at 300 K, are a=10.008 Å, b = 8,948 Å, c=5.523 Å, β= 107.59° [4]. This structure coincides with that of a mineral diopside [2].

3. Experiment
The neutron scattering experiments in the temperature range of 1.6-100 K were performed on the cold neutron powder diffractometer DMC [5] at Swiss spallation neutron source SINQ [6]. The sample was enclosed in a cylindrical vanadium container under helium atmosphere and mounted in a helium cryostat. The neutron wavelength used was λ = 2.4576 Å. The data has been corrected for absorption and was refined using the FULLPROF program package [7]. For the refinement of the magnetic structure at 1.6 K the paramagnetic data at 30 K was subtracted to obtain the pure magnetic diffraction pattern.

4. Results
Neutron diffraction is the method of choice for the determination of magnetic structures. In the paramagnetic state the experiment confirmed the monoclinic crystallographic structure. Below the critical temperature for magnetic order we find a large number of additional Bragg peaks due to long-range order of the magnetic moments (figure 2). The temperature dependence of the strongest magnetic peak is shown in figure 3, revealing the onset of magnetic order near 13 K. This is in accordance with the results of SQUID magnetic measurements [4]. The integrated neutron intensity increases with decreasing temperature as the square of the ordered moment and at the lowest temperature T = 1.6 K saturation is not yet reached.

A careful analysis of the experimental peak positions showed that the magnetic reflections cannot be described with a commensurate magnetic unit cell and immediately revealed the incommensurate nature of the magnetic order. We succeeded to refine the magnetic periodicity and the magnetic peaks can be indexed by the incommensurate propagation vector **k** = (0.3357(4), 0, 0.0814(3)) at 1.6 K.
We performed a systematic refinement of the arrangement of the magnetic moments at 1.6 K and it turned out that just two models are able to describe the incommensurate peak positions as well as the

observed magnetic intensities. Both models consist of antiferromagnetically coupled pairs of iron moments with incommensurate modulation along the propagation vector **k**. One model describes a sine-modulation of the moments, the other describes a helical modulation. The R-factor and χ^2 of the refinement clearly favor the second model which shows an overall better agreement with the data (figure1). Neutron powder diffraction experiments results are listed in table 1.

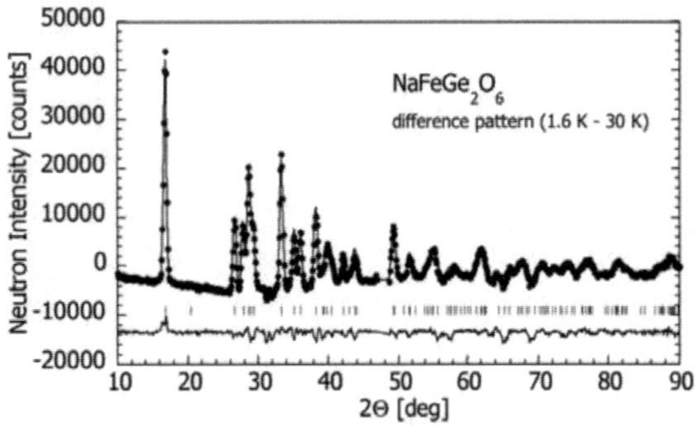

Figure 1. Difference (1.6 K – 30 K) neutron diffraction pattern measured on DMC at SINQ, showing the pure magnetic scattering only (•). The lines denote the model calculation and the deviation of experiment and refinement as described in the text.

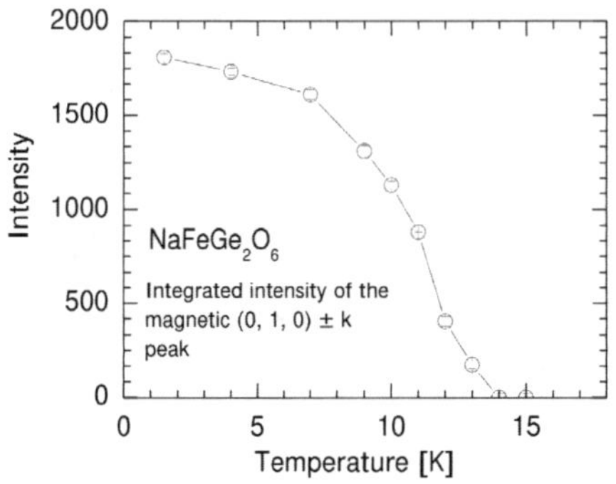

Figure 2. Integrated intensity of the strongest magnetic peak (0, 1, 0) ± **k** of *$NaFeGe_2O_6$* determined by neutron diffraction as a function of temperature, Néel temperature $T_N \approx 13$ K.

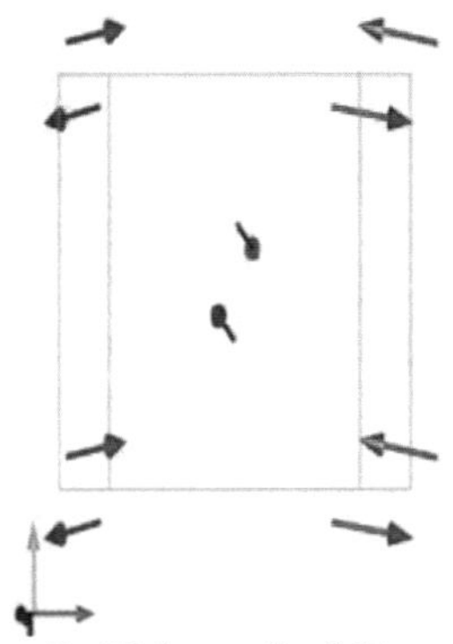

Figure 3. (Color online) Fe moments in one crystallographic unit cell, view along c (ab in plane). The magnetic structure of $NaFeGe_2O_6$ is built up by antiferromagnetically coupled pairs.

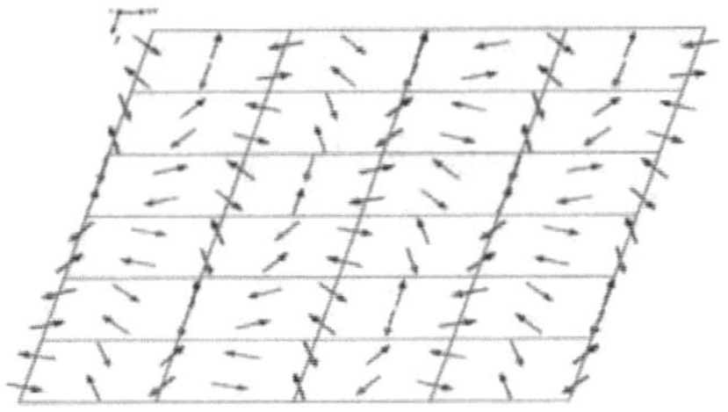

Figure 4. (Color online) Incommensurate modulation of the magnetic structure of $NaFeGe_2O_6$ (ac in plane).

Therefore we conclude that $NaFeGe_2O_6$ orders with an incommensurate helical modulation of the iron magnetic moments along **k** = (0.3357(4), 0, 0.0814(3)). At T = 1.6 K the ordered moment per Fe^{3+} ion is refined to be 2.55(1) μ_B. The refinement of the difference pattern (1.6 K – 30 K) with the helical model described above is shown in figure 1. In the difference pattern all peaks are due to magnetic order and our magnetic model nicely reproduces the experimental data with a large number of magnetic peaks for the full scattering angle range. The resulting pair arrangement of the moments in one chemical unit cell is shown in figure 3. The magnetic moments mainly lie in the ac plane and the

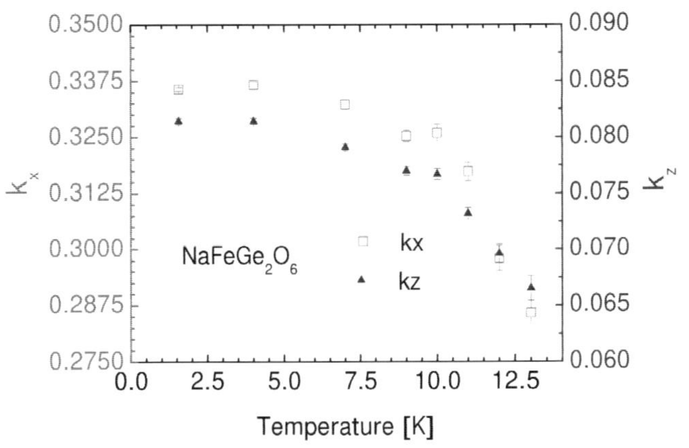

Figure 5. (color online) Temperature dependemce of the propagation vector k_x (left axis, red) and k_z (right axis, blue)

refinement only shows a small component along b due to a canting of ~10 degrees in respect to the ac-plane. The modulation over several chemical unit cells is shown in figure 4, displaying the incommensurate nature of the order in the ac plane. The errors in the refinement of the incommensurate vector in figure 5 as well as the magnetic moment do not include systematic errors and are therefore underestimated in the Rietveld refinement.

TABLE I. Results for a helical spin arrangement at 1.6K

Propagation vector	**k** = (0.3357(4), 0 , 0.0814(3))
Moment arrangement:	helical modulation of antiferromagnetically coupled pairs
ordered moment per Fe^{3+} ion	M = 2.55(1) μ_B
Plane of moments	close to a-c plane (small component along b)
Reliability factors of refinement	$R_p = 4.5$, $\chi^2 = 4.43$

Conclusions

The magnetic structure of the sodium iron germanate $NaFeGe_2O_6$ was determined by neutron diffraction. Below $T_N \approx 13$ K $NaFeGe_2O_6$ undergoes a phase transition to an incommensurate magnetic structure. Antiferromagnetically coupled pairs of Fe^{3+} moments show an incommensurate modulation according to the magnetic propagation vector **k** = (0.3357(4), 0, 0.0814(3)) at 1.6 K within the ac plane and no modulation along b. The k vector is slighly temperature dependent (c.f. figure 5).

Acknowledgments

Neutron beam time at the cold neutron powder diffractometer DMC [5] at the Swiss spallation neutron source SINQ, Paul Scherrer Institute, Villigen, Switzerland is greatfully acknowledged.

References

[1] Jodlauk S, Becker P, Mydosh J A, Khomskii D I, Lorenz T, Streltsov S V, Hezel D C and Bohaty L 2007 *J. Phys. Cond. Matter* **19** 432201
[2] Solovyova L and Bakakin V 1967 *Krystallographiya* **12** 591
[3] Katanin A and Irhin V 2007 *Usp.Phys.Nauk* **177**, 639
[4] Drokina T, Bajukov O, Petrakovskiy G, Velikanov D, Bovina A, Stepanov G and Ivanov D 2008 *Fiz. tv. tela (rus)*, **50**, 2050
[5] Fischer P, Keller L, Schefer J and Kohlbrecher J 2000 *Neutron News* **11** 19
[6] Fischer W E 1997 *Physica* **B 234-236**, 1202-1248
[7] Rodriguez-Cravajal J 1993 *Physica,* **B 192**, 55

Journal of Physics: Conference Series **251** (2010) 012017 doi:10.1088/1742-6596/251/1/012017

Neutron diffraction study of magnetic structures in single crystal Ho_2PdSi_3 in magnetic fields up to 5 T

Fei Tang[1], Peter Link[2], Matthias Frontzek[1], Jean-Michel Mignot[3], Jens-Uwe Hoffmann[4], Wolfgang Löser[5] and Michael Loewenhaupt[1]

[1] Institut für Festkörperphysik, TU Dresden, Dresden, Germany
[2] Forschungsneutronenquelle Heinz-Maier-Leibnitz, TU München, Garching, Germany
[3] Laboratoire Léon Brillouin, CEA-CNRS, CEA Saclay, France
[4] Helmholtz-Zentrum Berlin für Materialien and Energie GmbH, Berlin, Germany
[5] Leibniz-Institut für Festkörper- und Werkstoffforschung, Dresden, Germany

E-mail: [1] `ftang@ifp.phy.tu-dresden.de`

Abstract.
Due to the interplay between RKKY exchange interaction, crystalline-electric field (CEF) effects and geometric frustration due to the AlB_2-derived hexagonal crystal structure, the study of R_2PdSi_3 (R = rare earth) has been found to be a challenging field in rare earth magnetism. In this contribution we present the results of neutron diffraction experiments on a Ho_2PdSi_3 single crystal in the magnetically ordered state at $T \leq 1.6$ K. The compound orders antiferromagnetically in the basal plane with the propagation vector $\boldsymbol{\tau} = (1/7\text{-}\delta,\ 2\delta,\ 0)$ where $\delta \sim 0.01$ r.l.u. In magnetic fields applied along the $(1\bar{1}0)$ magnetic hard axis, the structure persists up to the highest measured field of 5 T. The dependence of the intensities as a function of field suggests that the $\boldsymbol{\tau}$ structure is a single-k multi-domain structure.

1. Introduction

The series of R_2PdSi_3 (R = rare earth) compounds have been a focus of interest for over 15 years[1]. The compounds crystallize in an AlB_2-derived hexagonal structure (P6/mmm) with a latent geometric frustration. The magnetic rare-earth ions occupy the Al positions of the AlB_2 structure while the non-magnetic Pd and Si atoms are assumed to be statistically distributed on the B positions[2]. A rich variety of magnetic phenomena from long-range antiferromagnetic order with propagation vectors in different directions to short-range magnetic correlations have been observed in the series[3]. This complex magnetic behavior results from the interplay between RKKY interaction, magneto-crystalline anisotropy based on crystal-electric field (CEF) effects and geometric frustration due to the AlB_2-derived hexagonal crystal structure.

A relatively weak magneto-crystalline anisotropy makes Ho_2PdSi_3 an interesting compound for the study of the delicate balance between the various interactions. The Stevens factor ($\alpha \sim$ -0.022) for Ho^{3+} suggests a magnetic easy axis in the basal plane which is also observed in high-temperature susceptibility data. However, in the magnetically ordered state the magnetic easy direction is the (001) direction. The exchange of the magnetic easy and hard axes is observed at around 50 K[4]. The Néel temperature T_N is 7.7 K, a second phase transition occurs at T_2 = 2.3 K [4].

Neutron diffraction measurements on powder Ho_2PdSi_3 samples were performed in zero field by Szytula et al.[2] proposing a magnetic structure with $\boldsymbol{c}$-axis oriented, sine-wave modulated magnetic moments with a propagation vector of (0.137, 0.0066, 0) at T = 1.5 K. In the present paper we present the results of neutron diffraction measurements on a single crystalline Ho_2PdSi_3 sample in zero field and in fields up to 5 T applied in the direction of the magnetic hard axis $(1\bar{1}0)$.

2. Experimental details

The single crystal preparation and characterization are described elsewhere[5, 6]. The neutron scattering experiments were carried out at the Helmholtz-Zentrum Berlin (formerly Hahn-Meitner Institut) with the E2 diffractometer and at Laboratoire Léon Brillouin on the 6T2 diffractometer with lifting counter option. The Ho_2PdSi_3 (m $\simeq$ 2.9 g) single crystal used in the studies is slightly irregular in shape with 6 oriented faces ground on a cylindrical sample of 9 mm length and 5 mm diameter. On both diffractometers a graphite monochromator was used with additional pyrolytic graphite filters to suppress $\lambda/2$ contaminations.

The E2 experiment used a standard orange-type cryostat capable of reaching 1.5 K base temperature. The E2 diffractometer employs a "banana"-type multidetector with 400 channels covering a 2θ angle of 80°. The sample is rotated around the axis perpendicular to the *HK*0 scattering plane in steps of 0.2°. The used neutron wavelength was 2.39 Å.

For the 6T2 experiment, the sample was mounted in a ^{4}He cryostat (base temperature $\sim$ 1.5 K) within a vertical magnet with the $(H, \bar{H}, 0)$ direction parallel to the field direction making the *HHL* plane the scattering plane. The diffractometer has a "classical" setup with a single detector mounted on a lifting arm to allow access to reflections off the scattering plane. It is worth noting that the $\mathbf{Q}$ resolution along the perpendicular direction is not as good as along directions in the scattering plane (normally the FWHM is about two times larger). The used neutron wavelength was 2.345 Å. The sample position was not very accurate in height and therefore produced an offset of the zero position in the reciprocal space.

3. Results and Discussion

3.1. The magnetic structure of $\boldsymbol{\tau}_\pm = (1/7\mp\delta, \pm2\delta, 0)$

Figure 1 shows the full reciprocal *HK*0 plane of Ho_2PdSi_3 at T = 1.5 K obtained on E2. Magnetic Bragg peaks appear on positions $\{\boldsymbol{G}\pm\boldsymbol{\tau}_0\}$ where $\boldsymbol{G}$ is a reciprocal lattice vector and $\boldsymbol{\tau}_0$ = (1/7, 0, 0). Closer examinations show that the magnetic reflections are split in directions perpendicular to the "main" propagation direction (*H*, 0, 0). Thus the actual propagation vectors are $\boldsymbol{\tau}_\pm = (1/7\mp\delta, \pm2\delta, 0)$. Shown in the upper left part of figure 2 is the centered Q scan measured on 6T2 at $\boldsymbol{Q}_0$ = (0, 0, 1) + (1/7, -1/7, 0) where the splitting of $\boldsymbol{\tau}_\pm$ is nicely resolved with δ $\delta \sim 0.01$ r.l.u. in $(\pm\delta, \pm\delta, 0)$. At $\boldsymbol{Q}_0$ = (0, 0, 1) + (0, -1/7, 0) (upper right in figure 2) it is more difficult to resolve the splitting because the scan direction is along the $\bar{2}10$ direction which is out of the scattering plane. The appearance of the higher harmonics at 3 $\boldsymbol{\tau}_\pm$ and 5 $\boldsymbol{\tau}_\pm$ in figure 1 indicates that the magnetic structure is not a simple sinusoidal-modulated structure. The temperature dependence of higher harmonics suggests that the squaring-up process is closely related to the transition at T_2 ($\sim$ 2 K), similar to the one observed in Er_2PdSi_3 compound[3].

3.2. Domain effect of the magnetic structure $\boldsymbol{\tau}_\pm = (1/7\mp\delta, \pm2\delta, 0)$

Due to the difficulty to resolve the two splits of $\boldsymbol{\tau}_\pm$, only the summed intensity of $\boldsymbol{\tau}_+$ and $\boldsymbol{\tau}_-$ can be compared. They are labeled as $\boldsymbol{\tau}_0^i$. With the 6T2 experiment, the field dependence of the $\boldsymbol{\tau}_0^i$ intensities has been carefully followed at T = 1.6 K from $\mu_0 H$ = 0 T to 5 T. Transverse scans along five (of the six) symmetric equivalent vectors of $\boldsymbol{\tau}_0$ that were accessible with the lifting counter were performed (see figure 3 for the relative orientations).

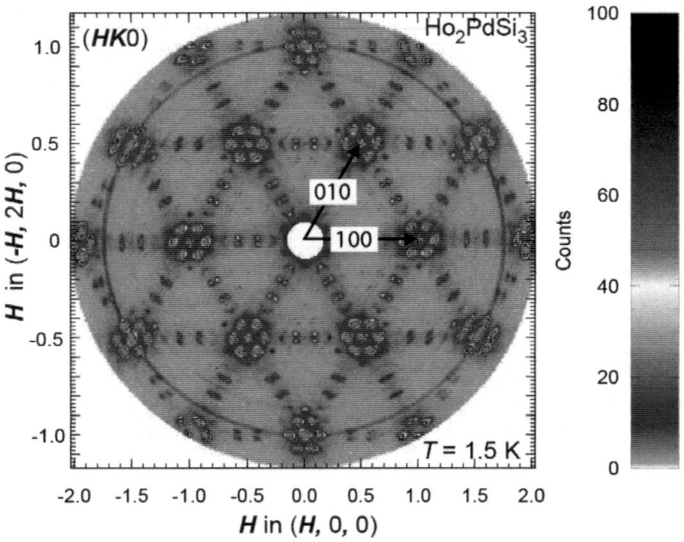

Figure 1. Full reciprocal $HK0$ plane of Ho_2PdSi_3 at T = 1.5 K obtained on E2. The reciprocal lattice directions (100) and (010) are marked with arrows. The two rings originate from the scattering on polycrystalline Al from the sample holder and the cryostat.

Shown in figure 2 is a comparison between τ_0^0 and τ_0^2 at 0 T and 5 T, respectively. The intensity of the τ_0^0 reflection exhibits a large increase in applied field while the τ_0^2 reflection is already oppressed at 5 T. In figure 4 the field dependence of the integrated intensities of all measured τ_0^i is shown. The clear trend of the increase of the intensity of τ_0^0 with the decrease of the other τ_0s suggests that the magnetic structure is a multi-domain structure. And the fact that the τ_0^0 intensity at 5 T is only slightly less than the sum of the $\tau_0^{0,1,2}$ intensities at zero field indicates a domain repopulation.

The magnetic structure is rather robust for fields along the $(1\bar{1}0)$ direction (persistent up to 5 T at least). This could be due to the fact that the antiferromagnetic couplings are between the magnetic components parallel to the **c**-axis. The application of a magnetic field perpendicular to the moment direction only cants the magnetic moment gradually.

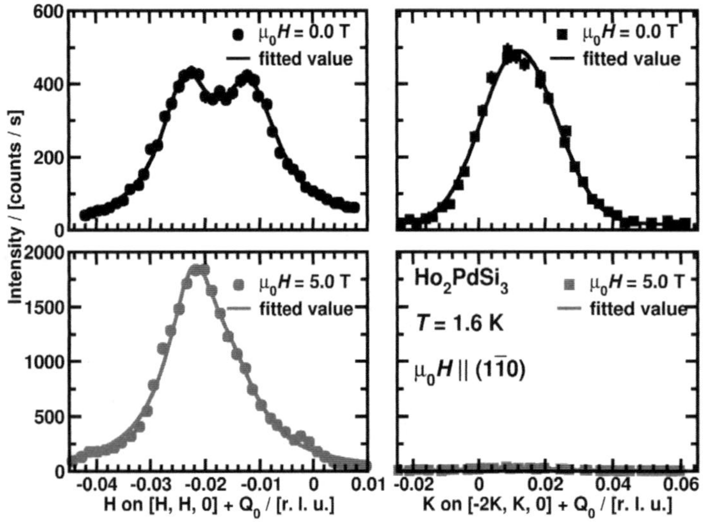

Figure 2. Centered Q scans of Ho_2PdSi_3 measured on 6T2 at T = 1.6 K at $\mathbf{Q_0}$ = (0, 0, 1)+$\tau_0^{0,2}$ where τ_0^0 = (1/7, -1/7, 0) (left frames) and τ_0^2 = (0, -1/7, 0) (right frames) in zero field (upper frames) and at 5 T (lower frames). The solid lines are the fitted values. For details see text.

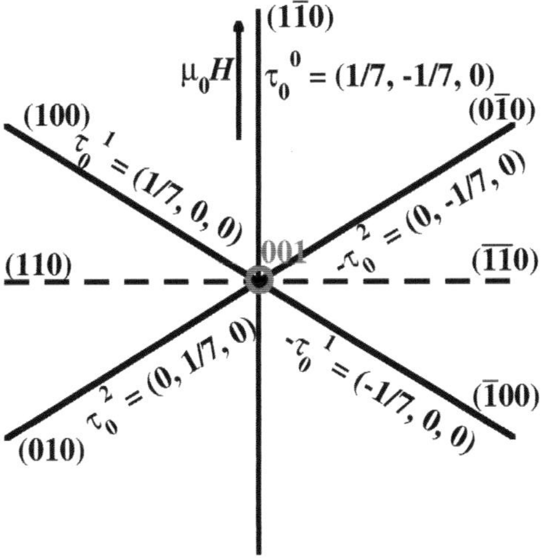

Figure 3. Directions in the hexagonal reciprocal plane: The magnetic field is parallel to the $(1\bar{1}0)$ direction; the symmetric equivalent directions for the vector $\boldsymbol{\tau}_0$ are labelled as $\boldsymbol{\tau}_0^i$.

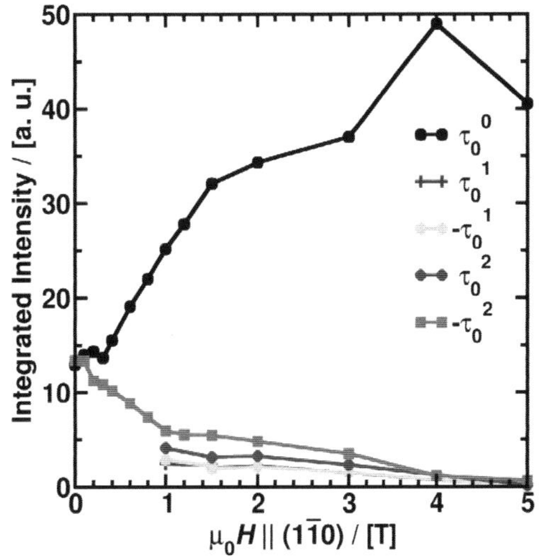

Figure 4. Field dependence of the summed intensities of propagation vectors $\boldsymbol{\tau}_{\pm}$ at different $\boldsymbol{\tau}_0^i$s of Ho_2PdSi_3 fitted from the HH scans shown in figure 2 at 1.6 K.

4. Conclusions

The field dependence of the magnetic structures of Ho_2PdSi_3 has been studied using neutron diffraction measurements on a single crystalline sample. Zero field measurements revealed propagation vectors of $\boldsymbol{\tau}_{\pm} = (1/7\mp\delta,\ \pm2\delta,\ 0)$ with $\delta \sim 0.01$ r.l.u. Furthermore, higher harmonics of $\boldsymbol{\tau}_{\pm}$ have been observed below the Néel temperature suggesting a squaring-up of the magnetic moments at low temperature. This is consistent with the theoretical observation that the sinusoidal modulated structure is generally not a ground state[7]. The field dependent measurements of the $\boldsymbol{\tau}_{\pm}$ structure show that it is a multi-domain structure. There are six domains, each related to one of the six symmetric equivalent directions of $\boldsymbol{\tau}_{\pm}$. The application of an external field along one of the six $\boldsymbol{\tau}_0^i$s ($1\bar{1}0$ here) suppresses the domains with other propagation vectors, leaving only the domains with propagation vectors ($\boldsymbol{\tau}_{\pm}^i$) most parallel to the field.

References

[1] Kotsanidis P A and Yakinthos J K 1990 *J. Magn. Magn. Mater.* **87** 199
[2] Szytula A, Hofmann M, , Penc B, Slaski M, Majumdar S, Sampathkumaran E V and Zygmuntand A 1999 *J. Magn. Magn. Mater.* **202** 365
[3] Frontzek M, Kreyssig A, , Doerr M, Schneidewind A, Hoffmann J U and Loewenhaupt M 2007 *J. Phys. : Condens. Matter* **19** 145276
[4] Frontzek M, Kreyssig A, , Doerr M, Rotter M, Behr G, Lser W, Mazilu I and Loewenhaupt M 2006 *J. Magn. Magn. Mater.* **301** 398
[5] Graw G, Bitterlich H, Löser W, Behr G, Fink J and Schultz L 2000 *J. Alloys and Compounds* **308** 193
[6] Mazilu I, Frontzek M, Löser W, Behr G, Teresiak A and Schultz L 2005 *J. Cryst. Growth* **275** e103
[7] Gignoux D and Schmitt D 1991 *J. Magn. Magn. Mater.* **100** 99

Magnetism and Magnetic Structure of $NdCr_2Si_2C$

Marcela Janatová[1], Jana Poltierová Vejpravová[1], Jan Prokleška[1], Vladimír Sechovský[1], Martin Diviš[1], Olivier Isnard[2]

[1]Charles University, Faculty of Mathematics and Physics, DCMP, 121 16 Prague 2, Czech Republic
[2]Institut Laue Langevin, and University J. Fourier Grenoble, 38042 Grenoble Cedex 9, France

Email: janatova@mag.mff.cuni.cz

Abstract. A $NdCr_2Si_2C$ polycrystal has been prepared and characterized by heat-capacity and magnetization, X-ray and neutron powder diffraction measurements. Ferromagnetism below T_C = 21 K with Nd magnetic moments aligned parallel to the *c*-axis has been confirmed. We also studied the electronic structure of $NdCr_2Si_2C$ by first-principles calculations, which predict the ordered magnetic moment both on the Nd and Cr sites. The magnetization and neutron diffraction data, however, do not allow resolving the question of existence of the Cr moment unambiguously.

1. Introduction

$NdCr_2Si_2C$ is a member of the recently described family of the RCr_2Si_2C compounds (R = a rare-earth element) crystallizing in the tetragonal $CeMg_2Si_2$-type structure with the space group *P4/mmm* [1-3]. In this structure, the atoms occupy the 1R in 1*a*: 0, 0, 0; 2Cr in 2*e*: 0, 1/2, 1/2; 2Si in 2*h*: 1/2, 1/2, z_{Si}; 1C in 1*b*: 0, 0, 1/2 crystallographic sites. According to Ref. [3], the R ions (R = Pr, Nd, Gd-Dy) order ferromagnetically at low temperatures ($T_C \leq 30$ K), and the Cr sublattice probably exhibits no magnetic ordering. Contrary, Moze et al. [4] reported that in the related RCr_2Si_2 compounds (*I4/mmm*), the Cr stable magnetic moments exist and order antiferromagnetically below 700 K.

In this paper, we report on magnetism in $NdCr_2Si_2C$ investigated by heat-capacity and magnetization measurements and by neutron powder diffraction (NPD). First-principles calculations based on the density functional theory (DFT) have been performed to investigate the Cr magnetism.

2. Experimental and computational

We have prepared a polycrystalline sample of $NdCr_2Si_2C$ by arc melting in a protective argon atmosphere. The sample was several times re-melted to improve the homogeneity. Prior to the sample preparation, the Nd metal had been refined by the solid-state electrotransport (SSE) [5]. The sample was characterized by X-ray powder diffraction (XRPD) and field-emission scanning electron microscopy (SEM). The heat capacity and magnetization measurements were performed with a PPMS 9T (Quantum Design) facility, using the standard Heat Capacity, ACMS and VSM Oven option, respectively, in the temperature range 2-900 K and magnetic fields up to 9 T. Neutron powder diffraction (NPD) measurements were performed in the temperature range 2-27 K and at the room temperature on the multidetector powder diffractometer D1B (λ = 2.52 Å) at ILL, Grenoble. The XRPD and NPD data were analyzed by the Rietveld method [6] using the program FULLPROF [7].

To obtain direct information about the ground-state electronic structure and magnetic properties, we performed first-principles theoretical calculations. The ground state electronic structure was calculated within the local spin density approximation (LSDA) and generalized gradient approximation (GGA) [8] using the full-potential WIEN2k code [9]. The localized Nd 4*f*-electron states were treated in the open-core approximation with the stable $4f^3$ atomic configuration.

3. Results and discussion

The XRPD and microprobe analyses confirmed the sample containing only the desired tetragonal $NdCr_2Si_2C$ phase with the lattice parameters a = 4.018 Å, c = 5.333 Å and z_{Si} = 0.232.

The temperature dependence of the specific heat is presented in Figure 1a. A sharp $C(T)$ anomaly is observed at about 21 K, which is attributed to the magnetic phase transition from paramagnetic to ferromagnetic state at the Curie temperature T_C. The derived magnetic entropy S_{mag}, which was calculated on the basis of the difference between the total specific heat measured for $NdCr_2Si_2C$ and corresponding non-magnetic analogue $LaCr_2Si_2C$, is presented in Figure 1b. Slightly above T_C, S_{mag} reaches a value of Rln2 expected for a doublet and, above 200 K, S_{mag} approaches the value of Rln10, which is expected for the Nd^{3+} ion (J = 9/2).

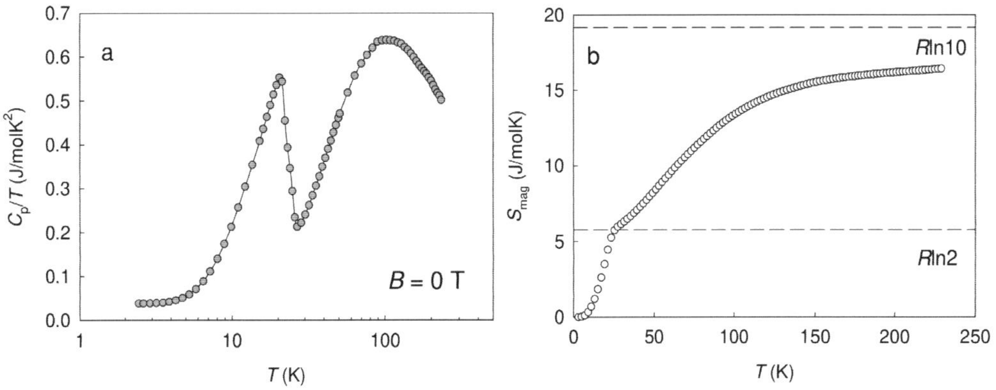

Figure 1. Temperature dependence of the specific heat as the *C/T* vs. *T* curve (a) and magnetic entropy (b).

Figure 2a shows the temperature dependence of the magnetization measured at various magnetic fields. The low-field ZFC and FC curves bifurcate approximately at T_C. The higher-field $M(T)$ curves coincide below 10 K suggesting rapid saturation of the magnetic moment at magnetic fields larger than 2 T. The plot of the inverse magnetic susceptibility is shown in Figure 2b. One can clearly see, that there are no high-temperature anomalies connected with the antiferromagnetic ordering of the Cr sublattice, which has been reported for the related RCr_2Si_2 compounds [4]. The high-temperature curve can be fitted using the modified Curie-Weiss law:

$$\chi = \chi_0 + \frac{C}{T - \theta_P}, C = \frac{N_a \mu_{eff}^2}{3k_B}.$$

We have obtained an effective magnetic moment value of 3.4(7) μ_B/f.u., a paramagnetic Curie temperature θ_p = (20 ± 3) K and χ_0 = 1.46 · 10^{-8} m^3/mol. The value of μ_{eff} is close to the theoretical effective moment of 3.62 μ_B/f.u. for Nd^{3+} ion.

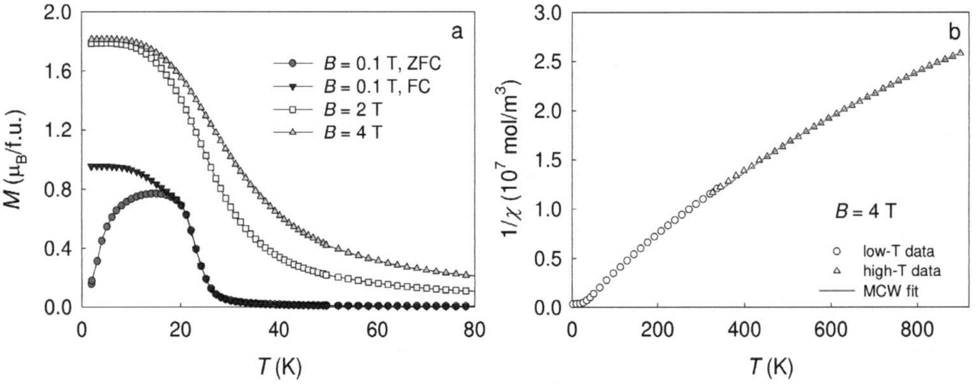

Figure 2. Temperature dependence of the magnetization (a) and inverse magnetic susceptibility (b).

The magnetization curves measured at selected temperatures are displayed in Figure 3. At 2 K, the compound exhibits a strong ferromagnetic component and rapid saturation already at 1 T with the saturation magnetic moment of about 1.86 μ_B/f.u. The hysteresis loop shows a coercivity of about 0.2 T and a remanence, M_r, of about 1.55 μ_B/f.u.. The value of M_r decreases with increasing temperature and vanishes at about 20 K (~ T_C).

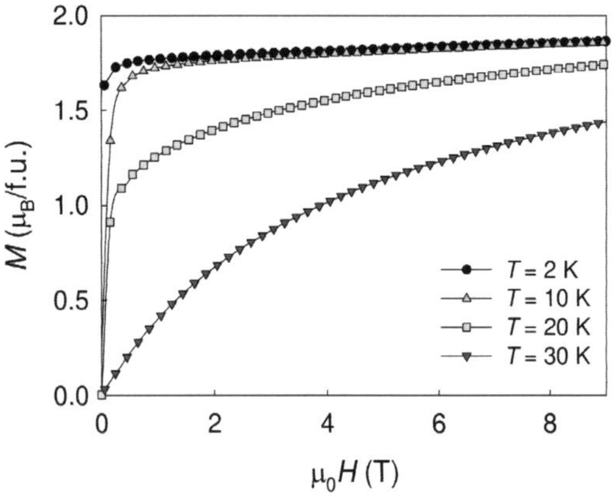

Figure 3. Magnetization isotherms of the $NdCr_2Si_2C$ measured at various temperatures.

The first-principles calculations suggest an ordered magnetic moment both on the Nd and Cr sites. The calculated value of the Cr moment depends on the method used (GGA or LSDA); it amounts about 0.89 μ_B using GGA [8]. The Cr moment is stabilized by the 5*d* states of Nd which are polarized by the localized Nd spin moment of about 3 μ_B and strongly hybridized with the Cr 3*d* states.

The NPD experiment was performed also in order to investigate the possibility of magnetic ordering of the Cr sublattice. The ferromagnetic ordering in the compound was confirmed by the temperature evolution of the NPD patterns (Figure 4a). Below T_C, we observed considerable additional intensities on top of nuclear reflections except for the (001), which indicates the ferromagnetic alignment of magnetic moments along the *c*-axis. Two different models were applied for the analysis of the magnetic diffraction data: 1) when assuming only the Nd moments, which are ordering ferromagnetically we obtain $\mu_{Nd} = (3.32 \pm 0.08)\ \mu_B$, 2) assuming the Nd moments ordering ferromagnetically and also the Cr moments ordered antiparallel to the Nd moments. Then the refined value of the magnetic moments are μ_{Nd} = (3.31 ± 0.08) μ_B and μ_{Cr} = (0.46 ± 0.05) μ_B. The refined values of the Nd magnetic moment for both the models are in good agreement with the theoretical value of the ordered moment of 3.27 μ_B/f.u for the free Nd^{3+} ion. Taking into account the results of our experiments and first-principles calculations, the second model corresponds to our calculated values. However, the two presented models have nearly the same R factor, i.e. the NPD data are not sufficient to resolve the question of existence of Cr magnetic moment unambiguously.

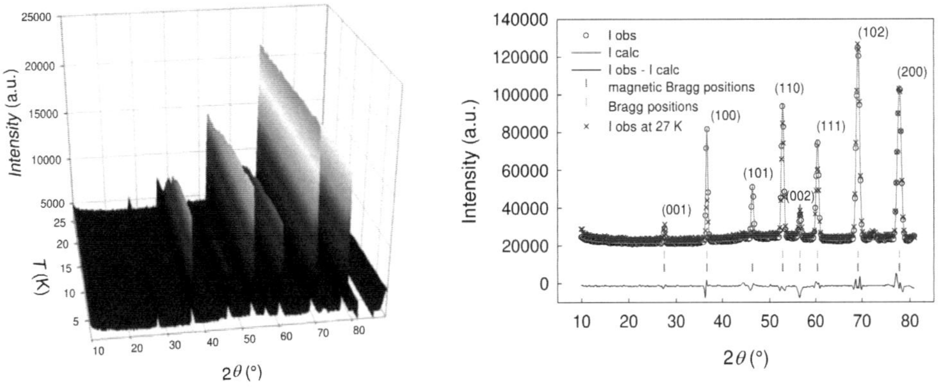

Figure 4. (a) The temperature evolution of the NPD patterns collected for $NdCr_2Si_2C$ between 5 and 27 K. (b) The NPD pattern at 2 and 27 K.

4. Conclusions

In conclusion, $NdCr_2Si_2C$ was studied experimentally by XRD, SEM, heat capacity, magnetization and NPD measurements on a polycrystal, and also by first-principles calculations. The magnetization data provide no indication of any antiferromagnetic ordering of the Cr sublattice at high temperatures. The first-principles calculations predict the ground-state ordered Cr moment. The low temperature magnetization and NPD data give no unambiguous answer to the question of the Cr moment. XMCD experiment on a single crystal is planned to investigate the existence of the Cr magnetic moment. Therefore, the polarized neutron diffraction and XMCD experiment on a single crystal are envisaged.

5. Acknowledgements

This work is a part of the research plan MSM0021620834 financed by the Ministry of Education of Czech Republic. The work of M.J. has been supported also by the Grant Agency of the Charles University no. 134309 and the Grant Agency of the Czech Republic no. 202/09/H041.

References

[1] Ch. Tang, S. Fan, M. Zhu, *J. Alloys Compd.* **299**, 1-4 (2000).
[2] M. Janatova, J. Vejpravova, M. Divis, V. Sechovsky, *Physica B* **403**, 2338-2343 (2008).
[3] V. Klosek, A. Vernière, B. Malaman, J. Tobola, S. Kaprzyk, *Phys. Rev. B* **78**, 104419 (2008).
[4] O. Moze, M. Hofmann, J.M. Cadogan, K.H.J. Buschow and D.H. Ryan, *Eur. Phys. J. B* **36**, 511 (2003).
[5] R. G. Jordan, *Contemp. Phys.* **15**, 375-400 (1974).
[6] H. M. Rietveld, *J. Appl. Crystallogr.* **2**, 65 (1969).
[7] J. Rodriguez-Carvajal, *Physica B* **192**, 5 (1993).
[8] J. P. Perdew, K. Burke, M. Ernzerhof, *Phys. Rev. Lett.* **77**, 3865 (1996).
[9] P. Blaha et al, *WIEN2k*, TU Wien, Austria, 2001, ISBN:3-9501031-1-2.
[10] M. Janatova, J. P. Vejpravova, M. Divis, *J. Appl. Phys.* **105**, 07E105 (2009).

Real-time observation of magnetic structural change in the multistep metamagnet $CeIr_3Si_2$

Kiyoichiro Motoya[1], **Yuji Muro**[1] **and Toshiro Takabatake**[2]
[1]Department of Physics, Faculty of Science and Technology, Tokyo University of Science, Noda 278-8510, Japan
[2]Department of Quantum Matter, ADSM, Hiroshima University, Higashi-Hiroshima 739-8530, Japan

E-mail: `motoya@ph.noda.tus.ac.jp`

Abstract. A ternary intermetallic compound $CeIr_3Si_2$ shows successive magnetic transitions at T_{N1}=4.1 K and T_{N2}=3.3 K. At $T < T_{N2}$ it shows three-step metamagnetic transitions below H=1.43 T. In this non-diluted compound a long-time variation of magnetic structure has been detected by means of time-resolved neutron scattering measurements. When a sample is rapidly cooled below T_{N2}, the magnetic Bragg peaks corresponding to the intermediate temperature phase ($T_{N2} < T < T_{N1}$) are observed. The amplitude of these Bragg peaks gradually decreases with time. On the other hand, another group of Bragg peaks corresponding to the low temperature phase ($T < T_{N2}$) gradually grow with time. The characteristic time for these variations follows the Arrhenius law with an activation energy E_a/k_B =4 K.

1. Introduction

Long-time variations of magnetic properties have been extensively investigated in the spin-glass study. These characteristic behavior of spin-glasses are regarded as due to the multi valley structure of the free energy which comes from random magnetic interactions. In a system without randomness or imperfections we have not expected to observe a time variation of the magnetic structure in an attainable time scale. We report the first observation of a long-time variation of the magnetic structure in a non-diluted magnet $CeIr_3Si_2$.

The first report of the magnetic property of this material was presented by Umarji and coworkers in 1987 [1]. In this work they reported the crystal structure of $CeIr_3Si_2$ as the $CeCo_3B_2$-type and no magnetic transition was observed down to 4.2 K. Recently the magnetic and transport properties of $CeIr_3Si_2$ were reinvestigated with polycrystalline sample by Muro and coworkers [2]. They found that $CeIr_3Si_2$ crystallizes not in the hexagonal $CeCo_3B_2$-type, but in the orthorhombic $ErRh_3Si_2$-type (Space Group *Imma*, No.74) that is a deformed structure of the $CeCo_3B_2$-type. A magnetic susceptibility and specific heat measurements revealed the existence of magnetic transitions at T_{N1}=4.1 K and T_{N2}=3.3 K. The isothermal magnetization curve at 1.9 K showed three-step metamagnetic transitions below H=1.43 T. The Curie-Weiss behavior of the magnetic susceptibility curve between 100 K and 300 K shows a paramagnetic Curie temperature $\theta_p = -32$ K and an effective magnetic moment μ_{eff}=2.57 μ_B, close to that expected from a Ce^{3+} free ion. The magnetic part of the electrical resistivity shows a maximum at T_{max} ~15 K accompanying a $-\ln T$ behavior between 20 and 50K. This shows that $CeIr_3Si_2$ is classified as a Kondo-lattice compound. Shigetoh and coworkers made magnetization, electrical resistivity and specific heat measurements under various magnetic field utilizing a single crystal sample [3]. These measurements revealed the easy-plane type anisotropic properties. The magnetization curves measured at T=0.3 K with a magnetic field along the b or c axis include a ferromagnetic component of 0.2-0.3 μ_B/Ce and show metamagnetic transitions. Two metamagnetic transitions were observed at H=0.68 T and 1.3 T when the magnetic field was applied along the b-axis. When the magnetic was field applied along the c-axis a transition at H=0.75 T was followed by a linear increase of megnetization up to a kink at H=3 T. These values of critical magnetic field agree with those measured with a polycrystalline sample [2].

We started a neutron scattering study of $CeIr_3Si_2$ to clarify the magnetic structures in various magnetic phases. During this study we found long-time variations of the amplitudes of magnetic Bragg peaks. In this report we present the results of time-resolved neutron scattering measurements in zero external field. We believe this is the first real-time observation of the magnetic structural change in a uniform magnetic system.

2. Experimental

A single crystal of $CeIr_3Si_2$ was grown by the Czochralski pulling method. Details of the procedure has been described in ref. 3. For the neutron scattering measurements a small piece (~2 mm × ~3 mm × ~3 mm) was cut from the ingot because of large neutron absorption of Ir atoms. The lattice parameters at room temperature are a=7.178 Å, b=9.726 Å and c=5.597 Å. The mosaic spread is ~0.3° (full width at half maximum, FWHM).

Neutron scattering experiments were conducted using the 4G and 5G triple-axis spectrometers installed at the JRR-3M reactor of JAEA-Tokai. Measurements were made by a double-axis mode using pyrolytic graphite crystals for a monochromator and a filter. The wave-length of the incident neutron and the horizontal beam collimation are 1.638 Å and $40' - 40' - 40'$, respectively. The sample was mounted with the [100] direction vertical in a closed-cycle ^{3}He gas refrigerator or a liquid ^{4}He cryostat. Time variations of neutron scattering patterns were measured after the sample was rapidly cooled from T=30 K up to the target temperature at zero external field. The time required to cool the sample was ~15 min.

3. Results and discussion

Preliminary measurements showed that in the intermediate temperature (IT) phase ($T_{N2} < T < T_{N1}$), magnetic Bragg peaks were observed at the (0, 1.375, 0.64) and equivalent reciprocal lattice points (IT phase signal). On cooling the sample below T_{N2} to the low temperature (LT) phase, these Bragg peaks decreased with time. Concomitantly, another group of Bragg peaks appeared and grew with time at the (0, 4/3, 2/3) and equivalent points (LT phase signal). These observations have shown that the magnetic structure of the IT phase is maintained in the rapidly cooled LT phase and it gradually transforms to the LT phase structure.

In the present study, we have observed the time variations by repeating two types of scans across the (0, 1.375, 0.64) and (0, 4/3, 2/3) points alternately for ~24 h at various temperatures. The measurements were made with scanning the spectrometer along the $(0, \xi, \xi/2)$ direction to resolve two kinds of peaks.

Figure 1 shows the scattering patterns around (0, 1.375, 0.64) taken at representative elapse times t after cooled to T=1.9 K. The amplitude decreases with increasing t and no appreciable signal was observed for $t > 500$ min. Figure 2 shows the scattering patterns around (0, 4/3, 2/3) taken at representative t at T=1.9 K. The amplitude of the signal increases with increasing t. For both of these measurements, no shift of the peak position with time was detected. The line shape of the scattering patterns fits a Gaussian function of the instrumental resolution.

Similar time variation measurements of the IT and LT phase signals were made at various temperatures. When the sample was cooled to the IT phase, the IT phase signal immediately appeared and no time variation was detected. Figures 3 and 4 show the time variations of

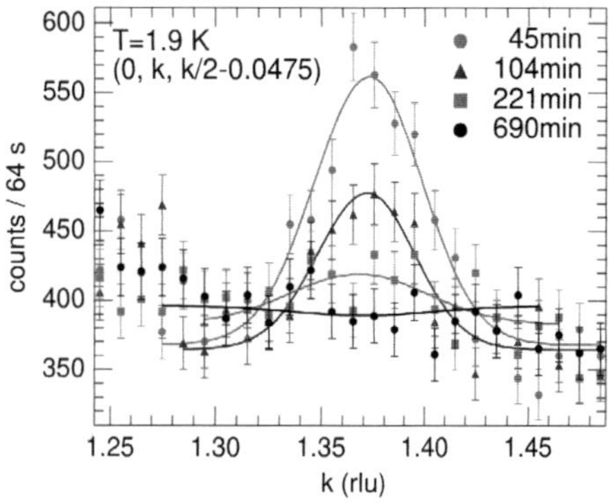

Figure 1. Time variation of neutron scattering patterns observed at T=1.9 K around the (0, 1.375, 0.64) reciprocal lattice point.

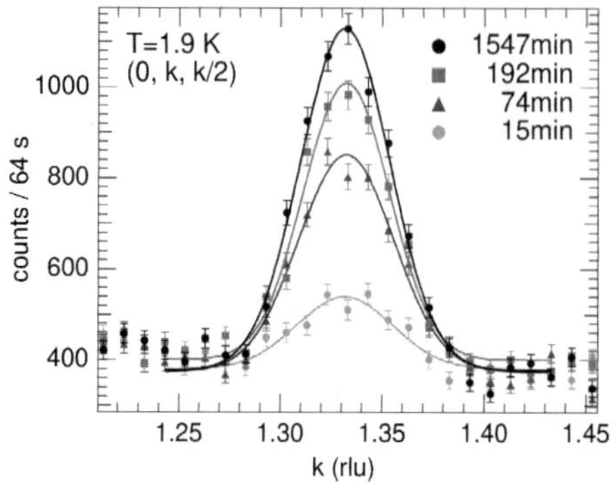

Figure 2. Time variation of neutron scattering patterns observed at T=1.9 K around the (0, 4/3, 2/3) reciprocal lattice point.

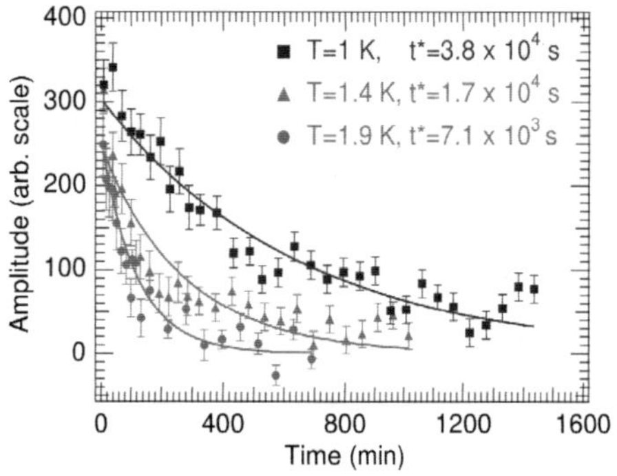

Figure 3. Time variations of the amplitude of the IT phase signal at various temperatures. Curves represent the results of the least squares fitting described in the text.

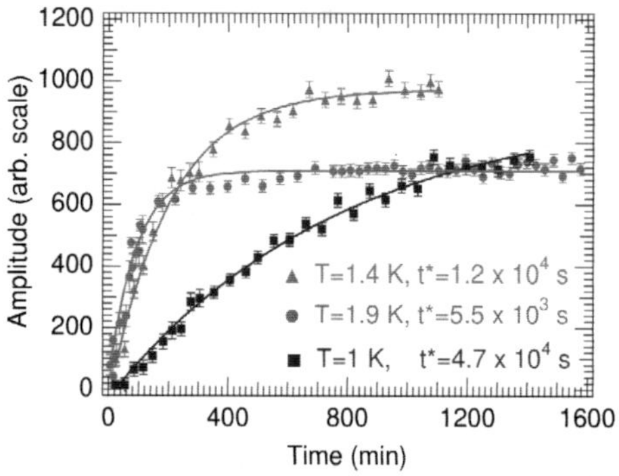

Figure 4. Time variations of the amplitude of the LT phase signal at various temperatures. Curves represent the results of the least squares fitting described in the text.

the IT and LT phase signals measured at various temperatures below T_{N2}. At T=0.7 K the IT-phase signal was observed without detectable change up to 10 h and no LT phase signal was observed. The time variations of the signal amplitude $A(t)$ are well expressed by simple exponential functions

$$A(t) = A_0 \exp(-\frac{t}{t^*}) \quad \text{and} \quad A(t) = A_0 \left[1 - \exp(-\frac{t}{t^*})\right]$$

for the IT and LT phase signals, respectively. In these equations A_0 and t^* are the constant value and the characteristic time for the amplitude variation, respectively. Curves in Figs. 3 and 4 show the results of the least squares fitting. For the LT phase signal, the asymptotic value decreases with increasing temperature. It is probably due to the normal temperature variation of the sub-lattice magnetization. Obtained values of t^* are posted in the figures. t^* values for the IT and LT phase signals are close for the same temperature.

We analyzed the results in terms of the Arrhenius model. Figure 5 shows the Arrhenius plot of t^* ($\log(1/t^*)$ vs $1/T$ plot). $1/t^*$ follows the Arrhenius law as shown by the straight guide line in the figure. The activation energy E_a of the time variation is E_a/k_B =4 K.

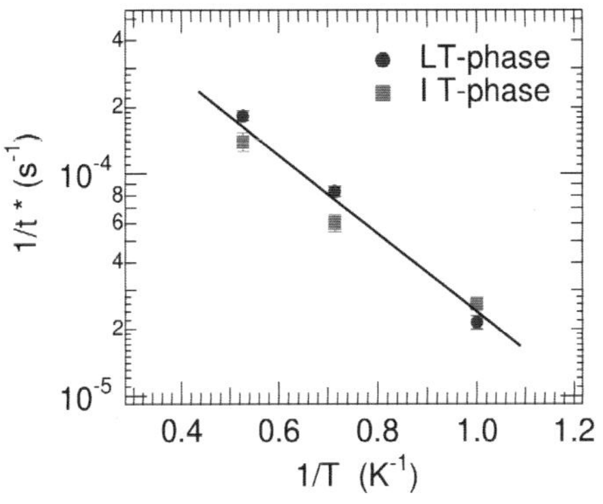

Figure 5. Arrhenius plot of 1/t* derived from the time variation of neutron scattering amplitudes of two kinds of signals.

We have shown that two groups of magnetic Bragg peaks corresponding to different types of magnetic structure coexist in rapidly cooled $CeIr_3Si_2$. The amplitude of the Bragg peaks corresponding to the IT phase structure decreases and that of the LT phase increases with time. These results show that the magnetic structure of the IT phase is maintained below T_{N2} and it gradually transforms to the LT phase structure. Although the magnetic structure in each phase has not been determined, we presume that the transition from the IT phase to the LT phase is basically an incommensurate to commensurate phase transition. We note that in the present measurements, only the amplitude of Bragg peaks varied with time. Neither the position nor the line width of Bragg peaks showed appreciable time variation. These results strongly suggest that we have observed the change of the volume fractions of two distinct magnetic regions with a commensurate and an incommensurate structures. The activation energy E_a deduced from the Arrhenius law corresponds to the energy barrier accompanied with the motion of the interface between two regions. In future experiments with high intensity apparatus, we will be able to observe the change of domain size at the early stage of the time evolution.

Acknowledgments

We thank the members of the Neutron Scattering Laboratory, Institute for Solid State Physics, the University of Tokyo for supporting our experiments. This work is partly supported by a Grant-in-Aid for Scientific Research from the Ministry of Education, Culture, Sports, Science, and Technology.

References

[1] Umarji A M, Dhar S K, Malik S K and Vijayaraghavan R 1987 *Phys. Rev.* B **36** 8929
[2] Muro Y, Ohno Y, Okada T and Motoya K 2007 *J. Mag. Mag. Mater.* **310** 389
[3] Shigetoh K, Ishida A, Ayabe Y, Onimaru T, Umeo K, Muro Y, Motoya K, Sera M and Takabatake T 2007 *Phys. Rev.* B **76** 184429

The Locked-in magnetic structure of $Er_3Cu_4Si_4$ below 1 K

J M Cadogan[1], D H Ryan[2] and L M D Cranswick[3]

[1] Department of Physics and Astronomy, University of Manitoba, Winnipeg, Manitoba, R3T 2N2, Canada
[2] Department of Physics, McGill University, Montreal, Québec, H3A 2T8, Canada
[3] Canadian Neutron Beam Centre, NRCC, Chalk River Laboratories, Chalk River, Ontario, K0J 1J0, Canada

E-mail: cadogan@physics.umanitoba.ca

Abstract. We have determined the magnetic structure of $Er_3Cu_4Si_4$ by high-resolution neutron powder diffraction down to 0.34 K. The magnetic ordering temperatures of the Er(2d) and Er(4e) sublattices are 14 K and 3.3 K, respectively. The Er(2d) order is commensurate down to 0.34 K, with a propagation vector $[0\ \frac{1}{2}\ 0]$ and a moment of 8.7(2) μ_B at 0.34 K. The Er(4e) order is initially short-range and incommensurate, with a propagation vector [0 0.876(5) 0] and a refined moment of only 4.4(2) μ_B at 1.5 K. Below about 1.4 K, the Er(4e) order starts to lock-in to a commensurate cell described by the propagation vector $[\frac{1}{3}\ \frac{1}{2}\ \frac{1}{2}]$ with the moments aligned in the crystal bc-plane, canted about 4° away from the b-axis. The lock-in is complete by 0.8 K and the Er(4e) moment is 8.8(2) μ_B at 0.34 K.

1. Introduction

The orthorhombic $R_3Cu_4X_4$ (X = Si, Ge, Sn) compounds show a complex variety of magnetic ordering behaviour. The R atoms occupy two crystallographically distinct sites in these compounds (2d and 4e), and the moments observed at the two R sites are generally quite different from each other. Often, the refined R moment is greatly reduced from the corresponding R^{3+} free-ion value. Furthermore, the ordering temperatures of the two R sites are usually different, as are the ordering directions.

Wawrzyńska *et al.* [1] carried out neutron powder diffraction on $Er_3Cu_4Si_4$ down to 1.5 K and they reported that the magnetic ordering temperatures for the Er(2d) and Er(4e) sites are 14 K and 3.3 K, respectively. At 10 K, the refined Er(2d) moment was 8.6(1) μ_B, ordered along the crystal c-axis, with a propagation vector $[0 \sim\frac{1}{2}\ 0]$. At 1.5 K, the Er(2d) order is virtually unchanged, with a refined Er(2d) moment of 9.5(1) μ_B, which is actually larger than the free-ion Er^{3+} moment of 9 μ_B. Curiously, the refined magnetic moment at the Er(4e) site at 1.5 K was only 4.4(2) μ_B, about one half of the free-ion value, lying in the crystal ac-plane, with a propagation vector [0 0.903(2) 0].

At the same time as the Wawrzyńska *et al.* [1] study, we published a study of the $R_3Cu_4X_4$ (X = Si, Ge, Sn) compounds [2] where we presented the results of our ^{166}Er and ^{119}Sn Mössbauer spectroscopy work, along with neutron powder diffraction on the silicide. ^{166}Er Mössbauer spectroscopy showed that at 2 K, the Er(2d) moment in $Er_3Cu_4Si_4$ is 8.5(1) μ_B and the Er(4e) moment is 7.1(1) μ_B. However, our neutron diffraction confirmed the findings of Wawrzyńska *et al.* [1] that the refined Er(4e) moment at 1.5 K is less than half the free-ion value. The apparent conflict in the Er(4e) measured moments reflects the *local* versus *extended* natures of the two measurement techniques. We showed that at 2.4 K, the Er(4e) order is incomplete and quite short-range, and we estimated the correlation length to be about 10 Å from the width of the broad feature in the neutron diffraction patterns.

In this paper we report neutron powder diffraction measurements on $Er_3Cu_4Si_4$ at temperatures down to 0.34 K in order to follow the ordering behaviour and the temperature dependence of the ordered moments on the two Er sites in detail. In particular, our aim was to determine if the Er(4e) moment eventually forms an extended magnetic structure with a long enough correlation for neutron diffraction to yield a free-ion moment for the Er(4e) sublattice, consistent with our ^{166}Er Mössbauer work [2].

2. Experimental Methods

The $Er_3Cu_4Si_4$ sample was prepared in a tri-arc furnace with a base pressure of less than 6×10^{-7} mbar. Stoichiometric amounts of the pure elements (Er (99.9%), Cu (99.99%), Si (99.999%)) were melted several times under pure (less than 1 ppm impurity) argon to ensure homogeneity. The resulting ingot was annealed under vacuum at 800°C for two weeks and water-quenched. Powder x-ray diffraction measurements were made at room temperature using Cu Kα radiation. Analysis confirms that the sample was primarily composed of the orthorhombic $Er_3Cu_4Si_4$ phase, ($Gd_3Cu_4Ge_4$-type structure [3], with the *Immm* space group #71). The sample also contained about 3 wt.% ErCuSi (hexagonal $P6_3/mmc$) impurity. The $Er_3Cu_4Si_4$ structure has two Er sites (2d and 4e), one Cu site (8n) and two Si sites (4f and 4h). The refined lattice parameters of $Er_3Cu_4Si_4$ at RT are a = 13.5597(4) Å, b = 6.4977(2) Å and c = 4.0890(1) Å.

Neutron powder diffraction experiments were carried out on the DUALSPEC C2 high-resolution diffractometer at the NRU reactor, Chalk River Laboratories. The neutron wavelength was 2.37164(14) Å. Diffraction patterns were obtained over the temperature range 0.34–295 K and all patterns were analysed using the Rietveld method and the FULLPROF/WinPLOTR program [4] [5]. For the neutron diffraction experiments, the sample was mixed with an approximately equal volume of pure (99.99 %) copper powder and hydraulically pressed into an OFHC copper sample holder in order to ensure proper thermalisation of the powder sample at the very low temperatures used here [6]. From 295 K to 4 K, the neutron diffraction data were collected with the sample loaded into a vanadium sample can in a Janis closed cycle fridge using Helium exchange gas. The lower temperature data from 20 K to 0.34 K were collected using an Oxford Heliox insert working with a standard Helium cryostat and a copper can [6].

3. Results

In figure 1 we show a comparison of the neutron powder diffraction patterns obtained on $Er_3Cu_4Si_4$ at 0.34 K, 1.42 K, 2.37 K, 6.67 K and 19.5 K. In figure 1 we also show an expanded-range comparison. At 19.5 K, the scattering is purely nuclear and the only peaks seen in the range $7° \leq 2\theta \leq 44°$ are the (200) at $2\theta = 20.1°$ and the (011) at $2\theta = 40.1°$.

The ordering of the Er(2d) sublattice is clear from the prominent ($0 \pm\frac{1}{2}$ 0) and ($1 \pm\frac{1}{2}$ 0) peaks, at $2\theta = 10.5°$ and 14.5°, respectively, and marked 'd' in the figure. There is also a clear magnetic peak from the Er(2d) sublattice at $2\theta = 22.7°$ the ($2 \pm\frac{1}{2}$ 0), also marked 'd' in the figure. The Er(4e) order sets in around 3 K but is superimposed on a broad background, reflecting the short-range ordering of the Er(4e) sublattice [2]. As the temperature is lowered, the Er(4e) long-range order is established and the broad background gives way to a set of peaks, two of which are marked 'e' in figure 1.

In figure 2 we show the temperature dependences of the integrated intensities of four of the Er(4e) magnetic peaks in the diffraction patterns of $Er_3Cu_4Si_4$. At about 1.4 K the magnetic

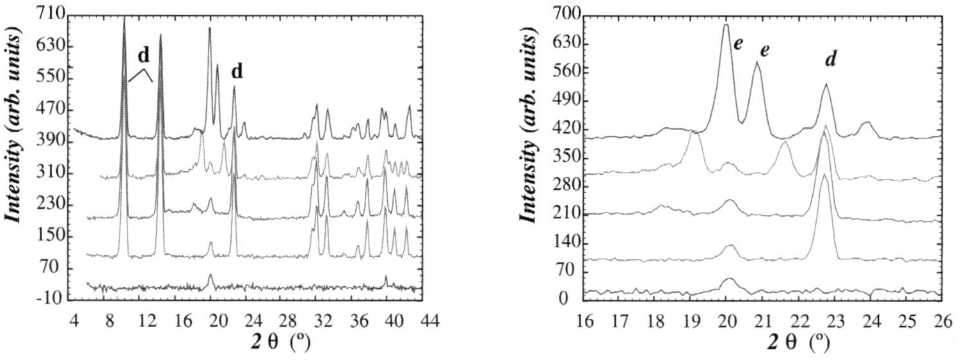

Figure 1. Comparison of the neutron powder diffraction patterns of $Er_3Cu_4Si_4$. From top to bottom, the temperatures are 0.34 K, 1.42 K, 2.37 K, 6.67 K and 19.5 K.

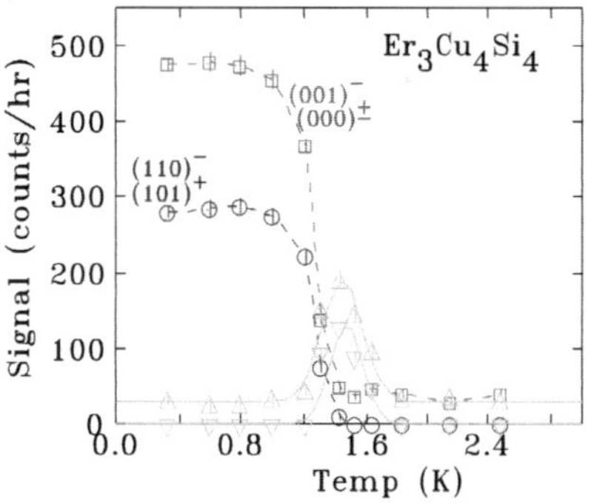

Figure 2. Temperature dependences of selected Er(4e) magnetic peaks in $Er_3Cu_4Si_4$.

order of the Er(4e) sublattice begins to change from the incommensurate, short-range phase towards a commensurate, long-range phase. The change is complete by 0.8 K.

Our refinement of the 0.34 K pattern is shown in figure 3. The standard refinement factors (%) are R(Bragg)=6.3, R(F)=3.9, R(mag-2d)=12.1 and R(mag-4e)=13.8. We find that that the Er(2d) order remains unchanged from that found at higher temperatures. The Er(2d) propagation vector ($\mathbf{q}_d$) is $[0\ \frac{1}{2}\ 0]$ with a refined Er(2d) moment of 8.7(1) μ_B. The Er(4e) order at 0.34 K has locked-in to a commensurate structure which is described by the propagation vector ($\mathbf{q}_e$) $[\frac{1}{3}\ \frac{1}{2}\ \frac{1}{2}]$ with the Er(4e) moments aligned in the crystal bc-plane, canted about 4° away from the b-axis. The Er(4e) moment at 0.34 K is 8.8(2) μ_B, i.e. the free-ion value.

The only question remaining now is whether the modulation of the Er(4e) order is square or sinusoidal. Our ^{166}Er Mössbauer work [2] strongly suggested that the Er(4e) order is a square-wave since we found no evidence of a distribution of ^{166}Er hyperfine field (and hence Er^{3+} moment). In figure 4 we show a small region of the fit to the 0.34 K diffraction pattern with the third-harmonic of the Er(4e) propagation vector included. For comparison, we show the same region without the third-harmonic. The effect of the Er(4e) third-harmonic is evident as a pronounced shoulder at $2\theta = 22.2°$ on the low-angle side of the (200)$\pm\ q_d$ magnetic peak at $2\theta = 22.7°$. Thus, the Er(4e) order is square-wave which is fully consistent with our Mössbauer work.

In conclusion, the magnetic structures of the Er(2d) and Er(4e) sublattices in $Er_3Cu_4Si_4$ are

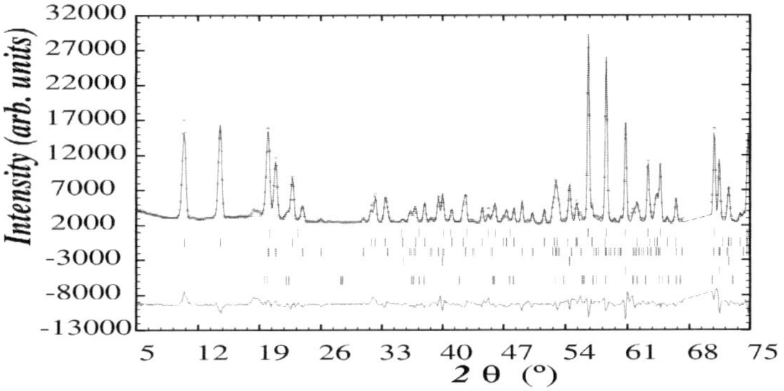

Figure 3. Neutron powder diffraction pattern at 0.34 K of $Er_3Cu_4Si_4$. A strong peak from the copper used in the sample mounting has been removed ($2\theta \sim 69°$).

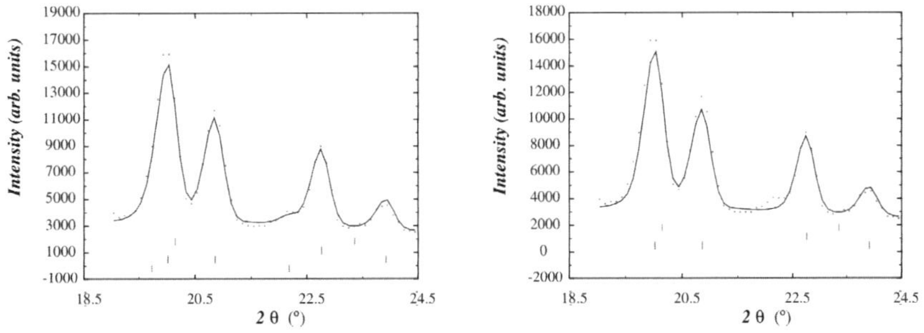

Figure 4. Section of the refined neutron powder diffraction pattern at 0.34 K of $Er_3Cu_4Si_4$ with and without the third-harmonic of the Er(4e) propagation vector included (Left and Right, respectively).

both commensurate (doubled and tripled, respectively) and long-range at 0.34 K, with both Er^{3+} magnetic moments attaining the free-ion value of 9 μ_B.

Acknowledgments
We are grateful for financial support from the Natural Sciences and Engineering Research Council of Canada and Fonds Québécois de la Recherche sur la Nature et les Technologies. We also thank the staff at Chalk River for their assistance during the neutron diffraction experiments. The $Er_3Cu_4Si_4$ sample was prepared by Robert Gagnon (McGill University). Finally, JMC acknowledges support from the Canada Research Chairs programme.

References

[1] E. Wawrzyńska, J. Hernández-Velasco, B. Penc and A. Szytula J. Magn. Magn. Mater. **280** 234 (2004).
[2] D.H. Ryan, J.M. Cadogan, R. Gagnon and I.P. Swainson, J. Phys. Condensed Matter **16** 3183 (2004).
[3] W. Rieger, Monat. f. Chemie **101**, 449–462 (1970); G. Hanel and H. Nowotny, Monat. f. Chemie **101**, 463–468 (1970).
[4] J. Rodríguez-Carvajal, Physica B **192**, 55 (1995)
[5] T. Roisnel and J. Rodríguez-Carvajal, Mater. Res. Forum **378–381**, 118 (2001).
[6] D.H. Ryan and I.P. Swainson, J. Appl. Cryst. **42** 43-47 (2009).

International Conference on Neutron Scattering 2009 IOP Publishing
Journal of Physics: Conference Series **251** (2010) 012021 doi:10.1088/1742-6596/251/1/012021

Potassium tantalate substrates for neutron experiments on antiferromagnetic perovskite films

H M Christen,[1,*] G J MacDougall,[2] H-S Kim,[1] D H Kim,[1,3] L A Boatner,[1] C J Callender Bennett,[1] J L Zarestky,[2] and S E Nagler[2]
[1] Oak Ridge National Laboratory, Materials Science and Technology Division, PO Box 2008, Oak Ridge, TN 37831, USA
[2] Oak Ridge National Laboratory, Neutron Scattering Science Division, PO Box 2008, Oak Ridge, TN 37831, USA
[3] Tulane University, Department of Physics and Engineering Physics, New Orleans, LA 70118, USA

E-mail: christenhm@ornl.gov

Abstract. For the study of antiferromagnetism in thin-film materials, neutron diffraction is a particularly important tool, especially since magnetometry experiments are often complicated by the substrate's strong diamagnetic or paramagnetic contribution. However, the substrate, by necessity, has a lattice parameter that is very similar to that of the film, and in most cases is over 1000 times more massive than the film. Therefore, even weak structural distortions in the substrate crystal may complicate the analysis of magnetic scattering from the film. Here we show that in contrast to most other perovskite substrates (including $SrTiO_3$, $LaAlO_3$, etc.), $KTaO_3$ provides a uniquely appropriate substrate platform for magnetic diffraction experiments on epitaxial oxide films.

1. Introduction

Epitaxial perovskite films have been studied extensively due to their interesting ferroelectric (FE), magnetic, and multiferroic properties [1]. Due to the possibility of forming precisely controlled interfaces and heterostructures [2,3], the coupling between dissimilar materials at well-defined length scales can be studied, with the goal of developing new materials whose properties are determined by effects occurring at the interfaces.

One of the mechanisms leading to a coupling of adjacent layers in a multi-layer structure is the exchange interaction between an antiferromagnetic (AFM) and a ferromagnetic (FM) material. Understanding and enhancing this coupling may provide a means to create composite materials with a strong magneto-electric effect at room temperature: combining a room-temperature multiferroic (AFM+FE) with a FM layer could yield macroscopic properties that, to date, have not been found in a single-phase material [4].

In order to understand and maximize the coupling between AFM and FM layers, the spin order in the AFM material needs to be known. While significant progress has been made in using x-ray dichroism techniques to probe magnetism in thin films [5], neutron diffraction remains the method-of-

[*] To whom any correspondence should be addressed

choice for a quantitative analysis of the entire volume of a film, and several studies have shown the power of neutron methods for understanding AFM in epitaxial perovskite layers [6].

Epitaxial growth of high-quality films typically requires a substrate material with a similar lattice parameter to that of the film, and thus, the film's nuclear scattering positions are necessarily close to those of the substrate. However, when indexed in the perovskite's pseudocubic notation, AFM magnetic diffraction occurs at half-integer positions [i.e. (½ ½ ½) for G-type and (0 0 ½) for the layered A-type], at which no nuclear diffraction from the substrate should occur if the substrate is cubic. However, as we show here, this is not the case for most of the commercially available perovskite substrates, and structural distortions yield nuclear diffraction from the substrate at positions that overlap with the AFM diffraction from the film. Considering that the typical substrate (0.5mm to 1mm thick) is 1000 times more massive than the actual material of interest (i.e. the film, with a thickness that is typically below 1 μm), this significantly complicates the data analysis. However, we find that $KTaO_3$ substrates remain cubic at all temperatures and, thus, are an ideal platform for such experiments.

2. Neutron diffraction of epitaxial $BiCrO_3$ films on $LaAlO_3$ and $SrTiO_3$

$BiCrO_3$ films are used here as a test system, since this AFM material [7] is readily grown by pulsed-laser deposition on various perovskite substrates [8], having a distorted perovskite structure that can be approximated as pseudo-cubic with a = b = c ≈ 3.90 Å [7]. Layers with thicknesses of 1.5 – 2 μm were deposited as described in our previous work [8], and the substrates were obtained commercially. Buffer layers [$SrRuO_3$ on $SrTiO_3$, (La,Ca)MnO_3 on $LaAlO_3$] with thicknesses of 4 – 8 nm were used since they led to better crystallinity of the $BiCrO_3$. These buffer layers are too thin to be observed in these neutron experiments and thus do not contribute to the observed scattering. Elastic neutron scattering measurements were performed at Oak Ridge National Laboratory's High Flux Isotope Reactor (HFIR) on beamline HB1A, using PG(002) monochromator and analyzer, collimations 48-48-40-136, and a fixed incident neutron energy of 14.64 meV.

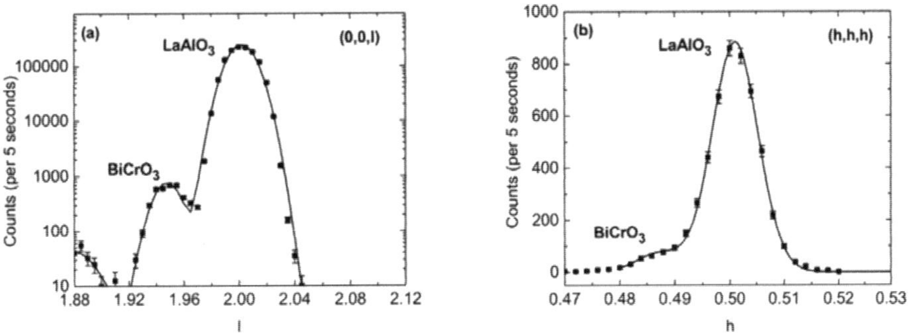

Figure 1. $BiCrO_3$ film on $LaAlO_3$ substrate. **a)** Radial scan through the film and substrate's (0 0 2) diffraction peak at 200K. **b)** Radial scan at the (½ ½ ½) position at 20K, showing the $LaAlO_3$ substrate's large contribution due to its non-cubic nature.

Figure 1a shows a radial scan through the film and the substrate (0 0 2) peaks. Note that pseudocubic notation is used throughout this manuscript with indices referenced according to the substrate lattice parameters). The weak film peak (h ≈ 1.95, c ≈ 3.90 Å) is clearly separated from the strong substrate reflection at h = 2. Figure 1b depicts a scan at the (½ ½ ½) position at 20 K (i.e. well below the Néel temperature of $BiCrO_3$, $T_N \approx 110$K). A strong substrate contribution, largely masking the diffraction from the thin film, is clearly observed as a consequence of this material's non-cubic structure ($R\bar{3}m$ [9], which can be approximated as pseudocubic below room temperature, where a = b = c = 3.788 Å and α ≈ 89.93°).

The lattice mismatch between the film and the substrate helps to distinguish between the different contributions, however, the weakness of the magnetic scattering combined with the large mass of the substrate make this distinction difficult. This therefore complicates the extraction of quantitative data. In the case of the better-matched (more appropriately chosen) $SrTiO_3$ substrate (Fig. 2), the lattice parameter difference between the film and the substrate (a = 3.905 Å at room temperature) is too small for a distinction to be made between the different components (Fig. 2a). More importantly, $SrTiO_3$ undergoes a structural phase transition at $T \approx 105$ K [10], which nearly coincides with the T_N of $BiCrO_3$. Below 105 K, $SrTiO_3$ is tetragonal and shows nuclear scattering at peak positions that overlap those of the magnetic scattering from $BiCrO_3$. As has been pointed out, some of this scattering persists at temperatures well above 105 K [11].

A number of other substrates are commercially available for the epitaxial growth of perovskite materials, including the orthorhombic perovskites $NdGaO_3$, $DyScO_3$, $GdScO_3$, the K_2NiF_4-structure materials $SrLaGaO_4$ and $SrLaAlO_4$, and the tetragonal alloy LSAT $[(LaAlO_3)_{0.3}(Sr_2AlTaO_6)_{0.7}]$. Similar experiment as those described above were also performed on LSAT substrates (data not shown), for which strong scattering was observed at those half-integer positions that lie close to the film's peaks for G-type AFM. Clearly, for the present studies, a material that is cubic across the entire temperature range is required.

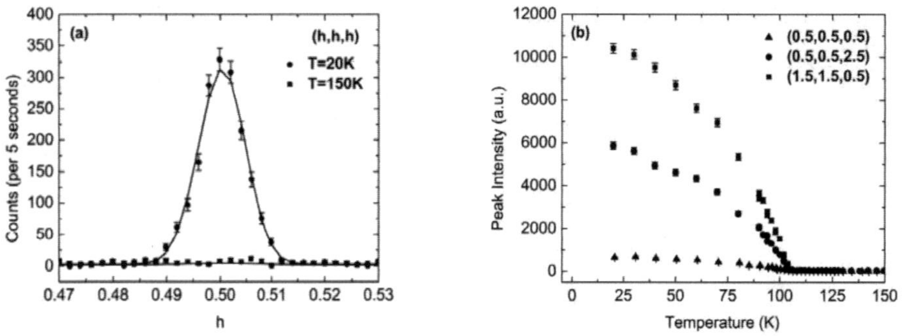

Figure 2. $BiCrO_3$ film on $SrTiO_3$ substrate. **a)** Radial scan at the (½ ½ ½) position at 150 K (above the $SrTiO_3$ structural phase transition and the $BiCrO_3$ Néel temperature) and at 20 K, where the film's (magnetic) and the substrate's (nuclear) contribution cannot be separated due to the small difference in lattice parameters. **b)** Temperature-dependence of the intensities of the (½ ½ ½), (½ ½ 5/2), (3/2 3/2 ½) and peaks, dominated by the $SrTiO_3$ structural phase transition at 105 K.

3. $BiCrO_3$ films on $KTaO_3$

$KTaO_3$ retains the prototype cubic perovskite symmetry down to the lowest temperatures investigated [10] and, therefore, promises to be an ideal substrate material for this work. It's room-temperature lattice parameter is 3.989 Å. $KTaO_3$ crystals were grown by spontaneous nucleation [12] and polished with a KOH-buffered solution of colloidal silica. $BiCrO_3$ films were grown as described above using a $SrRuO_3$ buffer layer.

Figure 3a shows a radial scan through the film and substrate (1 1 1) peak, allowing us to identify the positions where the corresponding half-integer peaks (resulting from structural distortions or magnetic scattering) are expected. As is seen in Fig. 3b, no intensity is detected at the substrate's (½ ½ ½) position. This result clearly allows us to determine *quantitatively* the scattering from the film. It is important to note that such quantitative experiments are, therefore, not only possible for films with lattice parameters that are well separated from those of the $KTaO_3$ substrate, but also for well-matched layers, due to the complete absence of substrate peaks at half-integer positions (experiments were also performed for (½ ½ 3/2), (½ ½ 5/2), (3/2 3/2 ½), and (5/2 3/2 ½), data not shown).

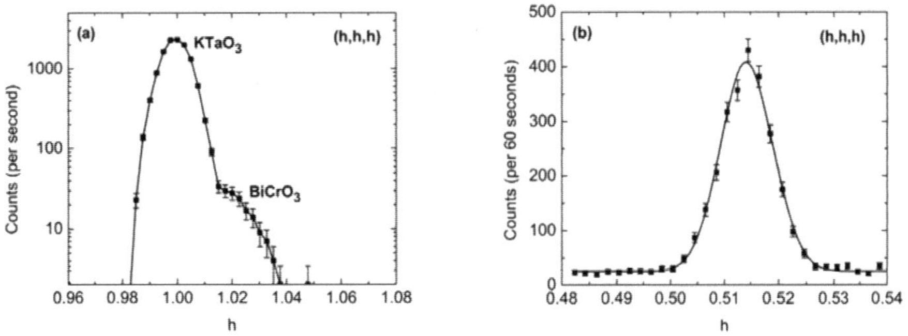

Figure 3. $BiCrO_3$ film on $KTaO_3$ substrate at 20 K. **a)** Radial scan at the (1 1 1) position showing the substrate at h = 1 and the film at h ≈ 1.03. **b)** Radial scan at the (½ ½ ½) position. Corresponding to the data in (a), the film's (½ ½ ½) peak is observed at h ≈ 0.515, while no intensity is seen at h = 0.5 (i.e., no substrate contribution).

4. Conclusions

Due to its perfect cubic symmetry at all investigated temperatures, $KTaO_3$ is an ideal substrate for the study of antiferromagnetism in epitaxial perovskite films. Our work demonstrates that quantitative analysis of the magnetic diffraction is possible for epitaxial layers, which will lead to a better understanding of the properties of thin-film materials that are being studied both for their fundamental properties as well as for a broad range of potential applications in oxide electronics and for sensors.

Acknowledgements

This work was supported by the Division of Materials Science and Engineering (HMC, HSK, DHK, LAB, CJCB) and the Division of Scientific User Facilities (GJM, JLZ, SEN), U.S. Department of Energy under contract with UT-Battelle, LLC.

References

[1] Rao C N R and Raveau B 1998 *Transition Metal Oxides* 2nd edn (New York: Wiley–VCH)
[2] Schlom D G, Haeni J H, Lettieri J, Theis C D, Tian W, Jiang J C, and Pan X Q 2001 *Mater. Sci. Eng. B* **87**, 282
[3] Lee H N, Christen H M, Chisholm M F, Rouleau C M, and Lowndes D H 2005 *Nature* **433**, 395
[4] Spaldin N A and Fiebig M 2005 *Science* **309** 391
[5] Arenholz E, van der Laan G, Yang F, Kemik N, Biegalski M B, Christen H M, and Takamura Y 2009 *Appl. Physl Lett.* **94** 072503
[6] Béa H, Bibes M, Petit S, Kreisel J, and Barthélémy A 2007 *Phil. Mag. Lett.* **87** 165
[7] Belik A A, Iikubo S, Kodama K, Igawa N, Shamoto S, and Takayama-Muromachi E 2008 *Chem. Mater.* **20** 3765
[8] Kim D H, Lee H N, Varela M, and Christen H M 2006 *Appl. Phys. Lett.* **89** 162904
[9] Howard C J, Kennedy B J, and Chakoumakos B C 2000 *J. Phys.: Condens. Matter* **12** 349
[10] Lines M E and Glass A M 1977 *Principles and Applications of Ferroelectrics and Related Materials* (Oxford: Clarendon Press)
[11] Cowley R A, Buyers W J L, and Dolling G 1969 *Sol. State Comm.* **7** 181
[12] Paulauskas I E, Katz J E, Jellison, Jr. G E, Lewis N S, Boatner L A, and Brown G M 2009 *J. Electrochem. Soc.* **156** B580

Journal of Physics: Conference Series **251** (2010) 012022 doi:10.1088/1742-6596/251/1/012022

Field-induced magnetic ordering in $ErNi_{1-x}Cu_xAl$

J. Prchal[a,1], M. Míšek[a], O. Isnard[b]

[a]Department of Condensed Matter Physics, Faculty of Mathematics and Physics, Charles University, Ke Karlovu 5, 12116 Prague 2, Czech Republic

[b]Institut Néel, CNRS and University J. Fourier Grenoble, 38042 Grenoble Cedex 9, France

[1]E-mail: prchal@karlov.mff.cuni.cz

Abstract. The RNi$_{1-x}$Cu$_x$Al pseudoternaries belong to group of *RTX* intermetallic compounds crystallizing in the hexagonal ZrNiAl type of structure. In the critical region of the substitution parameter $x \sim 0.8$ there was found a loss of a long-range magnetic order at the lowest studied temperatures for series with R = Tb, Dy, Er. As an explanation for this effect a competition of different ordering mechanisms for either *R*NiAl or *R*CuAl compounds was proposed. Nevertheless, the magnetization curves of the $ErNi_{0.2}Cu_{0.8}Al$ compound (within the critical concentration region) showed behaviour close to the parent ErCuAl, which was confirmed as a ferromagnet. Neutron diffraction on $ErNi_{0.2}Cu_{0.8}Al$ compound with applied magnetic field signifies appearing of the ferromagnetic order at the lowest applied magnetic field of 0.1 T. The magnitude of the ordered magnetic moment increases with increasing magnetic field in agreement with the magnetization curve, confirming tendency to ferromagnetic ordering or short-range magnetic order of ferromagnetic character in this critical region.

1. Introduction

The *R*NiAl and *R*CuAl compounds belong to a large group of 1:1:1 *RTX* compounds (where *R* stands for a rare-earth element, *T* for a transition metal and *X* for a *p*-metal element) crystallizing in the hexagonal ZrNiAl-type of structure and exhibiting a large variety of magnetic structures in the ordered state. Very surprising results have been observed for the $TbNi_{1-x}Cu_xAl$ series [1]. A surprising behaviour of the ordered magnetic moments on the position of rare-earth element *R* was found around specific concentration of the substituted Cu instead of Ni. Within past years we have performed structural and magnetic investigation for the $ErNi_{1-x}Cu_xAl$ and $DyNi_{1-x}Cu_xAl$ compounds including neutron powder diffraction. In this work we discuss our recent neutron-diffraction results on the $ErNi_{1-x}Cu_xAl$ series under external magnetic field in the context with above-mentioned work and previously studied Er- and Dy-based series [2,3].

In all studied systems the loss of long-range magnetic order (LRO) is observed at the same Cu-concentration range between $x = 0.6$ to 0.8. That means that the rare-earth element species does not play considerable role in this effect. It is thus only the number of 3*d*-electrons in (Ni,Cu) sublattice that is the decisive parameter. It may be conceived with an idea of competition between different mechanisms responsible for magnetic order in *R*NiAl or *R*CuAl, respectively: RKKY interaction (Rudeman-Kittel-Kasuya-Yosida interaction referring to a coupling mechanism between localized moments through polarization of the conduction electrons) and exchange interaction mediated by transition-metal free *d*-electron states [1].

We have performed a neutron diffraction experiment on $ErNi_{0.2}Cu_{0.8}Al$ compound with applied magnetic field in order to investigate magnetic structures induced by the magnetic field, and to clarify the similarity of the magnetization curves of the compound with lost LRO against the parent magnetically ordered one.

2. Experimental

The polycrystalline sample was prepared by arc-melting in a monoarc furnace under protection of argon atmosphere. The sample was then checked by X-ray diffraction for its quality. The sample was additionally checked for its phase composition by using a microprobe analysis using the JXA-733 electron microprobe microanalyzer. Pure Er, Ni, Cu and Al metals were used as standards. The weight concentrations were obtained from the measured relative intensities (i.e. ratios to the standards) using the ZAF method [4] and the SAMx program support [5]. Besides the main phase with 1:0.2:0.8:1 (Er:Ni:Cu:Al) stoichiometry (which corresponds to the nominal composition within accuracy of the used method) there were found also traces of pure Er and a small amount of foreign phase corresponding to composition $Er_{0.2}([Ni,Cu])_{0.5}Al_{0.3}$. No reflection corresponding to the pure erbium was observable in the X-ray pattern, but two small excess peaks are visible on the X-ray pattern that most probably correspond to the last mentioned foreign phase. We shall mention that also boundary compound ErCuAl contained similar amount of the same impurities that have no considerable influence on the observed magnetic properties. As it was seen later by neutron diffraction (correspondingly also at present investigation), there was no foreign peak visible on the diffraction pattern that does not correspond to the ZrNiAl-type hexagonal structure, as well as not to other propagation than the ferromagnetic (000) one, so we believe that the impurities do not influence magnetic behavior of our compound.

The neutron-diffraction experiment under magnetic field was performed at the Institute-Laue-Langevin Grenoble (ILL), using the D1B diffractometer working with a banana-type multidetector covering 80°. There was no collimation used only horizontal and vertical slits have been used before the cryostat in order to adjust the beam to the sample size. The a monochromatic beam of 2.52 Å has been selected from a thermal neutron beam by a highly oriented pyrolitic graphite monochromator (002 reflection), then to remove the contamination by higher harmonics a graphite filter has been used. The external magnetic field was used in the range up to 1.2 T.

The obtained diffraction data were analyzed by a whole-profile fitting - Rietveld refinement method, using the FullProf program [6] with corrections on instrumental abbreviations. The Fermi lengths and absorption coefficients were taken from Ref. [7].

3. Results and discussion

We have studied a sample with composition around the critical concentration of Cu (x = 0.8) under applied magnetic field to find out changes induced by the applied magnetic field. The diffraction patterns taken at zero magnetic field in the paramagnetic state and at the lowest temperature (T = 15 and 1.5 K) are in good agreement with the previous results [2]. Under applied magnetic field new magnetic intensities on the top of the nuclear reflections appear already with very small magnetic field of 0.1 T, as it can be seen on Fig. 1. The induced magnetic moment at field of 0.1 T is 4.3 μ_B (magnetic agreement factor $R_{Bragg}^{mag} = 9.8$).

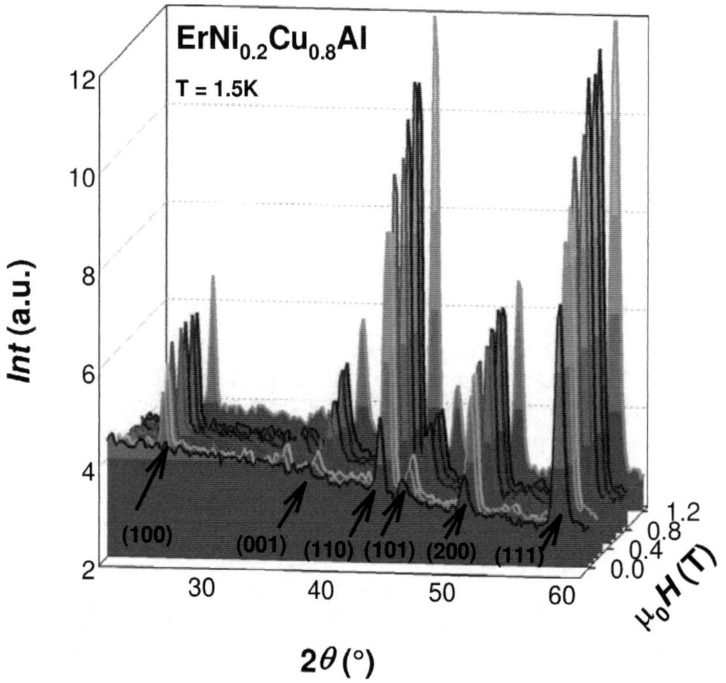

Figure 1. Field evolution of the neutron diffraction patterns at lowest temperature. One can see a rapid increase of diffracted intensity of most reflections already at the first nonzero field.

After application of the magnetic field of 0.35 T, the magnetic moment is 4.8 μ_B and under the magnetic field of 1.2 T the compound exhibits already the value of the ordered (ferro) magnetic moment of 6.8(4) μ_B ($R_{Bragg}^{mag} = 7.2$) what corresponds to the observed magnetic moment in the parent ErCuAl [8]. The diffraction pattern (displayed on Fig. 1) clearly shows onset of magnetic contribution on position of the nuclear reflections, especially the (100) and (001) reflections increase very rapidly from almost zero value intensity to quite intense peaks.

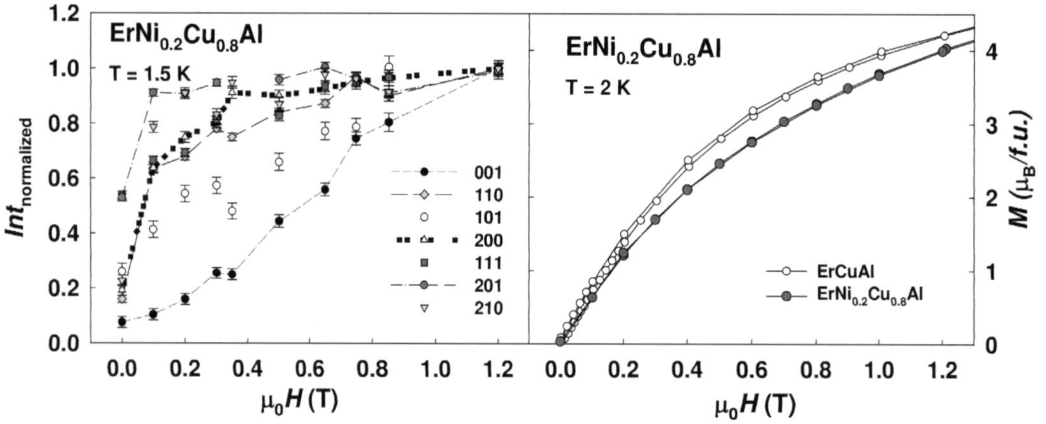

Figure 2. The field dependence of the intensities of selected reflections of the $ErNi_{0.2}Cu_{0.8}Al$ sample. Data are displayed for comparison together with magnetization curve in the same field range at similar temperature; also magnetization data of ErCuAl (taken from [8]) are displayed for comparison.

The evolution of intensities for several reflections with applied magnetic field at the lowest temperature of 1.5 K is displayed on the Fig. 2. A rapid onset of magnetic contribution to intensities of the reflections is visible on practically all of the reflections, representing onset of long-range magnetic ordering even at very small applied external field. The evolution of the integrated intensities (left part of Fig.2) is corresponding to the field evolution of magnetization (right part of Fig. 2), where for comparison also magnetization data of ErCuAl are added. Note that ErCuAl was confirmed by neutron diffraction as a simple ferromagnet [8].

Previous assumption of magnetic order on the short range scale is most probably confirmed as well as its ferromagnetic character although its character at zero magnetic fields cannot be by principle estimated by this method of applying even very small magnetic field. Short range magnetic order caused probably by competition of two different ordering mechanisms in the concentration region around $x = 0.8$ (as proposed by Ehlers [1]) can be very easily removed by applying quite small magnetic field (0.1 T) what leads to ferromagnetic order of the magnetic moments ordered up to this point locally. This finding explains the inconsistency between the bulk magnetic data (magnetization curves indicating ferromagnetic behaviour of the compound [9]) and results from the neutron diffraction experiment performed on these compounds previously at zero external field which did not discover any magnetic contribution down to the lowest used temperature, except for enhanced intensities at low angles [10]. Confirmation of the ferromagnetic-type ordering on the local scale without applied magnetic field is going to be confirmed by the μSR experiment in near future.

Acknowledgments

This work is a part of a research plan MSM0021620834 financed by the Ministry of Education of the Czech Republic and was also supported by the Grant Agency of the Czech Science Foundation under project No.202/07/P153. The authors also thank to ILL for providing the beamtime and the ILL staff for support during the experiment.

References

[1] Ehlers G, Ahlert D, Ritter C, Miekeley W and Maletta H 1997 *Europhys. Let.* **37(4)** 269
[2] Prchal J, Javorský P, Sechovský V, Dopita M, Isnard O and Jurek K 2004 *J. Magn. Magn. Mater.* **283** 34
[3] Prchal J, Javorský P, Detlefs B, Daniš S and Isnard O 2007 *J. Magn. Magn. Mater.* **310** e589
[4] Hulinský V and Jurek K *Zkoumání látek elektronovým svazkem* 1983 SNTL Prague
[5] KEVEX manual, SAMx manual
[6] Rodriguez-Carvajal J 1993 *Physica B* **192** 55
[7] Sears VF 1992 *Neutron News* **3** 26
[8] Javorský P, Burlet P, Ressouche E, Sechovský V, Michor H and Lapertot G 1996 *Physica B* **225** 230
[9] Prchal J, Javorský P, Vejpravová J, Dopita M, Rafaja D, Sechovský V, Jurek K, Šantavá E and Maryško M 2003 *Acta Physica Polonica B* **34** 1485
[10] Prchal J, Javorský P, Isnard O and Sechovský V 2004 *Physica B* **350** e159

Field induced phases in a GMR system TbNiSn

K. Prokeš[1], K. Katsumata[2], S. M. Shapiro[3] and Y. Isikawa[4]

[1]Helmholtz Centre Berlin for Materials and Energy, Glienicker Straße 100, D-14109 Berlin, Germany

[2]RIKEN SPring-8 Center, Harima Institute, Sayo, Hyogo 679-5148, Japan

[3]CMP/MS, Brookhaven National Laboratory, Upton, New York 11973-5000, USA

[4]Department of Physics, Toyama University, Toyama 930-8555, Japan

PACS : 75.25.+z, 75.30.Kz, 75.60.Ej
Keywords :

prokes@helmholtz-berlin.de

Abstract. Results of neutron diffraction measurements on TbNiSn in magnetic fields up to 6T are reported. This material shows successive phase transitions at T_1 = 7.6 K, T_2 = 6.0 K, T_3 = 2.0 K and T_N = 18.5 K in zero field, and a giant magnetoresistance effect (GMR) near four successive metamagnetic transitions at low temperatures. Propagation vectors and main signatures of the magnetic phases appearing at elevated fields applied along the b axis at low temperatures are determined. We find large hysteresis at several phase boundaries. In some cases a co-existence of magnetic structures is detected.

1.Introduction

In the last two decades materials have been found for which the canonical definition of the magnetoresistance does not apply. Instead of an increase with the increasing field a significant decreases of the electrical resistivity by as much as -90% with increasing fields, has been found in some intermetallic compounds [1] and magnetic multilayers [2]. Such phenomenon is referred to as a giant magnetoresistance (GMR) and is usually connected with a disappearance of superzone boundaries introduced into the system by an antiferromagnetic (AF) ordering. In the transition metal oxides [3], the change in the resistance with magnetic field can be much larger so the magnetoresistance is called a colossal magnetoresistance (CMR). The materials showing GMR or CMR have been a subject of intensive studies because of potential applications to magnetic recording devices.

About a decade ago, Hori and co-workers [4] found that the rare earth ternary compound TbNiSn exhibits a GMR. It forms in the orthorhombic structure with the space group Pnma with lattice constants: a= 7.146 Å, b= 4.445 Å and c= 7.663 Å [5]. This material exhibits AF order below T_N = 18.5 K [5,6] followed by three successive order-order magnetic phase transitions at T_1 = 7.6 K, T_2 = 6.0 K and T_3 = 2.0 K [6]. We denote the magnetic phases stable between T_N and T_1, T_1 and T_2, T_2 and T_3 and below T_3 as A, B, C and D, respectively. For a magnetic field applied along the easy b-axis at low temperatures, successive metamagnetic transitions occur at B_{c1} = 0.6 T, B_{c2} = 1.8 T, B_{c3} = 4.3 T and B_{c4} = 5.3 T leading to saturated magnetization of 8.1μ_B/Tb [4,7] as shown in Fig. 1. A GMR is observed near these critical fields [4].

Zero field neutron diffraction using powder samples [5] and single crystals [6,8] showed that the phase A is a sine-modulated magnetic structure with a wavevector q_A = (0.311, 0.325, 0.0) at 14 K followed by the B phase with q_B = (0.665, 0.330, 0.0) between $T_2 < T < T_1$ [6] and an antiphase-type structure (phase C) with q_C = (0.399, 0.344, 0.0) below T_2 [5]. Single crystal measurements [6,8] revealed that the C phase is not sine-wave modulated but squared-up one. We note that in different studies different, but equivalent, propagation vectors are utilized and that no report has been published regarding the ground state phase D stable below T_3.

2. Experimental

The single crystal of TbNiSn was grown by the Czochralski method using an induction furnace. Magnetization measurements were done with a MagLab VSM vibrating-sample magnetometer (Oxford Instruments, UK) and measured at low temperatures in fields along the principal directions up to 6 T. Neutron diffraction measurements were performed on the E4 diffractometer installed at the BER-II reactor at the Helmholtz Zentrum Berlin (formerly HMI). The incident-neutron wavelength λ= 2.44 Å was used. The E4 diffractometer was at the time of the experiment equipped with a single detector. When necessary, two λ/2 filters were used which leave a residual λ/2 contamination at a level of less than 10^{-4}. The external magnetic field generated by a 6 T horizontal magnet was applied parallel to the b* axis of the crystal which is the easy direction of the magnetization [8]. In this geometry, which is quite restrictive due to dead regions produced by the magnet coils, reflections of the (hk0) type were in horizontal scattering plane. The individual scan profiles were analyzed by fitting to Gaussian profiles and the appropriate scattering lengths and the standard Tb^{3+} magnetic form factor were used in the refinement. We also followed the intensities of few reachable nuclear and magnetic reflections as a function of temperature and magnetic field by monitoring the peak intensities.

3. Results and Discussion

In Fig. 1 we show the field dependence of the magnetization measured at 1.5 K with field applied along the b axis in continuously swept applied field. As can be seen, in the present measurement, the values of critical fields, defined as the average of mid-points of sweep-up and sweep-down branches, are B_{c1} = 0.6 T, B_{c2} = 2.1 T, B_{c3} = 4.3 T and B_{c4} = 5.4 T. In the present experiment some of the transitions appear to be quite hysteretic. In the previous measurement [7], however, no appreciable hysteresis was observed at any of the critical fields. We conclude therefore that this is a signature of a non-equilibrium state caused by non-zero field sweeps being 0.1T/min. Moreover, we observe an additional transition at 0.4 T, which we denote B'_{c1}. The saturation moment is achieved above the B_{c4} and amounts to about 8.1 μ_B/Tb. Very remarkable are the plateaus visible between the relevant critical fields showing magnetization values that are roughly 1/18, 1/9, 1/3 and 2/3 of the saturated value, respectively.

At 1.5 K in zero external field one anticipates a ground-state AF structure (phase D) in TbNiSn. A mesh scan in the vicinity of (2/3, 1/3, 0) reciprocal position shows a coexistence of two magnetic reflections that can be described by two different propagation vectors, namely by q_1 = (0.685, 1/3, 0) and q_2 = (0.6, 0.35, 0). This observation is new and suggests a non-trivial ground state of TbNiSn. This observation will be discussed in detail elsewhere [9]. We want only to note that in TbNiSn different magnetic structures compete.

Fig. 2 shows the field dependence of the diffracted intensity measured at the (2,0,0) reciprocal position. Because of the restrictive geometry of the horizontal magnet, this intensity

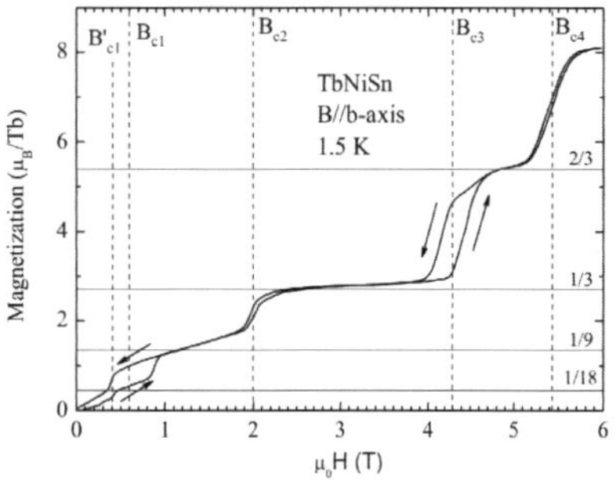

Fig. 1: The field dependence of the TbNiSn magnetization measured with field applied along the b axis. Critical fields of respective magnetic phase transitions are marked by dashed lines.

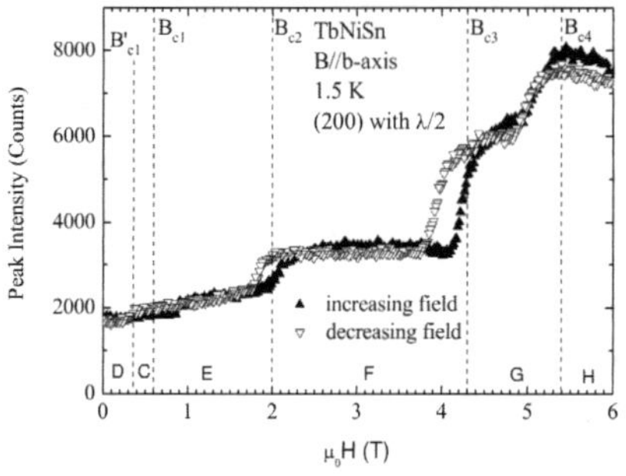

Fig. 2: Field dependence of the diffracted signal recorded at 1.5 K for (2, 0, 0) with the λ/2 filter removed. Various phases are indicated at the bottom of the figure.

was measured by removing the filters and using λ/2 wavelength with the spectrometer set at the forbidden (1,0,0) position. Although the scaling with other reflections is a more complicated, the calculation of the magnetic moment component assumed to be aligned along the b axis is independent of this fact. It appears that above B_{c4} = 5.3 T one deals indeed with the ferromagnetic component close to 9.0 μ_B/Tb. Also the relevant critical fields are in good agreement with the magnetizaion measurement shown in Fig. 1. The (0,2,0) reflection measured under identical conditions is absent for all the fields. This demonstrates that the Tb magnetic moments for all the field-induced magnetic structures are most probably oriented along the b axis. Let us denote the phases between B_{c1} and B_{c2} as phase E, between B_{c2} and B_{c3} as F and between B_{c3} and B_{c4} as G before arriving to the field-forced ferromagnetic phase H. The B'_{c1} is caused by a transition from the phase D to C that is stable up to B_{c1}. This phase is not discussed here.

In Figs. 3a and 3b we show the field dependence of the diffracted intensity measured at the (4/3, -1/3, 0) and (2/3, 0, 0) reciprocal positions, respectively. On increasing the field, the intensity measured at the first position shows a step-like increase at B_{c1}, then a slow decrease and an abrupt disappearance at B_{c3}. With decreasing field one observes a hysteresis around B_{c3} and clearly different behavior in the low-field limit. However, within several minutes the intensity decays to its original zero field value (see the inset of Fig. 3a). At the (2/3, 0, 0) position the signal appears with increasing field at B_{c2}, increases steadily and then sharply decreases at B_{c3}, followed by an increase with a maximum around 5 T and finally decreases and vanishes at B_{c4} = 5.4 T. With decreasing field this signal shows a significant hysteresis around B_{c2} and B_{c3} and vanishes in the low field region. The described observation is highly unusual as it suggests coexistence of different propagation vectors at fields between B_{c2} and B_{c3}. The first propagation vector is q_E = (1/3, 1/3, 0) which is a unique vector for the phase E and the other q_G = (2/3, 0, 0) for phase G. Phase F is described by q_E and q_G simultaneously. For all the field induced phases a ferromagnetic component also exists.

The symmetry group analysis considering the Pnma space group and propagation vectors q_E and q_G respectively allows Tb moments to be oriented either along the b axis or within the a-c plane for the q_G and to have all the Cartesian components for the vector q_E. Consideration of the ferromagnetic components leads to an approximate magnetic structures that are depicted to the Fig. 4. Although, in general, the main features are correct, some details remain still unclear. This applies especially to phase E. The spin arrangement shown in Fig. 4a would imply small non-zero intensity at the (0,2,0) position that is not observed. It is not clear at this moment whether this is due to domains or whether the proposed magnetic structure is incorrect. Due to complexity of the magnetic ordering in TbNiSn under influence of magnetic field, further study is needed.

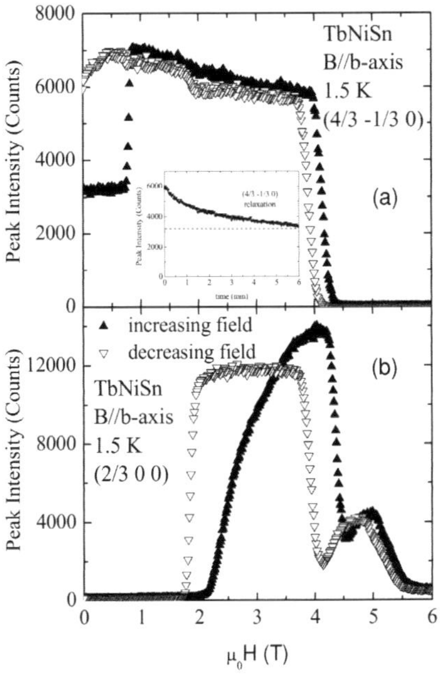

Fig. 3: The field dependence of the diffracted signal recorded at 1.5 K at the (4/3, -1/3, 0) (a) and (2/3, 0, 0) (b) reciprocal space positions. The time dependence of the (4/3, -1/3, 0) intensity in zero field after sweep down is shown in the inset of (a).

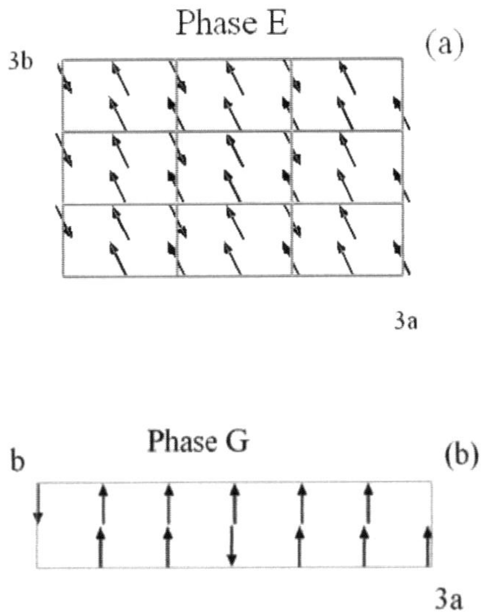

Fig. 4: Schematic representations of possible magnetic structures of TbNiSn belonging to the q_E = (1/3, 1/3, 0) (a) and q_G = (2/3, 0, 0) (b) propagation vectors. Only Tb magnetic moments are shown. In the case of the E phase Tb moments are confined to the a-c plane.

ACKNOWLEDGMENTS: Work performed at Brookhaven is supported by the U. S. Department of Energy under Contract No. DE-AC02-98CH10886. We also thank Y. Yamamoto and H. Hori for their contribution to the early stage of this work.

REFERENCES

[1] V. Sechovský, L. Havela, K. Prokeš, H. Nakotte, F. R. de Boer and E. Brück J. Appl. Phys. **76**, 6913 (1994).
[2] J.-Ph. Ansermet, J. Phys. : Condens. Matter **10**, 6027 (1998).
[3] M. Imada, A. Fujimori and Y. Tokura, Rev. Mod. Phys. **70**, 1039 (1998).
[4] H. Hori, A. Oki, M. Kurisu, M. Furusawa, Y. Andoh and K. Kindo, Phys. lett. A **224**, 293 (1997).
[5] J. K. Yakinthos and Ch. Routsi, J. Mag. Mag. Mat. **149**, 273 (1995).
[6] Y. Andoh, M. Kurisu and S. Kawano, Physica B **237-238**, 575 (1997).
[7] M. Kurisu, H. Hori, M. Furusawa, M. Miyake, Y. Andoh, I. Oguro, K. Kindo, T. Takeuchi and A. Yamagishi, Physica B **201**, 107 (1994).
[8] S. Kawano, Y. Andoh and M. Kurisu, J. Phys. Chern. Solids **60**, 1205 (1999).
[9] K. Katsumata,,K. Prokeš, S.M. Shapiro, Y. Isikawa, in preparation (2009).

Neutron diffraction on the layered Ising magnet $KEr(MoO_4)_2$

S Maťaš[1,2] , A Orendáčová[3] , K Prokeš[1], B. Klemke[1]

[1] Helmholtz-Zentrum für Materialien und Energie GmbH, Glienicker Straße 100, D-14109 Berlin, Germany

[2]Slovak Academy of Sciences, Watsonova 47, 040 01 Košice, Slovakia

[3]University of P. J. Šafárik, Park Angelinum 9, 040 01 Košice, Slovakia

slavomir.matas@helmholtz-berlin.de

Abstract. The magnetic and structural properties of the low-dimensional (2D) Ising magnet $KEr(MoO_4)_2$ are investigated in a temperature region from 70 mK to 8 K in zero field and in external magnetic fields up to 6 T. At 1.8 K, an additional scattering contribution appears at forbidden nuclear reflection positions when ramping an external magnetic field along the c*-axis, showing appreciable hysteresis. This is in agreement with already observed peculiarities in the EPR spectra, microwave absorption spectra and the differential magnetic susceptibility. The low temperature magnetic contribution to the scattering recorded at temperature of 70 mK is confronted with a zero field non-collinear four-lattice magnetic structure magnetic model suggested in the literature.

1. Introduction

The $KEr(MoO_4)_2$ belongs to a large family of *2D* compounds, the so-called double rare-earth molybdates $MR(MoO_4)_2$, where *R* stands for rare-earth and *M* for alkali metal. They crystallize in a variety of layered structures, exhibit strong magnetic anisotropy and are often close to the ideal *2D* Ising system. The magnetic susceptibility measurements taken on $KEr(MoO_4)_2$ system suggest [1-4] that the magnetic phase transition to an antiferromagnetic (AF) ordered state takes place at about T_N = 0.95 K. A non-collinear magnetic structure, in Ref. 1 and 2 named as *6AF-12*, with magnetic moments lying predominantly in the ac-plane but having an angle of about 9 degree with respect to the c axis direction has been suggested by Anders et al. to be the stable AF structure in zero field. This would imply the doubling of the a- and the b- axes. This structure, however, has been suggested on the basis of theoretical calculations and electron paramagnetic resonance experiments only. The magnetic bulk measurements by Khatsko et al. [4] suggested that the easy magnetization direction is parallel to the principal axis with the lattice constant 7.91 Å and also parallel to chains of the rare earth ions separated by distance of 3.957 Å. In Ref. 4 it was named as the b axis. However, the notation of Anders et al. in Ref. 2 is different. The suggested ground state of $KEr(MoO_4)_2$ and the field induced AF structures would suggest the easy magnetization axis being parallel to the c axis. Due to the ambiguity in literature we define and use throughout this contribution the following lattice constant values a = 5.063 Å, b = 18.25 Å and c = 7.917 Å. The ground state of free erbium is $^4I_{15/2}$ with degeneracy 2J+1, which splits by a crystal field into 8 Kramers doublets. The energy separation between the first excited state and the ground state of $\Delta \sim 1.6$ meV is much larger than the cooperative Jahn-Teller interaction energy $A \sim 0.5$ meV [5]. This prevents the spontaneous Jahn –Teller effect when cooling down in zero field. However, upon application of a rather moderate magnetic field along a particular direction above the magnetic phase transition, a first-order structural transition indicating the presence of a magnetically induced Jahn-Teller effect is observed [5]. This problem is rather appealing and motivated our neutron study together with the aim to determine the magnetic structure of $KEr(MoO_4)_2$ at low temperatures.

2. Experimental

A single crystal of $KEr(MoO_4)_2$ has been prepared by the flux method at the Institute of Low Temperature Physics in Kharkov. The sample preparation details are given in Ref. [7-9]. Because of the layered crystal structure, crystals of this material have the form of thin parallelepipeds. We have investigated a good quality crystal with dimensions 12 mm x 0.67 mm x 12 mm. The single crystal was oriented prior to the experiment using several strong nuclear reflections. It was glued on a copper sample holder, installed in a dilution refrigerator that in turn was placed inside a superconducting split-pair cryomagnet manufactured by Oxford Instruments. Reflections in the a*-b* plane were accessible. A vertical external magnetic field up to 6.5 T was applied parallel to the c*–axis. The integrated intensities were measured on the E4 diffractometer at the BER-II reactor at the Helmholtz Centre Berlin. An incident-neutron wavelength $\lambda = 2.44$ Å was used. The E4 diffractometer currently is equipped with a 20 cm x 20 cm position-sensitive detector. Two $\lambda/2$ filters imply a residual $\lambda/2$ contamination at a level of less than 10^{-4}. We have also followed the intensities of few reflections as a function of temperature and magnetic field. Each reflection was measured typically for about 15 minutes. The individual scan profiles were analyzed by fitting to Gaussian profiles. The appropriate scattering lengths and the standard Er^{3+} magnetic form factor were used in the refinement. The crystallographic and magnetic structures were refined by fitting to given models using the computer code Winplotr [6]. Temperature dependencies of magnetization and hysteresis loops measurements were performed in the temperature range between 2 K and 300 K and in magnetic fields up to 5.5 T on the commercial SQUID MPMS magnetometer manufactured by Quantum Design. The data were taken using both a standard and rotator setups. For our magnetic measurements we used the single crystal $KEr(MoO_4)_2$ of 3.89 mg weight cut along the principal a*- and c*- crystallographic directions.

3. Results and Discussion

3.1. *Magnetic susceptibility and magnetization*

Figure 1 shows the field dependence of the longitudinal magnetic moment for the $KEr(MoO_4)_2$ single crystal with magnetic field applied along the easy axis. The data were taken at different temperatures from 2 K up to 30 K. At 2 K, the magnetic moment saturates to a value of $\mu_s = 7.8\ \mu_B$ already in a low magnetic field of 2 T. As the temperature increases the saturation occurs at increasingly higher fields. At 30 K only a linear increase of longitudinal magnetic moment on the external magnetic field is observed. The magnetization curves displayed in the figure 1 show no hysteresis, however, there are a few exceptions in case of other crystal orientations when a very narrow hysteresis of the magnetization was observed. Our magnetization and susceptibility measurements (see figure 2) show that the hard magnetic axis corresponds to the b*- crystallographic direction and that the magnetic easy axis corresponds to the a*- crystallographic direction (a=5.063 Å), which is contradiction with already published results [1-4]. From the high temperature part of the magnetic susceptibility (20 K - 300 K) we estimated the effective magnetic moment of $KEr(MoO_4)_2$ to $\mu_{eff} = 9.8\ \mu_B$, $8.4\ \mu_B$ and $8.7\ \mu_B$ within an error $\pm 0.4\ \mu_B$ for the orientation B//a*-axis, B//b*-axis and B//c*-axis, respectively.

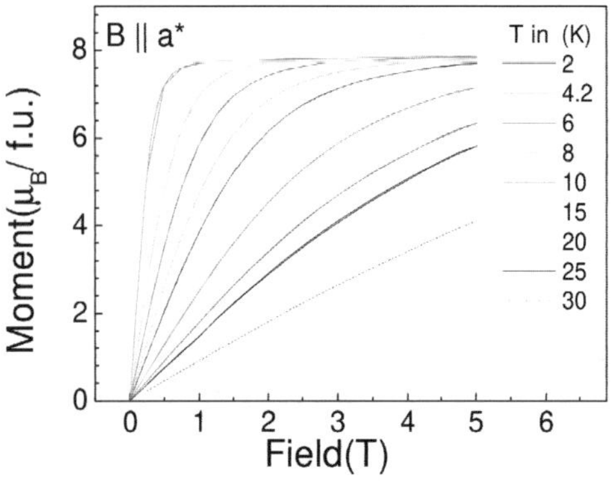

Figure 1 The field dependence of the longitudinal moment for a $KEr(MoO_4)_2$ single crystal at different temperatures in a magnetic field B||a*.

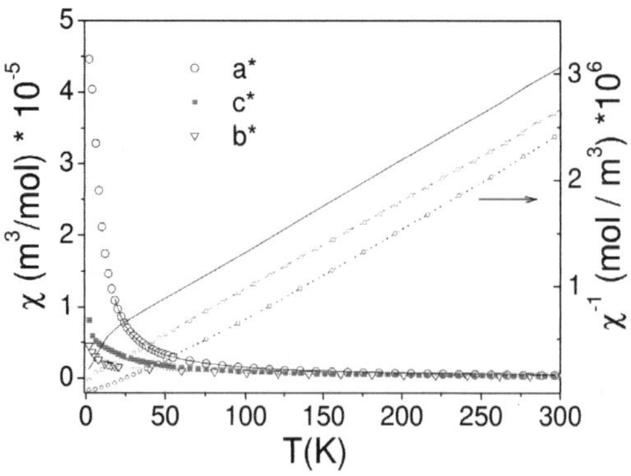

Figure 2 The temperature dependence of χ(T) and χ^{-1}(T) for a $KEr(MoO_4)_2$ single crystal measured in an external field of 1T for B||a*, B||b* , B||c*.

3.2. Neutron diffraction in zero field and in external magnetic field

At 70 mK, well below the magnetic phase transition temperature, the *6AF-12* zero field magnetic structure with magnetic moments lying in the a-c plane and having an angle about ±9 degree with the [001] direction has been suggested by Anders et al. [1,2]. This imposes the appearance of additional magnetic reflections described by propagation vector $\boldsymbol{\tau}$ = (½, ½, 0). However, we did not find any magnetic contribution at any of following positions (1.5, 1.5, 0), (1.5,-1.5, 0), (0.5, k, 0), k= 0.5, 1.5, 2.5, 3.5 and 4.5. Additional scans across the [h00] direction and across [hh0] direction did not show any additional magnetic contribution at positions outside nuclear reflections either. The missing signal calls for explanation: The critical field of the $KEr(MoO_4)_2$ system reported in the literature is of the order of ~70-100 mT, so it is high enough to rule out the influence of the remanent field of the cryomagnet. Hall-probe measurements have shown the remanent field of the magnet is at least one order of magnitude smaller.

Contrary to that, magnetic contributions have been identified at several forbidden nuclear reflection positions of the (0k0) type where k is odd; as (030), (050), (070), (090) and on top of allowed nuclear reflections. As the temperature increases the magnetic contribution at forbidden nuclear positions decreases, reference scans taken above the transition temperature at 1.8 K did not show any magnetic intensity. Due to the absence of reflections with half integer indices we discarded all wave vectors as (½, ½, 0), (½, 0, ½), (0, ½, ½), (½, 0, 0), (0, ½, 0) and (0, 0, ½,) from our symmetry considerations. The additional magnetic intensities observed at forbidden reflection positions, however, suggest that the wave vector ***k***= (0,0,0) should be relevant in describing the zero field magnetic structure. The symmetry analysis performed with help of Mody program [12] gives the possible magnetic ordering modes: (τ_1) G_y, (τ_2) A_y, (τ_3) G_xF_z, (τ_4) A_xC_z, (τ_5) F_xG_z, (τ_6) C_xA_z, (τ_7) F_y and (τ_8) C_y, where $F=S_1+S_2+S_3+S_4$, $G=S_1-S_2+S_3-S_4$, $A=S_1-S_2-S_3+S_4$, $C=S_1+S_2-S_3-S_4$. The best agreement between our magnetic model and the measured data was achieved for the τ_4 mode where two nonzero components of the magnetic moment, $A_xC_z = (+ - - +)_x (+ + - -)_z$ are located at the erbium atom positions *(1)*-(0, y, 0.25), *(2)*-(0, 1-y, 0.25), *(3)*-(0, 0.5-y, 0.25), *(4)*-(0, 1+y, 0.25), y= 0.008, respectively. Figure 3 shows a field dependence for the forbidden (030) reflection at 70 mK temperature. The integrated intensity for the (030) reflection quickly approaches a maximum (at about ≈ 0.9 T), it decreases in higher magnetic field (H>3.5 T). One can interpret the decrease of the intensity above 0.9 T with a picture where the magnetic moments initially have an angle with c*-direction and get aligned by the external magnetic field c*||B. They approach ferromagnetic ordering for which the extinction rules do not allow magnetic intensity at forbidden nuclear reflection positions. In figure 4 an interesting hysteresis of the integrated intensity for the (030) AF reflection at a temperature 1.8 K is shown. The quite unusual field dependence and hysteresis can be understood in terms of field induced lattice changes due to reversible Jahn-Teller processes and by the contribution from the field induced magnetic moment at the Erbium sites. Both contributions can be separated by a synchrotron experiment at 1.8K and in moderate external magnetic field.

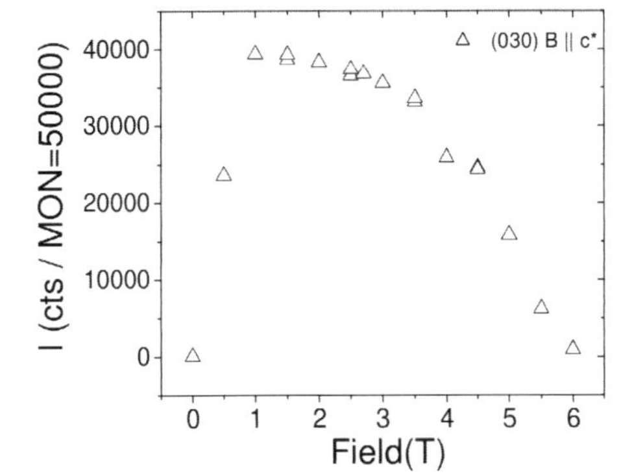

Figure 3 The field dependence of the integrated intensity at forbidden nuclear reflection (030) at 70 mK.

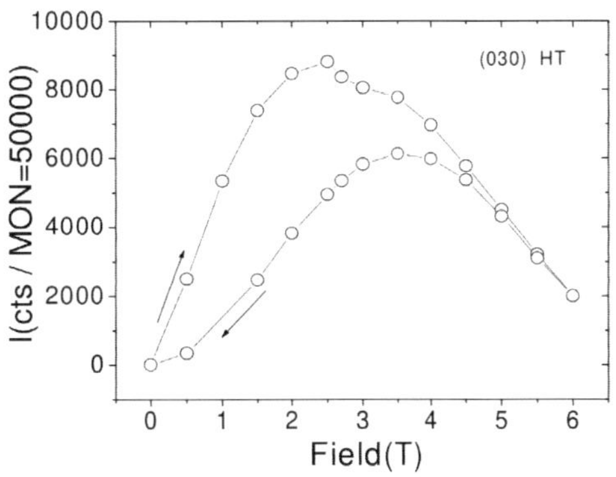

Figure 4 The field dependence of the integrated intensity at the (030) AF reflection at1.8 K. The lines are guides to the eyes.

In conclusion, the a-axis is the easy magnetic axis of $KEr(MoO_4)_2$. The neutron scattering experiment did not confirm the *6AF-12* zero field AF structure with wave vector $\boldsymbol{\tau} = (½, ½, 0)$ as suggested by Anders et al. [1,2]. In contrast, additional magnetic contributions were detected at positions of both allowed and forbidden nuclear reflections of type (0, k, 0) where k is odd number. Our magnetic model qualitatively describes the observed magnetic reflections at wave-vector $\boldsymbol{k}$=(0,0,0) within the symmetry mode τ_4 : A_xC_z.

References

[1] Anders A. G. et al. 1994 *Low Temp. Phys.* **20** (2) 105
[2] Anders A. G. et al. 1994 *Low Temp. Phys.* **20** (2) 137
[3] Khatsko E. N. et al. 2004 *Low Temp. Phys.* **30** (2) 133
[4] Khatsko E. N. et al. 1993 *Low Temp. Phys.* **19** (11) 1217
[5] Khatsko E. N. et. al. 1999 *Ferroelectrics* **233** 93
[6] Roisnel T. and Rodríguez-Carvajal J. 2001 *Materials Science Forum* **378** 118
[7] Dergachev K. G., Kobets M. I., Khatsko E. N. http://arxiv.org/abs/cond-mat/0601471
[8] Vinokurov V. A. and Rlevtsov P. V. 1972 *Kristallografia* **17** 127.
[9] Wanklyn B. M. and Wondre F. R. 1978 *Journal of Crystal Growth* **43** 93
[10] Kobets M. I., Khatsko E. N., Bagulja V. A. 2000 *Ukranian Journal of Physical Optics* **1** (1) 46
[11] Pashchenko V. A. et. al. 2000 *Phys. Rev.B*, **62** (2) 1197
[12] Sikora W., Białas F. and Pytlik L. 2004 *Journal of Applied Crystallography*, **37** 1015

Acknowledgement

We would like to thank A. G. Anders for providing the single crystal for the neutron scattering experiment. We also thank K. Kiefer and S. Gerischer for the support during our neutron measurements in HZ Berlin.

Comparison of thermal stability in *MAX* 211 and 312 phases

W K Pang[1], I M Low[1], B H O'Connor[1], A J Studer[2], V K Peterson[2], Z M Sun[3] and J -P Palmquist[4]

[1] Centre for Materials Research, Department of Imaging and Applied Physics, Curtin University of Technology, GPO Box U 1987, Perth WA, Australia.
[2] The Bragg Institute, ANSTO, PMB 1, Menai, NSW 2234, Australia
[3] Materials Research Institute for Sustainable Development, National Institute of Advanced Industrial Science and Technology (AIST), Nagoya 463-8560, Japan.
[4] Kanthal AB, P.O. Box 502, SE-734 27 Hallstahammar, Sweden.

E-mail: j.low@curtin.edu.au

Abstract. The susceptibility of four *MAX* phases (Ti_2AlC, Cr_2AlC, Ti_3AlC_2, and Ti_3SiC_2) to high-temperature thermal dissociation in vacuum has been investigated using *in-situ* neutron diffraction. In high vacuum, these phases decomposed above 1400°C through the sublimation of *M* and *A* elements, forming a surface coating of MC. The apparent activation energies for the decomposition of sintered Ti_3SiC_2, Ti_3AlC_2, and Ti_2AlC were determined to be 179.3, -71.9, and 85.7 kJ mol^{-1}, respectively. The spontaneous release of Ti_2AlC and TiC from de-intercalation during decomposition of Ti_3AlC_2 resulted in a negative activation energy.

1. Introduction

A group of layered ternary ceramics with the general formula $M_{n+1}AX_n$ (n = 1–3, *M* is an early transition metal, *A* is a group *A* element, and *X* is either carbon and/or nitrogen) were found to be excellent candidate materials for high temperature applications, due to the unique combination of both metallic and ceramic properties [1-6]. *MAX* phases are layered hexagonal (space group P6_3/mmc) Z = 2. In brief, the M_6X octahedra are edge sharing and are identical to the *NaCl*-type structure of the corresponding binary carbides. These M_6X octahedra are interleaved by close-packed planes of the *A* element [7-10]. Ternary carbides such as Ti_3AlC_2, Ti_3SiC_2, Ti_2AlC, and Cr_2AlC are common representatives of the *MAX* phases. The electrical and mechanical properties, and the oxidation of these four compounds has been well studied [9, 11-21], however, very little is published on their susceptibility to high-temperature phase dissociation in vacuum.

In this paper, the thermal stability and the corresponding phase dissociation of these ternary carbides in vacuum was investigated using high-temperature *in-situ* neutron diffraction. The apparent activation energies of phase decomposition were calculated. The characteristics of thermal stability in *MAX* 211 (Ti_2AlC and Cr_2AlC) and 312 phases (Ti_3AlC_2 and Ti_3SiC_2) are compared and discussed.

2. Experimental

The collection of high-temperature *in-situ* neutron diffraction data were collected using Wombat (the high-intensity neutron powder diffractometer) at the OPAL source in Australia.. Data were collected using neutrons with incident wavelength of λ = ~1.660 Å from 15 to 135° 2θ, with a step size of 0.125°, with the use of the oscillating tertiary collimator. Rietica 1.7.7 was used for phase identification and Rietveld refinement. The optimized parameters during refinement were background coefficients, zero-shift error, peak shape parameters, cell parameters, and anisotropic thermal factors. The residual values of the refinement, statistical reliability factor of Bragg (R_B), R-weighted pattern (R_{wp}), R-expected (R_{exp}), and the goodness-of-fit (χ^2), were evaluated. In Rietica, χ^2 is defined as the square of the ratio of R_{wp} to R_{exp}.

Fully-dense solid bars of hot-isostatically-pressed Ti_2AlC, Cr_2AlC, Ti_3SiC_2, and spark-plasma sintered Ti_3AlC_2 were used in this study [2, 16]. The temperature of sample environment was controlled by a closed cylindrical niobium vacuum furnace (10^{-6}-10^{-8} torr). The sample was held by vanadium wire and heated to 1000°C at a heating rate of 10°C/min and thereafter at 5°C/min to 1550°C. The dwell times at between 1000 and 1550°C are shown in Table 1. Diffraction patterns were collected every minute from starting to the end of the experiment.

Table 1. Heating schedule during the in-situ neutron diffraction study.

Temperature (°C)	Dwell time (min)	Temperature (°C)	Dwell time (min)
1000	30	1400	200
1100	30	1450	200
1200	30	1500	200
1300	30	1550	200

The decomposition rates at different temperatures were calculated and the corresponding apparent activation energies were determined in order to understand further the mechanism of high-temperature phase dissociation in vacuum.

3. Results and Discussion

3.1 Phase Transitions during Thermal Dissociation of MAX 211 and 312 Phases

The phase evolution of *MAX* phases and the relative phase abundances at various temperatures as revealed by in-situ neutron diffraction is shown in Figure 2. The samples used in this experiment were not single-phase with TiC as the most common impurity. The starting compositions of each sample are shown in Figure 2. For Ti_3SiC_2, it decomposed to TiC above 1300°C (Fig. 2a). A weight loss of ~ 4% was observed for decomposed Ti_3SiC_2 which may be attributed to the release of gaseous Ti and Si by sublimation during the decomposition process. These results agree with those reported by Low and co-workers [22-23]. For Ti_3AlC_2, its decomposition into TiC and Ti_2AlC as lower order or intermediate phase was observed at ≥ 1400°C (Fig. 2b). However, at higher temperatures, when compared to TiC, a smaller growth rate for Ti_2AlC may indicate that Ti_2AlC experienced further decomposition into TiC via the sublimation of Al and Ti, similar to decomposition of Ti_3SiC_2. In contrast to Ti_3AlC, no intermediate or lower order phase was observed for the decomposition of Ti_3SiC_2. This difference can be attributed to the fact that Ti_3SiC_2 is the only stable ternary phase in Ti-Si-C system. For Ti_2AlC (Fig. 2c), the formation of TiC results mainly from the decomposition of Ti_3AlC_2 as the impurity phase. Ti_2AlC would only decompose above 1450°C by sublimation of Ti and Al. Figure 2(d) shows the excellent stability of Cr_2AlC up to 1450°C where no observable phase dissociation was detected. However, it started to react with the vanadium wire and became unstable above 1450 °C. Like Ti_2AlC, it would only decompose above 1450°C via the sublimation of Cr and Al.

The existence of a lower order 211(Ti_2AlC) phase in Ti_3AlC_2 is believed to be responsible for its unique two-step decomposition behaviour when compared to Ti_2AlC, Cr_2AlC, Ti_3SiC_2 which do not possess such a lower order phase. As a consequence, the former undergoes an initial spontaneous decomposition to lower order Ti_2AlC phase, followed by eventual decomposition of Ti_2AlC through the sublimation of Al and Ti at higher temperature. On the contrary, the latter decompose at elevated temperature in vacuum through the direct sublimation of *M* and *A* elements with the concomitant release of a binary carbide (*MX*) as follows:

$$M_{n+1}AX_n \rightarrow nMX + A + M$$

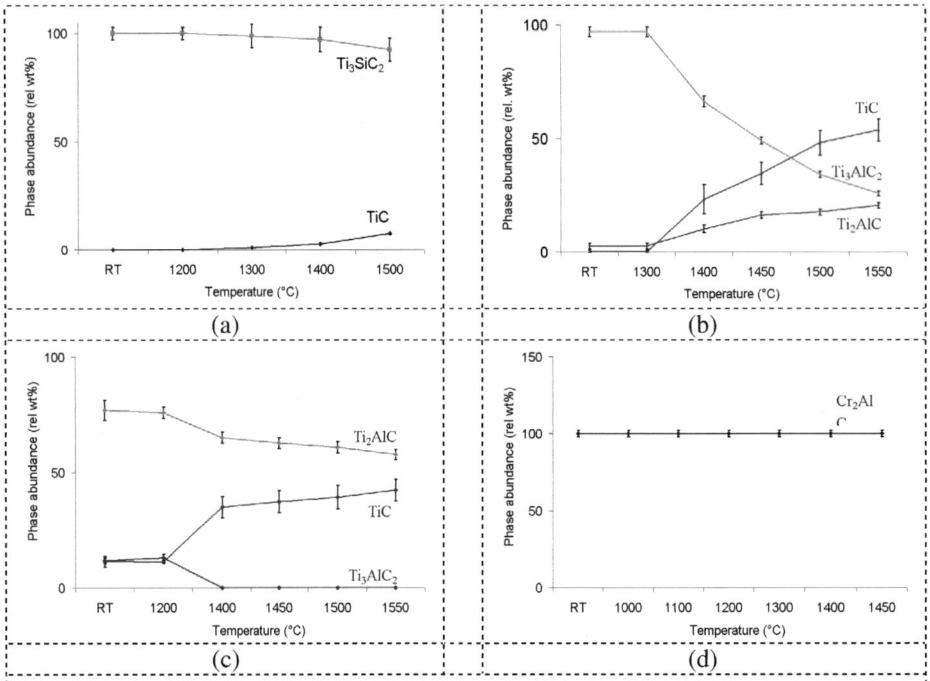

Figure 2. Phase abundance as a function of temperature for the decomposition of (a) Ti_3SiC_2, (b) Ti_3AlC_2, (c) Ti_2AlC, and (d) Cr_2AlC in vacuum.

3.2 Apparent Activation Energies During Isothermal Decomposition

During the isothermal decomposition of 211 and 312 phases at different temperatures, the decomposition rates were calculated by analyzing the relationship between the phase abundances and soaking time. These rates were used to determine the corresponding apparent activation energy of decomposition using an Arrhenius equation [24], $k = A\ \exp(-E\ /\ RT)$, where k is the reaction rate, E is activation energy, R is gas constant, T is absolute temperature and A is a constant.

The calculated apparent activation energies for the four *MAX* phases and the proposed reactions are summarized and listed in Table 1.

Table 1. Comparison of the decomposition of *MAX* 211 and 312 phases.

MAX phase	**Apparent activation energy**	**Proposed reactions**	
Ti_3SiC_2	179.3 **kJ mol^{-1}**	$Ti_3SiC_{2(s)} \rightarrow 2TiC_{(s)} + Ti_{(g)} + Si_{(g)}$	(1)
Ti_3AlC_2	-71.9 **kJ mol^{-1}**	$Ti_3AlC_{2(s)} \rightarrow Ti_2AlC_{(s)} + TiC_{(s)}$	(2a)
		$Ti_2AlC_{(s)} \rightarrow TiC_{(s)} + Ti_{(g)} + Al_{(g)}$	(2b)
Ti_2AlC	85.7 **kJ mol^{-1}**	$Ti_2AlC_{(s)} \rightarrow TiC_{(s)} + Ti_{(g)} + Al_{(g)}$	(3)
Cr_2AlC	Stable in vacuum up to 1450 °C.	$2Cr_2AlC_{(s)} \rightarrow Cr_3C_{2(s)} + Cr_{(g)} + 2Al_{(g)}$	(4)

A negative activation energy obtained for Ti_3AlC_2 can be attributed to the presence of weak Ti-Al bonds [25] which will favour spontaneous decomposition through the de-intercalation of TiC from the crystal structure (see Eq 2a). Subsequent decomposition of released Ti_2AlC at higher temperature (see

Equation 2b) would involve a small but positive activation energy. It follows that the high activation energy obtained for Ti_3SiC_2 can be attributed to much stronger bonding between Ti and Si, leading to better resistance against thermal dissociation *via* sublimation of Ti and Si from the crystal structure. The lower but positive activation energy observed for decomposition of Ti_2AlC (see Eq. 3) is related to weaker Ti-Al bonds and thus its susceptibility to sublimation in vacuum at elevated temperature.

4. Conclusions

The high-temperature thermal stability of *MAX* 211 and 312 phases in vacuum has been studied using in-situ neutron diffraction. Both 211 and 312 phases were susceptible to decomposition above 1400°C through sublimation of M and A elements, resulting in a surface coating of TiC being formed. The apparent activation energies for the decomposition of sintered Ti_3SiC_2, Ti_3AlC_2 and Ti_2AlC were determined to be 179.3, -71.9 and 85.7 kJ mol^{-1}, respectively. The negative activated energy for Ti_3AlC_2 can be attributed to the release of lower order Ti_2AlC and TiC from the crystal structure via spontaneous de-intercalation.

Acknowledgements

This work was funded by an ARC Discovery-Project grant (DP0664586) and an ARC Linkage-International grant (LX0774743) for one of us (IML). The collection of diffraction data was conducted at the OPAL research reactor in ANSTO with financial support from AINSE (08/329 & 09/606).

References

[1] Barsoum M W 2000 *Prog. Solid State Chem.* **28** 201.
[2] Barsoum M W, Brodkin D and El-Raghy T 1997 *Scripta Mater.* **36** 535.
[3] Barsoum M W, Buschow K H J, Robert W C, Merton C F, Bernard I, Edward J K, Subhash M and Patrick V in "Encyclopedia of Materials: Science and Technology" (Elsevier, Oxford, 2006) p. 1.
[4] Barsoum M W, Crossley A and Myhra S 2002 *J. Phys. Chem. Solids* **63** 2063.
[5] Barsoum M W and El-Raghy T 2001 *Am. Sci.* **89** 334.
[6] Barsoum M W, Buschow K H J, Robert W C, Merton C F, Bernard I, Edward J K, Subhash M and Patrick V in "Encyclopedia of Materials: Science and Technology" (Elsevier, Oxford, 2004) p. 1.
[7] Wu L, Chen J X, Liu M Y, Bao Y W and Zhou Y C 2009 *Wear* **266** 158.
[8] Wang J, Zhou Y, Lin Z and Hu C 2008 *Scripta Mater.* **58** 1043.
[9] Wang J, Zhou Y, Liao T, Zhang J and Lin Z 2008 *Scripta Mater.* **58** 227.
[10] Zhang H Z and Wang S Q 2007 *Acta Mater.* **55** 4645.
[11] Han J H, Hwang S S, Lee Dand Park S W 2008 *J. Eur. Ceram. Soc.* **28** 979.
[12] Yang C, Jin S, Liang B, Liu G, Duan L and Jia S 2009 *J. Alloys Compd.* **472** 79.
[13] Yang C, Jin S, Liang B, Liu G and Jia S 2009 *J. Mater. Process. Tech.* **209** 871.
[14] Lee D B, Nguyen T D, Han J H and Park S W 2007 *Corr. Sci.* **49** 3926.
[15] Lin Z J, Li M S, Wang J Y and Zhou Y C 2007 *Acta Mater.* **55** 6182.
[16] Tian W, Sun Z, Du Y and Hashimoto H 2008 *Mater. Lett.* **62** 3852.
[17] Fraczkiewicz M, Zhou A G and Barsoum M W 2006 *Acta Mater.* **54** 5261.
[18] Kisi E H, Crossley J A A, Myhra S and Barsoum M W 1998 *J. Phys. Chem.Solids* **59** 1437.
[19] Kooi B J, Poppen R J, Carvalho N J M, De Hosson J T M and Barsoum M W 2003 *Acta Mater.* **51** 2859.
[20] Khoptiar Y and Gotman I 2002 *Mater. Lett.* **57** 72.
[21] Wang P, Mei B C, Hong X L and Zhou W B 2007 *Trans. Nonferrous Metals Soc. Chi.* **17** 1001.
[22] Low I M 2004 *Mater. Lett.* **58** 927.
[23] Low I M, Oo Z and Prince K E 2007 *J. Am. Ceram. Soc.* **90** 2610.
[24] Riley D P, Kisi E H and Hansen T C 2008 *J. Am. Ceram. Soc.* **91** 3207.
[25] Xu X, Wu E, Du X, Tian Y, and He J 2008 *J. Phys. Chem. Solids* **69** 1356.

Journal of Physics: Conference Series **251** (2010) 012026 doi:10.1088/1742-6596/251/1/012026

A Generic Phase Diagram for R_2PdSi_3 (R = Heavy Rare Earth)?

M Frontzek[1], F Tang[1], P Link[2], A Schneidewind[1,2], J M Mignot[3], J U Hoffman[4] and M Loewenhaupt[1]

[1]Institut für Festkörperphysik, Technische Universität Dresden, D-01062 Dresden, Germany

[2]Technische Universität München, FRM II, D-85747 Garching, Germany

[3]Laboratoire Léon Brillouin, CE-Saclay, F-91191, Gif-sur-Yvette, France

[4]Helmholtz-Zentrum Berlin für Materialien und Energie GmbH, D-14109 Berlin, Germany

Frontzek@physik.tu-dresden.de

Abstract. The magnetic phase diagrams of hexagonal R_2PdSi_3 compounds have been found to feature an apparent similar phase when the field is applied along the respective magnetic easy axes. To elucidate this, neutron diffraction on single crystalline Tb_2PdSi_3 and Ho_2PdSi_3 in magnetic fields was performed. These two compounds have different magnetic structures in zero field. In Tb_2PdSi_3 the propagation vector is parallel to the hexagonal axis while in Ho_2PdSi_3 the propagation is in the basal plane. In both compounds the magnetic moment direction is perpendicular to the propagation vector. The application of a magnetic field along the respective easy axes destroys the zero field magnetic structure and leads to a ferrimagnetic phase with magnetic intensity on the same positions for both compounds. These findings and their connections to the crystallographic (super-)structure are presented and discussed.

1. Introduction

The R_2PdSi_3 (R = heavy rare earth) compounds crystallize in an AlB_2-derived hexagonal structure (P6/*mmm*). The magnetic rare-earth ions form layers which can be understood as triangular lattice with latent geometric frustration. These layers are intersected by layers of the non-magnetic Pd and Si atoms. The stacking of the layers is along the *c*-axis. Experimental evidence from single crystal neutron diffraction strongly suggests the presence of Pd-Si order at the B positions [1]. Additional structural reflections were found on (m/2, 0, n/8) and (m/2, m/2, n/8) positions with n, m as integers. No intensity was found on (0, 0, 1/2), (1/2, 1/2, 0) and (1/2, 1/2, 1/2) positions which would be expected from earlier published lattice constants by Kotsanidis et al. [2]. A detailed study of the crystallographic (super)-structure will be published elsewhere [3]. Here, only pre-published facts which are crucial for the understanding of the magnetic structure will be mentioned. The Pd-Si order doubles the lattice constant a of the primitive unit cell and the (super)-structure reflections can then be described by a propagation vector (0, 0, 1/8). In the following, we describe the structure in the picture of the primitive AlB_2 structure and refer to the Pd-Si order as crystallographic superstructure. A direct consequence of the doubling is the occurrence of two inequivalent rare earth positions. One of the rare earth positions (Wyckoff layer *1a*) keeps still the hexagonal on-site symmetry while the other position (Wyckoff layer *3f*) has monocline symmetry.

2. Phase diagrams

Ac-susceptibility measurements on single crystals grown in the Institut für Festkörper- und Werkstoffforschung, Dresden [4, 5] were analyzed for features like maxima, steps and kinks [6]. The position of these features in (*H*, *T*) space has been used to derive phase diagrams. Similar features for different parameters are connected with lines as is shown exemplarily in figure 1, the phase diagram of Gd_2PdSi_3 with the field applied along the (*H*, 0, 0)-direction. The low temperature (below T_N), low field (below H_{c1}) part of the phase diagram labelled AFM displays a variety of metamagnetic phases, also for other R_2PdSi_3 compounds (in most compounds also a second transition temperature T_2 is observed).

The FiM labelled phase in figure 1 is of interest since it has an intriguing behavior with the phase boundary increasing with higher fields towards higher temperatures. Similar phases are found in the phase diagrams for *R* = Gd, Tb, Dy and Ho when a field is applied along the magnetic easy direction. For the compounds with a small magneto-crystalline anisotropy like Ho_2PdSi_3 and Gd_2PdSi_3 this phase can be found also in the magnetic hard direction. Therefore, the existence of a generic phase diagram as schematically drawn in figure 2 is proposed for the R_2PdSi_3 (*R* = heavy rare earth) compounds.

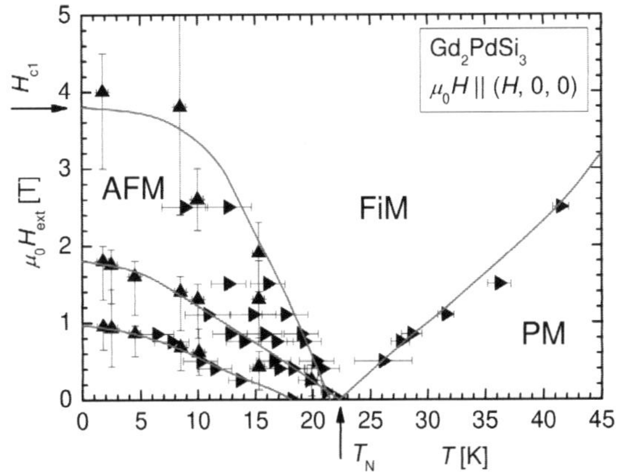

Figure 1: Magnetic phase diagram of Gd_2PdSi_3 derived from ac-susceptibility measurements. The critical magnetic field H_{c1} and the Néel temperature T_N are marked.

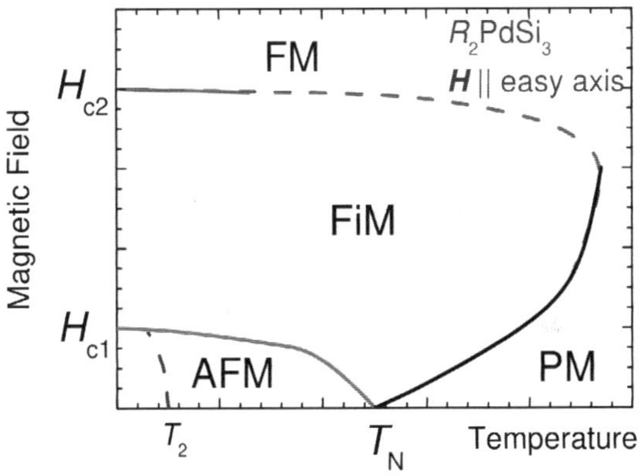

Figure 2: Proposal for a generic phase diagram for R_2PdSi_3 when a field is applied along the magnetic easy axis. For details see text.

The FiM phase might be interpreted as a ferromagnetic induced state. However, the critical field H_{c2} at T = 2 K above which the compound is in the saturated state is found at high fields (9.4 T for Gd_2PdSi_3; larger than 13 T for Tb_2PdSi_3 and Ho_2PdSi_3). Thus we are left with the question about the the nature of the FiM phase.

3. Neutron scattering in applied magnetic fields for Ho_2PdSi_3 and Tb_2PdSi_3

Neutron diffraction measurements were performed at the Helmholtz-Zentrum Berlin (HZB, formerly HMI) at the E2 diffractometer for Tb_2PdSi_3 (T_N = 23.6 K; H_{c1} = 3.8 T) and at the Laboratoire Leon Brillouin (LLB, Saclay) at the 6T2 diffractometer for Ho_2PdSi_3 (T_N = 7.7 K; H_{c1} = 0.8 T). Both experiments used a graphite monochromator with λ/2 filters at a wavelength of about 2.4 Å. The two investigated compounds are especially suited for the evaluation of a generic phase since their magnetic structures in zero field are completely different.

In zero field the magnetic structure of Tb_2PdSi_3 can be described with a propagation vector $\boldsymbol{\tau}_{Tb,1}$ = (0, 0, 1/16) and that of $Ho_2PdSi_3$with a propagation vector $\boldsymbol{\tau}_{Ho,1}$ = (1/7+δ, -2δ, 0). In both cases the propagation vector is perpendicular to the magnetic moment direction.
The zero field magnetic structure of Tb_2PdSi_3 is influenced by the crystallographic superstructure. Magnetic intensity is found not only for $\boldsymbol{G}+\boldsymbol{\tau}_{Tb,1}$ where $\boldsymbol{G}$ is a reciprocal lattice vector but also for $\boldsymbol{G}$+(1/2, 1/2, 0)+$\boldsymbol{\tau}_{Tb,1}$ according to the description of the crystallographic superstructure with the doubled basal plane. When a magnetic field is applied in the basal plane (in the magnetic easy direction) the zero field magnetic structure is gradually replaced by a magnetic structure with a ferromagnetic component and with an antiferromagnetic component with $\boldsymbol{\tau}_2$ = (0, 0, 1/8). This vector equals the propagation vector for the crystallographic superstructure. We therefore observe in applied fields a strong increase of intensity on the positions of the superstructure reflections. In figure 3 scans along $\boldsymbol{Q}$ = (1/2, 1/2, L) are shown for different temperatures and fields. The (nuclear) reflections due to the crystallographic superstructure are very weak and barely to be seen for the chosen scale (green curve). The magnetic reflections with $\boldsymbol{\tau}_{Tb,1}$ of the zero field magnetic structure (blue curve) disappear for fields above H_{c1}. They are replaced by reflections with $\boldsymbol{\tau}_2$ (red curve at the positions of the crystallographic superstructure). They exist in the field and temperature range as indicated for the FiM phase in figure 2. Below H_{c1} and T_N the magnetic phases with $\boldsymbol{\tau}_{Tb,1}$ and $\boldsymbol{\tau}_2$ coexist.

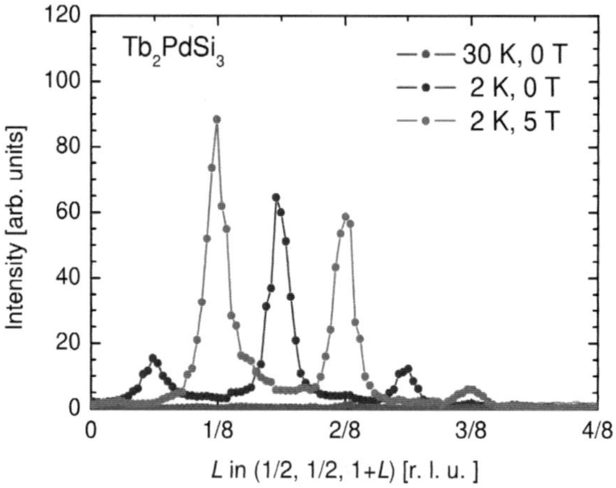

Figure 3: Diffractogram of Tb_2PdSi_3 along $\boldsymbol{Q}$ = (1/2, 1/2, L) for different fields and temperatures.

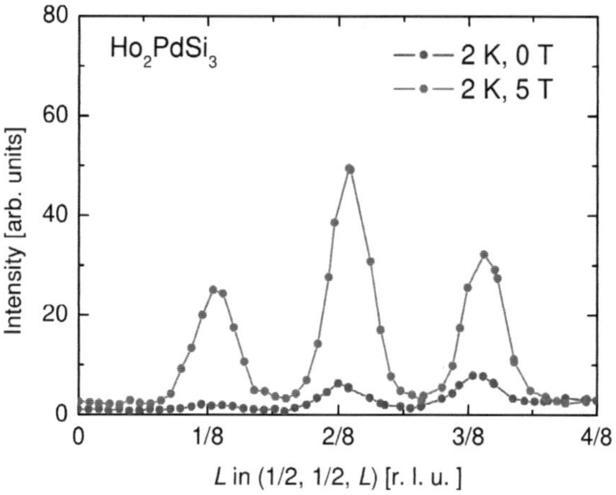

Figure 4: Diffractogram of Ho_2PdSi_3 along $\boldsymbol{Q}$ = (1/2, 1/2, L) for different fields and temperatures.

In Ho_2PdSi_3 the zero field magnetic structure with propagation vector $\boldsymbol{\tau}_{Ho,1}$ = (1/7+δ, -2δ, 0) has no magnetic intensity along the (1/2, 1/2, L) direction as shown in figure 4. Only (nuclear) reflections due to the crystallographic superstructure (green curve) are observed in zero field. However, applying a magnetic field larger than H_{c1} (0.8 T) in the magnetic easy direction leads not only to the destruction of the zero field magnetic structure (not shown here) but to the appearance of magnetic intensity on the positions $\boldsymbol{Q}$ = (1/2, 1/2, n/8) of the crystallographic superstructure. Again, these reflections exist in the field and temperature range as indicated for the FiM phase in figure 2.

4. Discussion and Conclusion

The application of a magnetic field along the easy magnetic directions in Tb_2PdSi_3 and Ho_2PdSi_3 leads for both compounds to a similar magnetic structure with a ferromagnetic component and an additional antiferromagnetic component. This magnetic structure can be identified with a magnetic phase named FiM in the magnetic phase diagram of these two compounds. Moreover, this phase has been identified in macroscopic measurements also for other R_2PdSi_3 compounds leading to the idea of a generic phase diagram. The FiM phase of other R_2PdSi_3 compounds still needs to be investigated more closely with neutrons to confirm if a similar magnetic structure connected to the FiM phase in Tb_2PdSi_3 and Ho_2PdSi_3 is also found there.

The nature of the strong connection between magnetic and crystallographic superstructure in R_2PdSi_3 is open for speculation. In the following two possible interpretations are discussed briefly. The possibility of the FiM phase to be the ferromagnetic induced state is discarded since the absolute value of the magnetization is still below $g\boldsymbol{J}$ in the FiM phase.

The first interpretation proposes a strong crystalline electric field (CEF) effect which is relevant when the magnetic structure due to the exchange interactions is destroyed by a magnetic field ($> H_{c1}$). Additionally, a pronounced difference of the CEF on the two different rare earth sites due to the crystallographic superstructure has to be assumed. The surrounding CEF defines the ground state wave function of the rare earth ion. Since the rare earth ions have different surroundings (either they are neighbored by Pd and Si or by Si alone) different ground states are imaginable. The FiM phase is then a true ferromagnetic phase but with different magnitude of the magnetic moments. The change of the magnetic intensity with the magnetic field can be explained by the changing of the mixing ratio of the wave functions. However, this model cannot explain the potential existence of a FiM phase for Gd_2PdSi_3. In addition, evidence for different magnetic moments in zero field due to different ground states is missing [7].

The second interpretation on the other hand assumes a strong exchange interaction. Then the CEF only defines the magneto-crystalline anisotropy (moment direction). Though strong exchange interactions between neighboring rare earth ions are present at higher temperatures the Néel temperature is diminished due to geometric frustration of the rare earth ions. Furthermore, the effect of geometric frustration is confined to layers of the rare earth ions. The application of a magnetic field lifts the geometric frustration in the layer and nearest neighbor exchange becomes relevant. The nearest neighbor exchange is antiferromagnetic and yields an antiferromagnetic in plane component following the $2*a$ basal plane. The sequence of this antiferromagnetic component along the $\boldsymbol{c}$-axis is modulated in the sequence of the Pd/Si layers. Then the crystallographic superstructure influences the exchange from rare earth layer to rare earth layer and consequently the resulting magnetic structure follows the modulation of the superstructure.

For a final interpretation of the nature of the similar magnetic structures found in Tb_2PdSi_3 and Ho_2PdSi_3 more measurements are needed. These are aimed to reveal the nature of the FiM phases in other R_2PdSi_3 compounds. The strong coupling of the magnetic properties to the crystallographic (super-)structure and the possibility to influence this coupling via an external magnetic field in the R_2PdSi_3 compounds make this series interesting for further studies.

References

[1] P. A. Kotsanidis, J. K. Yakinthos, E. Gamari-Seale, J. Magn. Magn. Mat. **87** (1990) 199
[2] M. Frontzek, A. Kreyssig, M. Doerr, J.-U. Hoffman, D. Hohlwein, H. Bitterlich, G. Behr, M. Loewenhaupt, Physica B **350** (2004) e187
[3] J. Dshemuchadse *to be published*
[4] G. Graw, H. Bitterlich, W. Löser, G. Behr, J. Fink, L. Schultz, J. Alloys Compounds **308** (2000) 193
[5] I. Mazilu, M. Frontzek, W. Löser, G. Behr, A. Teresiak, L. Schultz, J. Crystal Growth **275** (2005) e103
[6] M. Frontzek, A. Kreyssig, M. Doerr, M.Rotter, G. Behr, W. Löser, I. Mazilu, M. Loewenhaupt, J. Magn. Magn. Mat. 301 (2006) 398
[7] M. Frontzek, A. Kreyssig, M. Doerr, A. Schneidewind, J.-U. Hoffman, M. Loewenhaupt, J. Phys.: Condens. Matter 19 (2007) 145276

International Conference on Neutron Scattering 2009 IOP Publishing
Journal of Physics: Conference Series **251** (2010) 012027 doi:10.1088/1742-6596/251/1/012027

Structural Phase Transition in AuZn Alloys

Barry Winn[1,2], S M Shapiro[1], Jason C Lashley[3], Cyril Opeil[4] and William Ratcliff[5]

[1]Brookhaven National Laboratory, Upton, NY 11973-5000 USA
[2]Oak Ridge National Laboratory, Oak Ridge, TN 37381 USA
[3]Los Alamos National Laboratory, Los Alamos, NM 87545 USA
[4] Boston College, Chestnut Hill, MA 02467 USA
[5]National Institute of Standards and Technology, Gaithersburg, MD 20899 USA

E-mail: winnbl@ornl.gov

Abstract. Au_xZn_{1-x} alloys undergo a shape memory martensitic transformation whose temperature and nature (continuous or discontinuous) is strongly composition dependent. Neutron diffraction experiments were performed on single crystals of x=50 and 52 to explore the structural changes occurring at the transition temperature. A transverse modulation with wavevector q_0=(1/3,1/3,0) develops below the transition temperature, with no observable change in lattice parameter. However, the Bragg peak width shows a broadening suggesting an unresolved rhombohedral distortion similar to what has been observed in NiTi-Fe alloys.

1. Introduction

Shape memory properties occur in materials undergoing martensitic phase transformation that are normally first order transitions showing an abrupt change in the lattice parameters at the transformation temperature. The transformation is diffusionless, reversible, with a very small volume change. The structure of the high temperature austenite phase of most shape memory materials is cubic and the low temperature martensite phase has a lower symmetry, either tetragonal, monoclinic or triclinic structure. The occurrence of many domains in the martensite phase is the key element of the shape memory properties.

Recently it was noted that the intermetallic alloy, Au_xZn_{1-x} undergoes a nearly continuous phase transformation for the 50-50 composition, but is first order at off-stoichiometric compositions [1]. Near the transition temperature, T_M, which is composition dependent, a transverse modulation of the lattice develops with wavevector q_0=(1/3,1/3,0). Also, the transition temperature decreases with increasing pressure suggesting a quantum critical regime. [2].

Here we probe the structural transition in more detail in order to gain more insight into the low temperature structure and compare the behavior with the well studied Fe doped NiTi.

2. Experimental Details and Results

The single crystals of Au_xZn_{1-x} were prepared by fusion of elements in a Bridgman furnace. The x=50 sample was cut into a $1mm^3$ cube with faces along the cubic direction and had a crystal mosaic of about 0.8°. The x=52 crystal was smaller with a slightly larger mosaic. The neutron experiments were performed on the BT9 triple-axis spectrometer at the NIST research reactor using an incident energy of 14.7 meV with filters placed in the beam. Because of the small size of the sample collimation of 40' was used throughout the instrument.

Figure 1 shows the temperature dependence of the satellite peak intensity measured at q_0 for x=50 and x=52. For x=50 the behavior is nearly continuous and can be fit with a power law with critical exponent β=0.51, consistent with mean field theory [2]. For x=52 the transition temperature is shifted to lower temperatures and is more discontinuous.

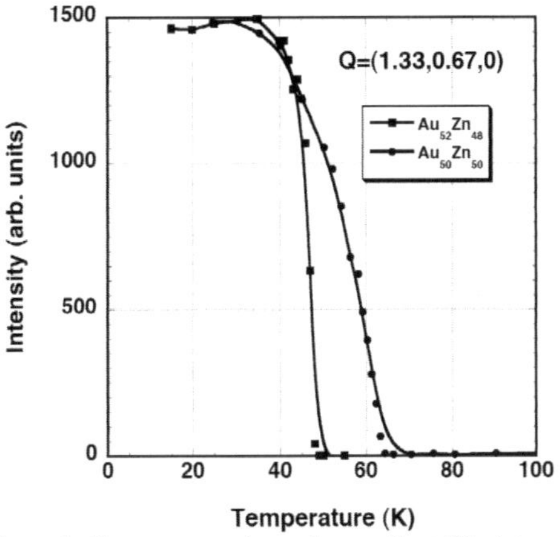

Figure 1. Temperature dependence of satellite intensity on heating for Au_xZn_{1-x}: x=50 and x=52.

The transition temperatures measured are consistent with those obtained from specific heat and resistivity measurements [3]. Also, the nature of the transition: nearly continuous for x=50 and discontinuous for x=52 are also consistent with other measurements. In our experiment we monitored the (1,1,0) Bragg peak in order to observe a change at the transition temperature. Figure 2 is a plot of the lattice parameter for x=52. Surprisingly, no anomaly is apparent near the transition temperature T_M=48K where the satellite intensity begins to increase. A similar observation was made for x=50.

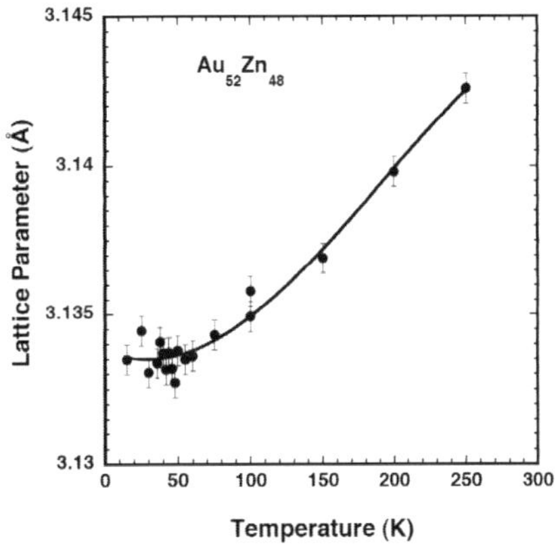

Figure 2. Lattice parameter for $Au_{52}Zn_{48}$ measured at the (1,1,0) position vs tempeature

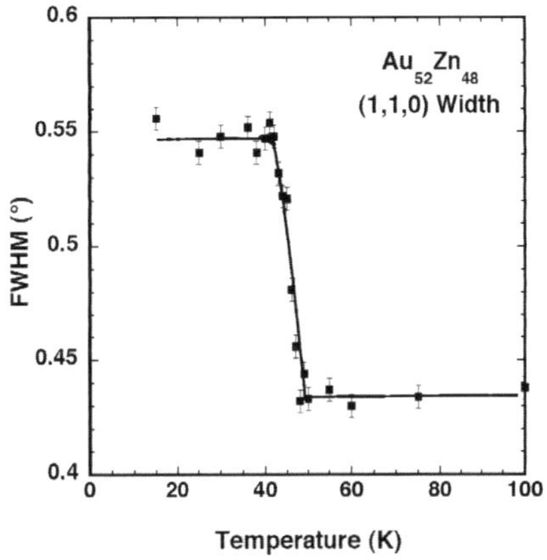

Figure 3. Angular width of longitudinal scan through (1,1,0) Bragg peak.

However, the width of the Bragg peak did change abruptly at the transition as shown in Fig. 3, which is the angular width of a longitudinal $\theta-2\theta$ scan.

Another observation is the presence of the satellites measured in the [HKK] zone for x=50 at equivalent q_1 = [1/3,1/3,1/3] satellite positions that appear below T_M. Mesh scans performed at 10K at several equivalent positions reveal an incommensurability that varied with satellite position. The temperature dependence of this incommensurability was not monitored to see if there was a lock-in.

3. Discussion

The high temperature phase of Au_xZn_{1-x} has the cubic CsCl structure. The low temperature phase was reported to be trigonal with $P3(C_3^1)$ space group [4]. The signature of this symmetry change is a splitting of the $(1,1,0)_C$ cubic Bragg peak into a $(1,1,2)_R$ and $(0,0,3)_R$ trigonal or R-phase Bragg peak. This is not happening in our samples (Fig. 2) and we conclude that the trigonal distortion is small and cannot be resolved, but shows up as broadening of the Bragg peak (Fig. 3)

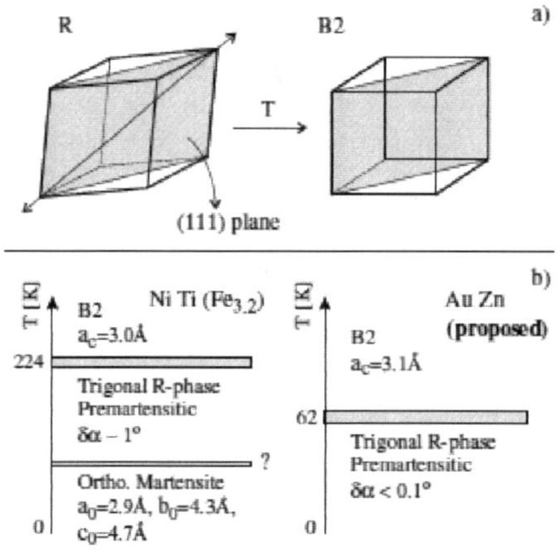

Figure 4. Schematic representation of (a) structural transformation from B2 to R phase and (b) comparison between AuZn and $NiTi\text{-}Fe_{3.2}$.

There is a large similarity between the phase transition in AuZn and that observed in $NiTi\text{-}Fe_x$ [5,6]. Figure 4 shows a schematic representation of the phase transition. In Fig. 4a the B2 phase is cubic and the R phase is obtained by effectively pulling along a [111] body diagonal. This causes a distortion and the angle between the cubic axis changes from 90° to 90-α°. If α is large enough the (1,1,0) Bragg peak will split as mentioned above. This occurs in $NiTi\text{-}Fe_x$ for x< 6 at. %. For $x \geq 6$ at.% no change in lattice parameter is observed and the transition appears continuous, precisely the behavior observed in Au_xZn_{1-x} for x=50. It is likely that $\alpha < 0.1°$ and shows up as a broadening of the Bragg peak rather than to a splitting. The comparison also extends to the presence of the q_1 type satellites observed in $NiTi\text{-}Fe_{3.2}$ in a detailed x-ray experiment [7]. In Ref. [7] study an incommensurability was observed for both q_0 and q_1 satellites but the incommensurability varied with the Brillouin zone measured. This precludes a simple condensed charge density and/or lattice-wave description of the transformation. Our experiments show similar features for $Au_{50}Zn_{50}$.

Based upon the comparison with $NiTi\text{-}Fe_x$ our picture of the transition in AuZn is that the low temperature phase is trigonal or rhombohedral with a very small distortion. It is similar to the R phase of $NiTi\text{-}Fe_{3.2}$ and can be considered as a premartensite phase (PM) where the martensite phase is suppressed. Also, like NiTi-Fe there are phonon anomalies [8] around the q_0 position [2,4] in the cubic parent phase that are precursors to the transformation to the PM phase.

More data are desired to probe the incommensurability of the low temperature phase and to explore in more detail the lattice dynamics of this interesting system.

Acknowledgements:
The work performed at Brookhaven, Los Alamos and Oak Ridge is supported by the US Department of Energy. We acknowledge the support of the National Institute of Standards and Technology, U.S. Department of Commerce, in providing the neutron research facilities used in this work.

References

[1] Darling T. W., et al., 2002 *Philos. Mag.* **82** 825
[2] Lashley J C, et al., 2008 *Phys. Rev. Lett.* **101** 135703
[3] McDonald R D, et al. 2005 *J. Phys. Condens. Matter* **17** L69
[4] Makita T, et al., 2005 *Physica B* **213&214** 430
[5] Choi, M –S, Fukuda T, Kakeshita T and Mori H, 2006 *Philos. Mag.* **86** 67
[6] Otsuka K and Ren X 2005 *Progr. In Mat. Science* **50** 511
[7] Shapiro S M, Noda Y, Fujii Y and Yamada Y, 1984 *Phys. Rev. B* **30** 4314
[8] Satija S K, Shapiro S M, Salamon M B and Wayman C M, 1984 *Phys. Rev. B* **29** 6031

Structure of $BiRe_2O_6$ re-investigated using single-crystal neutron Laue diffraction

N Sharma[1], T Söhnel[2], G J McIntyre[3], R Piltz[4] and C D Ling[1, 4]

[1] School of Chemistry, The University of Sydney, Sydney, NSW 2006, Australia

[2] Department of Chemistry, The University of Auckland, Auckland, New Zealand

[3] Institut Laue-Langevin, 6 rue Jules Horowitz, BP 156, 38042 Grenoble Cedex 9, France

[4] Bragg Institute, Australian Nuclear Science and Technology Organisation, PMB1, Menai, NSW 2234, Australia

n.sharma@chem.usyd.edu.au

Abstract. Single crystals of $BiRe_2O_6$ of typical volume 0.03 mm^3 were grown by chemical vapor transport and characterized using room-temperature single-crystal neutron diffraction in monoclinic *C2/c* symmetry with cell parameters a = 16.1178(11), b = 4.9235(3), c = 5.5278(3) Å and β = 92.475(5) ° (R_{all} = 11.39 wR_{all} = 7.97). The structure contains ordered layers of corner sharing units of Re_2O_{10} with Re-Re distances of 2.519(1) Å. The unit cell is doubled along c in comparison to previous studies based on X-ray diffraction data where the layer stacking was described as disorderd.

1. Introduction

Rhenium has been reported in oxidation states of 4^+ [1], $4^+/5^+$ [2-4], 5^+ [2,3], $5^+/6^+$ [5] and 7^+ [6,7] in ternary compounds. Re^{7+}-containing compounds are generally synthesized in atmospheric conditions and feature $5d^0$ electronic configuration of Re. The other oxidation states of Re, which contain $5d$ electrons, are synthesized under a variety of reducing conditions. The high-spin versions of these compounds are of interest for their magnetic properties, and in particular for what has been termed "unconventional rhenium magnetism." [8]

$BiRe^{5+}O_4$ [2,3], $BiRe^{4+/5+}{}_2O_6$ [3] and $Bi_3Re^{4+/5+}{}_3O_{11}$ [2] are known examples with unpaired d electrons on Re. $BiRe_2O_6$ [2,3] has a $KSbO_3$-related structure previously observed in $SbRe_2O_6$, [9] $ScRe_2O_6$ [8] and $Sb(Nb/Ta)TiO_6$. [10] These compounds all feature edge-sharing Re_2O_{10} octahedral dimers that form layers by sharing corners with other dimers. [5] The Re–Re bond within the dimers often results in metallic behavior and good electrical conductivity. [8] The Re–Re bond lengths vary from 2.415(1) Å in $La_4Re_6O_{19}$ [4] to 3.102(1) Å in $PbRe_2O_6$. [11]

Initial work on $BiRe_2O_6$ by Cheetham and Smith [3] proposed a monoclinic *C2/m* cell with a = 5.516(1), b = 4.906(1), c = 8.384(1) Å and β = 106.71(1) °. A disordered stacking of the Re–O layers was described. Some Weissenberg photographs contained weak reflections suggesting a doubling of the c axis, but they did not contain enough significant intensity to be used in refinements. Bi is found in distorted square pyramidal and distorted octahedral environments depending on the relative orientation of the Re–O layers.

Here, we have re-investigated the structure of $BiRe_2O_6$ by single-crystal neutron diffraction, after synthesizing sufficiently large crystals for the white-beam Laue technique used by the instruments VIVALDI at the Institut Laue-Langevin (ILL) Grenoble, France [12], and KOALA at the Open-Pool Australian Light-water reactor (OPAL), Lucas Heights, Australia.

2. Experimental

Chemical vapor transport was used to synthesize crystals of $BiRe_2O_6$. Bi_9ReO_{17} [7] and Re metal were pre-reacted in evacuated sealed quartz tubes for 4 days at 800 °C. This mixture was combined with the NH_4Br transport agent and placed at the source position of a quartz tube. The quartz tube was then evacuated and sealed. The source position of the tube was heated to 650 - 750 °C, while the sink position was maintained 50 °C lower in temperature. The reaction proceeded for 5 days.

Crystals were initially analyzed on a Bruker-Nonius APEX II X-ray diffractometer with Mo K_α radiation (λ = 0.71073 Å). Preliminary neutron Laue data were collected at room temperature on VIVALDI at the ILL, followed by the collection of a more complete data set on KOALA at OPAL. VIVALDI and KOALA use 'white' thermal-neutron beams (0.8 Å $\sim\!<\lambda\sim\!<$ 5.2 Å) with cylindrical image-plate detectors. The same 0.3×1.0×0.1 mm crystal mounted on a vanadium pin was used for data collection on both instruments.

Neutron Laue patterns were collected on KOALA in 12 exposures of 2 h each, separated by 12° rotations of the ϕ-axis (perpendicular to the neutron beam), providing good data redundancy and completeness. The patterns were indexed using the program LAUEGEN [13], and the reflections integrated using ARGONNE_BOXES [14] and normalised to a common wavelength using LAUENORM. [15] The reduced data contained 4322 reflections (2084 unique) in the ranges $-29 \leq h \leq 29$, $0 \leq k \leq 8$, $0 \leq l \leq 8$. Structural refinements were undertaken with the Jana2006 [16] package.

3. Results and Discussion

The fact that these crystals can only be obtained in small quantities by chemical vapor transport presents an unusual situation for structural analysis, where single-crystal neutron diffraction using VIVALDI and KOALA was feasible, but insufficient total quantities were available for neutron powder diffraction (after grinding up the crystals). The majority of the crystals tested on VIVALDI were laminated perpendicular to the stacking axis *a*, but a suitable candidate for full data collection was identified from the 10 tested.

The single-crystal neutron Laue patterns clearly showed reflections corresponding to a doubling of the *c* axis of the reported [3] unit cell. The standard setting for this enlarged cell (in which *a* and *c* are exchanged with respect to that of Cheetham *et al.* [3]) is monoclinic *C*2/*c* a = 16.1178(11), b = 4.9235(3), c = 5.5278(3) Å and β = 92.475(5) °. Using $SbRe_2O_6$ [9] as a starting model, the positions and independent anisotropic atomic displacement parameters (ADP's) could be refined for all atoms. Final statistics for 43 refined parameters were R_{all} = 11.39 %, R_{obs} = 7.08 %, wR_{all} = 9.97 %, and wR_{obs} = 7.82 %. Refined atomic parameters, ADP's, and bond valence sums (BVS) [17] are presented in table 1 and bond lengths and angles are presented in table 2. Figure 1 shows the refined structural model of $BiRe_2O_6$.

Table 1. Final refined fractional atomic coordinates, ADP's, and bond valence sums (BVS) for $BiRe_2O_6$ in *C*2/*c*, a = 16.1178(11), b = 4.9235(3), c = 5.5278(3) Å and β = 92.475(5) ° with Z = 4. Equivalent isotropic ADP's are given for all atoms.

Atom	*Wyckoff position*	*x*	*y*	*z*	$100 \times U_{eq}/U_{iso}$	BVS
Bi	4*a*	0.0	0.2735(2)	0.25	0.75(4)	3.35(1)
Re	8*b*	0.1804(1)	0.2356(2)	-0.1129(2)	0.22(2)	4.53(1)
O1	8*b*	0.2156(1)	0.4769(2)	0.1690(3)	0.62(4)	2.13(1)
O2	8*b*	0.1378(1)	-0.0300(3)	0.1087(3)	0.75(4)	1.80(1)
O3	8*b*	0.0699(1)	0.4044(3)	-0.0836(3)	0.81(4)	2.27(1)

Table 2. Selected metal–oxygen bond lengths (Å) in the refined structure of $BiRe_2O_6$.

Bi - O1	2.799(1) × 2
Bi - O3	2.270(1) × 2
Bi - O3	2.129(2) × 2
Re - Re	2.519(1)
Re - O1	2.014(1)
Re - O1	1.999(2)
Re - O1	1.947(2)
Re - O2	1.935(2)
Re - O2	1.926(2)
Re - O3	1.970(1)

Figure 1. The crystal structure of $BiRe_2O_6$ with Bi in light gray, edge-sharing octahedral Re_2O_{10} groups in dark gray and Re illustrated as gray spheres. The structure viewed along the **(a)** [010] and **(b)** [011] direction.

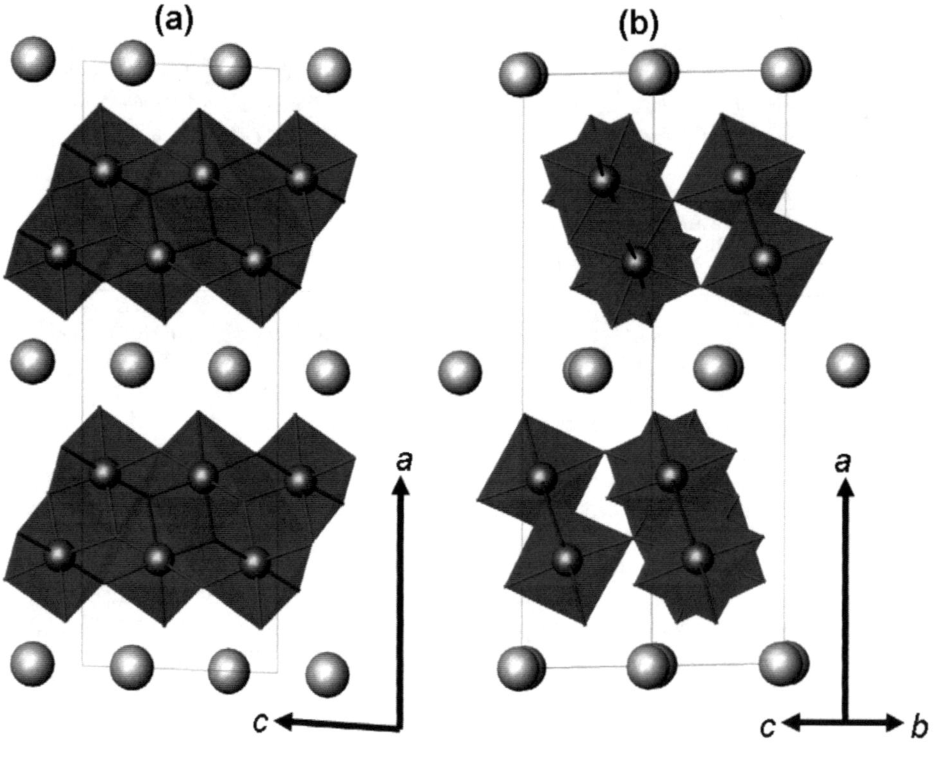

Our model shows the presence of ordered stacking of layers of Re_2O_{10} dimers to form two-dimensional layers. The Re–Re bonds in this structure are 2.519(1) Å, a value which is intermediate between those found in $SbRe_2O_6$ and $ScRe_2O_6$. [8,9] Re–O bonds range from 1.926(2) to 2.014(1) Å. The BVS's indicate chemically plausible coordination environments for all ions, including $Re^{4.5+}$, and the Bi–O distances and coordination environments (figure 2) are all reasonable.

Figure 2. The coordination environments of Bi and Re viewed along the [001] direction in the refined structure of $BiRe_2O_6$. Re - Re bond is also shown.

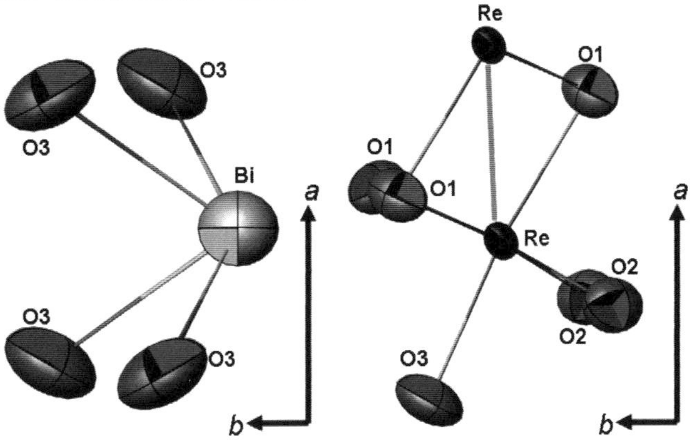

4. Conclusions

Single-crystal neutron Laue diffraction is a powerful technique for solving and refining the structures of compounds that require neutron (rather than X-ray) diffraction data, but for which it is not possible to obtain sufficient quantities for neutron powder diffraction. Such a situation arose in this case due to the use of chemical vapor transport synthesis. Data collected on VIVALDI and KOALA were used to identify and refine the ordered superstructure of $BiRe_2O_6$, in which Re_2O_{10} edge-sharing octahedral dimers are ordered into two-dimensional layers separated by Bi atoms with square pyramidal coordination.

Acknowledgements

This work was supported by the Australian Research Council – Discovery Projects (DP0666465) and the Australian Institute of Nuclear Science and Engineering Postgraduate Research Awards scheme. We acknowledge support to travel to the ILL by the Access to Major Research Facilities Programme.

References

[1] Waltersson K 1976 *Acta Cryst. B* **32** 1485.
[2] Cheetham A K and Smith A R R 1981 *Mat. Res. Bull.* **16** 7.
[3] Smith A R R and Cheetham A K 1979 *J. Solid State Chem.* **30** 345.
[4] Morrow N L and Katz L 1968 *Acta Cryst. B* **B24** 1466.
[5] Torardi C C and Sleight A W 1986 *J. Less-Common Metals* **116** 293.
[6] Cheetham A K and Smith A R R 1985 *Acta Cryst. B* **B41** 225; Crumpton T E, Mosselmans J F W, and Greaves C 2005 *J. Mater. Chem.* **15** 164.
[7] Sharma N, Withers R L, Knight K S, and Ling C D 2009 *J. Solid State Chem.* **Submitted**.
[8] Mikhailova D, Ehrenberg H, Miehe G, Trots D, Hess C, Schneider R, and Fuess H 2008 *J. Solid State Chem.* **181** 190.
[9] Watanabe H, Imoto H, and Tanaka H 1998 *J. Solid State Chem.* **138** 245.
[10] Ling C D, Thompson J G, Withers R L, and Schmid S 1996 *J. Solid State Chem.* **125** 19.
[11] Wentzell I, Fuess H, Bats J W, and Cheetham A K 1985 *Z. anorg. allg. Chem.* **528** 48.
[12] Mcintyre G J, Lemée-Cailleau M H, and Wilkinson C 2006 *Physica B* **385-386** 1055.
[13] Campbell J W, Hao G, Harding M M, Nguti N D, and Wilkinson C J 1998 *J. Appl. Crystallogr.* **31** 23.
[14] Wilkinson C, Khamis H W, Stansfield R F D, and Mcintyre G J 1988 *J. Appl. Crystallogr.* **21** 471.
[15] Campbell J W, Habash J, Helliwell J R, and Moffat K 1986 *Information Quarterly for Protein Crystallography* **18**.
[16] Petricek V, Dusek M, and Palatinus L, JANA2006 The crystallographic computing system, 2007 Institute of Physics, Praha, Czech Republic.
[17] Brese N E and O'Keeffe M 1991 *Acta Cryst.* **B47** 192.

Structure and Properties of Manganese–Substituted Bismuth Titanates

J Ting and B J Kennedy

School of Chemistry, the University of Sydney, Sydney NSW Australia, 2006

E-mail: j.ting@chem.usyd.edu.au

Abstract. Partial substitution of manganese for titanium in bismuth titanate is a potential method to introduce ferromagnetic properties to ferroelectric bismuth titanate. We have established that manganese-substituted bismuth titanate forms only at low manganese contents. As the manganese content is increased, secondary phases form. A combination of neutron and synchrotron X-ray powder diffraction has been used to determine the structure of manganese-substituted bismuth titanates with chemical composition $Bi_4Mn_xTi_{3-x}O_{12}$ ($0 \leq x \leq 1$ at 0.05 steps). Measurements between 4 K and 1073 K have been used to examine the phase transitions that they undergo. It was found the substitution of manganese into bismuth titanate increases the phase transition temperature to tetragonal by about 100 °C. Neutron diffraction and bulk magnetic susceptibility measurements demonstrate the absence of long-range magnetic cation ordering in selected examples.

1. Introduction

Bismuth titanate ($Bi_4Ti_3O_{12}$) is a well known ferroelectric material that has been extensively studied as a consequence of its potential commercial applications in non-volatile memories and electro-optical devices. The ferroelectric polarisation is a result of the interaction between the d^0 Ti^{4+} cation, the $6s^2$ lone pair electrons from bismuth Bi^{3+} in the "Bi_2O_2" layer and the bridging O^{2-} anion, Figure 1 [1].

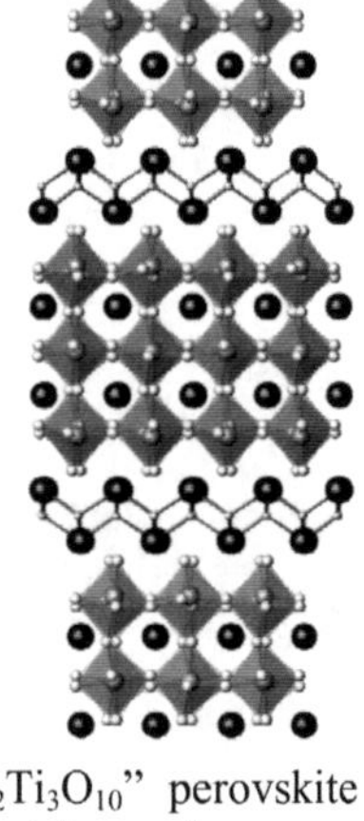

Figure 1. A schematic diagram of the Aurivillius oxide $Bi_4Ti_3O_{12}$. The black spheres denote bismuth and the white spheres denote oxygen. Bi–O bonds are drawn within the "Bi_2O_2" layers and the titanium cations are contained by the octahedrons representing TiO_6 units.

The "$Bi_2Ti_3O_{10}$" perovskite block in bismuth titanate offers the possibility of substitution by transition metals, such as manganese. Such substitution is an avenue to introduce magnetism to ferroelectric bismuth titanate and could result in a multiferroic material. Multiferroic materials are rare due to the usually conflicting requirement for a structurally symmetrical environment for the d^n magnetic cations and a structurally distorted environment stabilised by a d^0 cation.

2. Experimental

The series $Bi_4Mn_xTi_{3-x}O_{12}$, where $0 \leq x \leq 1$ at 0.05 intervals, was synthesised via a solid state reaction with a maximum temperature of 950 °C. For the x = 0, 0.10, 0.15 and 0.20, 10 g samples were prepared; 2 g samples were prepared for all other compositions. The reactions were monitored using Cu-K_α radiation on a Shimadzu X-6000 diffractometer.

Synchrotron X-ray powder diffraction data at $\lambda \approx 0.8$ Å was measured at the Australian National Beamline Facility at the Photon Factory in Tsukuba, Japan [2]. Neutron diffraction patterns were measured for selected samples using the high resolution powder diffractometer Echidna at ANSTO's OPAL facility at Lucas Heights [3]. Rietveld refinements of the diffraction patterns were done using the program RIETICA [4]. Magnetic susceptibilities were measured using a Quantum Design Physical Property Measurement System.

3. Results and discussion

Figure 2 shows the series of diffraction patterns for $Bi_4Mn_xTi_{3-x}O_{12}$ for $0 \leq x \leq 1$. Single phase samples were obtained for x = 0, 0.05 and 0.10. At x = 0.15, peaks from $Bi_{12}TiO_{20}$ were first observed, with the most prominent peak at d ~ 3.2 Å. The intensity of these peaks increases with increasing manganese content. At x = 0.55, a $Bi_2Mn_4O_{10}$ impurity was observed and this is seen to increase with increasing Mn contents. An example of a Rietveld refinement of the final compound in the series, $Bi_4MnTi_2O_{12}$ is shown in Figure 3. The diffraction pattern was refined using the three phases. While the parent $Bi_4Ti_3O_{12}$ adopts a monoclinic structure, the monoclinic distortion is reasonably small [1]. Approximating the β angle to be 90°, an orthorhombic crystal system with space group *B2cb* was used for the $Bi_4MnTi_2O_{12}$ phase.

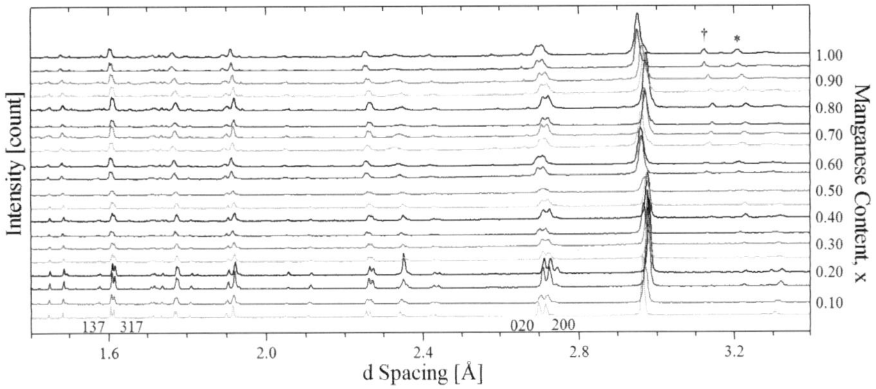

Figure 2. Portion of the diffraction patterns for $Bi_4Mn_xTi_{3-x}O_{12}$ where $0.05 \leq x \leq 1.00$. The * indicate the peak due to $Bi_{12}TiO_{20}$ and † indicates the peak due to $Bi_2Mn_4O_{10}$.

Splitting of the orthorhombic 200/020 and 317/137 peaks at 2.7 Å and 1.6 Å respectively is evident in Figure 2 and 3. As the manganese content increases, the splitting of these peaks decreases, suggesting a shift to a tetragonal structure. This was not, however, achieved by x = 1 as evident in Figure 3.

Variable temperature synchrotron diffraction patterns were measured for the x = 0.05 and 0.50 samples. Both show a first order phase transition from orthorhombic *B2cb* to paraelectric tetragonal *I4/mmm* at 800 °C and 775 °C for x = 0.05 and 0.50 respectively, Figure 4 and 5. This compares to a temperature of 670–690 °C found for $Bi_4Ti_3O_{12}$ [5,6]. This suggests that manganese increases the phase transition temperature, but this may not be a linear effect.

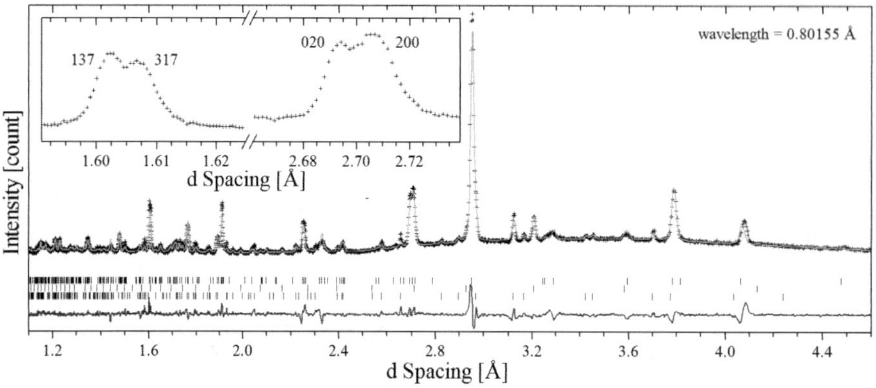

Figure 3. The Rietveld refinement of $Bi_4MnTi_2O_{12}$ using 3 phases. The first phase is orthorhombic *B2cb* $Bi_4MnTi_2O_{12}$ 72.1(7) wt%, the second is $Bi_{12}TiO_{20}$ 7.62(2) wt% and the third phase is $Bi_2Mn_4O_{10}$ 20.3(4) wt%. Inset, the resolved 200/020 and 317/137 peaks.

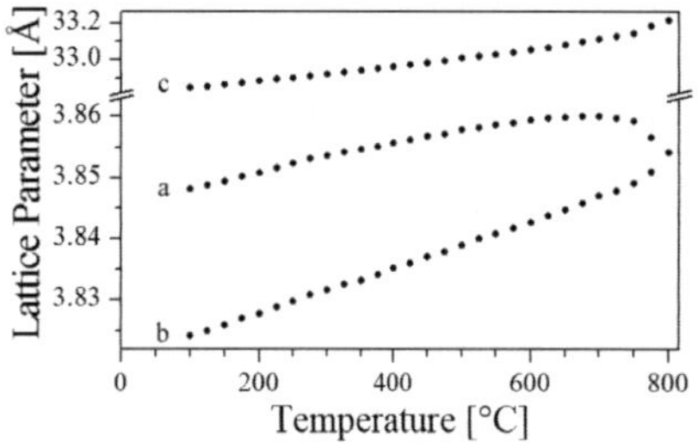

Figure 4. The lattice parameters for $Bi_4Mn_{0.05}Ti_{2.95}O_{12}$.

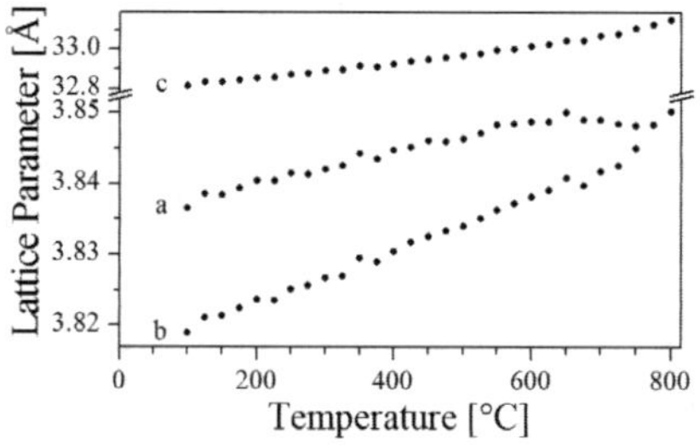

Figure 5. The lattice parameters for $Bi_4Mn_{0.50}Ti_{2.50}O_{12}$.

The inverse susceptibility against temperature for $x = 0.15$ is given in Figure 6 and the magnetisation against magnetic field is in Figure 7. No evidence for magnetic exchange was observed for this or the $x = 0.10$ sample, reflecting the small amount of manganese in these compounds. Higher values of x were not examined due to the existence of impurity phases.

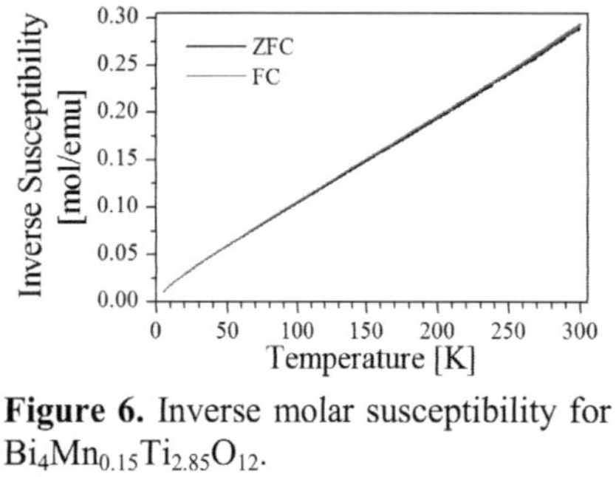

Figure 6. Inverse molar susceptibility for $Bi_4Mn_{0.15}Ti_{2.85}O_{12}$.

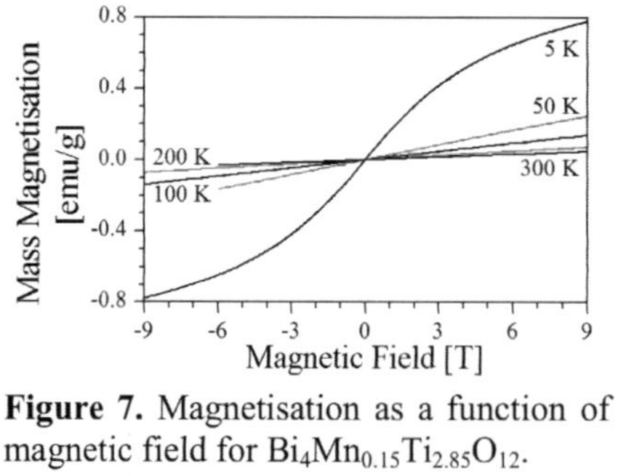

Figure 7. Magnetisation as a function of magnetic field for $Bi_4Mn_{0.15}Ti_{2.85}O_{12}$.

A comparison of the neutron diffraction pattern for $x = 0.20$ at 3 K and 300 K is in Figure 8 and shows no evidence of any structural or magnetic transition, indicating the lack of magnetic ordering in the low manganese content materials. One of the advantages of neutron diffraction is the ability to provide accurate structural information. However, given the negative scattering lengths of both Ti and Mn, -3.438 fm and -3.73 fm respectively, establishing the ordering of manganese in the perovskite block is difficult. Examination of the metal–oxygen bond lengths and measuring the distortion of the Ti/MnO_6 octahedral units using equation (1) shows that the inner but not the outer octahedral units

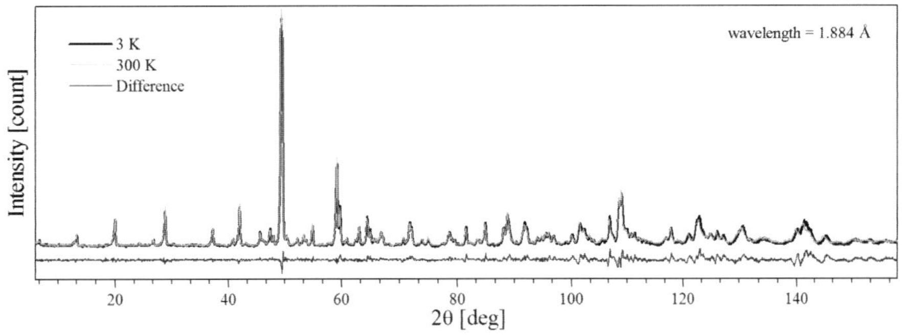

Figure 8. A comparison of the patterns for $Bi_4Mn_{0.20}Ti_{2.80}O_{12}$ collected at 3 K and 300 K. The 300 K pattern subtracted from the 3 K pattern is the difference profile.

become more symmetrical with the addition of manganese, Table 1 [7]. We propose that this is due to preferential occupancy by the manganese cations in the central octahedral units.

$$\Delta d = \frac{1}{6}\sum\left(\frac{\text{M-O}_{\text{obs}} - \text{M-O}_{\text{avg}}}{\text{M-O}_{\text{avg}}}\right)^2 \quad (1)$$

Table 1. Distortion of the Ti/MnO_6 octahedral units in the perovskite block.

	Outer	Inner
$Bi_4Ti_3O_{12}$	6.534×10^{-3}	2.318×10^{-3}
$Bi_4Mn_{0.10}Ti_{2.90}O_{12}$	6.324×10^{-3}	1.206×10^{-3}
$Bi_4Mn_{0.15}Ti_{2.85}O_{12}$	7.310×10^{-3}	0.412×10^{-3}
$Bi_4Mn_{0.20}Ti_{2.80}O_{12}$	6.708×10^{-3}	0.675×10^{-3}

4. Conclusion

The series of $Bi_4Mn_xTi_{3-x}O_{12}$ $0 \leq x \leq 1$ was synthesised to study their magnetic properties in the anticipation that the introduction of magnetic cations into ferroelectric bismuth titanate would result in a multiferroic material. Synchrotron X-ray diffraction shows that only low manganese content samples are single phase. Increasing the Mn content produces $Bi_{12}TiO_{20}$ and $Bi_2Mn_4O_{10}$ as secondary phases and decreases the orthorhombic distortion. Variable temperature measurements also show an increase in the ferroelectric–paraelectric phase transition temperature by about 100 °C upon Mn doping. Magnetic susceptibility measurements of the single phase samples show a lack of long range magnetic ordering and this is supported by low temperature neutron diffraction. From changes in the distortion of the octahedra, we suggest that the Mn occupies the inner octahedral units of the perovskite block.

References

[1] Rae A D, Thompson J G, Withers R and Willis A C 1990 *Acta. Cryst.* B **46** 474–87
[2] Sabine T M, Kennedy B J, Garrett R F, Foran G J and Cookson D J 1995 *J. Appl. Cryst.* **28** 513–7
[3] Liss K D, Hunter B A, Hagen M, Noakes T and Kennedy S J 2006 *Phys. B Cond. Matt.* **385-386** 1010–2
[4] Howard C J and Hunter B A 1998 *A Computer Program for Rietveld Analysis of X-ray and Neutron Powder Diffraction Patterns* Lucas Heights Research Laboratories, New South Wales, Australia pp 1–27
[5] Hervoches C H and Lightfoot P 1999 *Chem. Mater.* **11** 3359–64
[6] Zhou Q, Kennedy B J and Howard C J 2003 *Chem. Mater.* **15** 5025–8
[7] Alonso J A, Martínez-Lope M J, Casais M T and Fernández-Díaz M T 2000 *Inorg. Chem.* **39** 917–23

International Conference on Neutron Scattering 2009 IOP Publishing
Journal of Physics: Conference Series **251** (2010) 012030 doi:10.1088/1742-6596/251/1/012030

Crystallization phase diagram, the growth of large single crystals of bovine β-Lactoglobulin A

D Yagi[1], Y Ohnishi[2], I Tanaka[2] and N Niimura[3,4]

[1]Graduate School of Science and Engineering, Ibaraki University, Japan
[2]College of Engineering, Ibaraki University,Japan
[3]Frontier Research Center for Applied Atomic Sciences, Ibaraki University, Japan

[4]E-mail: niimura@mx.ibaraki.ac.jp

Abstract. A crystallization phase diagram defining the meta-stable region of bovine β-lactoglobulin A (β-Lg) was firstly determined by a dialysis method. We have succeeded in growing a large single crystal of β-Lg by selecting a crystal grown in this "meta-stable region" method described in the present paper. The quality of protein crystals was characterized quantitatively via rapid X-ray data collections, followed by the use of Wilson plots to analyze their resulting average B-factors.

1. Introduction
β-lactoglobulin A (β-Lg) is one of the model proteins for the study of amyloid fibril formation. [1] It is well-known that property changes of these proteins are closely related to a change of the hydrogen bonding between β-strands, and therefore it is important to know the distinctive features of H-bonds of such proteins for a better understanding the mechanism of amyloid fibril formation.

Neutron protein crystallography, in principle, provides the information of all the hydrogen atoms and the hydration structure around a macromolecule, and it is one of the most powerful techniques to investigate H-bonds in which the positions of hydrogen atoms are explicitly included. [2] However, one of the main disadvantages of neutron protein crystallography is that large crystals are needed: currently, the volume required for single-crystal work is usually on the order of 1 mm3. In this paper the crystallization phase diagram of β-Lg will be reported and a method of growing a large single crystal of β-Lg on the basis of this phase diagram will be demonstrated. Finally, the qualities of several crystals grown under various conditions in the phase diagram will be described.

2. Experiments, results and discussion

2.1. Sample preparation
Bovine β-lactoglobulin A (β-Lg) was purchased from Sigma Chemical Co. (L7880), and protein solution was prepared into Tris/HCl (pH 7.9) buffer without further purification. Ammonium sulfate was selected to be the crystallization (precipitating) agent. [3]

[4] Author to whom any correspondence should be addressed.

2.2. The crystallization phase diagram

A typical and general crystallization phase diagram is shown in Figure1, where Cp and Cc are the concentrations of the protein and the crystallization (precipitating) agent, respectively. The phase diagram is classified into three regions: unsaturated, meta-stable, and nucleation region. The particular crystallization phase diagram for β-Lg (shown in Figure 2) was determined by a dialysis method. A Lucite dialysis button, in which the volume of protein solution is 5 μl, was used. The molecular weight cutoff of the dialysis membrane was 10,000. Five protein solutions, with Cp of 4, 11, 15, 19 and 36 mg/ml, respectively, were prepared. For each concentration, four solutions were prepared in order to monitor reproducibility. The button containing the protein was immersed in reservoir solution, consisting of the crystallization agent and the buffer. Since protein solutions Cp of β-Lg higher than 40mg/ml become cloudy upon addition of ammonium sulfate, the maximum Cp was limited to 36 mg/ml in our experiments. Moreover it must be pointed out that the protein solution became translucent beyond about 1.5 M of Cc.

Firstly, the nucleation borderline between meta-stable region and nucleation region (dotted blue line in Figure 2) was determined as follows: The Cc was increased step by step. Each incremental step was 0.05 M and the protein solution in the button was kept for 1 week at T= 20°C, upon which it was noted whether crystallization had occurred or not.

Secondly, the solubility curve (solid red line in Figure 2) was determined by dissolving crystals in the button as follows: The Cc was decreased step by step, and the protein solution was kept in this condition for one day and it was observed whether crystals had completely dissolved or not. When the crystals were completely dissolved, a point (Cc, Cp) was marked in the phase diagram, whereas if crystals still remained, the Cc was decreased by one more step, -0.05M. It was experimentally confirmed that the solubility curve above 2.2 M of Cc comes very close to the low concentration limit of Cp (dotted red line in Figure 2). Crystals were put into protein-free solutions of ammonium sulfate between 2.2 M and 2.7 M of Cc, and kept there for two weeks. It was observed that only a tiny amount of dissolution of these crystals occurred under those high-precipitant concentrations, and when measured by UV ($A_{280}^{1\%} = 9.6$) the protein concentrations turned out to have values of 0.018 mg/ml and 0.021 mg/ml for 2.2 M and 2.7 M of ammonium sulphate, respectively. The resulting crystallization phase diagram for bovine β-lactoglobulin A is shown in Figure 2. The meta-stable region has been clearly determined. The crystallization phase diagram of β-Lg, (in which the protein concentration Cp was kept below 40 mg/ml) was first determined by the dialysis method in order to define meta-stable region. It was found that this meta-stable region was rather wide, a feature which is normally conducive to the growth of a large single crystal. We succeeded in growing a large single crystal in the meta-stable region as described in the next subsection.

2.3. Growth of a large single crystal

The phase diagram which is useful for growing a large single crystal of a protein is illustrated by following the gray arrows of Figure 1. A crystal starts growing in the nucleation region. When Cc is decreased into the meta-stable region, this crystal becomes a seed crystal and continues to grow in the meta-stable region (left gray arrow). As the crystal grows, the protein concentration remaining in solution decreases simultaneously (downwards gray arrow) and reaches the solubility curve (solid curved line) and crystal growth stops. Then Cc is increased to the right boundary (right gray arrow) of the meta-stable region and the crystal growth resumes. When this procedure is repeated, the protein molecules corresponding to the concentration differences between the initial and the final conditions are concentrated into the one crystal and a large crystal is expected to grow (Figure 3). This method has been applied to β-Lg on the basis of the initially determined crystallization phase diagram as shown schematically in Figure 2. The procedure of the Cc change and the photographs of the actual grown crystal are shown in Figure 3. The numerical Cc values and the actual period of days at each precipitant level are indicated in Table 1. In this experiment, 4mg/ml concentration of β-Lg was selected, because low concentration of Cp is easy to control nucleation in comparison with high concentration of Cp. The initial size of the crystal (0.1 mm x 0.1 mm x 0.1 mm) was grown to 1.0 mm x 0.5 mm x 0.4 mm in size eventually, and the crystal was irradiated by using a neutron diffractometer, BIX-4 at JRR-3 in JAEA, and data to 3.0Å resolution was obtained. It is worthy to note that during such a very long crystal-growing period no new nucleation had occurred, and only a single seeded crystal was observed to grow within the nucleation boundaries of the meta-stable region of the phase diagram, and it is proved that the method of growing a large single crystal is promising.

3. Conclusion

1) The crystallization phase diagram of bovine β-lactoglobulin A has been determined by the dialysis method.
2) In the meta-stable region, the large crystal was grown. The broadness of the meta-stable region might help to grow a large crystal, and it is important that in meta-stable region no nucleation occurs and only the seed introduced grows.

Acknowledgments
This work was supported in part by Grant-in-Aid for Scientific Research from the Ministry of Education, Culture, Sports, Science and Technology of Japan, and by a Research Grant funded by the Human Frontier Science Program.

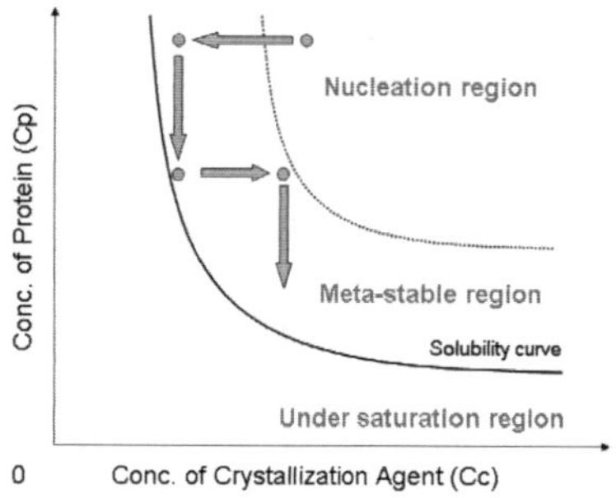

Figure 1. A typical and general crystallization phase diagram. The principle of crystallization of a large crystal is illustrated by the sequence of gray arrows.

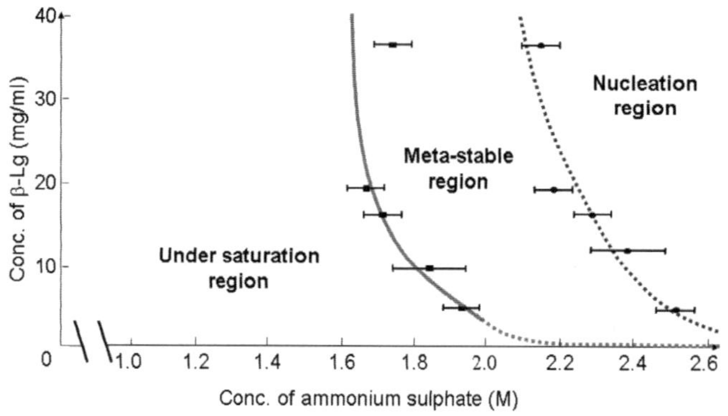

Figure 2. The obtained crystallization phase diagram of bovine β-lactoglobulin A. The meta-stable region is defined by the area between the dotted blue line and the red solid line, the latter of which is often referred to as the "solubility curve". The solubility curve at low Cp concentration has been determined by dissolution of crystals at defined Cc and shown in dotted red line.

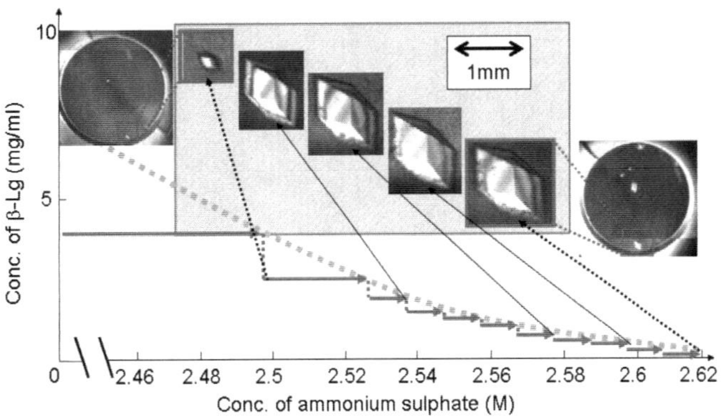

Figure 3. Photographs of a crystal grown stepwise in a systematic way by changing Cc, as indicated by the series of red arrows (also see Table 1 in the text).

Table 1. A period of days at each precipitant concentration level (Cc) for the crystal shown in Figure 3.

Cc (M)	2.50	2.53	2.54	2.55	2.56	2.57	2.58	2.59	2.60	2.61	2.62
Duration (Days)	92	10	3	11	11	9	5	6	7	29	14

References

[1] Dubay K F, Pawar A P, Chiti F, Zurdo J, Dobson C M, 2004 *J. Mol. Biol.* **341** 1317

[2] Niimura N, Arai S, Kurihara K, Chatake T, Tanaka I, and Bau R, 2006 *Cell. Mol. Life Sci.***63** 285

[3] Kesley M G, Oliveria V, Valente-Mesquita , Botelho M M, Sawyer L, Sergio, T. Ferreira. L T, and Polikarpov I, 2001 *Eur. J.Biochem.* **268** 477

Current status of Ibaraki biological crystal diffractometer iBIX -Several examples of the measurement -

K Kusaka[1], T Hosoya[1], I Tanaka[1], N Niimura[1], K Kurihara[2] and T Ohhara[2]

[1] Frontier Research Center for Applied Atomic Sciences, Ibaraki University
162-1 Shirakata, Tokai, Ibaraki, 319-1106, JAPAN
[2] J-PARC Center, Japan Atomic Energy Agency
2-4 Shirakata-Shirane, Tokai, Ibaraki, 319-1195, JAPAN

E-mail: kusakats@mx.ibaraki.ac.jp

Abstract. Since 2004, Ibaraki prefecture has constructed the TOF neutron biological diffractometer (iBIX) at J-PARC for industrial use. Since the end of 2008, Ibaraki University has operated iBIX in order to support the experiment of users and to improve the instruments. The diffractometer is designed to measure samples with their cell edges up to around 150Å. In the beginning of December in 2008, the basic optics, the support system of detectors and the three detectors were completed for the diffraction experiments. We have tried to measure the TOF diffraction data of several proteins and organic compounds in order to estimate the efficiency and characteristics of the diffcactometer. The TOF diffraction data of Ribonuclease A can be measured under the following conditions, beam power: 20kW, pulse repetition: 25Hz, range of wavelength: 0.5~4Å, exposure time: 18.7 hours. About one hundred Bragg reflections could be observed clearly on all detectors. The reflection of 1.58Å in minimum *d*-spacing which $I_{obs}/\sigma(I_{obs}) > 7$ was observed. This result implies that the efficiency of the iBIX will become 50~100 times higher than that of the present high performance diffracatometer BIX-3 in JAEA after 1MW operation of J-PARC.

1. Introduction

Hydrogen atoms and water molecules around proteins play a very important role in the stability of their three-dimensional structure and many physiological functions of them. Recently, the demand on the positional information of hydrogen atoms in a protein molecule has been increasing after many projects in the world which have analyzed thousands of protein structures. Neutron diffraction can provide an experimental method of directly identifying hydrogen atoms in proteins as a technique complementary to ultra-high-resolution X-ray diffraction.

Recently, development of neutron imaging plate (NIP) and the diffractometer equipped with NIP on reactor neutron sources make a breakthrough event in neutron protein crystallography and meet the demand on structural biology in life science field [1]. Although the demand on neutron structural biology in the current life science field has continuously increased, neutron crystallography still remains to this day a severely limited technique.

To meet the increasing demand on neutron structural biology, Ibaraki Prefectural Government in Japan has started to construct a new high performance TOF neutron single crystal diffractometer

(IBARAKI Biological Crystal Diffractometer: iBIX) at the 1MW pulsed neutron source in J-PARC for industrial use since 2004. The diffractometer is designed to measure samples with their cell edges up to around 150Å with a resolution up to 1.2Å in case of biological macromolecules, and a resolution up to 0.7Å in case of organic compounds. The efficiency of the diffractometer would be achieved more than 100 times better than the present high performance diffractometer, BIX-3 (JRR-3, JAEA, JAPAN) [2]. Since the end of 2008, Ibaraki University has operated iBIX in order to support the experiment of users and to improve the instrument. In the beginning of December in 2008, the basic optics, the support system of detectors and the three detectors (a two dimensional position sensitive detectors of ZnS:Ag/$^{10}B_2O_3$ scintillator with wavelength shift fiber system) [3]. In this paper, the current status of iBIX and several examples of measurement are presented.

2. Current status of iBIX

To realize the performance, iBIX has been installed on a coupled moderator which provides more intense peak and integrated intensity of neutrons flux but wider pulse shape than a decoupled one. The optic parameters (beam divergence and the distance between sample and detector: L_2) have been determined to consider the overlapping of Bragg spots by using the original simulation program[4]. A super-mirror guide tube of iBIX was designed based on the determined optic parameter [2]. The curved guide tube can protect samples from high energy γ-ray and neutrons without any T_0 chopper and the well designed tapered guide tube can serve to reduce the reflection number of neutrons on the surface of the mirror so that more neutrons arrive at the sample position. To realize high efficiency, iBIX requires a completely new detector system. A 2-dimensional position sensitive detector using ZnS:Ag/$^{10}B_2O_3$ scintillator with wavelength shift fiber system which has a high spatial resolution, a less dead area, a high counting rate and high efficiency were developed [3].

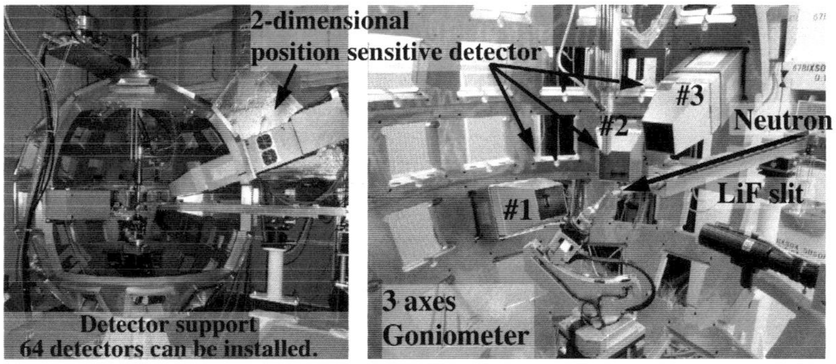

Figure 1. Diffractometer of iBIX.

Since 2004, iBIX has been started to construct at BL03, MLF, J-PARC. In March 2008, the basic optics, the beam line shielding and the experimental shielding house were completed for the commissioning of the beam line. In May 2008, we started commissioning of iBIX under a trial operation. On the 31st of May, we could observe the first neutrons with current-TOF detectors. Spatial- and TOF-profiles of the direct beam at the sample position were measured by using a NIP and a ^{3}He detector. These results showed that a feature of the direct beam is as expected from the simulation based on the design of the supper mirror guide tube. At the beginning of December in 2008, the support system of detectors, which is radial arrangement of detectors, and the three detectors (2θ for the centre of detective area: #1=95.6°, #2=139.0°, #3=156.6°, respectively, A total solid angle subtended by a sample: 1.953%) were installed in the experimental shielding house for the diffraction experiment (Fig.1). Since the end of December in 2008, iBIX has been opened to users.

3. Several examples of diffraction by iBIX

Since the end of December in 2008, we have started to measure the TOF diffraction pattern of several proteins and organic compounds in order to estimate the efficiency and characteristics of iBIX.

3.1. Results of several examples for proteins and organic compounds.

The results of several examples of TOF diffraction pattern measured by iBIX were shown in Table 1. As the curved guide tube of iBIX can protect samples from high energy γ-ray and neutrons, we can select intermediate band of wavelength by adjusting the phase delay of the tail-cutter installed near the moderator. It is possible to select the range of wavelength from 0.5~8Å with about 4Å band width of wavelength. The optimum ranges of wavelength were selected for each sample. In case of organic compounds, minimum *d*-spacing less than 0.5Å could be achieved in PW12. In case of proteins, minimum *d*-spacing 1.58Å could be achieved in Ribonuclease A (RNase A).

Table 1. Results of several examples of TOF diffraction pattern measured by iBIX

Sample	Crystal Volume (mm^3)	Cell Volume ($10^3 Å^3$)	Power (kW)	Exp. Time (h)	Wave length (Å)	d_{min} (Å)
Na_2UMP*[1]	9	12	20	10.0	0.5~4	0.70
PW12*[2]	40	12	20	20.8	0.5~4	0.47
BIPO*[3]	0.052	17	20	13.5	3~7	3.56
Trypsin+BPTI*[4]	2.25	783	20	16.0	4~8	3.80
RNase A*[5]	12.7	63	20	18.7	0.5~4	1.58
Thrombin+BIV*[6]	6	395	20	31.2	4~8	2.73

*[1,2,3]Organic compound, *[4,5,6]protein, *[1]sample provided by Prof. Sugawara, *[2]data given by Prof. Ozeki, *[3]data given by Prof. Kawano, *[6]data given by Dr. Yamada

3.2. The results of TOF-diffraction pattern of RNaseA – the highest resolution of protein samples –

One of the measured crystals is RNase A (RNase A, Crystal Volume=12.5mm^{3,} *a*=30.4Å, *b*=38.6Å, *c*=53.4Å, *β*=105.8°). Measurement conditions are as follow, beam power: 20kW, pulse repetition: 25Hz, range of wavelength: 0.5~4Å (1st frame) and exposure time: 18.7 hours. About one hundred Bragg reflections could be observed clearly on all detectors. The reflection of minimum *d*-spacing in the TOF diffraction pattern of RNase A is shown in Fig. 2. The *d*-spacing of the reflection is 1.58Å and $I_{obs}/\sigma(I_{obs}) > 7$. Although there were some counting fluctuations in the data, a clear Bragg reflection were observed. Indexing of the TOF diffraction spots will be carried out by using new data reduction software "STARGazer" developed for iBIX [5]. It is expected that we can recognize a reflection with a smaller *d*-spacing than 1.58Å on the diffraction pattern of RNase A after indexing.

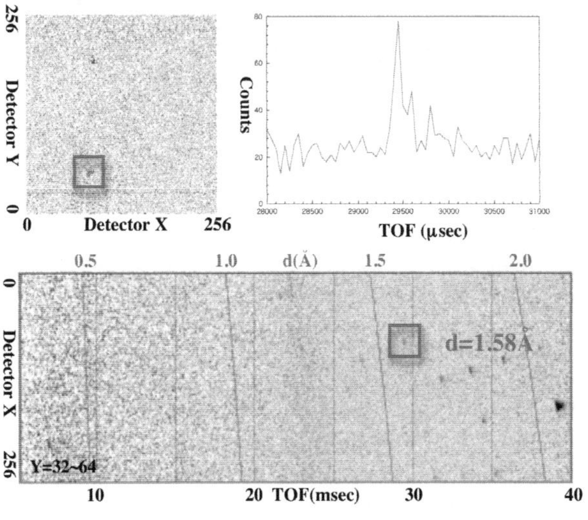

Figure 2. X-TOF slice map, X-Y slice map and TOF-profile of RNase A measured by iBIX with the detector #2 which is arranged at 2θ=137°

3.3. Estimation of measurement efficiency for iBIX under 1MW operation of J-PARC

The measurement efficiency of iBIX under 1MW operation of J-PARC could be estimated from the results of diffraction pattern of RNase A as mentioned above section. This sample could provide high quality diffraction pattern which has minimum *d*-spacing 1.58Å in measurement time 18.7 hours per a setting of crystal orientation even under the operation of low accelerator power in J-PARC (20kW).

The overall completeness of Bragg reflections was calculated using the original simulation program [4] in case of the diffractometer with 64 detectors and 10 measurement settings. The result of overall completeness is achieved over 90% in minimum *d*-spacing 1.2Å. We can calculate total number of crystal settings for each number of detectors based on the simulation result (Table 2). The exposure time for one crystal setting was also estimated simply from the gain of accelerator power (Table 2). Consequently, the progress for measurement efficiency of iBIX according to increasing the accelerator power and the total number of detectors was estimated and summarized in Table 2.

Under 1MW operation of the accelerator and 30 detectors available, we can measure the full data set of the RNaseA sample which has a standard volume of 1mm^3 in about 4 days. The estimation of the measurement efficiency as mentioned above was assumed that Laue group of crystal is $\bar{1}$. If Laue symmetry of a measurement sample becames higher, we can expect that the total measurement time becomes shorter. Generally, the resolution becomes worse according to the increase of the unit cell volume. The total measurement time of the sample of a large unit cell volume becomes longer than that of the sample of a small unit cell volume.

Currently, this estimation implies that the measurement efficiency of the iBIX will become about 50~100 times better than that of the present high performance diffractometer BIX-3 after 1MW operation of J-PARC. More accurate estimation must be curried out after the accelerator power increases.

Table 2. Measurement efficiency of iBIX estimated from TOF diffraction data of RNase A (Crystal Volume: 12.5mm^3, a=30.4Å, b=38.6Å, c=53.4Å, β=105.8°)

Date	Power (kW)	No. of detectors	Exp. time (h/frame)	Total no. of setting	Total exp. Time (days/sample)
Feb., 2008~	20	3	18.7	10*(64/3)=213	166
Apr., 2009~	20	14	18.7	10*(64/14)=46	35.8
2012?~	1000	14	18.7/50=0.37	10*(64/14)=46	0.72
2013?~	1000	30	18.7/50=0.37	10*(64/30)=21	0.33
Crystal volume = 1mm^3 (Standard volume for iBIX), Total exposure time= 4days					

References

[1] Niimra N and Bau R 2008 Acta Cryst. A **64** 12
[2] Tanaka I, Kusaka K, omoyori K, Niimura N, Ohhara T, Kurihara K, Hosoya T and Ozeki T 2009 NIMA A **600** 161
[3] Hosoya T, Nakamura T, Katagiri M, Tirumachi A,Ebine M,Soyama K 2009 NIMA A **600** 217
[4] Kusaka K,Ohhara T, Tanaka I, Niimura N, Ozeki T, Kurihara K, Aizawa K, Morii Y, Arai M, Ebagta K, Takano Y 2006 Physiva B **385-286** 1062
[5] Ohhara T, Kusaka K, Hosoya T, Kurihara K, Tomoyori K,Niimura N, Tanaka I, Suzuki J, Nakatani T,Otomo T, Matsuoka S, Tomita K, Nishimaki Y, Ajima T, Ryuuhuku S 2009 NIMA **600** 195

A preliminary neutron diffraction analysis of Achromobacter protease I

Yuki Ohnishi[1], Takeharu Masaki[2], Taro Yamada[3], Kazuo Kurihara[4], Ichiro Tanaka[1], Nobuo Niimura[3]*

[1]Faculty of Engineering, Ibaraki University, 4-12-1, Nakanarusawa, Hitachi, Ibaraki 316-8511, Japan.
[2]Faculty of Agriculture, Ibaraki University, 3-21-1 Chuo, Ami, Ibaraki 300-0393, Japan.
[3]Frontier Research Center of Applied Atomic Sciences, Ibaraki University, 162-1 Shirakata, Tokai, Naka, Ibaraki 319-1106, Japan.
[4]Quantum Beam Science Directirate, Japan Atomic Energy Agency, 2-4 Shirakata-shirane, Tokai, Naka, Ibaraki 319-1195.

E-mail:niimura@mx.ibaraki.ac.jp

Abstract.Achromobacter protease I (API, E.C. 3.4.21.50) is one of the serine proteases produced by *Achromobacter lyticus* M497-1. API is distinct from the other tripsin type protease in its lysine specificity. The neutron structure analysis of catalytic triad with Trp169 and His210 was presented. His57 was double protonated and formed hydrogen bonds to Ser194O^{γ} and Asp113$O^{\delta 1}$, $O^{\delta 2}$.

1. Introduction

Achromobacter protease I (API, E.C. 3.4.21.50) is one of the serine proteases isolated from soil bacterium, *Achromobacter lyticus* M487-1 [1, 2]. API specifically hydrolyzes the peptide bonds at the carboxyl side of lysine residues. When API is compared with tripsin type protease, API has several characteristics such as lysine specificity, a higher peptidase activity, a wide pH optimum ranging from 8.5 to 10.7 and stability against denaturation with 4M urea and 0.1% sodium dodesyl sulfate, respectively [3]. From the X-ray structure analysis of API (protein data bank ID: 1arb), imidazole-indole stacks between Trp169 and His210, which is closed to catalytic triad residues, contributes to reduces a space at the active site [4]. This structure seems to be a high catalytic potentiality of lysine specificity. To elucidate these results in detail by observing protons, hydrogen bonds and hydration structure of the protein, we have carried out neutron diffraction experiment. In this study, structure of API was determined by neutron diffraction data. The preliminary results of neutron structure analysis of the catalytic triad with Trp169-His210 will be presented.

* To whom any correspondence should be addressed.

2. Materials and Methods

API was purified as reported previously [2]. Purified API was stored at 277K and concentrated to 15.3 mg/mL before crystallization.

Crystallization was performed by vapor diffusion method followed by modified macro-seeding technique which was used as a large crystal growth [5]. The neutron diffraction experiment was carried out at room temperature by the single-crystal diffractometer BIX-4, which is installed at the JRR-3 reactor of the JAEA [6]. A step scanning method with an interval of 0.3° was used to collect data and the exposure time was 2 hours per frame. The intensities of Bragg reflections were integrated and scaled by the program *DENZO* and *SCALEPACK* [7]. The neutron crystal structure refinement was performed by using *PHENIX* for joint XN refinement [8] and *Coot* [9] for model building. In order to obtain the initial model for neutron structure analysis, an X-ray structure analysis was carried out by using another crystal which was grown under the same conditions and the X-ray diffraction data were collected by MacScience DIP2000 equipped with x-ray generator M06CEXG-SRA using Cu rotating anode. The neutron and X-ray data collection and joint XN refinement statistics were summarized in Table 1.

3. Results

Catalytic triad with Trp169 and His210

Nuclear density map of the catalytic triad with Trp168 and His210 is shown in Figure 1. His57 was doubly protonated as deuterium atom, and hydrogen bonds were observed between His57$D^{\varepsilon 2}$ and Ser194O^{γ}, His57$D^{\delta 1}$ and Asp113$O^{\delta 1}$, and His57$D^{\delta 1}$ and Asp113$O^{\delta 2}$. The distances of these hydrogen bonds are 2.1, 2.0 and 2.6 Å, respectively.

In this report, we have demonstrated the preliminary neutron diffraction analysis of API. The structure of catalytic triad involving Trp169 and His210 has determined. The appropriate resolution of the data was 2.0 Å, because the completeness of 20-4.0 Å, 4.0-2.0 Å and 2.0-1.86 Å shells are 96.4%, 69.0% and 20.1%, respectively, after the recent re-processing of data. For this refinement, however, 1.86 Å neutron data was used to obtain the fine Fourier map. The further refinement is under way.

Acknowledgments

This work was supported in part by a Grant of Regional Consortium by Kanto Economical and Industry, a "Grant-in-aid for scientific research" from the Ministry of Education, Culture, Sports, Science and Technology of Japan, a common-use facility of Japan Atomic Energy Agency, and a research grant funded by the Human Frontier Science Program.

Table 1. Data collection and refinement statistics

	Neutron	X-ray
Space group	*P1*	*P1*
Unit cell (Å)	a = 39.755, b = 40.792, c = 43.991 α = 64.83, β = 66.07, γ = 73.41	a = 39.755, b = 40.792, c = 43.991 α = 64.83, β = 66.07, γ = 73.41
Beam port/radiation	1G, JRR-3	MacScience M06XCEXG-SRA
Wavelength (Å)	2.60	1.54
Diffractometer	BIX-4	DIP2020
Exposure time (min/fr)	118	5
Total frames	823	390
Resolution (Å)	80 - 1.86	80 - 1.90
Observed reflection	18557	58252
Unique reflection	11068	15308
I/σ (***I***)	3.9 (2.2)	21.6 (5.7)
Completeness (%)	58.9 (19.5)	88.4 (79.8)
$^aR_{merge}$	16.7 (27.3)	6.1 (16.3)
^{b}*Rcrysta/Rfree (%)*	22.90 / 28.34	14.71 / 20.56

^{a} $R_{merge} = \sum | I - \langle I \rangle | / \sum I$, where I is the intensity of a reflection and $\langle I \rangle$ is the average intensity.
^{b}R=R_{free}=$\sum||F_o|-|F_c||/\sum|F_o|$ calculated for the reflections of working and test sets, respectively.
Values in parentheses are for the highest resolution shell.

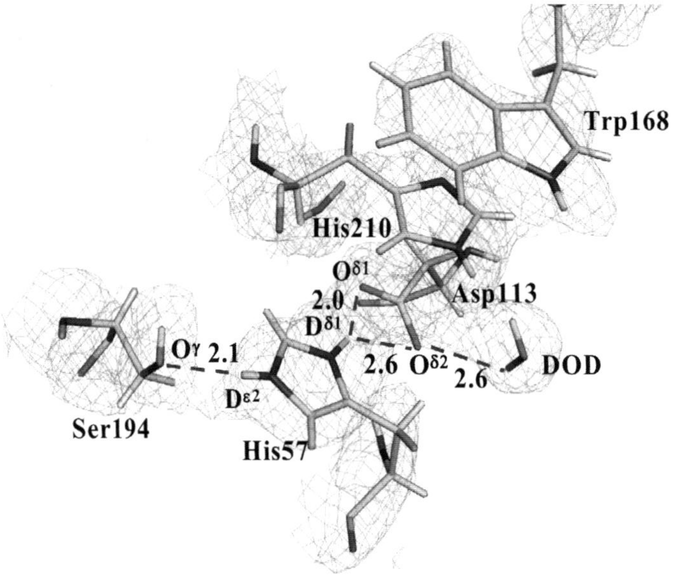

Figure 1. Nuclear density map for catalytic triad of API with π-π stacking Trp168 and His210. Map colour of $2|F_o|$-$|F_c|$ for model is grey.

References

[1] Masaki T, Nakamura K, Isono M, Soejima M 1978 *Agric. Biol. Chem.* **42**, 1443-1445.
[2] Masaki T, Fujihashi T, Nakamura K, Soejima M 1981 *Biochim. Biophys. Acta.* **660**, 51-55.
[3] Masaki T, Tanabe M, Nakamura K, Soejima M 1981 *Biosci. Biotechnol. Biochem.* **56**, 1604-1607.
[4] Shiraki K, Norioka S, Li S, Sakiyama F 2002 *J. Biochem.* **131**. 213-218.
[5] Ohnishi Y, Masaki T, Yamada T, Tanaka I, Niimura N (To be published)
[6] Kurihara K, Tanaka I, Refai-Muslih M, Ostermann A, Niimura N 2004 *J. Synchr. Rad.* **11**, 68-71.
[7] Otwinowski, Z. & Minor, W. (1997). *Methods Enzymol*. **276**, 307–326.
[8] Adams P D, Mustyakimov M, Afonine P V, Langan P 2009 *Acta Cryst.* **D65**, 567-573.
[9] Emsley P, Cowtan K, 2004 *Acta Crystallogr. D.* **60**, 2126-2132.

International Conference on Neutron Scattering 2009 IOP Publishing
Journal of Physics: Conference Series **251** (2010) 012033 doi:10.1088/1742-6596/251/1/012033

A preliminary neutron crystallography on the trypsin-bovine pancreatic trypsin inhibitor complex

K Kawamura[1], T Yamada[2], K Kurihara[3], T Tamada[3], R Kuroki[3], I Tanaka[4,2], H Takahashi[6] and N Niimura[2,5]

[1]Institute of Applied Beam Science, Graduate school of Science and Engineering, Ibaraki University, 4-12-1 Naka-Narusawa, Hitachi, Ibaraki, 316-8511, Japan
[2]Frontier research center for applied atomic sciences, Ibaraki University, Ibaraki Quantum Beam Center, 162-1 Shirakata, Tokai, Ibaraki 319-1195, Japan
[3]Molecular Structural Biology Group, Quantum Beam Science Directorate, Japan Atomic Energy Agency, 2-4 Shirakata-Shirane, Tokai, Ibaraki 319-1195,Japan
[4]Department of Materials Science, the College of Engineering, Ibaraki University, 4-12-1 Naka-Narusawa, Hitachi, Ibaraki 316-8511, Japan
[6]Institute of Applied Beam Science, Ibaraki University, 4-12-1 Naka-Narusawa, Hitachi, Ibaraki, 316-8511, Japan

E-mail: niimura@mx.ibaraki.ac.jp

Abstract. Trypsin is one of serine proteases. BPTI (Bovine Pancreatic Trypsin Inhibitor) is a protein inhibitor, which binds trypsin tightly and inhibits cleavage of peptide bonds. X-ray structure determination of trypsin-BPTI complex could make clear the overview of the active site. However, information of hydrogen atoms related to catalytic mechanism has not been satisfied. In this study, the trypsin-BPTI complex structure has been determined by neutron diffraction data at 2.0 Å resolution. Deuterium atoms of catalytic triad, hydration structures in the binding pocket of trypsin and hydrogen bonds were observed. We would like to discuss details of hydrogen bonds in the interface between trypsin and BPTI and the adjacent water molecules including hydrogen atoms involved in the enzymatic reaction.

1. Introduction

Hydrogen atoms of dissociable groups and the hydrate structures, deeply related to many chemical reactions in biological systems of several proteins, are directly observed by neutron diffraction experiments. In the active site, trypsin has three dissociable amino acids of His57, Asp102 and Ser195 (called catalytic triad), recognizes specifically the basic amino acid residues Lys or Arg and catalyzes the hydrolysis of peptide bond. The catalytic mechanism is explained by hydrogen bonds related to proton transfer among residues in the triad. The overview is as follows: Asp102 increases the basicity of the $N^{\varepsilon 2}$ of the imidazole of His57 by forming hydrogen bonding between the COO^- of Asp102 and $N^{\delta 1}H$ of the imidazole of His57. His57 then abstracts a proton from Ser195OH, thereby promoting its nucleophilic attack on the carbonyl carbon of the scissile peptide bond. The nucleophilic attack of Ser195O^{γ} results in formation of a tetrahedral intermediate of the peptide. In the tetrahedral

[5] Author to whom any correspondence should be addressed.

intermediate, the carbonyl oxygen ion of the scissile peptide forms hydrogen bonds with the backbone NH groups of Gly193 and Ser195 [1]. To make clear the reason, it may be interesting to reveal these positions of hydrogen atoms in the active site. Also, to prevent cell destruction by autolysis of trypsin, pancreatic trypsin inhibitors invariably exists within organisms. Bovine pancreatic tryspin inhibitor (BPTI) has high inhibition effect for trypsin [2] [3]. The 58-residue BPTI has Lys15 (P1) whose side chain interacts electrostatically with Asp189 of specific binding site in trypsin. Several extensive studies of trypsin-BPTI complex have been performed at the atomic level [4]. To date, structures of trypsin-MIP (monoisopropyl ester phoshonic acid) complex [5][16] and BPTI alone [6] have been determined by neutron protein crystallography, but that of the trypsin-BPTI complex has not yet been determined. In this study, the trypsin-BPTI complex structure was determined by neutron diffraction data, and hydrogen atoms of the catalytic triad, hydration structures in the binding pocket of trypsin and hydrogen bonds were observed.

2. Materials and Methods

Recombinant bovine β-trypsin from corn and BPTI (trypsin inhibitor from bovine pancreas, type I-P) were purchased from Acris Antibody GmbH and Sigma-Aldrich, respectively. The BPTI was mixed with bovine β-trypsin in a 1.25:1 molar ratio [7] in a solution of 20 m*M* HEPES buffer (pH 7.5). The protein concentration was adjusted 55 mg ml^{-1}. Crystallization was carried out by the sitting-drop method. The droplet was contained 1:1 mixture of protein solution and reservoir solution. The reservoir solution was 1.45 *M* ammonium sulfate, 200 m*M* HEPES buffer (pH 7.5), and 10 m*M* calcium chloride. The crystal grew to 2.2 mm^3 (2.0 mm × 1.1 mm × 1.0 mm) for about 2 weeks. Thereafter, it was soaked in a D_2O solution at pD 7.9 for two weeks.

The neutron diffraction experiment was carried out at room temperature with the single-crystal diffractometer BIX-4, which is installed at the JRR-3 reactor of the JAEA [8]. A step scanning method with an interval of 0.3° was used to collect data and the exposure time was 4 hours per frame. The intensities of reflections were integrated and scaled by the program *DENZO* and *SCALEPACK* [9]. The statistics for the data reduction are summarized in Table 1. This neutron experiment was performed under the Common-Use Facility Program of JAEA (2008A-A06).

The neutron crystal structure refinement was carried out by a version of *CNS* [10] called *nCNS* modified for neutron refinement [14] and the graphics program was *COOT* [11]. To obtain an initial model for refinement, we solved the structure from X-Ray data. This data was collected from another crystal grown the same condition soaked in D_2O solution, refined by *REFMAC* of *CCP4* packages [12] (initial model is PDB ID: 2PTC) [13]. In our neutron structure analysis, we included reflection data from 80 to 2.0 Å resolution in the neutron map calculation. The protonation or deprotonation states of dissociable groups were determined by difference Fourier maps for a model with omitted partial deuterium atoms.

3. Results and discussion

As the result of neutron structure refinement, the $R_{crystal}$ is 20.71 %, and R_{free} is 25.48 % at resolution of 2.0 Å. The deuterium atom between His57$N^{\varepsilon2}$ and Ser195O^{γ} was observed as shown in Figure 1(a) and the initial stage of deprotonation from the serine was confirmed. The nuclear density map of the specific binding area of the BPTI is indicated in Figure 1(b). The refinement statistics for neutron structure of trypsin-BPTI complex are listed in Table1.

3.1. Catalytic triad

A hydrogen atom binding His57$N^{\delta1}$ was exchanged for deuterium. The His57$D^{\delta1}$- Asp102$O^{\delta1}$ and -Asp102$O^{\delta2}$ distances were 2.5 Å and 2.0 Å, respectively. A deuterium atom was identified between His57$N^{\varepsilon2}$ and Ser195O^{γ}. The distance between His57$N^{\varepsilon2}$-Ser195O^{γ} was 2.6 Å and the distances from the observed nuclear density map peak position of the deuterium atom to His57$N^{\varepsilon2}$ and Ser195O^{γ} were about 1.4 Å and 1.2 Å, respectively. These hydrogen bonds are short distance, which might facilitate the proton transfer during the catalytic mechanism. [15].

3.2. Specific binding site S1 Asp189 and P1 Lys15

In the X-ray result of trypsin-BPTI complex, two water molecules (oxygen atoms) have been assigned around ε-amine of Lys15, but the hydrogen atoms of the water molecules were identified. The hydrogen atoms of these water molecules were determined by neutron crystal structure analysis, and the exact hydrogen bonds scheme (donor and acceptor relations) were assigned clearly in this binding area as show in Figure 1(b). These hydrogen bond lengths indicated in Figure 1(b) were about 1.9 Å. The water molecules were locked tightly by the hydrogen bond.

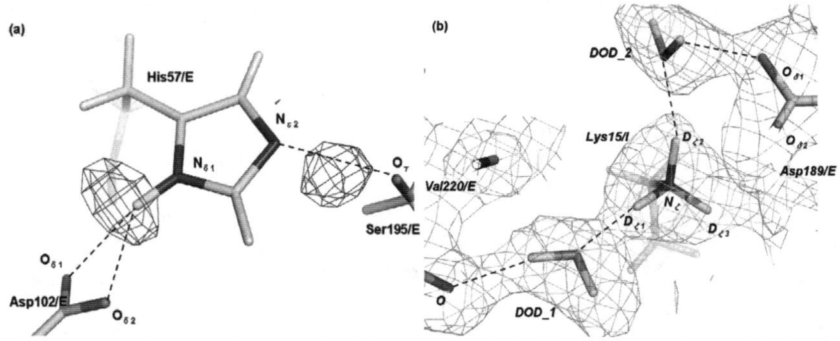

Figure 1. Nuclear density map for (a) catalytic triad of trypisn and (b) specific binding site Asp189, Lys15 of inhibitor and two deuterium waters. Map colour of $|Fo|$-$|Fc|$ map (3.0 σ) for model omitted deuterium atoms of catalytic triad is blue in Fig (a). $2|Fo|$-$|Fc|$ map (1.5 σ) for model included all hydrogen and deuterium atom is green colour in Fig (b). Hydrogen bonding is located with dot-line in this figure (a) and (b).

3.3. Waters are present in trypsin-BPTI binding area

Although the refinement is still underway, we could discuss the presence of water into the active site because the trypsin-BPTI complex crystal used in neutron diffraction experiment is soaked in D_2O solution. In general, it has been thought that water molecules cannot enter the active site of trypsin because the BPTI-binding loop is in close contact with the trypsin-binding area [1] [13]. In this study, hydrogen atoms in side chains of the catalytic triad were exchanged to deuterium atoms (occupancy of His57 Dδ1 was 1.0) and also waters in specific binding area were exchanged to deuterium waters, which means that water molecules invaded in active site.

Table 1. Statistic data collection, reduction and refinement

Data collection		Refinement	
Space group	*I*222	Resolution (Å)	80-2.0
Unit cell parameter (Å)		$R_{crystal}$ (%)	20.7
a	75.8	R_{free} [b] (%)	25.5
b	85.6	R_{free} test set (%)	10
c	123.1		
Beam port/radiation	1G, JRR-3		
Wavelength (Å)	2.6		
Diffractometer	BIX-4		
Exposure time (h/fr)	4		
Total frames	275		
Resolution (Å)	80-2.0		
Observed reflection	48,815		
Unique reflection	23,822		
I/σ (***I***)	4.8 (2.7)		
Completeness (%)	86.7 (80.1)		
R_{merge} [a]	0.162(0.248)		

[a] $R_{merge} = \sum | I - \langle I \rangle | / \sum I$, where I is the intensity of a reflection and $\langle I \rangle$ is the average intensity.
[b] $R = R_{free} = \sum||F_o| - |F_c|| / \sum|F_o|$ calculated for the reflections of working and test sets, respectively.
Values in parentheses are for the highest resolution shell.

References

[1] Hedstrom, L., (2002). *Chem. Rev.* **102**, 4501-4523

[2] Vincent, J.P. & Lazdunski, M. (1972). *Biochemistry*, **11**, 2967-2977

[3] Quast, U., Engel, J., Steffen, E., Tschesche, H. & Kupfer, S. (1978). *Biochemistry*, **17**, 1675-1682

[4] Grzesiak, A. Helland, R. SmalaÊs, O.A., Krowarsch, D., Dadlez, M. and Otlewski, J. (2000). *J. Mol. Biol.***301**, 205-217

[5] Kossiakoff, A.A. & Spencer, S.A. (1980). *Nature.* **288**, 414–416

[6] Wlodawer, A., Walt, J., Huber, R., Sjölin, L. (1984). *J. Mol. Biol.* **180**, 301-29

[7] Hanson, W.M., Domek, G.J., Horvath, M.P., Goldenberg, D.P. (2007). *J. Mol. Biol.* **366**, 230-243

[8] Kurihara, K., Tanaka, I., Refai-Muslih, M., Ostermann, A and Niimura, N. (2004). *J. Synchr. Rad* **11**, 68-71

[9] Otwinowski, Z. & Minor, W. (1997). *Methods Enzymol.* **276**, 307–326

[10] Brünger, A.T., Adams PD, Clore GM, DeLano WL, Gros P, Grosse-Kunstleve RW, Jiang J-S, Kuszewski J, Nilges M, Pannu NS, Read RJ, Rice LM, Simonson T, Warren GL. (1998) *Acta Cryst D,* **54,** 905–921.

[11] Emsley, P., and Cowtan, K. (2004) *Acta Crystallogr. D. Biol. Crystallogr.* **60**,2126– 2132

[12] Collaborative Computational Project, Number 4. (1994) *Acta Crystallogr., Sect. D* **50**, 760-767

[13] Marquart, M., Walter, J., Deisenhofer, J., Bode, W., & Huber, R. (1983). *Acta Cryst.* **B39**, 480-490

[14] Adams, P.D., Mustyakimov, M., Afonine, P.V., Langan, P.,(2009). *Acta Crystallogr D.*, **65**,567-573.

[15] Yamaguchi, S.,Kamikubo, H., Kurihara, K., Kuroki, R., Niimura, N., Shimizu, N., Yamazaki, Y. and Kataoka, M. (2009) *Proc. Nat. Acad. Sci.* **106**, 440-444

[16] Kossiakoff, A.A. & Spencer, S.A. (1981). *Biochemistry.* **20**, 6462-6474

Neutron scattering study of a membrane phase miscibility gap: Coexistence of L_3 "sponge" and L_α Lamellar phases

W A Hamilton[1,†] **and L Porcar**[2]

[1]Bragg Institute, Australian Nuclear Science and Technology Organisation, Lucas Heights, Australia

[2]Large Scale Structures Group, Institut Laue-Langevin, Grenoble, France

Email: Hamilton.WA@gmail.com

Abstract. We report on a small angle neutron scattering (SANS) study of a temperature driven first order phase transition in a 25wt% solution of the surfactant AOT (Sodium Di-2-ethylhexyl Sulfosuccinate) in 1.5wt% heavy brine between an isotropic L_3 "sponge" state at 27°C and a stacked lamellar L_α monophase 55°C. The prominent scattering features of these phases are correlation peaks due to the mean passage size of the L_3 sponge and the L_α stacking separation. This ratio of the monophase peak positions $Q_\alpha/Q_3 \approx 1.3$, is consistent with previous observations in this and similar systems. In the present study we tracked this system through the intermediate $L_3 + L_\alpha$ biphasic miscibility gap. There the initial appearance of the L_α peak at 33.25°C was at a scattering vector some 23% higher than the final high temperature monophase value. During coexistence both L_3 and L_α phase peak positions decreased linearly with increasing temperature maintaining a roughly constant ratio Q_α/Q_3 ~1.6-1.7. Two phase fits to the scattering data and application of scaling law predictions allow us to obtain local L_3 phase volume fractions in the biphasic region and make preliminary determinations of the structural accomodations necessitated by phase coexistence in this system's miscibility gap.

1. AOT membrane phases and scaling laws

AOT (Sodium Di-2-ethylhexyl Sulfosuccinate) is a double tailed anionic surfactant which displays a variety of self-assembled mesophases in aqueous solution[1]. Over a range of concentrations and temperatures in brine solution AOT molecules sell-assemble into bilayer membranes with the surfactants headgroup on the exterior shielding the hydrophobic tails from the solvent. Among the mesophases that may be formed by arrangement of these membranes are the familiar anisotropic L_α lamellar phase in which membranes are ordered as separated stacks and the isotropic L_3 bicontinuous "sponge" phase in which the membranes are connected by random passages to form an isotropic network that spans the solution[2]. These topological differences result in distinct material properties: L_α phases are relatively viscous and optically birefringent, while L_3 phases are free flowing and optically inactive. In small angle neutron and x-ray scattering (SANS and SAXS) measurements the L_α phase gives rise to sharp quasi-Bragg scattering peak due to the stacking separation, while for L_3 a broader correlation peak characteristic of the mean passage size is observed. These morphologies are

† To whom any correspondence should be addressed

often found adjacent to one another in the phase diagram, with the phase behavior being driven by the intrinsic curvature of the bilayer, which may vary with temperature (as in the current case), pressure, membrane or solvent composition.

Over a broad range of dilution these phases follow a "sheet" scaling equation where the membrane volume fraction ϕ is proportional to the membrane thickness δ divided by the characterstic length scale of the structure d, to a constant of proportionality C depending on their morphology. For rigid flat membranes stacked with period d we can see from Figure 1 that $C\equiv1$. Surfactant bilayers are not totally rigid, so ripples diminish the projected area normal to stacking and membranes are more distant for a given ϕ so C is a little greater than unity. Similarly the L_3 phase may be considered as a distorted version of the Shwartz "P" regular cubic structure which has a constant negative Gaussian sheet curvature. For this structure with d as the passage width a value $C\approx1.486$ has been estimated numerically[3]. For sponge phases the width of their scattering correlation peak indicates a greater variation of local curvature and passage size so we may expect C to be somewhat greater than 1.5.

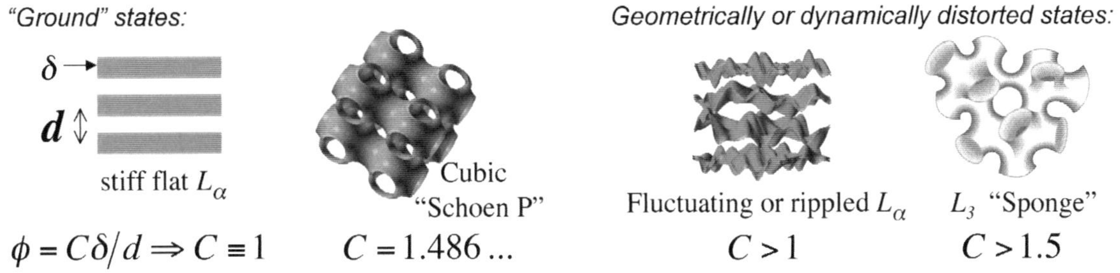

Figure 1. Membrane morphology and sheet equation coefficent of proportionality.

2. Scattering measurements over the L_3 to L_α phase transition miscibilty gap

In this paper we briefly report on a small angle neutron scattering (SANS) study of 25wt% solution of the surfactant AOT in 1.5wt% heavy brine between 27°C and 55°C over which range decreasing intrinsic curvature of the surfactant bilayer drives the system from an initial L_3 state to a final L_α state. In an intermediate miscibility gap region microphase regions of the L_α phase nucleate within the L_3 phase and we find coexistence of the two morphologies. Our measurements were taken on the SAND time-of-flight (TOF) SANS instrument at the Intense Pulsed Neutron Source (IPNS) at Argonne National Laboratory (ANL). SANS measurements of 30m duration were taken on a 1mm thick sample in a quartz cell in 4°C steps (controlled to about 0.1°C), and in finer steps over over the range ~31–34°C where the onset of the transition was expected from observed optical birefringence.

Figure 2(a) shows SANS over the full scattering vector Q range measured (to about 1.0 Å^{-1}) for the L_3 and L_α monophases at 27°C and 55°C. At 27°C the broad L_3 correlation peak at Q=0.062 Å^{-1} corresponds to a mean passage size of 101 Å. At 55°C the narrower quasi-Bragg L_α peak at Q=0.079 Å^{-1} indicates a membrane stacking separation of 79 Å. This ratio of monophase peak positions and inversely correlation lengths $Q_\alpha/Q_3 = d_3/d_\alpha = 1.27$ is consistent with previous observations in this and similar systems (for instance [4] and [5]). Figure 2(b) shows the evolution of the scattering in the correlation peak region. As the temperature was increased we observed no significant change until the biphasic boundary is crossed, signalled in our data by the first detectable appearance of an L_α peak at 33.25°C at Q=0.103 Å^{-1} some 23% higher than the final monophase value at 55°C and indicating an initial L_α separation of only 61 Å. Both peaks then move to lower Q, with the L_3 peak weakening, to become undetectable in our 55°C data, as the L_α peak becomes stronger.

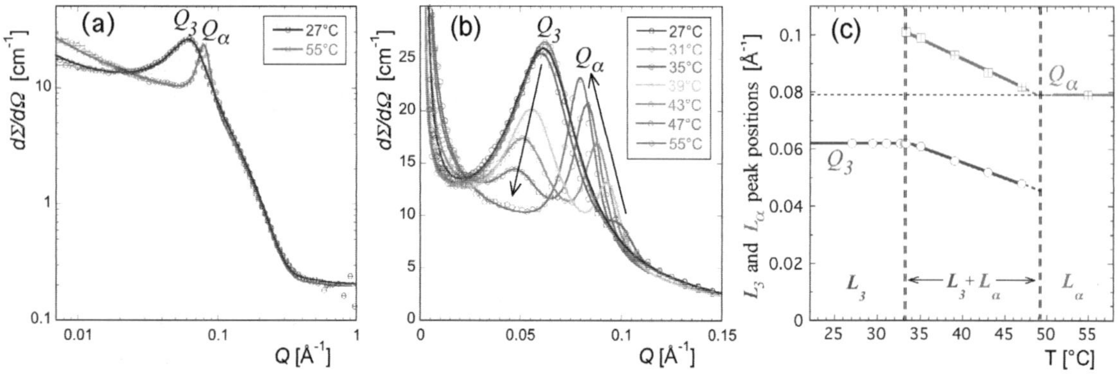

Figure 2. Data and fits for 25wt% AOT in 1.5wt% heavy brine. **(a)** SANS for 27°C L_3 and 55°C L_α monophases. **(b)** SANS as transition proceeds. **(c)** Q_3 and Q_α peak position versus temperature.

3. Scattering data fitting

Fits to our SANS measurements shown were obtained using the scattering function:

$$\begin{aligned}\frac{d\Sigma}{d\Omega}[Q] &= NV^2(\Delta\rho)^2 P[Q]\, S[Q] + I_{bg} \\ &\cong \left(\frac{\pi\phi\sigma^2(\Delta\rho)^2}{(t_o+t_h)}\right)\left(\frac{4\left(1-\cos[Qt_o]\exp\left[-Q^2\delta_o^2/2\right]\right)}{Q^2\left(Q^2\sigma^2+2\exp\left(-Q^2\sigma^2/6\right)\right)}\right)\left(1+\frac{A_3}{Q}+\frac{B_3}{1+(Q-Q_3)^2\xi_3^2}+\frac{B_\alpha}{1+(Q-Q_\alpha)^2\xi_\alpha^2}\right)+I_{bg}\end{aligned} \tag{1}$$

The first two terms in bracesof the second line capture in an analytic form[6] the number density N, scattering power (as the square of volume multiplied by the neutron contrast) and form factor (P) of randomly oriented discoids of membrane of rms radius σ, of contrast (tail) thickness t_o with and bilayer surface (headgroup) thickness t_h with a rms variation in thickness δ_o, which are arranged in mesophases described by the third term, a structure factor (S). This coherent scattering sits on an incoherent background I_{bg} (0.19 cm^{-1} from our fits). For measurements on a single sample the first braced term is a constant scaling parameter. From our fits we found t_o=15.5±0.5 Å with δ_o, about one quarter of this thickness ~3.8 Å, typical of surfactant membranes[7]. This second form factor term dominates at higher Q, where we observe a shoulder at Q~0.18 Å^{-1} as the scattering changes from ~Q^{-2} behavior characteristic of membrane scattering to the ~Q^{-4} dependence of surface scattering (see Figure 2(a)). This feature and a slight shift to lower Q with increasing temperature was reproduced by a gradual increase in the form factor disc radius σ from 11.0 Å at 27°C to 12.7 Å at 55°C, as we might expect going from the highly curved L_3 to the flatter L_α phase.

In our structure factor[8] the L_3 and L_α peaks are modelled as Lorentzians at Q_3 and Q_α with correlation lengths ξ_3 and ξ_α. The A_3/Q is term is intended to fit long range "in-out" scattering attributed the L_3 membrane's division of the solvent volume[9]. This term was adequate to fit the obvious increase in low Q scattering as the transition proceeds adequately, due perhaps to L_α phase domain formation, over this Q range. Our fits to this series were all averaged over resolution of the reduced TOF SANS data (see horizontal error bars in Figure 2(a)), which maintains a constant ratio of variance to Q (i.e. σ_Q^2/Q~0.0003 Å^{-1}) but does not significantly broaden any scattering features[10]. Some parameters obtained from our data series fits are presented in Table 1.

4. Preliminary results and interpretation

Figure 3(a) graphs the variation in L_3 and L_α peak positions Q_3 and Q_α peak positions showing their linear variation with temperature over the miscibility gap. Extrapolating this variation of Q_α to it's monophase limit at 55°C we may determine that the upper boundary of L_3+L_α coexistence ocurrs at 49°C. In the L_3+L_α region the ratio Q_α/Q_3 gradually increases from 1.61 to 1.71 between 33.25°C and the last measurement in this region at 47°C. We note that this is close to the "ground state" value for Q_α/Q_3 we might infer from Section 1. Assuming the scaling relation holds for L_3 over the biphasic region, which seems reasonable as the normalized correlation width of the L_3 structure factor peak is constant at $(\xi_3 Q_3)^{-1}$~0.30 over it, we may extract the local volume fraction within the L_3 phase domains, ϕ_3[T]. This will simply be proportional to $1/d_3[T] \propto Q_3[T]$. We present ϕ_3[T] in Table 1 - normalized to the L_3 monophase value at 27°C. In principle this value can be used to determine the "tie-line" boundaries of L_3+L_α coexistence versus temperature in the system phase diagram over a range near the current volume fraction. Another value which may be extracted is the effective "matching angle" $\theta \equiv \sin^{-1}(d_\alpha/d_3)$ between the mesophase bilayer periodicities at their microphase boundaries. We see that this angle is ~37° over the biphasic region in our system. This is rather lower than the ~50–60° deduced by Quillet *et al.*[4] for L_3+L_α coexistence in the CetylPyridinium Chloride/Hexanol membrane system. It is possible that this difference and the behavior of the biphasic ratio of Q_α/Q_3 may be due to the greater stiffness of and local curvature constraints on bilayers formed by AOT as a double tailed surfactant[11].

Table 1. Structure factor fitting parameters and derived local sponge volume fraction and phase contact angle values obtained from our data for the L_3 to L_α phase transition over the measured range. The Lorentzian widths of correlation peaks ξ_3^{-1} and ξ_α^{-1} have been normalized to peak position values.

T [°C]	A_3 [cm^{-1}]	B_3 [cm^{-1}]	Q_3 [Å^{-1}]	$(\xi_3 Q_3)^{-1}$	$\phi_3[T]/\phi_3[27°C]$ $\approx Q_3[T]/Q_3[27°C]$	B_α [cm^{-1}]	Q_α [Å^{-1}]	$(\xi_\alpha Q_\alpha)^{-1}$	Q_α/Q_3 $= d_3/d_\alpha$	θ [°] $\equiv \sin^{-1}(d_\alpha/d_3)$
27	0.008	3.5	0.062	0.34	1.00	0.0	-	-	-	-
31	0.004	3.5	0.064	0.29	1.02	0.0	-	-	-	-
33.25	0.004	3.5	0.064	0.28	1.02	0.2	0.103	0.05	1.61	38.4
35	0.005	3.2	0.063	0.29	1.00	0.7	0.100	0.057	1.61	38.4
39	0.006	2.1	0.057	0.28	0.92	1.9	0.094	0.068	1.65	37.3
43	0.009	1.4	0.053	0.30	0.85	2.6	0.089	0.079	1.68	36.5
47	0.014	0.8	0.049	0.3	0.80	3.3	0.084	0.089	1.71	35.7
55	0.016	0.0	-	-	-	3.7	0.08	0.099	-	-

Acknowledgements
This research sponsored by the Office of Basic Energy Sciences, U.S. Department of Energy under contracts DE-AC05-00OR22725 with Oak Ridge National Laboratory(ORNL) and W-31-109-ENG–38 with ANL. The help given of IPNS and SAND personnel was greatly appreciated. Fits to the data are from the MIRROR program developed by J B Hayter and W A Hamilton (1992–2005).

References

[1] O Ghosh and C A Miller 1987 *J. Phys. Chem.* **91** 4528
[2] G Porte 1999 From giant micelles to fluid membranes *Soft Matter Physics* ed M Daoud and C E Williams (Berlin,Springer-Verlag) chapter 5 p 155
[3] I S Barnes, S T Hyde, B W Ninham, P J Derian, M Drifford, G G Warr and T Zemb 1988 *Prog. Colloid Polym. Sci.* **1** 358
[4] C Quillet, C Blanc and M Kléman 1996 *Phys. Rev. Lett.* **77** 522
[5] W A Hamilton, L Porcar, P D Butler and G G Warr 2002 *J. Chem. Phys.* **116** 8533
[6] L Porcar, W A Hamilton, P D Butler and G G Warr 2003 Langmuir **19** 10779
[7] F. Nallet, R. Laversanne, and D. Roux, J. Phys. II **3**, 487 ~1993
[8] Modified from: N Lei, C R Safinya, D Roux, and K S Liang 1997 *Phys. Rev. E* **56** 608
[9] D Roux and C Coulon 1992 *J. Phys. Chem.* **96** 4174
[10] R P Hjelm 1987 *J.Appl. Cryst.* **20** 273
SAND web page http://www.pns.anl/instruments/sand/ (download 29 April 2003)
[11] O D Lavrentovich, C Quillet and M Kléman 1997 *J. Chem. Phys.* **101** 420

Spin-echo small-angle neutron scattering (SESANS) measurements of needle-like crystallites of gelator compounds

Pieter-Jan C J J Coumou[1], Aurelie M A Brizard[2], Jan H van Esch[2], Ignatz M de Schepper[1] and Wim G Bouwman[1]

[1]Delft University of Technology, Faculty of Applied Sciences, Department Radiation, Radionuclides and Reactors, Mekelweg 15, NL-2629 JB Delft, the Netherlands
[2]Delft University of Technology, Faculty of Applied Sciences, Department of Chemical Technology, Julianalaan 136, NL-2628 BL Delft, the Netherlands

E-mail: `w.g.bouwman@tudelft.nl`

Abstract. From dibenzoyl cystine, a low molecular weight gelator, we have prepared needle shaped crystals at relatively high concentrations. For the first time SESANS measurements are performed on objects with this geometry. From the measurements the average diameter can be seen directly. From a more careful analysis the width distribution is determined. The gel phase itself prepared at lower concentrations did not show any signal, in contrast to what one observes with conventional SANS. This shows the complementarity of SESANS and SANS.

1. Introduction

Spin-echo small-angle neutron scattering (SESANS) is a relatively new neutron scattering technique [1, 2, 3] to determine the scattering length density structure of materials [4] in the length range between 30 nm and 20 μm. This is derived from the polarisation as a function of the spin-echo length. Up till now the technique has mainly been applied to nearly spherical particles, as colloids [5, 6], casein micelles and aggregates [7], emulsions [8, 9] and granular materials [10, 11].

In this article we present the first measurements on clearly anisotropic particles, needle shaped crystals of gelator molecules. Hydrogels (gels of aqueous solutions) are well known, because of their applications in food, cosmetics and biomedical use. Dibenzoyl cystine (DBC) (see Fig. 1) is such a low molecular weight gelator [12]. From this material we have prepared needle shaped crystals at relatively high concentrations. From the SESANS measurements the average diameter can be seen directly. From a more careful analysis the width distribution is determined.

The gel phase itself prepared at lower concentrations did not show any signal, in contrast to what one observes with conventional SANS [13].

2. Experiment

The standard method for the preparation of a hydro-gel based on low molecular weight compounds is by completely dissolving the gelator in water. Then the compounds self-assemble into fibrils. A common way to dissolve the gelator in water is by heating the solvent. A gel is

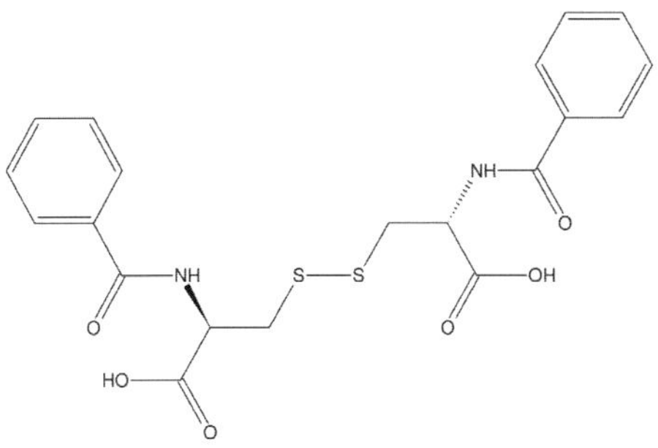

Figure 1. Structural formula of dibenzoyl cystine (DBC).

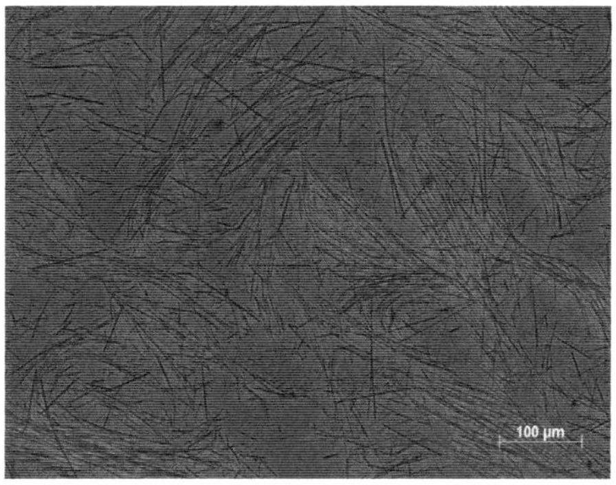

Figure 2. Optical microscopy pictures of a DBC 25 mM gel sample. The length of the fibres is in the order of magnitude of 100 μm, the width is in the order of 1 μm.

formed when the solution cools down again. The temperature for solubility for DBC in water is above the atmospheric boiling point of water, which means that the sample has to be heated under pressure. A solvent temperature of 140°C is high enough to dissolve the gelator DBC. The gels should be prepared directly in the quartz cuvettes, since gels cannot be transferred without breaking the gel, which would affect the original micro structure. An autoclave was used to increase the boiling point of water, and thus to avoid evaporation of the solvent during sample preparation (DBC dissolution). The autoclave was put in the oven for two and a half hours at 140°C and then cooled at room temperature. The cooling process takes several hours. Seven DBC gel samples (concentrations ranging from 10 to 50 mM) were prepared with the use of the autoclave.

Optical microscopy pictures (Fig. 2) have been taken for the 25 mM DBC gel. Long needle shaped crystals can be seen. The length of the fibres is in the order of magnitude of 100 μm. The width is in the order of 1 μm. With the resolution of the microscope it was impossible to determine the width more precisely.

SESANS measures the polarisation as a function of the real-space parameter (spin-echo length) z after normalisation with the empty beam

$$P(z) = e^{\Sigma_t (G(z)-1)}, \tag{1}$$

where Σ_t is the average number of times a neutron scatters when traversing the sample. $G(z)$ is the projection of the scattering length density correlation function $\gamma(r)$ over a distance r. The projection of $\gamma(r)$ along the neutron axis, taken to be x, is

$$G(z) = \frac{2}{\xi}\int_z^\infty \frac{\gamma(r)r}{\sqrt{r^2 - z^2}}dr, \tag{2}$$

where we have the correlation length along the neutron beam axis

$$\xi = 2\int_0^\infty \gamma(r)dr. \tag{3}$$

The correlation function $\gamma(r, D)$ for a straight and infinitely long cylinder with a diameter D is [14]

$$\gamma(r, D) = 1 - \frac{{}_2F_1\left(\frac{1}{2}, \frac{3}{2}; 3; \frac{r^2}{D^2}\right) r^3}{4D^3} - \frac{{}_2F_1\left(-\frac{1}{2}, \frac{3}{2}; 2; \frac{r^2}{D^2}\right) r}{D}, \tag{4}$$

for $0 \leq r < D$ and

$$\gamma(r, D) = 1 - \frac{{}_2F_1\left(\frac{1}{2}, \frac{3}{2}; 3; \frac{D^2}{r^2}\right) D^2}{4r^2} - {}_2F_1\left(-\frac{1}{2}, \frac{3}{2}; 2; \frac{D^2}{r^2}\right), \tag{5}$$

for $D < r < \infty$.

Where ${}_2F_1$ is the Hypergeometric function. We are unable to solve the projection $G(z)$ analytically but we have used Eq. 2 to calculate it [4]. The correlation function $\gamma(r)$ and $G(z)$ show a levelling decay saturating, not surprising, around D as a function of the spin echo length.

Polydispersity can be taken into account by weighing the projected correlation function with for example the normalised log-normal distribution function $f(D)$

$$G(z) = \int_0^\infty G(z, D) f(D) dD. \tag{6}$$

We choose the log-normal distribution, since it is one of the most frequently occurring distribution functions for sizes in nature. However, we are not sensitive to the fine details of the distribution. A normal distribution would give a similar description of the measured data.

SESANS was performed at the neutron source at the Delft University of Technology. A neutron beam, polarised by a set of supermirrors, with a flux of 10^3 neutrons/s at the sample position and a wavelength $\lambda = 0.21$ nm passed through the sample. More technical details can be found elsewhere [3]. Under ideal conditions, the set-up allows for analysis of data up to spin-echo lengths of 20 μm, although the spin-echo length is restricted to 10 μm in the present experiment. The sample cuvettes had a thickness of 10 mm.

3. Results

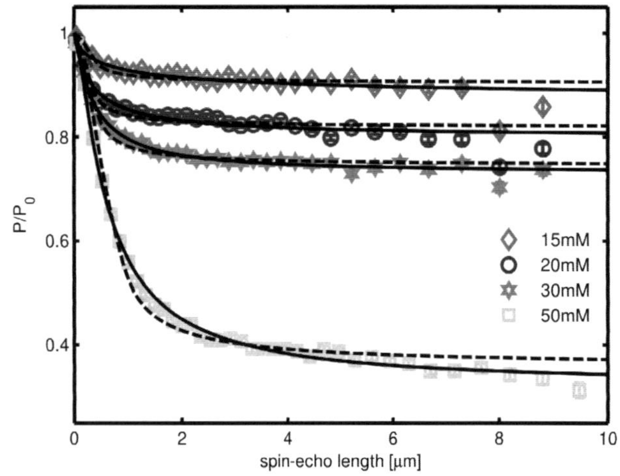

Figure 3. SESANS measurements of DBC-gels prepared with increasing concentrations. The symbols represent the data, the dashed curve represents a fit with the model for the infinitely long cylinder with diameter D, the drawn curves represent the fits with a log-normal distribution.

The SESANS measurements are shown in Fig. 3. The depolarisation increases with concentration. The curves have nearly the same shape: first a strong decay in polarisation and then a longer decaying tail. This is a clear signature of anisotropic particles. First all data were fit with the infinitely long cylinder model (the diameter D and Σ_t were the 2 fitting parameters) , which gave a fair description of the data. The diameter for the lowest three concentrations was $D = 0.50 \pm 0.07$ μm and for the 50 mM sample $D = 1.1 \pm 0.1$ μm. A second fit was done taking a log-normal distribution for the diameter of the cylinders (having as a third fit parameter the width of the distribution). This gave a better fit for the highest concentration samples for the decaying tail. The mean diameter found remained the same as with the first fit.

The gel phase itself prepared at lower concentrations did not show any signal, in contrast to what one observes with conventional SANS [13]. The features of the gel phase are apparently to small to be observed with SESANS. This shows the complementarity of SESANS and SANS.

4. Summary

DBC molecules dissolved at high enough concentrations form needle shaped crystals. The strong preference to crystallise in one direction is not surprising, since the gelator molecules have tendency to stack preferentially in one dimension. With optical microscopy we determined the length to be ≈ 100 μm. With SESANS on the other hand it is possible to determine the width to be $\approx 0.5 - 1$ μm. This is the first example of SESANS measurements on needle shaped particles.

Acknowledgements

We acknowledge Chris P. Duif for assistance with the SESANS measurements.

References

[1] Rekveldt M T 1996 *Nucl. Instr. & Methods in Phys. Res. B* **114** 366–370
[2] Gähler R, Golub R, Habicht K, Keller T and Felber J 1996 *Physica B* **229** 1–17
[3] Rekveldt M, Plomp J, Bouwman W G, Kraan W H, Grigoriev S and Blaauw M 2005 *Review of Scientific Instruments* **76** 033901
[4] Andersson R, van Heijkamp L F, de Schepper I M and Bouwman W G 2008 *Journal of Applied Crystallography* **41** 868–885
[5] Krouglov T, Bouwman W G, Plomp J, Rekveldt M T, Vroege G J, Petukhov A V and Thies-Weesie D M E 2003 *Journal of Applied Crystallography* **36** 1417–1423
[6] Kruglov T, Bouwman W G, de Schepper I M and Rekveldt M T 2005 *Physica B* **356** 218–222
[7] Tromp H and Bouwman W G 2007 *Food Hydrocolloids* **21** 154–158
[8] Bot A, Duval F P and Bouwman W G 2007 *Food Hydrocolloids* **21** 844–854
[9] Goddeeris C, Cuppo F, Reynaers H, Bouwman W and Van den Mooter G 2006 *International journal of pharmaceutics* **312** 187–195
[10] Andersson R, Bouwman W G, Luding S and de Schepper I M 2008 *Granular Matter* **10** 407–414
[11] Andersson R, Bouwman W G, Luding S and de Schepper I M 2008 *Physical Review E* **77** 051303
[12] Menger F, Yamasaki Y, Catlin K and Nishimi T 1995 *Angewandte Chemie-International edition in English* **34** 585–586
[13] Willemen H M, Marcelis A T M, lter E J R, Bouwman W G, Deme B and Terech P 2004 *Langmuir* **20** 2075–2080
[14] Gille W 2003 *Powder Technology* **138** 124–131

Diffusion of domains on nanometer sized vesicle

Y Sakuma, N Urakami[1], Y Ogata[1], M Nagao[2,3], S Komura[4], T Kawakatsu[5] and M Imai

Dept. of Physics, Ochanomizu Univ., Tokyo112-8610, Japan
[1]Dept. of Physics and Information Sciences, Yamaguchi Univ., Yamaguchi 753-8512, Japan
[2]Cyclotron Facility, Indiana Univ., Bloomington, IN 47408-1398, USA
[3]Center for Neutron Research, National Institute of Standards and Technology, Gaithersburg, MD 20899-6102, USA
[4]Dept. of Chemistry, Tokyo Metropolitan Univ., Tokyo 192-0397, Japan
[5]Dept. of Physics, Tohoku Univ., Sendai 980-8578, Japan

sakuma@sweetcream.phys.ocha.ac.jp

Abstract. We have investigated the diffusion coefficient of nano-meter-sized domains on a vesicle composed of saturated phospholipids, unsaturated phospholipids and cholesterols by means of neutron spin echo spectroscopy. The obtained diffusion coefficient was examined from a viewpoint of the hydrodynamic model of Brownian objects in a fluid membrane.

1. Introduction

In recent years, a lateral phase separation in cell membranes so-called lipid raft has gathered much attention in relation to biological functionalities [1]. Lipid rafts are liquid domains rich in cholesterol, sphingolipid and specific proteins, which might play important roles in the biological processes. Since the biochemical functions may be diffusion controlled [2], the lateral diffusion of lipid rafts in bilayers is of biological interest.

In order to understand the physiochemical nature of lipid rafts, the model giant unilamellar vesicles (GUVs) consisting of saturated lipids, unsaturated lipids and cholesterol have been investigated, because a lateral phase separation of the lipids into a liquid-ordered (L_o) phase and a liquid-disordered (L_d) phase might be responsible for the formation of rafts in cell membranes. One of the advantages of the model GUV systems is that the domains have the size in micrometer length scale, which makes it possible to observe the phase separation directly by a fluorescence microscopy. Taking this advantage, Cicuta *et al.* [3] estimated the diffusion coefficients of micrometer-sized liquid domains. They found that the liquid-ordered domains diffuse in the liquid-disordered matrix via Brownian motion and the diffusion coefficient, D, obeys a $D \propto 1/r$ law in the micrometer length scale, where r denotes the domain radius.

Based on the hydrodynamic continuum theory, Saffman and Delbrück [4] derived the size dependence of the diffusion coefficient of an object undergoing Brownian motion in the membrane. For the case of $r < \eta'/\eta_w$ (η': 2D membrane viscosity and η_w: 3D bulk viscosity of surrounding aqueous phase), the diffusion coefficient is expressed by [4]

$$D(r) = \frac{k_B T}{4\pi\eta''}\left[\ln\left(\frac{\eta''}{\eta_w}\frac{1}{r}\right) - \gamma + \frac{1}{2}\right], \qquad (1)$$

where k_B is the Boltzmann constant, T is the temperature and γ is the Euler constant, 0.5772. The 3D membrane viscosity, η''_{3D}, is given by $\eta''_{3D} \cong \eta''/h$, where h is the bilayer thickness. Using typical values for the membranes immersed in water, $\eta_w \approx 10^{-3}$ Ns/m^2, $\eta''_{3D} \approx 10^{-1}$ Ns/m^2 [5] and $h \approx 5 \times 10^{-9}$ m, we obtain $\eta''/\eta_w \cong 500$ nm. Thus this logarithmic scaling law is valid for diffusing objects with nanometer-length scale. On the other hand, for the diffusing domains with micrometer-length scale, the diffusion coefficient is expressed by [6]

$$D(r) = \frac{k_B T}{16\eta_w}\frac{1}{r}, \qquad (2)$$

which well explains the experimental results reported by Cicuta *et al.*

Since the lipid rafts are believed to have the diameter of 10-100 nm [7], the nanometer-sized domains should obey the $D \sim \ln(1/r)$ law. In order to confirm the scaling laws, it is desirable to measure the diffusion coefficients of objects in the nanometer scale range. In this study we investigated the dynamics of nanometer-sized domains using a contrast matching technique for neutron spin echo (NSE) and small unilamellar vesicles (SUVs).

2. Experiments

2.1. Samples

1,2-dioleoyl-sn-glycero-3-phosphocholine (DOPC) (> 99 % purity), 1,2-dipalmitoyl-sn-glycero-3-phosphocholine (*h*-DPPC) (> 99 % purity) and 1,2-dipalmitoyl-d62-sn-glycero-3-phosphocholine (*d*-DPPC) (> 99 % purity) were obtained in a powder form from Avanti Polar Lipid, Inc. Cholesterol (Chol) (> 99 % purity) was purchased from Sigma-Aldrich Co. All lipids were used without further purification and stored 253 K until use [8].

2.2. Sample preparations

GUVs were prepared by a gentle hydration method [9]. The obtained GUV suspension was sonicated using an ultrasonic homogenizer with 23kHz frequency (Branson, Sonifire Model 150) for ~20 min. After the sonication the suspension became transparent and we obtained the SUVs.

2.3. Measurements

In this study we performed small angle neutron scattering (SANS) experiments under two experimental conditions, one is a film contrast condition where we can obtain whole SUV shape and the other is a contrast matching condition where we can estimate the heterogeneity in the SUVs. The SANS measurements were performed using the SANS-U instrument of ISSP, The University of Tokyo at JRR-3M of the JAEA [10]. In order to obtain the dynamical structure factor of nanometer-sized domains we performed NSE experiments. Since the scattering intensity is low for the matching contrast samples, in this work we concentrated the NSE measurement at q = 0.07 Å^{-1} and 285 K in order to maximize statistics. The NSE experiments were performed using the NG5-NSE spectrometer at the NIST Center for Neutron Research [11].

3. Results and Discussion

First, we characterized the ternary SUVs with the composition of DPPC/DOPC/Chol = 4/4/2 by SANS profiles under the film contrast condition [12]. The obtained SANS profile is well described by the model scattering function for the spherical shell structure. From the fitting we extracted mean outer radius of vesicles R_{out} = (11.2±0.1) nm, the bilayer thickness h = (4.2±0.6) nm and the size polydispersity ($p^2 = \langle R_{in}^2 \rangle / \langle R_{in} \rangle^2 - 1$) p = (0.32±0.02), and these values were independent of the temperature.

The static structure of the nanometer-sized domains on a SUV was examined by the contrast matching technique for SANS. We adjusted the mean scattering length density (SLD) of the homogeneous ternary membrane $\rho_{d\text{-mix}}$ to that of the solvent ρ_{solv}. The obtained SANS profiles for the SUV at the contrast matching condition are shown in figure 1(a). At 333 K in the homogeneous one phase region, we observed no significant scattering from the vesicles, indicating that the lipids are homogeneously mixed in the one phase region and each SUV has the same composition. When we decreased the temperature to the coexisting two-phase region (below 303 K), a scattering maximum due to the phase separation was observed in the vicinity of $q \cong 0.02$ Å^{-1} [12, 13, 14]. With decreasing temperature to 285 K, the scattering maximum increases its intensity with keeping the maximum position. We analyzed the observed profiles using a multi-domain model (*e.g.* two domain model is shown in figure 1(b)). The model scattering function of the multi-domain model was calculated by a Monte Carlo simulation technique [12], where we assumed that each domain has the same size and composition. The model scattering function with two domains was well described the observed profile as shown in figure (a), which gave $\theta_c = 35°$ (corresponding to $r \sim 8$ nm).

We examined the dynamics of nanometer-sized domains on the SUV from the intermediate scattering function $S(q,t)/S(q,0)$ obtained by the NSE measurement. Since the diffusion of domains on a vesicle is composed of two modes, diffusion of the center of gravity of the SUV and that of domains on the SUV, we assumed that the intermediate scattering function of the phase separated SUVs under the matching contrast is expressed by

$$S(q,t)/S(q,0) = A\exp[-D_0q^2t] + (1-A)\exp[-(D_0+D_d)q^2t] \quad (3)$$

where D_0 and D_d are the diffusion coefficients of the whole SUV and the nanometer-sized lipid domain on a SUV, respectively, and A is a numerical constant. Here we estimated D_0 from the $S(q,t)/S(q,0)$ of the ternary SUVs at the film contrast condition. The intermediate scattering functions were well described by single exponential functions $S(q,t)/S(q,0) = \exp(-\Gamma t)$. The obtained relaxation rates, Γ, at various q has the relation $\Gamma = D_0q^2$ and gives the diffusion coefficient of whole SUV, $D_0 = (1.31\pm0.01) \times 10^{-11}$ m^2/s at 285 K. This value agrees with the diffusion coefficient $D_{SE} = 1.14 \times 10^{-11}$ m^2/s calculated by the Stokes-Einstein relation using $R = 11.2$ nm.

The $S(q,t)/S(q,0)$ of the ternary SUV at the matching contrast condition is shown in figure 2 with that at the film contrast condition. The difference between $S(q,t)/S(q,0)$ of the film contrast and that of the matching contrast is attributed to the diffusion of domains in an SUV and the fitting using equation (3) gives the diffusion coefficient of the domains with $r = 8$ nm as $D_d = 8.8 \times 10^{-12}$ m^2/s.

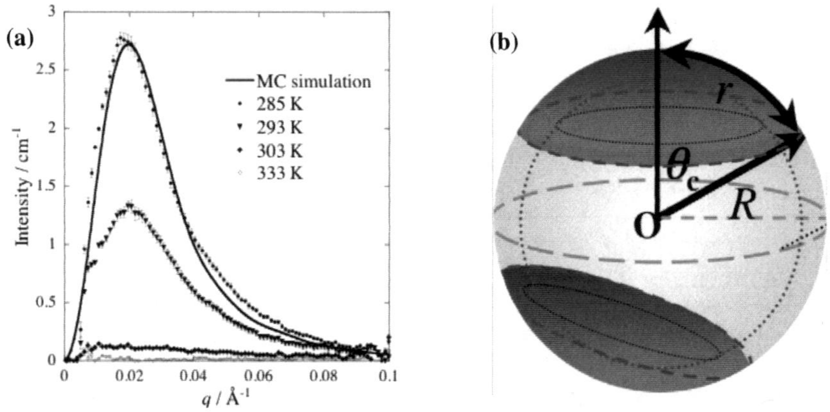

Figure 1. **(a)**Temperature dependence of SANS profiles of binary SUVs at matching contrast condition. Fitting result is shown by solid line. **(b)** Geometry of two-domain model for Monte Carlo simulations.

In figure 3 we plot the diffusion coefficients of liquid-ordered domains in the liquid-disordered matrix obtained by the NSE measurements (nanometer-scale) and the microscope observation (micrometer-scale) [3] as a function of the radius of the domains. The broken and the solid lines in figure 3 show the $1/r$ (equation (2)) and the $\ln(1/r)$ (equation (1)) scaling laws, respectively, where we used $\eta_w \approx 10^{-3}$ Ns/m^2 and $\eta'_{3D} \approx 10^{-1}$ Ns/m^2. The obtained diffusion coefficient was intermediate between the $1/r$ and $\ln(1/r)$ scaling law, indicating the necessity of further experiments to examine the Saffman and Delbrück law quantitatively. On this issue we will address in the forthcoming paper.

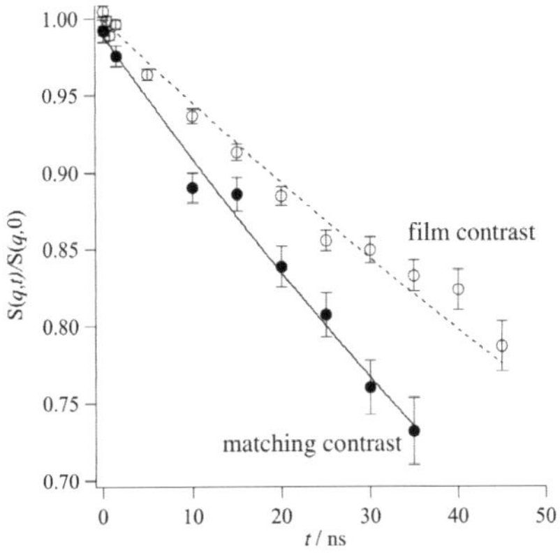

Figure 2. Intermediate scattering functions at $q = 0.07 Å^{-1}$ obtained from NSE experiments. Open circles and closed circles represent the film and the matching contrast experiments, respectively.

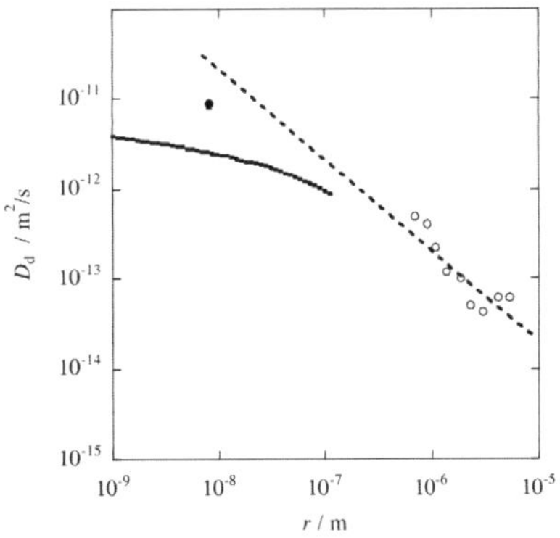

Figure 3. Diffusion coefficients of domains as a function of domain radius with $1/r$ (broken line) and the $\ln(1/r)$ (solid line) scaling laws. Closed circle and open circles indicate the data from this work and Cicuta *et al.* [3], respectively.

References

[1] Simons K and Ikonen E 1997 *Nature* **387** 569
[2] Hackenbrock C R 1981 *Trends Biochem. Sci.* **6** 151
[3] Cicuta P, Keller S L and Veatch S L 2007 *J. Phys. Chem. B.* **111** 3328
[4] Saffman P and Delbrück M 1975 *Proc. Natl. Acad. Sci.* **72** 3111
[5] Peters R and Cherry R J 1982 *Proc. Natl. Acad. Sci.* **79** 4317
[6] Hughes B D, Pailthorpe B A and White L R 1981 *J. Fluid Mech.* **110** 349
[7] Plowman S J, Muncke C, Parton R G and Hancock J F 2005 *Proc. Natl. Acad. Sci.* **102** 15500
[8] Certain commercial equipment or materials are identified in this paper to foster understanding. Such identification does not imply recommendation by the National Institute of Standards and Technology, nor does it imply that the materials or equipment identified are necessarily the best available for the purpose.
[9] Reeves J P and Dowben R M 1969 *J. Cell Physiol.* **73** 49
[10] Ito Y, Imai M and Takahashi S 1995 *Physica B* **213&214** 889
[11] Monkenbusch M, Schätzler R and Richter D 1997 *Nuclear Instruments and Methods in Physics Research Section A* **399** 301
[12] Masui T, Urakami N and Imai M, 2008 *Eur. Phys. J. E* **27** 379
[13] Pencer J, Mills T, Anghel V, Krueger S, Epand R M and Katsaras J 2005 *Eur. Phys. J. E* **18** 447
[14] Pencer J, Jackson A, Kucerka N, Nieh M-P and Katsaras J 2008 *Eur. Biophys. J.* **37** 665

Journal of Physics: Conference Series **251** (2010) 012037 doi:10.1088/1742-6596/251/1/012037

How lidocaine influence the bilayer thickness and bending elasticity of biomembranes

Zheng Yi[1], Michihiro Nagao[2,3], Dobrin P Bossev[4]

[1]UT/ORNL Center for Molecular Biophysics, Oak Ridge National Laboratory, Oak Ridge, TN 37831, USA
[2]Indiana University Cyclotron Facility, Indiana University, Bloomington, IN 47408, USA
[3]Center for Neutron Research, National Institute of Standards and Technology, Gaithersburg, MD 20899, USA
[2]Physics Department, Indiana University, Bloomington, IN 47405, USA

Email: zhyi@indiana.edu

Abstract. We have studied how local anesthetics influence the structural and dynamical properties of model bio-membranes. The measurements of small-angle neutron scattering have been performed on 1,2-dimyristoyl-*sn*-glycero-3-phosphocholine (DMPC) unilamellar vesicles with different concentrations of lidocaine in D_2O to determine the bilayer thickness as a function of the lidocaine concentration. The neutron-spin echo spectroscopy (NSE) has been used to study the influence of lidocaine on the bending elasticity of DMPC bilayers in fluid crystal phase (L_α) and the ripple gel (P_β') phase.

1. Introduction

Local anesthetics (LAs) are known to produce loss of sensation to pain in the certain area of the body without the loss of consciousness [1]. However, the mechanism of their action is still unclear. The current consensus is that the site of action of LAs is within the cellular membrane of the neuron: LAs produce a blockade of nerve impulse by blocking Na^+ voltage-gated ion channels of nerve membranes and slowing the rate of depolarization [2-4]. In order to block the sodium channel the molecules of LAs directly bind to the membrane transport proteins affecting their function [3-6]. Moreover, recent experiments by a number of methods [7-9] have demonstrated that LAs also interact with the lipid membranes and alter the organizational properties of membranes. Such changes on biological membranes could interfere with lipid-protein interactions, and lead to protein conformational changes with a reflection on their activity. In our study, unilamellar vesicles (ULV) of DMPC, one of the most extensively studied ester-linked phospholipids with regard to structural and thermodynamic properties, mixed with different concentrations of lidocaine, the most clinically used LAs, were studied. Differential scanning calorimetry (DSC) has been used to investigate the effect of lidocaine on the phase behaviour of DMPC bilayers in deuterated water solutions. We have used small angle neutron scattering (SANS) to investigate the influence of LAs on the bilayer thickness of DMPC bilayers in aqueous solutions. Furthermore, the bending elasticities (κ_c) of DMPC ULV bilayers with different concentrations of lidocaine in different phases were measured by Neutron Spin Echo (NSE) technique. The intermediate scattering function acquired through NSE was explained by the Zilman-Granek model [10, 11] for 2-D membranes.

2. Materials and methods

Synthetic 1,2-dimyristoyl-*sn*-glycero-3-phosphocholine (DMPC) was purchased from Avanti® Polar Lipids, (Alabaster, AL) and used without further purification. 2-diethylamino-N-(2,6-dimethylphenyl)acetamide (lidocaine) hydrochloride monohydrate was purchased from Sigma Chemicals (St. Louis, MO) in the solid form. D_2O (99 % atomic percent) was purchased from Cambridge Isotope Labs, (Andover, MA). DMPC Bilayers of unilamellar vesicles were prepared by extrusion method using a polycarbonate filter with pore diameter of 1000 Å [12]. A lipid concentration of a mass fraction of 1.0 % in D2O was used for SANS, 2.0 % for NSE and 10.0% for DSC measurement. The averaged radii of the vesicles in 30 °C, which are about 450 Å, were characterized by dynamic light scattering (DLS).

DSC was used to investigate the effect of LAs on the phase-transition temperatures of DMPC bilayer membrane. The thermotropic phase transition of multilamellar vesicles of molar ratio (*R*) of lidocaine/DMPC = 0, 0.2, 0.5, 0.7, 1.0 and 1.5 (lipid concentration in D_2O of a mass fraction of 10 %) was determined by TA instruments Q1000 differential scanning calorimeter (New castle, DE). The temperature scanning was from 30 °C to 0 °C, then returning back 30 °C with the temperature accuracy of ± 0.1 °C. The heating and cooling scanning rates were 1 °C per min.

SANS measurements were performed at NG3 30m SANS instrument located at NIST Center for Neutron Research (NCNR) [13]. Neutrons of wavelength $\lambda = 8.4$ Å with a wavelength spread of $\Delta\lambda/\lambda = 11$ % were used. Data were collected with a two-dimensional detector at three different sample-to detector distances (1.3m, 4.0m and 13.7m) in order to span the range of scattering vectors q from 2×10^{-3} to 4×10^{-1} Å^{-1}. Samples were contained in standard, 1-mm-path-length quartz cells. The data were corrected for instrumental and empty cell backgrounds using the data reduction macros of Igor Pro supplied by NIST [15]. Samples of R = 0, 0.2, 0.5, 2.0 and 5.0 were measured at 30 °C in L_α phase, above the main transition temperature of DMPC. The temperature was controlled by the water circulation with an accuracy of ± 0.1 °C. The measured scattering intensity of DMPC ULV bilayers were fitted as polydispersed spherical vesicles by single-strip model [14-16].

NSE spectrometer measures in the time domain, the real part of the intermediate scattering function, $I(q,t)$. In Zilman-Granek model for 2-D membranes, the time decay of $I(q,t)$ originating from thermal undulations of isolated single membranes can be predicted to exhibit stretched exponential decay:

$$I(q,t) = I(q,0)e^{-(\Gamma t)^{2/3}} \tag{1}$$

where Γ is the relaxation rate, and related to the bending elasticity as:

$$\Gamma = 0.025\gamma_k \left(\frac{k_B T}{\kappa_c}\right)^{\frac{1}{2}} \frac{k_B T}{\eta} q^3 \tag{2}$$

where η is a viscosity of a solvent, and a factor γ_k originates from averaging over the angle between the wave vector and the plaquette surface normal in the calculation of $I(q, t)/I(q, 0)$. The parameter γ_k unifies when $\kappa_c >> k_B T$. We use three times the value of average solvent ($\eta = 3\ \eta_{\text{solvent}}$) for viscosity η when taking the local dissipation at the membrane into consideration [17-19]. Our data were taken on the spectrometer located on the NG5 guide of NCNR. Neutrons with wavelengths of 8 Å and 11.6 Å were used. The range of momentum transfer q was from 0.03 Å^{-1} to 0.12 Å^{-1}. The reduced NSE data were analyzed using NCNR program DAVE. The temperature dependence of the bending elasticity was measured for samples with molar ratio R of Lidocaine and DMPC = 0, 0.5 and 2.0 at several temperatures from 10 °C to 30 °C.

3. Results and discussion

DSC measurement shows that the presence of lidocaine shifts the main transition temperature (T_m) of DMPC bilayers and depresses and broadens its endothermic peak. T_m of DMPC bilayers (red triangles) as a function of R was plotted in Figure 1. These scans show that the presence of lidocaine progressively depresses the main transition temperature and enthalpy in an approximately quadratic dependence. At the highest R = 1.5, the concentration of lidocaine in D_2O went over to about 200 mmol/kg and the main transition temperature was depressed to as low as 18.8 °C. This anesthetic-induced depression of main transition temperature of DMPC bilayers is an evidence of the change in membrane organization by LAs. It is necessary to investigate the structural change of DMPC model bio-membranes by the presence of LAs.

From SANS result, Bilayer thicknesses d of DMPC bilayers in various R were calculated and the dependence of d (black circles) on R is also plotted in Figure 1. Bilayer thickness of vesicles d decreases with R in quadratic dependence. The decrease is in agreement with the conclusions of the previous NMR study of phosphatidylcholines-tetracaine hydrochloride interaction.

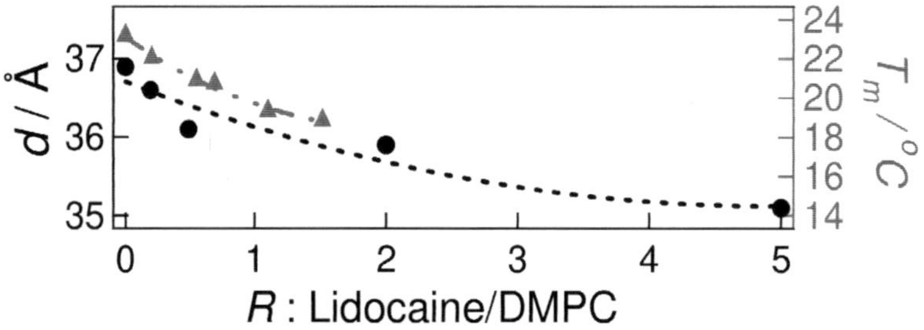

Figure 1. Transition temperature T_m (red triangles) and bilayer thickness d (black circles)of DMPC bilayers as a function of R

LAs molecules penetrate into the lipid bilayers: the hydrophilic parts of LAs mostly remain in phospholipid head-group level, and their lipophilic parts insert little in the bilayer hydrophobic region [20]. The depth of the insertion depends on polarity of lipid head groups and pH [21, 22]. The molecules of LAs with shorter length (lidocaine length ≈ 9.4 Å) loose the packing of the lipids with longer length (DMPC length ≈ 25Å), and induce the lateral membrane expansion. The intercalation of LAs provides more inter-lipid spaces for lipid hydrophobic chains, and then results in the decrease of bilayer thickness. Moreover, lateral membrane expansion provides hydrophobic tails more space to swing, and then disorders the hydrophobic parts of bilayers. This disordering in hydrophobic parts causes the decrease of the main transition temperature.

Table 1. Bending elasticity κ_c/K_BT of DMPC bilayers with different R in D_2O

R	T (°C)					
	10	15	20	23	25	30
0			151.0±6.10	120.7±7.25	20.4±0.68	16.7±0.48
0.5			119.5±7.67	68.2±3.34	31.8±1.17	35.9±1.23
2.0	114.4±10.46	49.8±3.46	69.4±4.70	31.0±1.40	33.8±1.46	

In our previous research [19] we have examined the effect of temperature on bending elasticity κ_c of pure lipid ULVs in D_2O: At T >> T_m, κ_c of pure DMPC bilayers is independent of temperature, and the temperature has a minimal effect on the properties of the lipid bilayer in L_α phase; when the temperature is approaching the vicinity of the main transition temperature, κ_c become slightly larger [23]; When temperature goes across the main transition temperature and into P_β' phase, the κ_c rapidly increases to 6-10 times of κ_c in L_α phase. In present study, bending elasticities of DMPC bilayer with R=0, 0.5 and 2.0 were calculated by equation (2) and shown in Table 1. In the vicinity of the main transition temperature of DMPC bilayers (data at 25 and 30 °C), κ_c fluctuates from 16.7k_BT for R=0 to 33.8 k_BT for R=2.0. The effect of lidocaine on κ_c in L_α phase isn't as obvious as its effect on transition temperature. In the discussion of our SANS result, we confirmed that the intercalation of lidocaine causes more inter-lipid spaces for lipid hydrophobic chains, so that the tail-tail interaction decreased.

However, the interaction of headgroups increased because of the possible closer packing in hydrophilic parts of DMPC and lidocaine. Under the dual effects of hydrophobic and hydrophilic groups, κ_c change of DMPC ULVs with lidocaine in L_α phase is negligible. At T_m=23 °C, κ_c of sample of R=0 was 6 times of its κ_c in L_α phase, however for samples of R=0.5, and 2.0, κ_c were only 1-3 times of their κ_c in L_α phase. The results of NSE also confirmed the decrease of T_m by lidocaine. The presence of lidocaine disordered the arrangement of hydrophobic parts of phospholipid bilayers, thus prevent "gelling" at lower temperatures.

4. Conclusions

We have used DSC, SANS and NSE to measure the influence of lidocaine on the structural and dynamic properties of DMPC phospholipid bilayers. Our results examined that the molecules of LAs loose the packing of the lipids, and induce the lateral membrane expansion. The intercalation of LAs molecules provides more inter-lipid spaces for lipid hydrophobic chains, and finally results in the decrease of bilayer thickness, decrease of the main transition temperature from L_α to P_β' phase and doubled bending elasticity in L_α phase. It is suggested that the increase of the head area and the bending elasticity makes the membrane-protein interaction larger, and possibly the appearance of the blockage of ion channels.

The identification of any commercial product or trade name does not imply endorsement or recommendation by the National Institute of Standards and Technology.

References

[1] Braun H 1914 *Local Anesthesia: Its Scientific Basis and Practical Use*
[2] Lee A G 1976 *Nature* **262** 545-548
[3] S.Wang, C. Nau, G.K. Wang, 2000 *Biophys. J.* **79** 1379-1387.
[4] M.F. Sheets, D. A. Hanck, 2003 *J. Gen. Physiol.* **121** 163-175.
[5] C. Nau, G.K. Wang, 2004 *J. membrane Biol.* **201** 1-8.
[6] M. Gregory, H.A. Fozzard, 2005 *Mol. Pharmacol.* **68** 1611-1622.
[7] Earnest J P, Limbacher H P, Mcnamee M G et al. 1986 *Biophys. J.* **49** A594
[8] Kuroda Y, Nasu H, Fujiwara Y et al. 2000 *J. of membrane Biol.* **177** 117-128
[9] Uhrikova D, Rapp G, Yaradaikin S, et al. 2004 *Biochemistry* **109** 361-373
[10] Zilman A and Granek R 1996 *Phys. Rev. Lett.* **77** 4788
[11] Zilman A and Granek R 2002 *Chem. Phys.* **284** 195
[12] Nayar R, Hope M J and Cullis P R 1989 *Biochim. Biophys. Acta* **986** 200-206
[13] Glinka C J, et al. 1998 *J. Appl. Cryst.* **31** 430
[14] Kline S R 2006 *J. Appl. Cryst.* **39** 6 895
[15] Kiselev M, Zemlyanaya E, Aswal V and neubert R 2006 *Eur. Biophys J.* **35** 477-493
[16] Kucerka N, Nagle J, Feller S and Balgavy P 2004 *Phys. Rev. E* **69** 051903
[17] Farrago B et al. 1995 *Physica B* **213** 712
[18] Takeda T et al. 1999 *J. Phys. Chem. Solids* **60** 1375
[19] Yi Z, Bossev D P and Nagao M 2009 *J. of Phys: Condens. Matter* **21** 155104
[20] Smith I, Auger M and jarrell H 1991 *Annals of the New York of Academy of Sci.* **625** 668-684
[21] Boulanger Y, Schreier S and Smith I C P 1981 *Biochemistry* **20** 6824-6830
[22] Hogberg C, maliniak A and Lyubartsev A 2007 *Biophys. Chem.* **125** 416-424
[23] Seto H, Yamada NL., Nagao M, Hishida M. and Takeda T. 2008 *Euro. Phys. J. E.* **26** 217

International Conference on Neutron Scattering 2009 IOP Publishing
Journal of Physics: Conference Series **251** (2010) 012038 doi:10.1088/1742-6596/251/1/012038

Cholesterol in unusual places

N Kučerka[1], M P Nieh[1], D Marquardt[2], T A Harroun[2], S R Wassall[3] and J Katsaras[1,2]

[1]Canadian Neutron Beam Centre, National Research Council, Chalk River, ON K0J 1J0, Canada

[2]Department of Physics, Brock University, St. Catharines, ON L2S 3A1, Canada

[3]Department of Physics, Indiana University-Purdue University Indianapolis, Indianapolis, Indiana 46202-3273, USA

John.Katsaras@nrc.gc.ca / Norbert.Kucerka@nrc.gc.ca

Abstract. Cholesterol is an essential component of mammalian cells, and is required for building and maintaining cell membranes, regulating their fluidity, and possibly acting as an antioxidant. Cholesterol has also been implicated in cell signaling processes, where it has been suggested that it triggers the formation of lipid rafts in the plasma membrane. Aside from cholesterol's physiological roles, what is also becoming clear is its poor affinity for lipids with unsaturated fatty acids as opposed to saturated lipids, such as sphingomyelin with which it forms rafts. We previously reported the location of cholesterol in membranes with varying degrees of acyl chain unsaturation as determined by neutron diffraction studies (Harroun *et al* 2006 *Biochemistry* **45,** 1227; Harroun *et al* 2008 *Biochemistry* **47,** 7090). In bilayers composed of phosphatidylcholine (PC) molecules with a saturated acyl chain at the *sn*-1 position or a monounsaturated acyl chain at both *sn*-1 and *sn*-2 positions, cholesterol was found in its much-accepted "upright" position. However, in dipolyunsaturated 1,2-diarachidonyl phosphatidylcholine (20:4-20:4PC) membranes the molecule was found sequestered in the center of the bilayers. In further experiments, mixing 1-palmitoyl-2-oleoyl phosphatidylcholine (16:0-18:1PC) with 20:4-20:4PC resulted in cholesterol reverting to its upright orientation at approximately 40 mol% 16:0-18:1 PC. Interestingly, the same effect was achieved with only 5 mol% 1,2-dimyristoyl phosphatidylchoile (14:0-14:0PC).

1. Introduction

Polyunsaturated fatty acids (PUFAs) constitute a biologically influential group of molecules. High levels, sometimes exceeding 50 mol %, are found in the phospholipids of specialized membranes where their depletion impairs function [1]. In retinal membranes, for instance, PUFAs are particularly abundant, and nearly 30% of phosphatidylcholines (PCs) isolated from bovine rod outer segments are dipolyunsaturated [2]. Dietary consumption of PUFAs is also known to elevate the modest concentration, usually less than 10 mol %, of PUFA-containing phospholipids in the plasma membrane alleviating a number of chronic conditions [3]. While interest in the topic has centered on the omega-3 class of PUFA lipids, it has also spanned a variety of human health issues, including

PUFA-associated effects on protein signaling in inflammation and cancer [4], arteriosclerosis [5], and suppressive effects on the immune system [6].

Cholesterol is an essential component of mammalian cells and is either obtained from foods of animal origin (e.g., milk, cheese, meat, eggs, etc.) or synthesized in the endoplasmic reticulum [7]. It is required for building and maintaining cell membranes, regulates their fluidity, and may act as an antioxidant [8]. Recently, cholesterol has also been implicated in cell signaling processes, where researchers have suggested that it is involved in the formation of lipid rafts in the plasma membrane [9, 10], and has also been found to reduce the permeability of the plasma membrane to sodium and hydrogen ions [11].

The interaction of cholesterol with the fatty acids of phospholipids plays a crucial role in modulating molecular organization within membranes [7]. Its interaction with saturated fatty acid (SFA)-containing PCs is well characterized. The introduction of the rigid steroid moiety into homoacid-disaturated PC membranes disrupts the regular packing of chains in the gel or solid ordered (so) phase and restricts the reorientation of the fatty acid chains in the liquid crystalline or liquid-disordered (ld) phase [12]. The differential between the phases is smeared out until a liquid-ordered (lo) phase - characterized by rapid reorientation but high conformational order - is formed over a wide range of temperatures at concentrations >16 mol % cholesterol. Excess sterol is expelled when the content exceeds >50 mol % [13]. Within the membrane, the 3β-hydroxyl group of cholesterol locates just below the aqueous interface [14] and the steroid moiety rotates rapidly about the long molecular axis that wobbles through a narrow range of angles slightly tilted relative to the bilayer normal [15]. A similar behavior is exhibited by heteroacid-saturated-monounsaturated PC [16-18]. However, the interaction of cholesterol with polyunsaturated fatty acid (PUFA)-containing phospholipids for which the sterol has diminished affinity is much less understood and has only recently begun to receive attention.

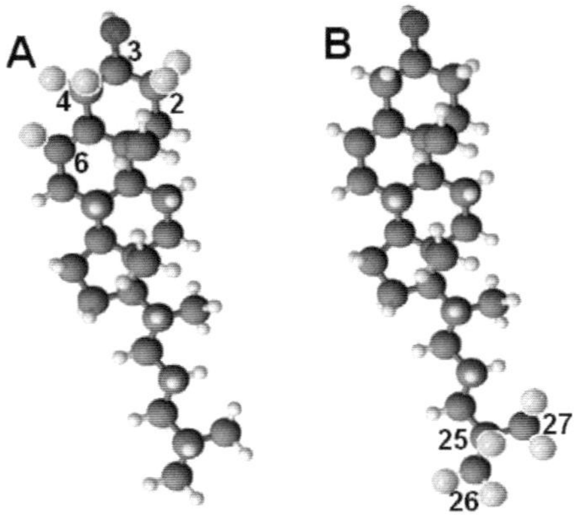

Figure 1. Deuterium label positions (yellow) of cholesterol. A) Headgroup-labeled cholesterol involving hydrogens [2,2,3,4,4,6-$^{2}H_{6}$] near the hydroxyl group on the steroid moiety. B) Acyl tail-labeled cholesterol [25,26,26,26,27,27,27-$^{2}H_{7}$]. Not all of the deuterons are visible.

2. Experimental details

Neutron diffraction data were recorded at the Canadian Neutron Beam Centre's N5 beamline, located at the National Research Universal (NRU) reactor (Chalk River, ON), using 2.37 Å wavelength λ neutrons. The appropriate wavelength neutrons were selected by the (002) reflection of a pyrolytic graphite (PG) monochromator, and a PG filter was used to eliminate higher order (i.e., $\lambda/2$, etc.) reflections. The methods of sample preparation and neutron data analysis follow exactly those described previously [19, 20]. Besides unlabeled cholesterol, experiments included 10 mol% of either "headgroup" [2,2,3,4,4,6-$^{2}H_{6}$] (Fig. 1A) or "tail" [25,26,26,26,27,27-$^{2}H_{7}$] (Fig. 1B) labeled cholesterol.

3. Results

Figure 2A shows difference (deuterated minus non deuterated cholesterol samples) scattering length density (SLD) profiles of bilayers with varying degree of acyl chain unsaturation containing 10 mol% of unlabeled or headgroup-labeled [2,2,3,4,4,6-$^2H_6$] cholesterol (Fig. 1A). For bilayers other than dipolyunsaturated 1,2-diarachidonyl phosphatidylcholine (20:4-20:4PC) the difference is dominated by a single pair of Gaussian shaped peaks symmetrically disposed on either side of the origin. On the other hand, for 20:4-20:4PC bilayers the difference SLD profile is described by a single Gaussian centered at the origin. These peaks designate where the center of mass of the six deuterated sites on the labeled sterol sits within each membrane. The solid line is a single Gaussian function fit to the difference data, which is used to determine the location and width. The appearance of a single Gaussian (20:4-20:4PC bilayers) is indicative of a single population for the label. This result clearly establishes that the cholesterol molecule has undergone a major reorientation within the dipolyunsaturated membrane, such that its hydroxyl group is now located in the middle of the bilayer instead of near the aqueous interface as is usually the case. The question that this surprising observation raises is whether the sterol lies flat between monolayers or, less likely, has become inverted in the bilayer. To answer this question we followed up with additional neutron experiments employing tail-labeled [25,26,26,26,27,27-$^2H_7$] cholesterol.

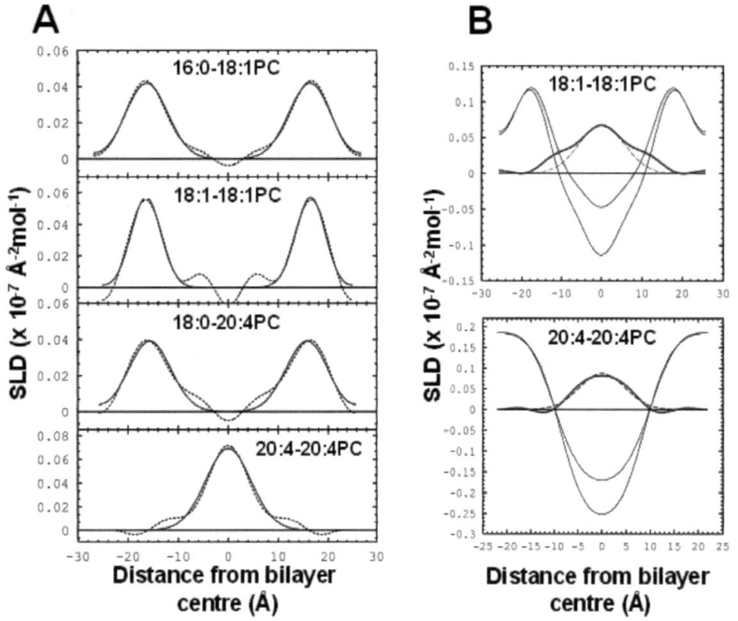

Figure 2. SLD difference profiles between labeled and unlabeled samples. A) Difference profiles for bilayers with varying levels of acyl chain unsaturation, containing 10 mol% of unlabeled or headgroup-labeled [2,2,3,4,4,6-$^2H_6$] cholesterol. B) Difference profiles for 18:1-18:1PC and 20:4-20:4PC bilayers containing 10 mol% of unlabeled or tail-labeled [25,26,26,26,27,27-$^2H_7$] cholesterol. The data were fit with single Gaussians. For details, the reader is referred to Refs. [19] and [20].

Figure 2B shows the difference SLD profiles of 1,2-dioleoyl phosphatidylcholine (18:1-18:1PC) and 20:4-20:4PC bilayers containing 10 mol % unlabeled cholesterol or [25,26,26,26,27,27,27-$^2H_7$] cholesterol. For both 18:1-18:1PC and 20:4-20:4PC bilayers there is a single peak at the center of the bilayer, indicating that cholesterol's tail is in the bilayer centre. In the case of 20:4-20:4PC, the entire sterol is now solvated at the same depth in the lipid acyl chain matrix, clear evidence that cholesterol lies flat in the midplane of 20:4-20:4PC bilayers.

3.1. Doping of 20:4-20:4PC/10 mol% cholesterol bilayers

Recently, we doped 20:4-20:4PC/10 mol% cholesterol bilayers with either 1-palmitoyl-2-oleoyl phosphatidylcholine (16:0-18:1PC) or 1,2-dimyristoyl phosphatidylcholine (14:0-14:0PC) lipids.

Preliminary data indicate that cholesterol can be induced to revert to its nominal upright orientation in 20:4-20:4PC bilayers when doped with 50 mol% 16:0-18:1PC (data not shown). However, only 5 mol% of 14:0-14:0PC is needed to accomplish the same effect (Fig. 3).

These results provide further evidence how different lipid species may affect the transmembrane, as well as the lateral distribution of cholesterol. For example, in plasma membranes, sphingolipids are primarily located in the outer monolayer [21], whereas PUFAs are preferentially incorporated into phospholipids, such as phosphatidylethanolamine, that are more abundant in the inner leaflet [22]. In light of the present results, it is conceivable that the presence of PUFA in the inner leaflet can enhance the transfer of cholesterol to the outer layer, potentially modifying raft composition and function.

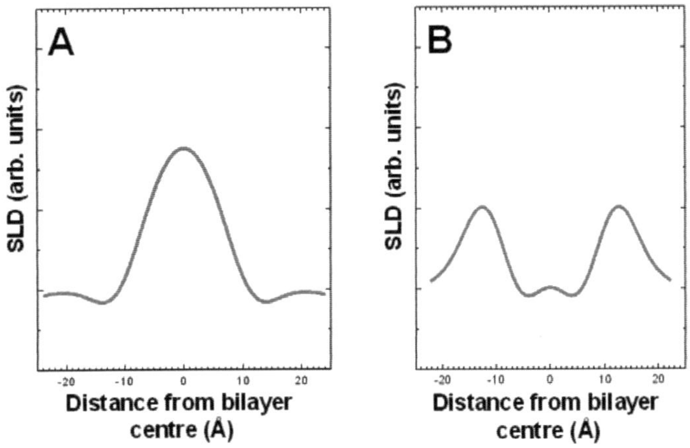

Figure 3. SLD difference profiles between headgoup-labeled and unlabeled 10 mol% cholesterol samples. A) Pure 20:4-20:4PC bilayers. B) 95% 20:4-20:4PC and 5% 14:0-14:0PC bilayers. Only 5 mol% 14:0-14:0PC is needed to flip the cholesterol in its nominal upright position.

References

[1] Salem, N Jr *et al* 1986 *Health Effects of Polyunsaturated Fatty Acids in Seafoods* Simopoulos A. P *et al* Eds (New York: Academic Press)
[2] Aveldano M I 1989 *Colloq. INSERM* **195** 87
[3] Stillwell W and Wassall S R 2003 *Chem. Phys. Lipids* **126** 1
[4] Siddiqui R A *et al* 2007 *Nutr. Clin. Pract.* **22** 74
[5] Kratz M 2005 *Handbook of Experimental Pharmacology* von Eckardstein A Ed. vol. 170 (New York: Springer)
[6] Shaikh, S R and Edidin M 2006 *Am. J. Clin. Nutr.* **84** 1277
[7] Finegold L 1993 *Cholesterol in Membrane Models* (Boca Raton FL: CRC Press)
[8] Smith L L 1991 *Free Radical Biol. Med.* **11,** 47
[9] Papanikolaou A *et al* 2005 *J. Biol. Chem.* **28** 26406
[10] Petrie R J *et al* 2000 *J. Immunol.* **165** 1220
[11] Haines T H 2001 *Prog. Lipid Res.* **40** 299
[12] Vist M R and Davis J H 1990 *Biochemistry* **29** 451
[13] Huang J *et al* 1999 *Biochim. Biophys. Acta* **1417** 89
[14] Léonard A *et al* 2001 *Langmuir* **17** 2019
[15] Marsan *et al* 1999 *Biophys. J.* **76** 351
[16] Thewalt J L and Bloom M 1992 *Biophys. J.* **63** 1176
[17] Worcester D L and Franks N P 1976 *J. Mol. Biol.* **100** 359
[18] Brzustowicz M R *et al* 1999 *FEBS Lett.* **451** 197
[19] Harroun *et al* 2006 *Biochemistry* **45** 1227
[20] Harroun *et al* 2008 *Biochemistry* **47** 7090
[21] Brown D A and London E 2000 *J. Biol. Chem.* **275** 17221
[22] Knapp H R *et al* 1994 *J. Lipid Res.* **35** 1283

International Conference on Neutron Scattering 2009 IOP Publishing
Journal of Physics: Conference Series **251** (2010) 012039 doi:10.1088/1742-6596/251/1/012039

Neutron diffraction and Vitamin E

T. A. Harroun,[1] D. Marquardt,[2] J. Katsaras[3] and J. Atkinson[2]

[1]Department of Physics, Brock University, St. Catharines, Ontario, L2S 3A1 Canada
[2]Department of Chemistry, Brock University, St. Catharines, Ontario, L2S 3A1 Canada
[3]Canadian Neutron Beam Centre, Building 459, Station 18, Chalk River, Ontario K0J 1J0, Canada

E-mail: tharroun@brocku.ca

Abstract. It is generally accepted that neutron diffraction from model membrane systems is an effective biophysical technique for determining membrane structure. Here we describe an example of how deuterium labelling can elucidate the location of specific membrane soluble molecules, including a brief discussion of the technique itself. We show that deuterium labelled α-tocopherol sits upright in the bilayer, as might be expected, but at very different locations within the bilayer, depending on the degree of lipid chain unsaturation.

1. Introduction

Despite its discovery over 80 years ago, the biological role of vitamin E remains a subject of tremendous controversy and confusion, with proposed theories being put forward with diametrically opposed views of its molecular function *in vivo* [1, 2].

Vitamin E occurs in two family groups, the tocopherols and tocotrienols, and there are two well established observations acknowledged by all competing theories. First, that α-tocopherol is the only member of the vitamin E family that is preferentially retained in all mammalian tissues, regardless of the amount of the four tocopherol or tocotrienol types found in an animal's diet. The discovery of the ubiquitous α-tocopherol transfer protein (αTTP) shows *how* this is accomplished, but the evolutionary reasons as to *why* α-tocopherol is selected in preference to the other family members remains a mystery.

The second observation is that the antioxidant properties of α-tocopherol in the test-tube is very well established. In fact, the fat-soluble α-tocopherol combined with water soluble ascorbic acid is arguably among the best anti-oxidant additives that can be used in food and chemical preservation.

In contrast, however, to these two observations, *in vivo* experiments on α-tocopherol have produced many confusing and contradictory results. Notably, vitamin E has been implicated in the regulation of cellular processes seemingly far removed from any antioxidant activity, including apoptosis and gene regulation [1, 2]. A majority of the biophysical studies regarding α-tocopherol have been on saturated acyl-chain model membranes, hardly a relevant model for determining whether α-tocopherol is really a lipid anti-oxidant.

We have begun a series of biophysical experiments based on neutron diffraction and isotopic/isomorphic labelling hoping to see whether α-tocopherol *could* act as an antioxidant protector of unsaturated lipids. To this end we were fortunate to have several milligrams of deuterium labelled α-tocopherol, as shown in Fig. 1. These were labelled either by $^2H_3$ on the methyl of the 5 carbon on the chromanol ring, or $^2H_2$ on the 9' carbon along the acyl chain.

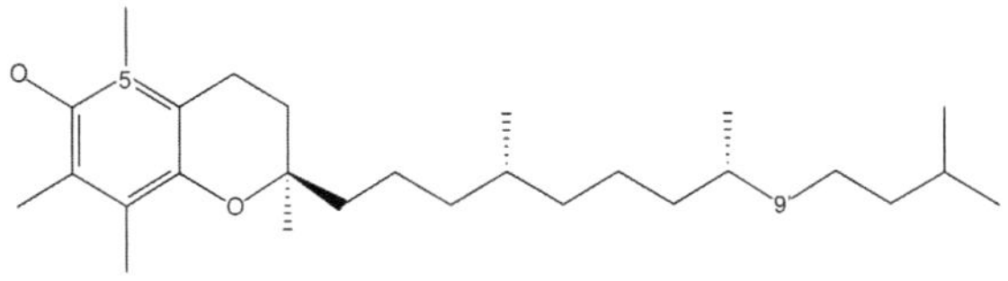

Figure 1. α-tocopherol, with deuterium label centres indicated by 5 and 9'.

Using aligned multi-lamellar stacks in hetero and homo-monounsaturated phosphocholine lipids, we have found the sample mass and time-averaged mass distribution of these labels within the membrane. From this data we hope to formulate whether, at these locations in the lipid bilayer, α-tocopherol is more capable of being a signalling molecule or a proper anti-oxidant.

2. Method of membrane diffraction and deuterium labelling

The entire preparation of aligned multilayer samples was carried out in a nitrogen-filled glove-box. A total of 12 mg of phospholipid with 5 mol % α-tocopherol were dissolved in chloroform-trifluoroethanol (3:1). The solution was deposited on a silicon crystal substrate, and the solvent evaporated while gently rocking the sample. This produces well aligned lamellar samples in a reproducible manner. The samples were then placed in a vacuum for $\sim$1 h to remove traces of the solvent. The samples were then sealed into sample holders and equilibrated in a humid nitrogen atmosphere at room temperature for several hours.

Samples were hydrated either; 1) at varying humidities using saturated salt solutions of KCl (84 % RH) and KNO_3 (94 % RH) or K_2SO_4 (97 % RH) with 8 mol % ^2H_2O, or 2) at a fixed humidity of 94 % RH with 70, 16, 8 and 0 mol % ^2H_2O.

Neutron diffraction data were taken on the N5 and D3 beam-lines at the Canadian Neutron Beam Centre (Chalk River, Ontario, Canada), using 2.37 wavelength neutrons selected by the (002) reflection of a pyrolytic graphite (PG) monochromator, while a PG filter was used to eliminate higher order (i.e., $\lambda/2$, etc.) reflections. Typically 5 to 6 pseudo-Bragg peaks were recorded, meaning that the reconstructed unit cell has a crystallographic resolution of 8-10 Å.

The bilayer profile was reconstructed by Fourier synthesis following the method outlined in [3]. The method records the integrated area of the Bragg peaks for each order I_h, and corrects for neutron absorption ($C_{\rm abs}$), geometry of beam and sample width ($C_{\rm flux}$), and the Lorentz factor ($C_{\rm Lor}$), which are given by

$$C_{\rm abs} = \frac{\alpha}{1-e^{-\alpha}}, \quad \alpha = \frac{2\mu t}{\sin\theta}, \tag{1}$$

$$C_{\rm flux} = \frac{1}{\sin\theta}, \tag{2}$$

$$C_{\rm Lorentz} = \sin(2\theta). \tag{3}$$

Here, μ is the absorption coefficient, and t is the sample thickness, calculated alongside the zeroth order Bragg peak F_0 (the scattering length density [SLD] of the entire unit cell, per mole of sample), using the total neutron cross section (rather than just the coherent scattering cross section), and some simple assumptions – the assumptions being the chemical composition (assuming 10% water by mass), and secondly, a mass density of 1 g/cm^3, which is likely correct to 10% with little effect on the final results.

Phasing of the structure factors was done either by 1) fitting a single continuous smooth form factor F_h versus scattering vector q to the swelling series of 8 mol % ^2H_2O data [4], or

2) by fitting straight lines to F_h as a function of the ^2H_2O percentage for each Bragg order h, expecting opposite signed slopes for even and odd values of h [5]. The first method has the advantage of being able to extrapolate and smooth out differences in the measured unit cell size, while the second method enables the difficult phasing of higher order peaks $h \geq 5$.

The SLD profile $\rho(z)$ is calculated from the Fourier transform of the structure factors according to,

$$\rho(z) = F_0 + 2 \sum_{h=1}^{h_{max}} F_h \cos(2\pi z h/d) , \qquad (4)$$

where z is the distance along the bilayer normal, $z = 0$ is defined as the centre of the bilayer.

The SLD difference between labelled (L) and unlabelled (U) samples can also be calculated with Equation 4 using the difference between the structure factors $F_h = F_h^{\rm L} - F_h^{\rm U}$, with one important caveat: The structure factors for the labelled and unlabelled experiments must be placed on the same relative scale. If they can, then the difference SLD profile is simply the centre of mass of the isotopic/isomorphic substitution label, with all other molecular components subtracted away. Unfortunately, since these are two different samples, unknown experimental factors mean $F_h^{\rm L}$ and $F_h^{\rm U}$ are, in fact, on different scales.

Placing these data on a similar, if not an absolute, scale is the heart of the method of neutron diffraction with isotopic and isomorphic labelling. We have followed closely the discussion and methods of this topic from Han et al. [6] and our earlier work on deuterium labelled cholesterol [3]. In the end we have adopted a scaling method similar to Wiener and White [7, 8], but we achieve the same results by several scaling methods: 1. Equating $\rho^{\rm L}(z)$ and $\rho^{\rm U}(z)$ at one point z which results in a set of simultaneous equations in h with scaling factors and label parameters as unknowns. 2. Equating $\rho^{\rm L}(z)$ and $\rho^{\rm U}(z)$ at two points, results in an analytical result for scaling factors. 3. Scale inter-bilayer water profiles to be the same, which places $\rho^{\rm L}(z)$ and $\rho^{\rm U}(z)$ on the same relative scale.

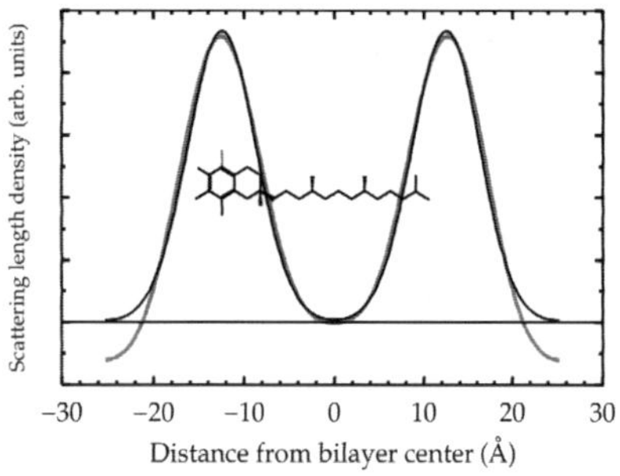

Figure 2. The deuterium mass distribution of labelled α-tocopherol at the C5a methyl (cf. Fig. 1), incorporated into 18:1-18:1 PC bilayers.

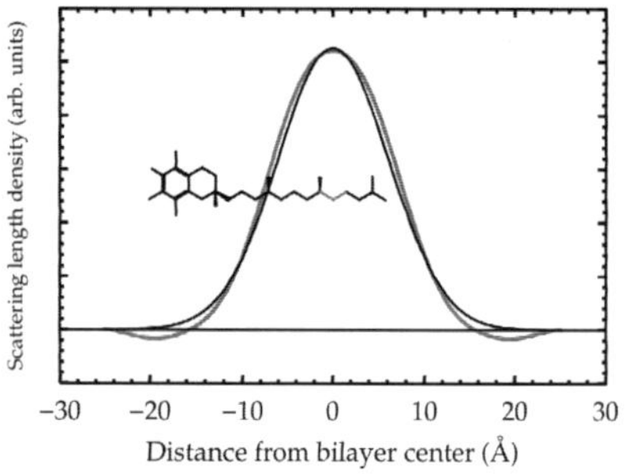

Figure 3. Similar to Fig 2, only for the C9' acyl chain label. The chemical diagram is to orient the viewer to the rough location of tocopherol.

3. Results

Figure 2 shows the the scattering length density (SLD) distribution of α-tocopherol labelled with three ^{2}H on the methyl attached to carbon 5 of the chromanol ring, as measured in a 18:1-18:1 phosphocholine membrane. The lighter curve is the label distribution, and the overlapping dark curve is the fit using a single Gaussian function. The unit cell contains one bilayer, and the origin of the abscissa is at the centre. The inter-lamellar water layers are located at the edges of the unit cell.

The data places the reactive hydroxyl group at 13 Å from the bilayer centre. This is at, or perhaps just below, the hydrophobic/hydrophilic interface, nearly at the same depth as the phospholipid glycerol backbone ester.

The location of the distal end of the α-tocopherol is shown in Figure 3. In this case, the deuterium label resides on the 9' carbon of the tocopherols acyl chain. The distribution is also well fitted by a single Gaussian function fixed at the $z = 0$ Å, and with a standard deviation width of 8.6 Å. In fact, the difference can also be fit by a narrower Gaussian whose centre lies within $z \lesssim 4$ Å, since the mirror distribution appearing on the opposing bilayer leaflet overlaps this Gaussian distribution, creating a much broader single distribution. This is because of the unit cell's mirror symmetry, where atoms in one half of the unit cell are reflected across the centre (and edges) of the unit cell.

Figures 2 and 3 together show that α-tocopherol stands "upright" in the bilayer, the hydroxyl group just above the depth of the first carbon in the 18:1-18:1 PC acyl chain. The chain of the tocopherol is highly disordered, unlike the chemical schematic shown in the figures, since the C9' carbon, only 3/4 along the chain, is localized to the centre of the bilayer. If the chain were extended it would be interdigitated into the opposing bilayer leaflet, but the additional branched methyls make an all-*trans* configuration highly unlikely due to higher gauche probability at the *tert*-carbons.

The location of the C5a-methyl $^2H_3$ label was also measured in the hetero-unsaturated lipid of 16:0-18:1 PC. Shown in Figure 4, the carbon 5 methyl is surprisingly found to reside much higher in the bilayer, around the choline groups. The C9' label was also found to reside at the centre of the bilayer as in Fig. 3 (data not shown). Tocopherol's curled chain spends most of its time at the center of the bilayer.

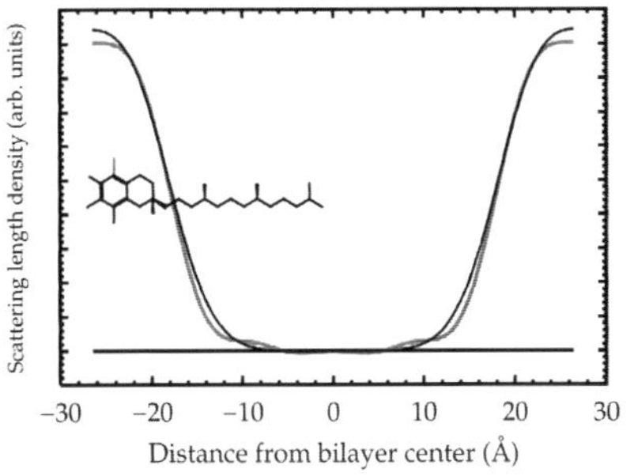

Figure 4. The deuterium mass distribution of labelled α-tocopherol at the C5a methyl (cf. Fig. 1), incorporated into 16:0-18:1 PC bilayers. The lighter line is the data, the darker line a Gaussian fit.

This unexpected result shows that the degree of lipid acyl chain unsaturation has a tremendous effect on the location of tocopherol in the bilayer. When all the lipid chains are unstaturated, the molecule sinks lower into the bilayer. This might indicate that tocopherol prefers to position itself near oxidizable bonds, which it would do to fulfil its anti-oxidant purpose.

References

[1] Traber M G and Atkinson J 2007 *Free Radical Biology and Medicine* **43** 4–15
[2] Azzi A 2007 *Free Radical Biology and Medicine* **43** 16–21
[3] Harroun T, Katsaras J and Wassall S 2008 *Biochemistry* **47** 7090–7096
[4] Darkes M J M and Bradshaw J P 2000 *Acta Cryst.* **D56** 48–54
[5] Worcester D L and Franks N P 1976 *J. Mol. Biol.* **100** 359–378
[6] Han X, Mihailascu M and Hristova K 2006 *Biophys. J.* **91** 3736–3747
[7] Wiener M C and White S H 1991 *Biophys. J.* **59** 162–173
[8] Wiener M C and White S H 1991 *Biophys. J.* **59** 174–185

New Promising Hydride Based on the Cu-Li-Mg System

M H Braga[1,2,3], A Acatrinei[1], M Hartl[1], S Vogel[1], Th Proffen[1] and L Daemen[1]
[1] LANSCE, Los Alamos National Laboratory, mail stop: H805, NM, 87545, USA
[2] CEMUC, Engineering Physics Department, FEUP, Porto University, R. Dr. Roberto Frias, s/n, 4200-465, Porto, Portugal

E-mail: mbraga@lanl.gov

Abstract. We investigated the ternary Cu-Li-Mg system, in particular the $CuLi_xMg_{2-x}$ (x = 0.08) for hydrogen storage. Instead of crystallizing in an orthorhombic phase, as $CuMg_2$, this phase presents a hexagonal structure very similar to that of $NiMg_2$ and $NiMg_2H_{0.3}$. In this work we will discuss the structure of $CuLi_xMg_{2-x}$ by the analysis of the neutron scattering data and first principles calculations. The first results for a hydride (deuteride) phase will also mentioned since preliminary studies at LANSCE showed that $CuLi_xMg_{2-x}$ might absorb approximately 5.3 to 6 wt% of H at an equilibrium pressure of approximately 27 bar at 200 °C. If these results are confirmed in future work, this will mean that, not only $CuLi_xMg_{2-x}$ absorbs a considerable amount of hydrogen (close to DOE's expectations for hydrogen storage materials), but also will probably release it at a temperature in the range of 50 to 150 °C, where applications are easier to develop. Hence it should be possible to use this alloy with fuel cells or in batteries. Another important observation is that cycling has a strong effect on the structure of the hydride.

1. Introduction

Efficient hydrogen storage remains a major technological obstacle toward the development of a hydrogen-based energy economy. Hydrogen stored as a liquid or gas occupies more space than in a suitable metal hydride or other hydrogen storage materials. Electric and hybrid vehicles already use metal hydrides as the negative electrode (instead of cadmium) in their batteries. In fact, metal hydrides are already being widely used in consumer electronics rechargeable batteries advertised as lasting three times longer than alkaline batteries. In Ni-MH, M is an intermetallic compound. The most common one has the AB_5 type, where A is a rare-earth mixture (e.g., La, Ce, Nd, Pr) and B is Ni, Co, Mn, and/or Al. Note that $LaNi_5H_6$, which is commonly used in batteries, stores only a maximum of 1.4 wt% of H. Another disadvantage of these materials for hydrogen storage is that lanthanides tend to be expensive and heavy.

We are currently investigating the Cu-Li-Mg system. The lighter and cheaper metals and our recent discovery that hydrogen can be reversibly stored in these compounds make it a very attractive alternative to lanthanide-based systems.

$CuMg_2$ has an orthorhombic crystal structure and does not form a hydride: it reacts reversibly with hydrogen to produce Cu_2Mg and MgH_2 [1]. However, $CuLi_xMg_{2-x}$ (x = 0.08) has a hexagonal crystal structure [2] (ICSD database [3]), just like $NiMg_2$, a compound known for its hydrogen storage properties. $NiMg_2$ absorbs up to 3.6 wt% H (of the hydride's weight), at 1 bar and 282 °C (555 K) [4].

[3] To whom any correspondence should be addressed.

In spite of the fact that the percentage of hydrogen absorbed by $NiMg_2$ is enough to propitiate practical applications, the temperature at which the alloy desorbs hydrogen is much too high for practical applications.
A comparison between the phase diagrams of the systems Cu-Mg and Ni-Mg shows that these binary systems form compounds with similar stoichiometry. $NiMg_2$ is formed by peritectic reaction of the elements at 759 °C (1032 K) and $CuMg_2$ by congruent melting at 568 °C (841 K). The presence of Li lowers even further the melting point of $CuMg_2$ [5].
Since the energy of formation of the hydride is related to that of the primary alloy [6], it was hypothesized that $CuLi_xMg_{2-x}$ might also be a hydrogen storage material similar to $NiMg_2$. Presumably, its advantage would be that it would release hydrogen at a lower temperature (possibly close to room temperature) [7].
Preliminary studies at the Los Alamos Neutron Scattering Center (LANSCE) showed that $CuLi_xMg_{2-x}$ might absorb approximately 5.3 wt% H for an equilibrium pressure of approx. 27 bar at 200 °C - a pressure and temperature regime easily attainable for practical applications. If these results are confirmed, this will mean that, not only $CuLi_xMg_{2-x}$ absorbs a considerable amount of hydrogen (near the expectations of the US Department of Energy for this type of hydrogen storage materials), but also will probably release it at a temperature in the range of 50 to 150 °C, where applications are easier to develop. Hence it should be possible to use this alloy with fuel cells or in batteries and hydrogen storage devices.

2. Sample preparation

The Cu-Li-Mg samples were prepared from the pure elements with a target composition of $CuLi_{0.10}Mg_{1.90}$. They were prepared by mixing stoichiometric amounts of Cu (electrolytic, 99.99% purity, 325 mesh), Mg (99.8% purity, 200 mesh, Alfa Aesar), and small pieces of Li (99% purity, Alfa Aesar). Because of the large vapor pressure of Mg, even below its melting point, the reagents were sealed in a stainless steel crucible in a dry box (He atmosphere). This eliminated possible reagent loss. The samples were heated in a tube furnace with a stirring device to ensure proper mixing of the heterogeneous starting mixture and complete dispersion of Li in the sample. Different reaction temperatures and times were used (from 450 °C for a day to 1200 °C for 1-2 h). Regardless of reaction conditions, the samples invariably contained Cu_2Mg, $CuMg_2$, or both. Nonetheless, we obtained final products containing up to 81.0 wt% (75.6 at%) of $CuLi_xMg_{2-x}$. Since the structure of Cu_2Mg and $CuMg_2$ is known, this complication translated merely in the refinement of two additional phases in the neutron powder diffraction pattern.

3. Hydrogen absorption experiments

A sample with 70.1 at% of $CuLi_{0.08}Mg_{2-0.08}$, 22.4 at% of $CuMg_2$ and 7.5 at% of Cu_2Mg was grinded to obtain a powder with particle size of the order of 37 μm. The sample was then studied in a HPVA high pressure absorption analyzer. The latter gas adsorption analyzers are designed to obtain high pressure adsorption isotherms of gases, such as hydrogen, using the static volumetric method. The equilibrium pressure of hydrogen versus the ratio between the number of atoms of H and the number of atoms of the other elements present in the $CuLi_{0.08}Mg_{2-0.08}$ was plotted in figure 1 a). It was also plotted in figure 1 b) the equilibrium pressure of hydrogen versus the ratio between the weight percentage of H and the weight percentage of H + $CuLi_{0.08}Mg_{2-0.08}$. Results show an isothermal curve that is much different of those obtained by Reilly and Wiswall for $NiMg_2$ [8] and $CuMg_2$ [1]. Even at a lower temperature, the difference between plateau pressures is higher than 21 atm. The latter mean that $CuLi_{0.08}Mg_{2-0.08}$ will release hydrogen at a much lower temperature. On the other hand, it seems that $CuLi_{0.08}Mg_{2-0.08}$ will absorb hydrogen corresponding to the formation of a hydride with $H/(Cu+Li+Mg) > 1.33$ (probably equal to 2 or higher).

4. Neutron scattering in deuteride and hydride samples

Time-of-flight (TOF) neutron diffraction data were collected at low and room temperatures on

NPDF and at room and high temperatures at HIPPO neutron diffractometers at LANSCE.

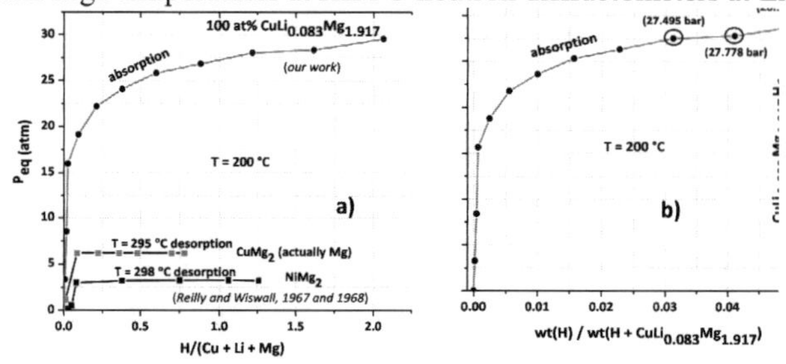

Figure 1. Absorption isothermal (at 200 °C). Equilibrium pressure of H as a function of: a) atoms of H/(Cu + Li + Mg) b) wt(H)/wt(H + $CuLi_{0.08}Mg_{1.92}$). It was considered that at 200 °C only $CuLi_{0.08}Mg_{2-0.08}$ will absorb hydrogen, since $CuMg_2$ will only start absorbing hydrogen at T ≥ 270 °C and Cu_2Mg at an even higher temperature.

5. Neutron scattering in deuteride and hydride samples

Time-of-flight (TOF) neutron diffraction data were collected at low and room temperatures on NPDF and at room and high temperatures at HIPPO neutron diffractometers at LANSCE.

5.1. Neutron diffraction

Neutron diffraction was firstly performed in order to characterize the phase $CuLi_xMg_{2-x}$. Rietveld refinements using GSAS [9] and Pair Distribution Function (PDF) refinements using PDFgui [10] were performed to characterize the phases present in the samples.

Two different measurements of the samples are shown in figure 2 and 3 for two different temperatures.

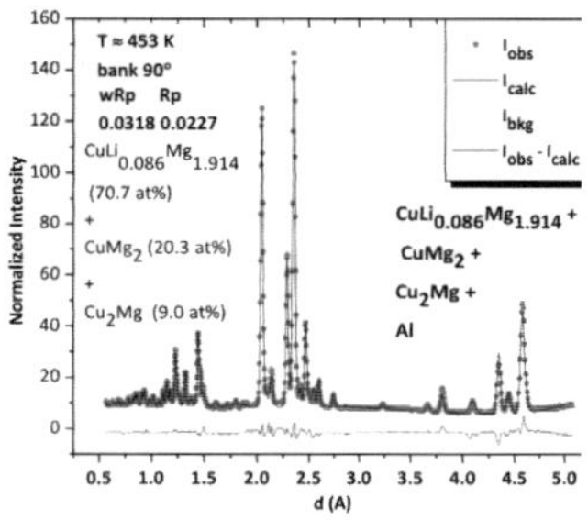

Figure 2. Refined neutron diffraction at T = 453 K. The sample contained $CuLi_{0.08}Mg_{1.92}$, $CuMg_2$ and Cu_2Mg; sample holder was of Al and the measurements were done on HIPPO.

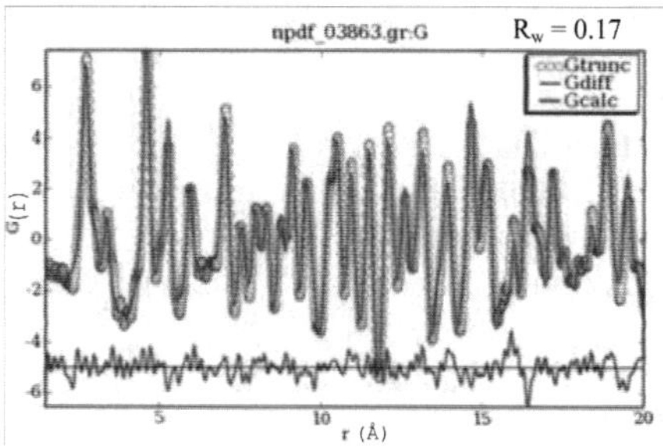

Figure 3. PDF fit after a NPDF measurement at T = 150 K. Sample containing $CuLi_{0.08}Mg_{1.92}$, $CuMg_2$ and Cu_2Mg.

Results obtained show that Li will substitute Mg in (1/2, 0, z) forming a hexagonal $P6_222$ compound $CuLi_{0.08}Mg_{1.92}$ with lattice parameters, a = b = 5.250 (1) Å and c = 13.621(1) Å (at T = 300 K).

5.2. Inelastic neutron scattering

A sample with 75.6 at% of $CuLi_{0.08}Mg_{1.92}$ and 24.4 at % of Cu_2Mg was loaded with H_2 at 200 °C and at different pressures and loading times of the order of several hours, before collecting a neutron vibrational spectrum (treatments were done on the same sample after each measurement). The sample was loaded:

- under 515 psi of H_2 pressure for 1 hour once (yellow curve),
- twice under 1450 psi of H_2 for 6 hours and once for 21 h (green curve),
- once for 17 h and 25 min. and twice for 3 hours (blue curve).

All data were collected at 10 K.

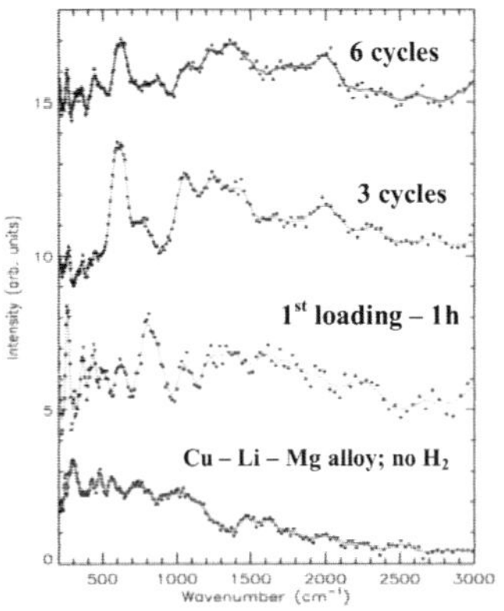

Figure 4. Inelastic neutron scattering (INS) results after cycling a sample with 75.6 at% of $CuLi_{0.08}Mg_{1.92}$ and 24.4 at % of Cu_2Mg with H_2.

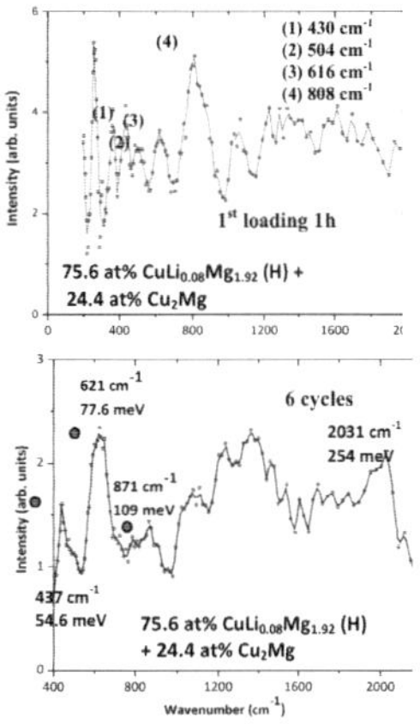

Figure 5. INS results for two different structures of the $CuLi_{0.08}Mg_{1.92}H_z$ hydride. Results show different structures depending on the number of cycles.

INS results show that cycling has a strong effect on structure and ion distribution. These results show that during the first loading the structure was monoclinic with the Cu (bonded to H) ion occupying C_1 sites like $[NiH_4]^{4-}$ ion in $NiMg_2H_4$ (monoclinic) [11], and after 6 cycles tetragonal with Cu (bonded to H) ion occupying C_{4v} sites like $[CoH_5]^{4-}$ ion in $CoMg_2H_5$ (tetragonal) [12].

Summary

This work presents a new compound, never studied before as a hydrogen storage material. In future work more absorption/desorption experiments will be done as well as neutron diffraction experiments, after deuterium absorption, to determine the structure of the hydride/deuteride phase. It is also possible that, as in the case of $CuMg_2$, MgH_2 will be formed upon hydrogen absorption.

Acknowledgments

MH Braga acknowledges FCT, Portugal, for the SFRH/BSAB/791/2008 grant and PTDC/CTM/099461/2008 project. LANL is operated by Los Alamos Nat. Sec. LLC under DOE contract.

6. References

[1] Reilly J J and Wiswall R W 1967 *Inorg. Chem.* **6**(12) 2220
[2] Braga M H, Ferreira J and Malheiros L F 2007 *J. Alloys Comp.* **436** 278
[3] Hewat P 2007 Inorganic Crystal Structure Database **156880**
[4] Schlapbach L and Zuttel A 2001 *Nature* **414** 353
[5] Braga M H, Ferreira J, Malheiros L F and M. Hamalainen 2007 *Z. Kristallog.* suppl. **26** 299
[6] Zuttel A 2007 *Private Communication* Slovenia
[7] Braga M H, Malheiros L F WO2007046017 2007
[8] Reilly J J and Wiswall R W 1968 *Inorg. Chem.* **7**(11) 2254
[9] Larson A C and von Dreele R B, *GSAS Generalized Structure Analysis System*, LANSCE, 2004
[10] Farrow C L, Juhas P, Liu J W, Bryndin D, Bozin E S, Bloch J, Proffen T and Billinge S J L 2007 *J. Phys.: Condens. Matter* **19** 335219
[11] Parker S F, Williams K P J, Smith T, Bortz M, Bertheville B and Yvon K 2002 *Phys. Chem. Chem. Phys.* **4** 1732
[12] Parker S F, Jayasooriya U A, Sprunt J C, Bortz M and Yvon K 1998 J. Chem. Soc. Faraday Trans. **94**(17) 2595

International Conference on Neutron Scattering 2009 IOP Publishing
Journal of Physics: Conference Series **251** (2010) 012041 doi:10.1088/1742-6596/251/1/012041

Investigation of detergent effects on the solution structure of spinach Light Harvesting Complex II

Mateus B. Cardoso, Dmitriy Smolensky, William T. Heller*, Hugh O'Neill*

Center for Structural Molecular Biology, Chemical Sciences Division, Oak Ridge National Laboratory, Oak Ridge, Tennessee 37831

Emails : hellerwt@ornl.gov; oneillhm@ornl.gov

Abstract. The properties of spinach light harvesting complex II (LHC II), stabilized in the detergents Triton X-100 (TX100) and n-Octyl-β-D-Glucoside (BOG), were investigated by small-angle neutron scattering (SANS). The LHC II-BOG scattering curve overlaid well with the theoretical scattering curve generated from the crystal structure of LHC II indicating that the protein preparation was in its native functional state. On the other hand, the simulated LHC II curve deviated significantly from the LHC II-TX100 experimental data. Analysis by circular dichroism spectroscopy supported the SANS analysis and showed that LHC II-TX100 is inactivated. This investigation has implications for extracting and stabilizing photosynthetic membrane proteins for the development of biohybrid photoconversion devices.

1. Introduction

Light harvesting complex II (LHC II) carries out the first step in photosynthesis by directing photons to the reaction centers of the photosynthetic apparatus of higher plants and algae.[1] It also plays a structural role by maintaining the supramolecular organization in the thylakoid membranes and is involved in the regulation of photosynthesis by dissipating excess excitation energy and controlling the activity of Photosystems I and II. The structures of spinach and pea LHC II have been resolved by X-ray crystallography at a resolution of 2.5 Å and 2.72 Å, respectively.[2, 3] Each subunit of the monomeric polypeptide chain is composed of ~232 amino acids with 3 membrane spanning helices that bind and orient 14 Chlorophylls (Chl) and 4 carotenoids. These studies have identified the position and orientation of the 8 Chl a, 6 Chl b and 4 carotenoids per monomer.

Functional LHC II trimers are typically isolated and purified by partial solubilization of thylakoid membranes using mild non-ionic detergents such as Triton X-100 (TX100), n-octyl-β-D-glucoside (BOG), and n-dodecyl-β-D-maltoside.[4-6] The properties of purified LHC II stabilized in detergent solution have been investigated using techniques such as fluorescence and circular dichroism (CD) spectroscopies.[4, 6] These studies provide excellent characterization of the spectroscopic properties of the detergent-associated LHC II but do not provide direct structural information about this system.

The development of photoconversion devices based on natural photosynthetic proteins, such as LHC II, requires the purification of structurally intact and functional membrane proteins and complexes. Protein-detergent complexes are ideally suited for study by SANS with contrast variation because proteins and detergents have inherently different scattering length densities, making contrast variation possible without isotopic labeling of the sample. In this report, we present a comparison of the effects of TX100 and BOG on the structure of LHC II isolated from spinach leaves using circular dichroism spectroscopy and small-angle neutron scattering (SANS).

2. Materials and Methods

2.1. LHC II purification

LHC II was purified from market spinach leaves using two different approaches. The method for isolation of the thylakoids and their solubilization in 0.8% (w/v) TX100 was the same in both procedures. For the LHC II-BOG preparation, the solubilized membranes were fractionated by density gradient ultracentrifugation over a continuous 0.1 to 1.0 M sucrose gradient followed by separation of free Chl from LHC II by Mg^{2+}/K^{+} cation-induced appression.[7] The LHC II containing pellet was dissolved in 10 mM Tris HCl pH 7.6 containing 1.8% BOG, 15.5 or 100 % D_2O, at a protein concentration of 3.95 mg/ml. The LHC II-TX100 preparation was purified as described earlier.[8] The sample for SANS analysis was dissolved in 10 mM Tris HCl pH 7.6 containing 1.7 % TX100, in 25% or 100 % D_2O, and 0.93 mg protein /ml. CD spectra were recorded using a Jasco 810 CD spectropolarimeter at 25°C.

2.2. Small-angle neutron scattering (SANS)

SANS experiments were carried out using the Bio-SANS instrument at the High Flux Isotope Reactor of Oak Ridge National Laboratory.[9] Scattering data were recorded for scattering vectors (q) $0.008 < q < 0.27$ Å^{-1} ($q = (4\pi/\lambda)\sin(\theta/2)$, λ is the neutron wavelength and θ is the scattering angle) using 6 Å neutrons with a wavelength spread ($\Delta\lambda/\lambda$) of 0.15. The measurements were performed at room temperature in 1.0 mm path quartz cuvettes. The raw data were corrected for detector sensitivity, solvent background and transmission. The radius of gyration (R_g) was calculated by using the Guinier approximation.[10] The theoretical scattering profile of the crystal structure of LHC II was calculated using the program ORNL_SAS.[11] The quality of the fit to the data was evaluated with the reduced χ^2 parameter.

3. Results

The properties of purified LHC II-TX100 were investigated using SANS. Figures 1A and B show the scattering profiles of LHC II-TX100 and free detergent in 75% and 25% D_2O, respectively. The free micelle and LHC II-TX100 curves overlay well in the high q region but deviate at low q. The absence of a Guinier region in the LHC II-TX100 curves in both 25% and 75% D_2O suggests that larger order aggregates are present in the sample. However, in the case of the free micelles, R_g was determined to be 23.7±0.2 Å (inset of Figure 1A). A theoretical scattering curve, calculated from the crystal structure of plant LHC II[3] using ORNL_SAS, was compared to the LHC II-TX100 curve in 25% D_2O, which is close to the contrast match point for TX100 (19.8 % D_2O[12]). It showed that there are significant differences between the simulated scattering curve and the experimental data ($\chi^2 = 4.6$, Figure 1B). The experimental and theoretical scattering curves deviate for $q < 0.02$ Å^{-1} and the experimental curve does not have a linear Guinier region indicative of monodisperse particles, but instead suggests that larger order aggregates are present in the LHC II preparation. At larger scattering vectors ($q > 0.05$ Å^{-1}), the measured profile is consistent with a structure that is smaller than the LHC II crystal structure. The low q data support the possibility that the LHC II is not in a native conformation. The scattering at high q is mainly dominated by TX100 micelles, which are present in solution at a considerable excess.

SANS measurements were also performed for LHC II-BOG detergent solution dissolved in 100% and 15.5% D_2O, and BOG detergent in 100 % D_2O (Figure 2A).[7] The scattering profile of LHC II in BOG dissolved in 15.5 % D_2O, the calculated contrast match point for the detergent, was compared to the theoretical LHC II scattering curve. The counting statistics in the experimental data at high q values were poor due to the low contrast between LHC II and the solvent and also the relatively low concentration of LHC II. In contrast to the fit with the LHC II-TX100 data, the simulated LHC II crystal structure scattering curve (shown as a solid red line) overlaid well with the profile of LHC II-BOG curve (χ^2 = 0.38, Figure 2B) suggesting that the protein is in its native folded state. Guinier fitting gives an R_g of 28.3 ± 1.6 Å.[7] A Guinier region is also clearly visible in the LHC II-BOG data in 100% D_2O, demonstrating that there are no larger order aggregates present.

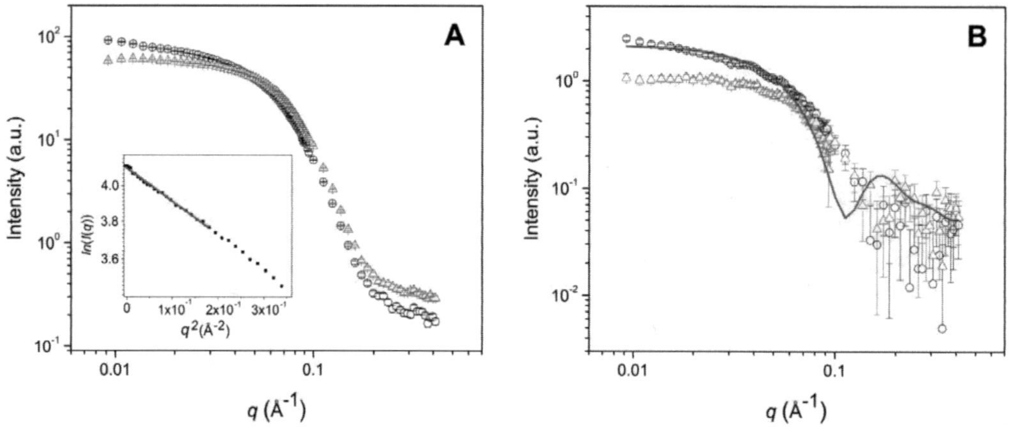

Figure 1. SANS data of LHC II-TX100 (open circles) and TX100 micelles (open triangles) in (A) 75% D_2O and (B) 25% D_2O. Theoretical scattering profile of LHC II crystal structure is presented as the solid red line. Panel A (inset): Guinier plot of TX100 in 75% D_2O.

The circular dichroism (CD) spectrum of LHC II in the visible region results from the excitonic coupling of adjacent Chls and is very sensitive to changes in the conformational state of the trimeric complex. The CD spectrum of LHC-BOG is shown in Figure 2B and is similar to previously reported spectra of the protein complex.[4, 13] The shoulders at 648 nm and 478 nm and a broad positive signal at 412 nm indicate that LHC II is in its native trimeric state. In comparison, no CD signal was observed

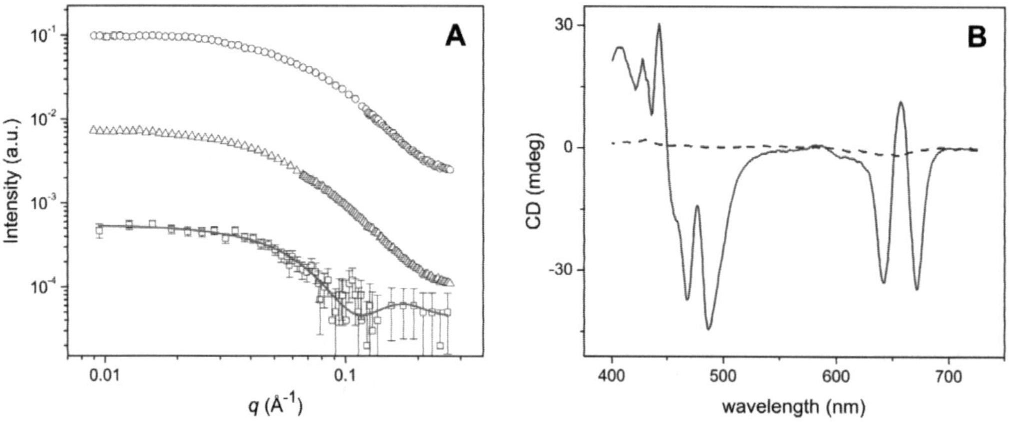

Figure 2. (A) SANS data of BOG in 100 % D_2O (open circles) and LHC II dissolved in 1.8% (w/v) BOG in 100 % D_2O (open triangles) and 15.5 % D_2O (opens squares). Theoretical scattering profile of LHC II crystal structure is presented as the solid red line. (B) Circular dichroism spectra of LHC II dissolved in 1.8% (w/v) BOG (solid green line) and in 1.7% (w/v) Triton X-100 (dashed blue line) at a detergent to Chl ratio of 40:1 after 48h incubation.

for LHC II dissolved in TX100 (dashed blue line) after 48 h. This indicates that the excitonic couplings between the Chls are lost, as would occur if the Chl dissociated from the protein as result of unfolding.

Conclusions

The molecular components of the photosynthetic apparatus such as the photosystems and their associated light harvesting antennae have properties that to date have not been emulated by synthetic approaches. In order to take advantage of their exquisite properties for the development of biohybrid photoconversion devices it is necessary to develop approaches for extraction of these transmembrane protein complexes in their functional state. This study investigated the effect of TX100 and BOG on the stabilization of LHC II. Our SANS analysis shows that the LHC II-TX100 preparation is in a non-native conformation that is inconsistent with an intact trimeric complex, but instead supports the presence of larger order aggregates in the preparation. The existence of a non-native conformation of LHC II is supported by circular dichrosim spectroscopy, which demonstrated that LHC II is inactive in TX100 detergent after 48h incubation. In contrast, LHC II is monodisperse in BOG detergent and maintains its native trimeric conformation under the conditions examined in this study. This investigation provides insight into the effects of detergents on the solution structure of LHC II.

Acknowledgements

M.B.C thanks Capes-Brazil for the support. This work was supported by the ORNL Laboratory Director's Research Development Program. Research at Oak Ridge National Laboratory's Center for Structural Molecular Biology (CSMB) is supported by the Office of Biological and Environmental Research, using facilities supported by the U. S. Department of Energy, managed by UT-Battelle, LLC under contract No.DE-AC05-00OR22725.

References

[1] Forester, T., *Discussion Faraday Transactions* **1959,** *27*, 7.

[2] Standfuss, R.; van Scheltinga, A. C. T.; Lamborghini, M.; Kuhlbrandt, W., *Embo Journal* **2005,** *24* (5), 919-928.

[3] Liu, Z. F.; Yan, H. C.; Wang, K. B.; Kuang, T. Y.; Zhang, J. P.; Gui, L. L.; An, X. M.; Chang, W. R., *Nature* **2004,** *428* (6980), 287-292.

[4] Ide, J. P.; Klug, D. R.; Kuhlbrandt, W.; Giorgi, L. B.; Porter, G., *Biochimica Et Biophysica Acta* **1987,** *893* (2), 349-364.

[5] Mullineaux, C. W.; Pascal, A. A.; Horton, P.; Holzwarth, A. R., *Biochimica et Biophysica Acta - Bioenergetics* **1993,** *1141* (1), 23-28.

[6] Peterman, E. J. G.; Hobe, S.; Calkoen, F.; van Grondelle, R.; Paulsen, H.; van Amerongen, H., *Biochimica et Biophysica Acta - Bioenergetics* **1996,** *1273* (2), 171-174.

[7] Cardoso, M. B.; Smolensky, D.; Heller, W. T.; O'Neill, H., *Journal of Physical Chemistry B* **DOI:** 10.1021/jp905050b.

[8] Simidjiev, I.; Barzda, V.; Mustardy, L.; Garab, G., *Analytical Biochemistry* **1997,** *250* (2), 169-175.

[9] Lynn, G. W.; Heller, W.; Urban, V.; Wignall, G. D.; Weiss, K.; Myles, D. A. A., *Physica B: Condensed Matter* **2006,** *385-386* (Part 2), 880-882.

[10] Guinier, A.; Fournet, G., *Small-Angle Scattering of X-Rays*. John Wiley & Sons, London, UK.: 1955.

[11] Tjioe, E.; Heller, W. T., *Journal of Applied Crystallography* **2007,** *40* (4), 782-785.

[12] Oberdisse, J., *Physical Chemistry Chemical Physics* **2004,** *6* (7), 1557-1561.

[13] Lambrev, P. H.; Varkonyi, Z.; Krumova, S.; Kovacs, L.; Miloslavina, Y.; Holzwarth, A. R.; Garab, G., *Biochimica et Biophysica Acta - Bioenergetics* **2007,** *1767* (6), 847-853.

SANS with contrast variation study of the bacteriorhodopsin-octyl glucoside complex

Yiming Mo and William T. Heller[1]

Center for Structural Molecular Biology and Chemical Sciences Division, Oak Ridge National Laboratory, Oak Ridge, TN 37831

E-mail: hellerwt@ornl.gov

Abstract. Membrane proteins (MPs), which play vital roles in trans-membrane trafficking and signalling between cells and their external environment, comprise a major fraction of the expressed proteomes of many organisms. MP production for biophysical characterization requires detergents for extracting MPs from their native membrane and to solubilize the MP in solution for purification and study. In a proper detergent solution, the detergent-associated MPs retain their native fold and oligomerization state, key requirements for biophysical characterization and crystallization. SANS with contrast variation was performed to characterize BR in complex with OG to better understand the MP-detergent complex. Contrast variation makes it possible to not only probe the conformation of the entire structure but also investigate the conformation of the polypeptide chain within the BR-OG complex. The BR-OG SANS contrast variation series is not consistent with a compact structure, such as a trimeric BR complex surrounded by a belt of detergent. The data strongly suggest that the protein is partially unfolded through its association with the detergent micelles.

1. Introduction

Membrane proteins (MPs) perform many vital cellular functions including membrane transport and trans-membrane signaling and make up roughly 30% of the expressed proteome. MPs are also the targets for many pharmaceuticals. The study of MPs continues to pose considerable technical challenges that result from their insolubility due to their hydrophobic character and the complex character of their natural environment. As a specific example, the number of unique atomic-resolution MP structures in the PDB, being less than 200 as of April, 2009 [1, 2], is dwarfed by the number of unique structures of soluble proteins. In spite of the challenges, the study of MPs continues to be driven by their biological importance.

Detergents are employed to both extract the MP from the native membrane and to maintain it in solution for further study. The traditionally assumed picture of a detergent-solubilized MP is that of a properly-folded MP in its native oligomerization state with the membrane-spanning, hydrophobic segments encapsulated in a belt of detergent [3], which is supported by crystal structures of membrane proteins showing detergent molecules clearly associated with membrane-spanning domains. For instance, the crystal structure of the OmpF porin from *E. coli*, a beta-barrel MP, has a single molecule thick ring of ordered detergent molecules encompassing the transmembrane surface of the structure

[1] To whom any correspondence should be addressed.

[4]. High-resolution structures of alpha-helical MPs, such as bacteriorhodopsin (BR) crystallized with detergent [5] and lipids [6] also support the existence of a detergent belt.

A recent SAXS study of BR in complex with octyl glucoside (OG), the detergent used to obtain the atomic-resolution structure [5], found that the protein contained significant alpha-helical secondary structure while displaying SAXS data consistent with unfolded structure [7]. The data demonstrated that the traditional model of a detergent-solubilized MP in solution is not always accurate, a problematic possibility for functional studies of MP-detergent complexes. To better understand the nature of this structure, small-angle neutron scattering (SANS) with contrast variation was used to probe the BR-OG complex. The results demonstrate that the structure of the BR in the complex is not consistent with a properly folded complex consistent with the expected, native trimeric structure.

2. Materials and Methods

2.1. SANS Data Collection

The BR-OG complex was prepared as previously described [7]. BR was dissolved (final 5 mg/mL) in a buffer containing 1% OG, 20mM PBS (pH =7.6), NaCl 100mM, and 0.02% NaN_3, at different H_2O/D_2O ratios. SANS data were collected using the CG-2 General Purpose SANS instrument at the High Flux Isotope Reactor of Oak Ridge National Laboratory under cooperative agreement with the Center for Structural Molecular Biology, operator of the Bio-SANS instrument [8]. Two sample-to-detector distances of 2.36 m and 8.36 m were employed. The wavelength, λ, was set to 4.75 Å, with a wavelength spread, $\Delta\lambda/\lambda$, of 0.15. Data reduction followed standard procedures to correct for detector sensitivity, dark current and buffer solution background. Data were azimuthally averaged to produce $I(q)$ *vs.* q, where $q = (4\pi sin\theta)/\lambda$, where 2θ is the scattering angle from the incident beam. Data from the two detector distances were merged using software developed at the National Institutes of Standards and Technology [9].

2.2. SANS Data Analysis

Data were analyzed according to Debye [10] for the radius of gyration, R_g. The "beads-on-a-string" model for the structure factor, $S(q)$ [11, 12], was also applied to the data. This model accounts for correlations between flexibly linked particles, such as micelles along the length of an unfolded protein, and provides the fractal dimension of the structure, D, and the correlation length between particles, ξ. A form factor, $F(q)$, of a core-shell ellipsoid of revolution [13] was used to model the detergent micelles using scattering length densities calculated from the chemical structure of OG. The micelle structural parameters (ellipsoidal semi axes and shell thickness), D and ξ were free parameters in the data fitting. The OG micelle structural parameters were allowed to range around previously determined values [14].

3. Results

The SANS contrast variation series of data collected for the BR-OG system is shown in Figure 1. The data do not indicate that the BR-OG complex is a compact particle. The data collected in high D_2O content solutions, where both the protein and detergent contribute to the scattering, are linear at low q on the log-log plot, suggestive of a fractal character. The data set collected at 30% D_2O shows very little signal, indicating that the scattering length density of the solution is very near to the average of the BR-OG complex.

The match point of OG, as calculated from the chemical structure of the molecule, is 18.7%. Therefore, the SANS data collected in 15% D_2O is dominated by the scattering from the BR. The R_g determined from the data set using the Debye model [10], is 152 ± 11Å when fit to a maximum q of 0.05 $Å^{-1}$ to avoid the nosie at higher q-values, which is a much larger structure than is expected for the native trimeric form of BR [5] embedded in a belt of detergent. The forward scatter, $I(0)$, of the 15% D_2O data indicates that the structure consists of ~7-8 BR polypeptide chains, presumably in an extended structure that links multiple OG micelles. The results of fitting the BR-OG data collected in

94% D_2O are shown in Figure 2. The data set is fit very well by the "beads-on-a-string"of the model [11,12]. The OG micelle determined from the fitting is an oblate core-shell ellipsoid 75 Å in diameter and 50 Å thick. The fractal dimension of the *S(q)* is 1.23 ± 0.02, consistent with an open structure [11,12] and ζ is 30.1 ± 1.0 Å.

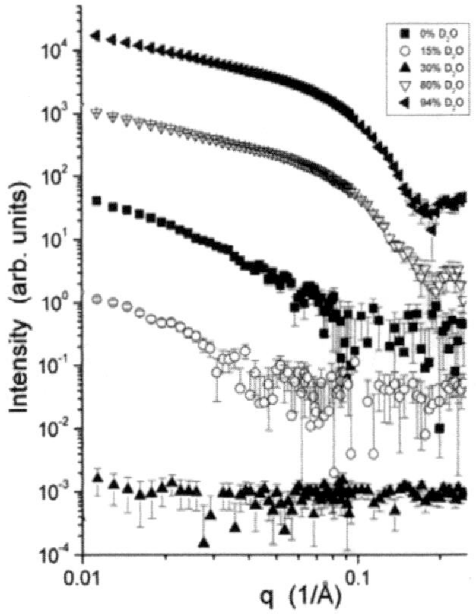

Figure 1. SANS contrast variation series data collected for the BR-OG complex. The curves have been offset for clarity.

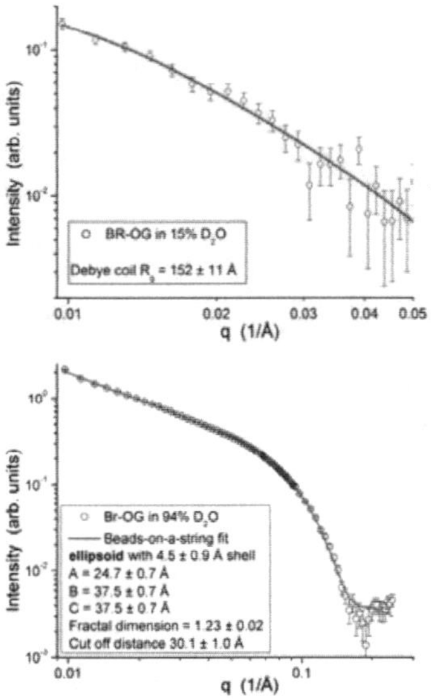

Figure 2. (top) Fit of the BR-OG SANS data in 15% D_2O for R_g according to Debye [10]. **(bottom)** Fit of the 94% D2O data using the "beads-on-a-string" model [11, 12].

4. Discussion

The SANS contrast variation series collected of the BR-OG complex demonstrates that the complex is not a compact structure. Instead, the BR within the complex appears to be both partially unfolded and in a non-native oligomerization state. The data collected at high scattering length density contrast suggest that the BR-OG complex forms an interconnected network. The data collected near the OG contrast match point clearly indicate that the protein is both not in a compact conformation and that the oligomerization state of BR is not the expected trimer [5].

The "beads-on-a-string" model agrees very well with the data, yet ζ is quite low. There are only ~150 OG monomers per BR polypeptide chain in the solution, which is very near the number of bound OG monomers per BR polypeptide chain found in a previous SANS study of the complex (~121 OG per BR) [15]. In a study of soluble proteins in complex with sodium dodecyl sulfate, ζ decreased significantly as the detergent-to-protein ratio decreased [16]. The critical micelle concentration of OG (~0.6% w/v) is also very near the concentration used (1% w/v), making it possible that not all of the OG are associated with the BR. The size of the micelle found by the "beads-on-a-string" fitting suggests that the OG has aggregated into larger structures interconnected by several protein chains.

The result is consistent with a previous SAXS study of the BR-OG complex by this group [7], but does not agree with a recent study of the complex by SANS [15]. The primary difference between the two experiments performed by the two groups is the use of BR from a commercial source in the present study versus the expressed and freshly purified BR studied by Santonicola and coworkers. Santonicola and coworkers concluded that the BR-OG complex was relatively compact. Close inspection of the published data [15] suggests that the scattering objects are not strictly monodisperse, or may not be compact. In particular, the BR-OG SANS curve presented in the upper panel of Figure 6 of reference [15] does not strictly have a slope of zero at low *q*, as it should for monodisperse, compact particles. Further, the Guinier plot shown in Figure 4 of reference [15] is not strictly linear.

The effect is subtle, and the *I(0)* determined from the data supports a predominantly monodisperse sample. Other groups [17, 18] have also concluded that MP-detergent complexes are compact based on scattering data that appear to suggest an unfolded state similar to that observed in the present study. In contrast, the expressed and purified Ste2p G-protein coupled receptor was found to adopt the same partially-unfolded state as BR study while still showing significant alpha-helical secondary structure content [7]. The photosystem I complex purified from spinach was found to be at least partially unfolded while still retaining some level of functionality, having a SAXS profile that was not consistent with the crystal structure of the multisubunit complex embedded in a belt of detergent [19].

5. Conclusions

The present work demonstrates that commercially-obtained BR resolubilized by OG does not result in a correctly folded structure or the native oligomerization state, in contrast to other results that suggest that purified BR adopts a compact conformation in OG. Such a state would undoubtedly have consequences for the interpretation of functional studies or other biophysical characterizations of commercially-obtained BR. SANS can play an important role in the study of membrane proteins by making it possible to determine the true solution conformation and oligomerization state of a detergent solubilized membrane protein prior to investigation by other experimental methods.

6. Acknowledgements

Research sponsored by the Laboratory Directed Research and Development Program of Oak Ridge National Laboratory (ORNL), managed by UT-Battelle, LLC under contract No.DE-AC05-00OR22725 for the U. S. Department of Energy. This Research at Oak Ridge National Laboratory's High Flux Isotope Reactor was sponsored by the Scientific User Facilities Division, Office of Basic Energy Sciences, U. S. Department of Energy.

References

[1] White S H 2004 Protein Sci. **13**, 1948
[2] http://blanco.biomol.uci.edu/Membrane_Proteins_xtal.html
[3] Gohon Y and Popot J L 2003 *Curr. Opin. Colloid Interface Sci.* **8**, 15.
[4] Cowan S W, Garavito R M, Jansonius J N, Jenkins J A, Karlsson R, Konig N, Pai E F, Pauptit R A, Rizkallah P J, Rosenbusch J P, Rummel G and Schirmer T 1995 *Structure* **3** 1041.
[5] Belrhali H, Nollert P, Royant A, Menzel C, Rosenbusch J P, Landau E M and Pebay-Peyroula, E 1999 *Structure* **7**, 909.
[6] Luecke H, Schobert B, Richter H T, Cartailler J P and Lanyi J K 1999 *J. Mol. Biol.* **291**, 899.
[7] Mo Y, Lee B-K, Ankner J F, Becker J M and Heller W T 2008 *J. Phys. Chem.* B **112**, 13349.
[8] Lynn G W, Heller W, Urban V, Wignall G D, Weiss K and Myles D A A 2006 *Physica* B **385-386**, 880.
[9] Kline S R 2006 *J. Appl. Crystallogr.* **39**, 895.
[10] Debye P 1947 *J. Phys. Colloid Chem.* **51**, 18.
[11] Chen S H and Teixeira J 1986 *Phys. Rev. Lett.* **57**, 2583.
[12] Teixeira J 1988 *J. Appl. Crystallogr.* **21**, 781.
[13] Pedersen J S 1997 *Adv. Colloid. Interface Sci.* **70**, 171.
[14] Lipfert J, Columbus L, Chu V B, Lesley S A and Doniach S 2007 *J. Phys. Chem.* B **111**, 12427.
[15] Santonicola M G, Lenhoff A M and Kaler E W 2008 *Biophys. J.* **94**, 3647.
[16] Guo X H, Zhao N M, Chen S H and Teixeira J 1990 *Biopolymers* **29**, 335.
[17] Du L C, Weng Y X, Hong X G, Xian D C and Katsumi K 2006 *Chin. Phys. Lett.* **23**, 1861.
[18] Tang K H, Guo H, Yi W, Tsai M D and Wang P G 2007 *Biochemistry* **46**, 11744.
[19] O'Neill H, Heller W T, Helton K E, Urban V S and Greenbaum E 2007 *J. Phys. Chem.* B **111**, 4211.

The need to revisit lipid areas

N Kučerka[1,2], J Gallová[2], D Uhríková[2], P Balgavý[2] and J Katsaras[1,3,4]

[1]Canadian Neutron Beam Centre, National Research Council, Chalk River, Ontario K0J 1P0, Canada

[2]Department of Physical Chemistry of Drugs, Faculty of Pharmacy, Comenius University, Odbojárov 10, 832 32 Bratislava, Slovakia

[3]Department of Physics, Brock University, St. Catharines, Ontario L2S 3A1, Canada

[4]Guelph-Waterloo Physics Institute and Biophysics Interdepartmental Group, University of Guelph, Guelph, Ontario N1G 2W1, Canada

norbert.kucerka@nrc.gc.ca / john.katsaras@nrc.gc.ca

Abstract. We have studied the structural properties of lipid bilayers made up of monounsaturated phosphatidylcholines (i.e. diCn:1PC, where n=14, 16, 18, 20, 22 and 24). High-resolution x-ray scattering data were analyzed in conjunction with contrast varied neutron scattering data, using the recently developed technique by Kučerka et al. [Biophys. J. **95,** 2356 (2008)]. Analyses of the data show that with increasing n lipid bilayers do not thicken in a linear fashion, as is often assumed, but quadratically, and that lipid area assumes a maximum value for n~18 bilayers. More importantly, compared to previous data our results strongly suggest that lipid areas are smaller by about 10%. This observation highlights the need to revisit lipid areas as they are extensively used in molecular dynamics simulations and for calibrating their force fields.

1. Introduction

The complex dynamics exhibited by biological membranes - characteristic of amphiphilic systems - are highly dependent on the membrane's various structural parameters. It should therefore not come as a surprise that accurate structural data regarding membrane components are important in determining specific biomembrane functions. One such datum is a lipid's lateral area, which is commonly understood to influence lipid-lipid and lipid-protein interactions, and which plays a central role in the outcome of molecular dynamics (MD) simulations.

Despite their importance, published lipid areas have been relatively scarce, and for the most part, inconsistent [1]. Noteworthy are the discrepancies between lipid areas as determined by x-ray and neutron scattering - arguably two of the most widely used experimental techniques in structural biology [2]. On the other hand, these inconsistencies have also been highlighted by the disparate results arising from MD simulations using different force fields. For example, MD simulations based on CHARMM potentials are performed at non-zero surface tension in order to agree with x-ray scattering data [3], while GROMOS potentials do not seem to require this additional "tweaking" [4].

Since the MD force fields are considered to be "well tuned" if they are able to reproduce experimental data, there is clearly much work that needs to be done in order to reconcile simulations and experiment.

Recently, we have developed a model for calculating scattering density profiles (SDP) from the simultaneous analyses of x-ray and neutron scattering data [5]. By appropriately parsing a lipid molecule and simultaneously analyzing the different "contrast" data (i.e. x-ray and different deuteration neutron scattering data), a more precise bilayer structure can be determined.

2. Results and Discussion

The SDP model is graphically shown in Figure 1. Briefly, the component groups in the SDP model are chosen on the basis that each group has the same functional form for all of the different contrast conditions. For example, carbonyl and glycerol groups are described by a single Gaussian (CG), the phosphate and part of choline (CH_2CH_2N) by another Gaussian (PCN), and the remaining choline ($3xCH_3$) by yet another Gaussian ($CholCH_3$). In effect, three Gaussians are used to describe the lipid headgroup. The error function represents the total hydrocarbon region (i.e. CH_2, CH and CH_3 groups). The CH and CH_3 groups are each described by a single Gaussian, which are then subtracted from the error function to obtain the CH_2 distribution. The water distribution is not defined by any particular function, rather it is calculated based on the "complementarity" requirement, whereby all of the probabilities add up to one. In this way, the model satisfies spatial conservation while capturing all of the features of the different SDPs [5].

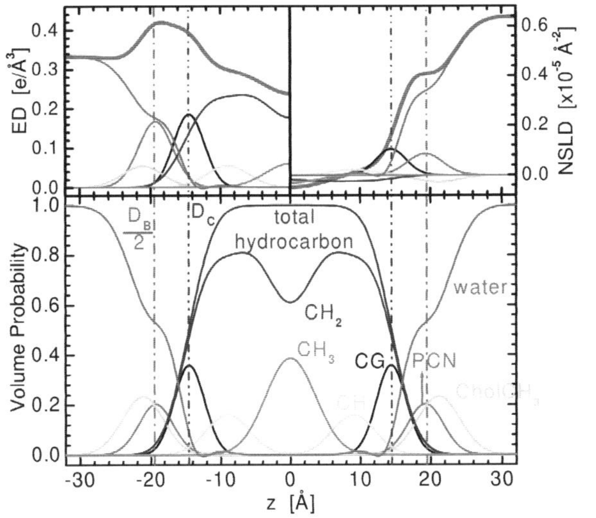

Figure 1: SDP model representation of a diC18:1PC bilayer. The top left panel shows electron densities, while the top right panel depicts neutron scattering length densities of the various component distributions, including the total scattering density (thick gray lines). The bottom panel shows volume probability distributions, where the total probability is equal to 1 at each point across the bilayer.

From Figure 1 it is obvious that the neutron and x-ray techniques are sensitive to different parts of the bilayer. For example, in the case of x-rays, the electron dense phosphate groups contrast very well with the low electron dense hydrocarbon region. Thus, x-ray data are well suited for the refinement of the lipid headgroups and hydrocarbon chains. On the other hand, the high neutron scattering length density of D_2O, permits neutron scattering to accurately determine the total bilayer thickness and consequently, lipid area.

Bilayers made up of the diCn:1PC, where n=14, 16, 18, 20, 22 and 24, were prepared in the form of unilamellar vesicles (ULVs) of diameter ~600 Å [6]. High-resolution small-angle x-ray scattering (SAXS) data were obtained over a q range [$4\pi/\lambda \sin(\theta/2)$, where λ is the wavelength and θ is the scattering angle] from 0.06 $Å^{-1}$ to 0.65 $Å^{-1}$. Small-angle neutron scattering (SANS) data were collected on the same lipids, but dispersed in 100% D_2O. Additional contrasts of 70% and 50% D_2O were used for n=14, 18 and 22 bilayers.

One of the important structural parameters when considering the hydrophobic matching of lipids and proteins is the hydrocarbon chain thickness, D_C. The SDP model defines D_C at the center of the error function, i.e. the hydrocarbon chain distribution (see Figure 1). Another important parameter is the total bilayer thickness (D_B) that we define as the Gibbs dividing surface for the water region (i.e. $D_B/2$) (please see Ref. [5] for details). As was mentioned, D_B is robustly determined by neutron scattering and leads to the determination of lipid area (A) - assuming that lipid volume is already known [7].

The pioneering work of Lewis and Engelman [8] studied three of these lipids (n=18, 22, and 24) using SAXS and pauci-lamellar vesicles (PLVs). When comparing to their data, we find our D_B values to be ~2 Å larger, while following the similar trend, and our lipid areas are smaller by as much as 6 Å^2. A recent x-ray study [9] has supported the Lewis and Engelman lipid area results for n=18 and 22 bilayers, while neutron scattering data have suggested much smaller areas [10, 11]. This inconsistency between SANS and SAXS data was thought to be the result of rudimentary models used to analyze SANS data. However, it was recently realized [5] that SAXS is better applied in determining the internal structure of lipid bilayers, but that it can underestimate D_B and A by as much as 10%, a task possibly more appropriate for SANS. However, by combining both techniques in one analysis the various bilayer parameters can be better determined.

Figure 2 shows the dependency of the various structural parameters as a function of hydrocarbon chain length. Both thicknesses (i.e. D_C and D_B) increase almost linearly with n, exhibiting a small, but not negligible quadratic behavior. Perhaps a more surprising result is how lipid area changes as a function of n (Figure 2). First it increases and then it decreases, deviating from the accepted monotonic behavior with a maximum near n~18. Although surprising at first, this behavior is consistent with the effect of the double bond position. For n=14 to 18 lipids the double bond is at the 9-cis position, while for n=20, 22 and 24 lipids the double bond is at position 11-cis, 13-cis and 15-cis, respectively. Coarse grained bilayer simulations reproduced this trend and qualitatively predicted the effects of double bond position on lipid area (S. J. Marrink, personal communication). As was discussed by Karlovská et al. [12], increasing the hydrocarbon chain length results in increased van der Waals attraction resulting in an ordering of the hydrocarbon chains, and thus reducing area per lipid. However, lipid chain disorder also depends on the position of the double bond, having the most effect when the double bond is located in the middle of the hydrocarbon chain [13]. Importantly, this indicates that lipid area, the result of a fine balance between intrabilayer forces, is a good indicator of lateral interactions within the bilayer.

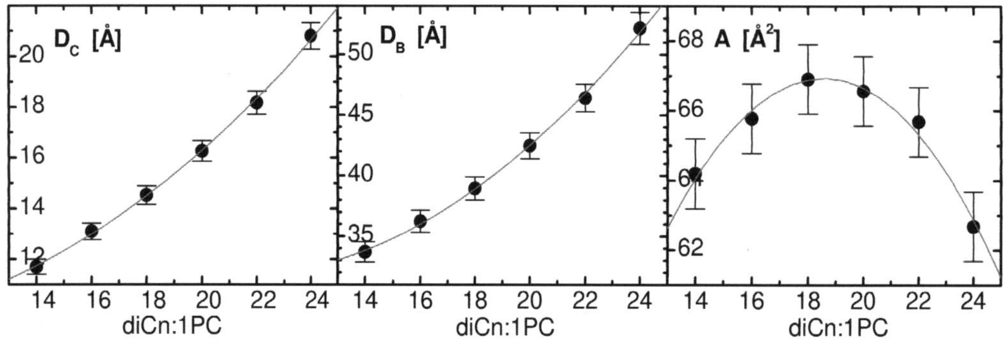

Figure 2: Bilayer structural parameters obtained by the simultaneous analysis of x-ray and neutron scattering data. Structural parameters D_c, D_B and A are plotted as a function of chain length, n.

3. Conclusions

In conclusion, recent developments in structural biophysics have allowed for a more precise determination of biomembrane properties. Through the simultaneous analysis of high resolution x-ray and neutron scattering data, we are able to robustly obtain bilayer structural parameters for the various di-monounsaturated phospholipids. Our results show a quadratic behavior of bilayer thicknesses as a function of n, and an area per lipid maximum near n=18 hydrocarbon chains. The present results can be used to better understand biomembrane-protein interactions, which are well known to depend on bilayer thickness and area per lipid. Perhaps more importantly, these results should serve as the foundation for fine-tuning the force field parameters of MD simulations, for which accurately known lipid areas are central. We encourage the MD simulators to compare their simulations results not only against x-ray scattering data, but to also include neutron data that are more sensitive to D_B and consequently A.

Acknowledgements

We thank Dr. Arthur Woll for assistance with the SAXS setup at the CHESS G-line (supported by the NSF & NIH/NIGMS via NSF award DMR-0225180) and Dr. José Teixeira's help with SANS experiments at the LLB PAXE spectrometer (supported by the EC FP6 via contract RII3-CT-2003-505925). This work was supported by the Dubna JINR 07-4-1069-09/2011 project and by the VEGA 1/0295/08 grant.

References

[1] Nagle J F and Tristram-Nagle S 2000 *Biochim. Biophys. Acta* **1469** 159
[2] Kučerka N *et al* 2007 *Curr. Opin. Coll. Int. Sci.* **12** 17
[3] Klauda J B *et al* 2006 *Biophys. J.* **90** 2796
[4] Pan J *et al* 2009 *Biochim. Biophys. Acta.* doi:10.1016/j.bbamem.2009.02.013
[5] Kučerka N *et al* 2008 *Biophys. J.* **95** 2356
[6] Kučerka N *et al* 2007 *Langmuir* **23** 1292
[7] Uhríková D *et al* 2007 *Chem. Phys. Lipids* **145** 97
[8] Lewis B A and Engelman D M 1983 *J. Mol. Biol.* **166** 211
[9] Kučerka N *et al* 2005 *J. Membr. Biol.* **208** 193
[10] Gallová J *et al* 2008 *Biochim. Biophys. Acta* **1778** 2627
[11] Kučerka N *et al* 2007 *Eur. Phys. J. E Soft. Matter* **23** 247
[12] Karlovská J *et al* 2006 *Biophys. Chem.* **119** 69
[13] Martinez-Seara H *et al* 2007 *J. Phys. Chem. B* **111** 11162

Effects of cutting and specimen size on neutron measurement of residual stresses

Law M, Luzin V, Kirstein O (Australian Nuclear Science and Technology Organisation)

ANSTO, PMB1, Menai, NSW, 2234, Australia

michael.law@ansto.gov.au

Abstract. To perform neutron residual stress measurements it is often necessary to cut samples to a manageable size. The effects of cutting a girth welded pipe were investigated with analytical methods and finite element analysis. The effect of cutting on measured stresses was calculated. A simplified method of modelling residual stresses in welds, "chill modelling", is introduced.

In ring slitting a cut is made in the axial direction and the deformation is maeesured. The change in elastic stress can be calculated and added to neutron diffraction measurements made on a cut ring to calculate the original stresses. Residual stress measurements were performed to validate the ring slitting correction using ANSTO's residual stress diffractometer Kowari.

1. Introduction

Many samples need to be reduced in size before making neutron measurements. The effect of cutting on residual stresses is explored in this paper, the cutting and analysis is performed on a number of welds in pipes.

There are 2 types of macroscopic residual stresses acting on the girth weld, "self-equilibrating" and "bending"[1]. Typical "self-equilibrating" stresses are in the weld direction; these are equilibrated by lower compressive stresses across a large area of parent material. Bending stresses may be caused by a "tourniquet" of shrunk weld metal around the weld. Cutting and drilling changes the local stress field. These effects reduce with distance in accordance with Saint-Venant's principle [2], a qualitative description of the observation that localised stresses do not extend beyond a certain distance from that area. This distance is considered to be a few "characteristic lengths" of the sample itself, a function not only of the plate thickness but also weld width.

Three cutting operations were analysed (figure 1):-"Length reduction" of the length of pipe containing the girth weld (cut 1). "Ring-slitting" cuts this ring axially, which may cause ring opening or ring closing (cut 2). "Coupon cutting" is making another axial cut to reduce the sample to a quarter segment approximately (cut 3).

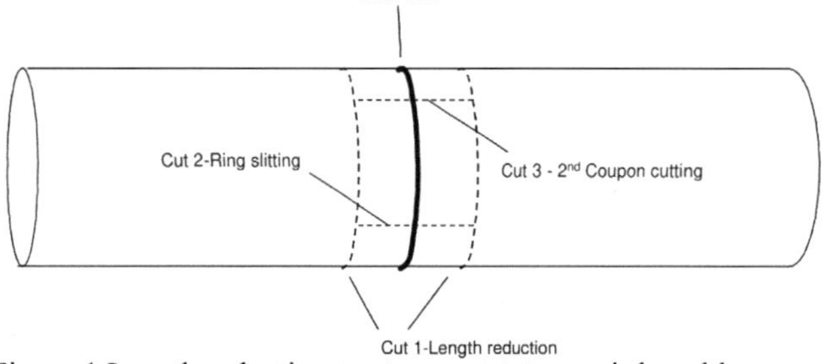

Figure 1 Length reduction to pipe containing a girth weld

Modelling was carried out by analytical and finite element methods at 3 common diameter to thickness ratios (D/t, where D is the diameter and t is the thickness) of 20, 40, and 100.

The sample size requirement may come from table size, slit configuration, weight considerations, or the size sample that be conveniently shipped.

2. Length Reduction

As the pipe length containing the weld is reduced, the tensile stresses along the girth weld will cause increasing inwards deflection at the weld. The deflection may be calculated [3] for a radial line loading at the cut end of the pipe (figure 2). The deflection is calculated at the cut end; it will be greater than a pipe where the load is in the centre, as rotation is possible in the cut end case. Using this conservative analytical approach, the deflections become constant (figure 2) where the length from the weld to the cut is greater than $3.5\sqrt{(Dt)}$. This is measured from the weld to the cut surface, and is doubled when cut either side of a weld.

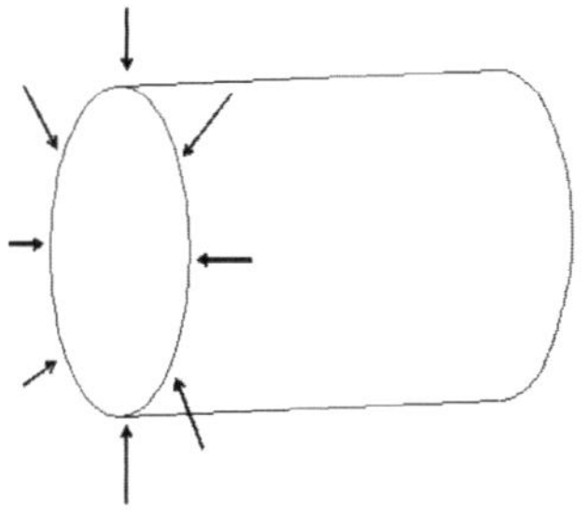

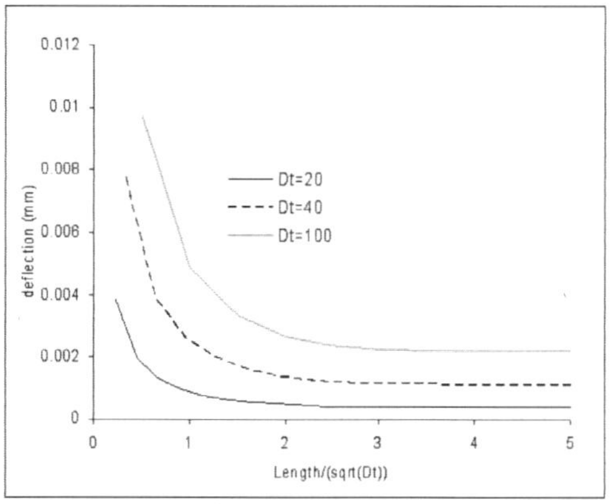

Figure 2 Pipe with line loading on free end and analytical results, from [2].

Rather than attempting full thermo-mechanical weld modelling, a complex procedure which is only now becoming feasible, a simplified method known as chill modelling is employed. Residual stresses were simulated with by cooling a weld section by 500° C, producing hoop stresses which were above yield. This procedure generates local-scale residual stress distributions typical, but not identical to, real welds but is considered representative of the global stresses and sufficient for this study. The stresses were normalised against the maximum tensile stress in each case (as each case had a different D/t ratio and slightly different weld profile, the maximum stresses are different), and against $\sqrt{(Dt)}$. It was assumed that the weld is in the centre. The maximum stress obtained is within 5% of the global stress value at lengths of > $3.5\sqrt{(Dt)}$ (figure 4).

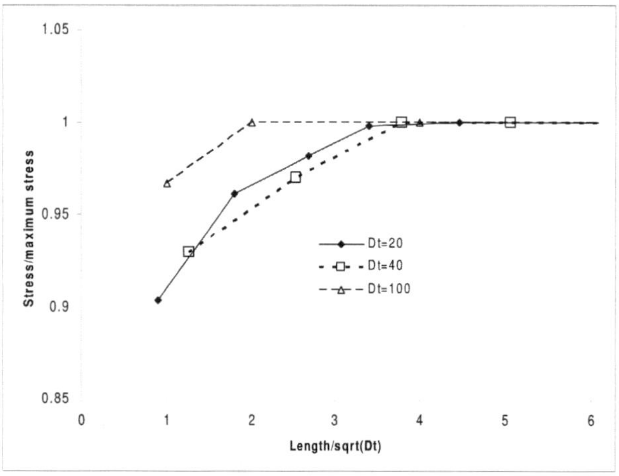

Figure 3 Normalized hoop stress vs. normalized distance from weld.

3 Ring Slitting

In ring slitting a cut is made in the axial direction and the change in circumference is measured and the elastic stress can be calculated and added to neutron diffraction results. The technique can only assess hoop stresses, and assumes that the hoop residual stresses are uniform in the ring. Ring openings of 35 mm or more have been seen in the field, and one case of a ring closing of 8mm has been characterised by neutron methods [4].

The bending stress is given by $\sigma = Et\,(1/D_o - 1/D_f)$ where D_o, D_f are the original and final mid-wall diameters, t is the wall thickness, and E is the Young's Modulus [5]. Results from the ring slitting and neutron methods give different information; the ring slitting method shows any imbalance in the total hoop direction bending moment, while neutron diffraction derived stresses are direct measurements in any direction at discrete locations.

4 Neutron residual stress measurements

Residual stress measurements were performed to validate the ring slitting correction detailed above using ANSTO's residual stress diffractometer Kowari [[6]]. Stresses in a ring of pipe were measured before and after slitting. Because the basic principles of this technique are well known [[7],[8],[9]] only details specific to this measurement will be reported. A monochromatic beam with $\lambda = 1.666$ Å and diffraction from Si{400} planes were used in this analysis. A nominal gauge volume of 2 x 2 x 2 mm^3 was used to perform through-thickness scans at 5 depths. A section of the pipe was cut into a 6 mm thick plate to derive a d-zero values (the unstressed inter-atomic spacing). Stresses were calculated with this d-zero value.

5 Neutron measurement results

The neutron results for the ring before and after slitting are shown in figure 6, along with linear fits to the data. These fits provide a simple method of stress linearization, which is separating the membrane stress (intercept with the midpoint) and the bending stress (which is the slope times the distance from the centre). At the outer surface the bending stresses are -6.9 MPa for the original ring, and 27.4 MPa for the cut sample. Thus a stress change of -34.3 MPa occurs at the outer surface during cutting.

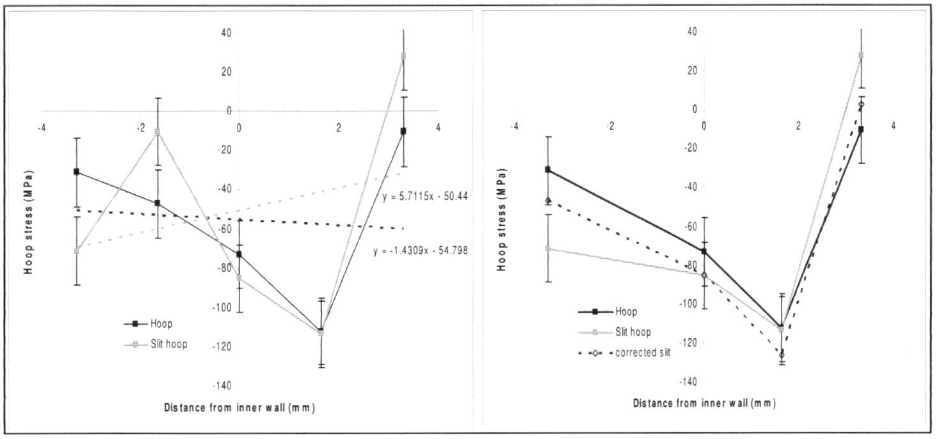

Figure 6a Original and cut residual stresses. Figure 6b Original, cut, and corrected residual stresses. For clarity, data at 3.15 mm has been removed error bars are only are shown on original measurements, the cut and corrected error bars are similar in magnitude.

After slitting an average ring opening of 7.97 mm was measured, equivalent to a stress change of 23.4 MPa. The corrected stresses are shown in figure 6b. The calculated average stress change is less than that seen in the neutron results. The most likely reason for the difference is a localised region showing strong ring-closing behaviour, such as the seam weld. Another issue is the small magnitude of the bending stresses in this sample, approximately the same as the error bars in the neutron measurements. Cases of residual stresses calculated from ring opening have been seen of up to 200 MPa.

6 Conclusions

Analytical methods may exist to asses stress change on cutting for simple geometry and global stress distributions, more complex geometries and localised stress fields (such as welds) may require finite element methods. A method of calculating any stress relaxation that occurs during ring-slitting is described; these stresses can be added to neutron diffraction residual stress measurements made on a cut ring to calculate the stresses before cutting. This assumes that uniform stresses exist in the sample, a condition that is often not met.

"Bending' type residual stresses may show large changes on cutting, while the "self-equilibrating" residual stresses are often little affected on cutting (except at the cut face).

References

1 Dong P, Brust F "Welding Residual Stresses and Effects on Fracture in Pressure Vessel and Piping Components: A Millennium Review and Beyond" J. Pressure vessel Technology August 2000, Vol. 122 pp 329-338
2 de Saint-Venant B., "Memoire sur la torsion des primes", Mem. Des Savants Etrangers, Vol. 14
3 Roark's formulas for stress and strain, 1989, 6th edition, McGraw-Hill, table 29, case 1
4 Clapham L et al "Characterization of texture and residual stress in a section of 610 mm pipeline steel" NDT & E V28, pp 73-82, 1995
5 Timoshenko S, Lessels J "Applied Elasticity" Westinghouse Technical Night School Press York, Pa, USA 1925
6 Kirstein O, Brule A, Nguyen H, Luzin V, Mat. Sc. For. Vols,571-572 (2008) pp 213-217
7 Allen A, Hutchings MT, Windsor CG, Andreani C, "Neutron diffraction methods for the study of residual stress fields," Adv. Phys. 34, 445-473, 1985
8 Measurement of Residual and Applied Stress Using Neutron Diffraction, eds. M. T. Hutchings and A. D. Krawitz; NATO ASI Series 216E, Kluwer Academic Publishers, Dordrecht / Boston / London, 1992.
9 Prask H.J, Brand PC "Residual Stress Determination by Means of Neutron Diffraction,", in Neutrons in Research and Industry, Proc. of Int. Conf. on Neutrons in Research and Industry, ed: G. Vourvopoulos, SPIE Proceedings Series, Vol. 2867, pp. 106-115 (1997)

Measurement of Residual Stress in a Welded Branch Connection and Effects on Fracture Behaviour

Law M, Luzin V, Kirstein O (Australian Nuclear Science and Technology Organisation)

ANSTO, PMB1, Menai, NSW, 2234, Australia

michael.law@ansto.gov.au

Abstract. The branch analysed in this paper was not post weld heat treated, resulting in significant residual stresses. Assessment codes assume these to be at, or close to, yield. An integrity assessment of a welded branch connection was carried out using these high assumed residual stresses.

The weld then had residual stresses determined by neutron diffraction, performed using ANSTO's residual stress diffractometer, Kowari. The maximum measured residual stress (290 MPa or 60% of yield) was much lower than the yield value assumed by assessment codes. Reanalysing with the actual residual stresses almost doubled the critical crack size, increasing the safety of the connection.

1 Introduction

Gas pipelines have connections to feed new facilities, especially gas-fired power stations. The connections are made on in-service pipelines by hot-tapping, welding directly onto a pressurised pipeline. The branch is welded directly to the 'live' pipeline and then reinforced with 2 half sleeves cut from the pipe.

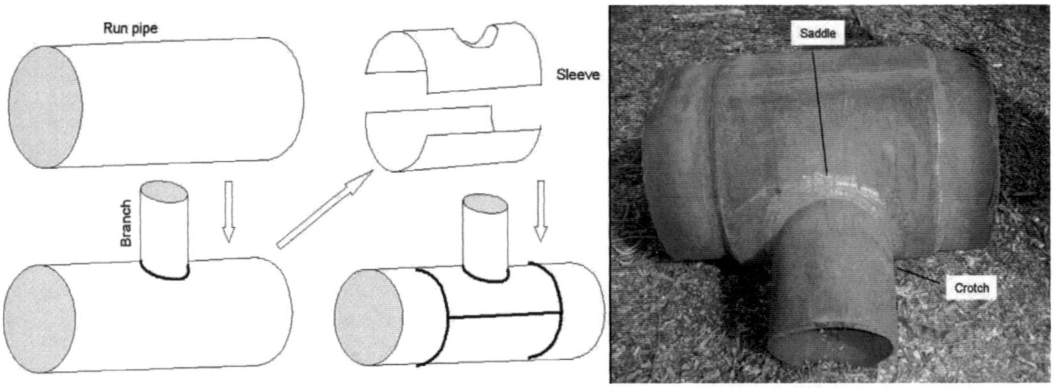

Figure 1 Pipe, branch welded on, components of reinforcing sleeve, and completed branch connection, welds are shown with bold lines.

The welding process generates significant residual stresses which may contribute to premature failure. Residual stress will generally reduce the critical flaw size for crack growth. Methods of measuring residual stress include destructive and non-destructive techniques such as Sachs boring and ring slitting, ultrasonic and magnetic methods, and diffraction methods using x-rays or neutrons. Neutron methods have great penetrating power, giving information to a depth of approximately 30 mm in steels. Reliable fracture mechanics assessments require accurately determined residual stresses. However, the conservative assessment of brittle fracture in codes such as BS 7910 [1], coupled with a high level of assumed residual stress (At yield strength, BS 7910-7.2.4.1) leads to very small critical crack sizes.

2 Component material and geometry

The connection is made from X70 ferritic steel with a minimum yield strength of 483 MPa. The reinforcement sleeve is in contact with the outer surface of the run pipe, and is welded to the run pipe at the ends of the sleeve, and also to the branch. By symmetry, a quarter model was used for all models (figure 2a). The residual stresses were measured where they are expected to be highest, where the connection is stiffest at the saddle. The integrity analysis was performed where the operational stresses are highest, conservatively using the residual stress from the saddle.

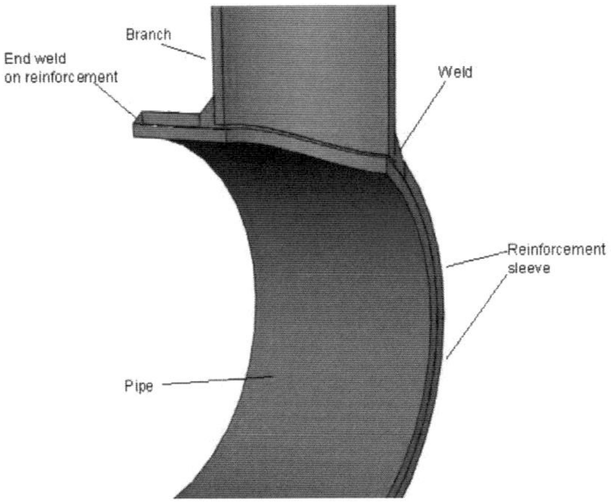

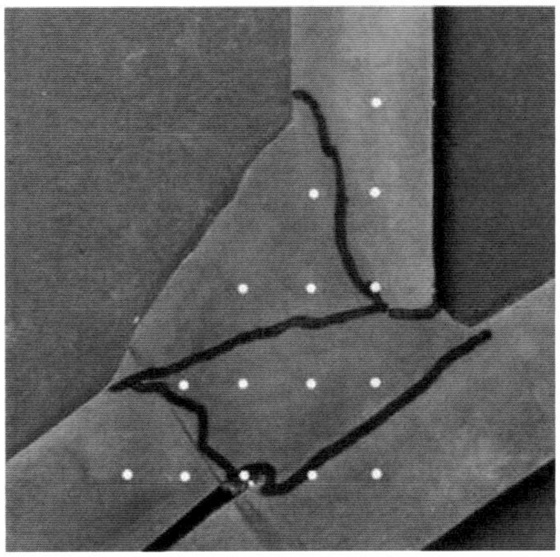

Figure 2a Quarter model of branch showing arrangement of components.
Figure 2b Measurement points over welded region.

3 Neutron residual stress measurements

The residual stress measurements were performed on the Kowari strain scanning instrument at ANSTO. Because the basic principles of this technique are well known [2,3,4] only details specific to this measurement will be reported. Through-thickness scans were performed across the weld and parent. A monochromatic beam with $\lambda = 1.666$ Å and diffraction from Si{400} planes were used to analyse the Fe{211} plane. A nominal gauge volume of 3 x 3 x 3 mm^3 was used to perform scans at 15 positions (figure 2b), the measurements were all at least 6 mm from the surface. A section through the weld and parent was cut into a 6 mm thick plate to derive d_0 values across the weld, heat affected zone (HAZ) and parent. The d_0 values were measured at the same 15 positions as the strains were measured, the values for the parent, HAZ and weld metal varied by less than 0.3%.

4 Analysis results

Residual stresses were measured in the saddle, the connection had not been hydrotested. The maximum stresses (290 MPa) were along the weld direction (figure 5a, b, c) with lower stresses both in the direction of the branch and in a direction normal to both the run and branch pipes.

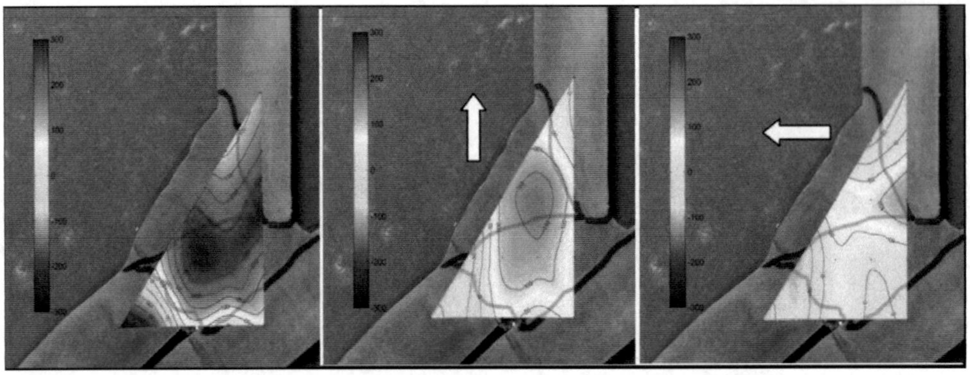

Figure 3a Stresses in direction of weld (out of page), maximum stress is 290 MPa.

Figure 3b Stresses in direction of branch (vertical), maximum stress is 187MPa.

Figure 3c Stresses normal to branch (horizontal), maximum stress is 138 MPa.

5 Integrity modelling

The branch can fail by plastic collapse or by brittle fracture, these failure mechanisms are modelled separately. The plastic collapse is not affected by residual stress, and occurs at a higher pressure than brittle fracture, and thus is not considered further. The brittle fracture analysis considered a flaw through the entire weld at the position of highest stress under operation, aligned axially to the run pipe situated at the weld crotch.

The toughness value of the weld was conservatively assumed to be a minimum of 40J Charpy V value at the minimum expected service temperature, this was converted to a K_{IC} value of 74 MPa√ using Appendix J.2.1 in BS7910-2005. Stresses were taken through the weld at the operating pressure of 16.6 MPa to input into the British Nuclear R6 integrity code [5] which is based on BS 7910-2005.

If this vessel were to be hydrotested the residual stresses would be reduced. The reduction is more significant at the crotch, where the weld is parallel to the major hoop stress. Thus the residual stresses were measured at the saddle. The stresses are taken at the crotch, where pressure induced stresses are highest. Combining these two stresses is a conservative method of assessment,

The stress values are based on the 5 greatest measurements within the weld, with a mean of 259 MPa and a standard deviation 29.9 MPa. Four estimates of the residual stress were used in the analysis:

- The mean value of 259 MPa.
- The maximum measured value of 290 MPa.
- A very conservative estimate of residual stress is the mean plus 3 standard deviations, giving a value of 349 MPa. If the values in the weld are normally distributed, and the area sampled is representative of the weld stresses, this ensures a probability of 0.135% that a residual stress will be above this value.
- The yield strength value assumed by BS 7910 of 483 MPa.

A single fracture mechanics case is presented, that of a through wall cracking equivalent to the full section weld defect in crotch. This is a very conservative model as weld cracking will only penetrate the outer sleeve (a possible 14 mm deep crack, not a full thickness 28 mm crack). The inner sleeve has higher toughness (80 J vs. 40J) and once cracking moves from the weld into the parent metal, residual stresses are much lower. Additionally, the stresses reduce away from the branch weld, while the analysis assumes they are constant. These conservative results are shown in figure 9.

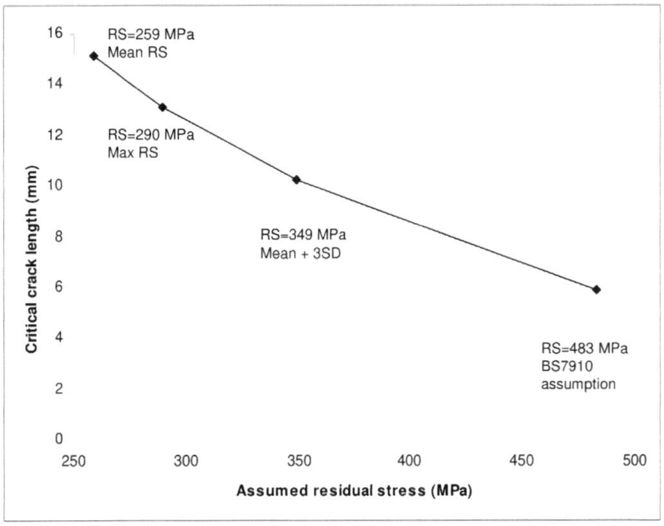

Figure 4 Critical crack size with different residual stress values.

6 Conclusions

The design of a branch connection has been analysed and has been found to be fit for service. The residual stresses were measured by neutron diffraction using ANSTO's residual stress diffractometer, Kowari. The branch was shown to be safe with the large residual stresses assumed in BS7910. The maximum stresses occurred in the direction where the connection was stiffest, along the weld. Transverse stresses were lower due to lower constraint.

Reanalysing with the actual residual stresses significantly increased the critical crack size using very conservative estimates of residual stress based on these measured values.

7 References

[1] BS 7910:2005 Guide to methods for assessing the acceptability of flaws in metallic structures
[2] Allen A, Hutchings MT, Windsor CG, Andreani C, "Neutron diffraction methods for the study of residual stress fields," Adv. Phys. 34, 445-473, 1985
[3] Measurement of Residual and Applied Stress Using Neutron Diffraction, eds. M. T. Hutchings and A. D. Krawitz; NATO ASI Series 216E, Kluwer Academic Publishers, Dordrecht / Boston / London, 1992.
[4] Prask H.J, Brand PC "Residual Stress Determination by Means of Neutron Diffraction,", in Neutrons in Research and Industry, Proc. of Int. Conf. on Neutrons in Research and Industry, ed: G. Vourvopoulos, SPIE Proceedings Series, Vol. 2867, pp. 106-115 (1997)
[5] R-Code R6.CODE, User Guide, Nuclear Electric, Gloucester, UK, 1996

Effect of alloying Ni-Mn-Ga with Cobalt on thermal and structural properties

K Rolfs[1,2], R C Wimpory[1], W Petry[2], R Schneider[1]
[1] Helmholtz-Centre Berlin for Materials and Energy, Glienicker Str.100, 14109 Berlin, Germany
[2] Technical University Munich, Physics Department E13, James-Franck-Str.1 85747 Garching, Germany

katharina.rolfs@helmholtz-berlin.de

Abstract. Materials showing a large magnetic field induced strain can potentially substitute giant magnetostrictive materials as well as piezoelectrical ceramics in actuating devices. However the magnetic shape memory alloys face several problems for an industrial application. Besides the well known brittleness the temperature-range is still limited due to the structural and magnetic phase transition around 75°C. By alloying one of the most common systems Ni-Mn-Ga with 4.8 at% and 6 at% Cobalt the Martensite temperature was increased up to 422K and the Curie temperature up to 432K. Neutron diffraction experiments were performed on these Ni-Mn-Ga-Co single crystalline samples at room temperature, which show the existence of two different non-modulated martensitic structures. Dependent on the composition, a tetragonal and an orthorhombic structure in single crystalline samples were verified. Furthermore intermartenstic phase transitions above room temperature were determined in several samples alloyed with 6 at% cobalt. A temperature dependent measurement of the structure of $Ni_{44.6}Mn_{29.1}Ga_{20.1}Co_{6.2}$ showed an orthorhombic phase from room temperature up to 379K changing to a tetragonal intermartensitic phase before reaching the austenitic phase above 389K.

1. INTRODUCTION

Materials, which show a magnetic field induced strain (MFIS) a magnitude higher than piezo- or magnetostrictive materials became a research topic of high interest for applications in sensors or actuators. One of the most promising materials is Ni-Mn-Ga, due to their MFIS of up to 6.8% in low magnetic fields i.e. less than 0.5T [1]. However the brittleness and the low phase transition temperatures limit the application. The replacement of Nickel by other elements like rare earth-elements, iron or aluminum leads to an improvement of the mechanical properties. The systems are less brittle which results in a longer lifetime [2]. Beside the brittleness, the thermal, structural and mechanical properties are affected by the alloying. Iron replacing nickel for example increases the Curie temperature while the substitution of manganese increases the Martensite temperature [3]. The same behavior has been expected in the case of cobalt as the substitute. Like in Ni-Mn-Ga – single crystals, the composition also influences the martensitic structure [4]. The results of the thermal and structural properties of Ni-Mn-Ga-Co single crystals with two different amounts of cobalt and changing Mn/Ga – ratio will be presented here.

2. Experimental

Using the Bridgeman-like technique Slag Remelting and Encapsulation (SLARE) [5] two single crystalline rods with the weighted compositions $Ni_{45}Mn_{31}Ga_{19.2}Co_{4.8}$ and $Ni_{44}Mn_{31}Ga_{19}Co_6$ were grown. The rods were cut into several 6x4x3mm samples along the 100 and 010 direction in the cubic phase.

The composition of the samples was verified by a calibrated Energy Dispersive X-Ray spectrometer (EDS), while the structural and magnetic phase transition temperatures were determined by Differential Scanning Calorimetry (DSC) cycling the temperature from 30 to 250°C one time with 10K/min and afterwards with 5K/min.

The martensitic structures of selected Ni-Mn-Ga-Co – single crystals were determined by neutron diffraction at the Instruments E3 with a perfectly bents silicon monochromator with a wavelength of 1.468A of the Helmholtz-Center Berlin of Materials and Energy [6] at room temperature and temperatures up to 400K.

3. Results

3.1 *Thermal properties of Cobalt alloyed Ni-Mn-Ga single crystals*

The high vapor pressure of manganese causes a concentration gradient developing during the growth process along the growth axis of the single crystalline rod. The composition of the samples cut out of the rod along the growth axis shows a deviation of manganese and gallium of more than 3 at%, whereby the amount of manganese increases with the decrease of gallium (see Tab.1). Nickel and cobalt are well distributed over the whole rod. The concentration of both elements remains nearly constant in the determined samples.

Tab.1 Composition of several Ni-Mn-Ga-Co cut out of two different single crystalline rods with the weighted compositions $Ni_{45}Mn_{31}Ga_{19.2}Co_{4.8}$ and $Ni_{44}Mn_{31}Ga_{19}Co_6$

Sample	Co [at%]	Ni [at%]	Mn [at%]	Ga [at%]	Mn/Ga
$Ni_{45}Mn_{31}Ga_{19.2}Co_{4.8}$					
1-A	5.0	45.7	29.4	19.9	1.477
1-B	4.8	45.4	30.2	19.6	1.540
1-C	4.8	46.0	30.6	18.6	1.645
1-D	4.9	45.1	30.9	19.1	1.618
1-E	4.9	44.7	31.7	18.7	1.695
$Ni_{44}Mn_{31}Ga_{19}Co_6$					
2-A	5.8	45.0	27.5	21.7	1.264
2-B	6.0	45.2	27.6	21.2	1.299
2-C	6.2	44.9	28.2	20.7	1.362
2-D	6.2	44.6	29.1	20.1	1.443
2-E	6.0	44.6	29.5	19.9	1.481

Due to this concentration gradient along the rod, the samples show different structural phase transitions temperatures. Samples containing approximately 4.8% cobalt show a Curie-temperature (T_C) of around 420K, while Samples with the higher cobalt content of around 6 at-% exhibit a higher T_C of approximately 435K independent from the Mn/Ga ratio (see Fig.1)

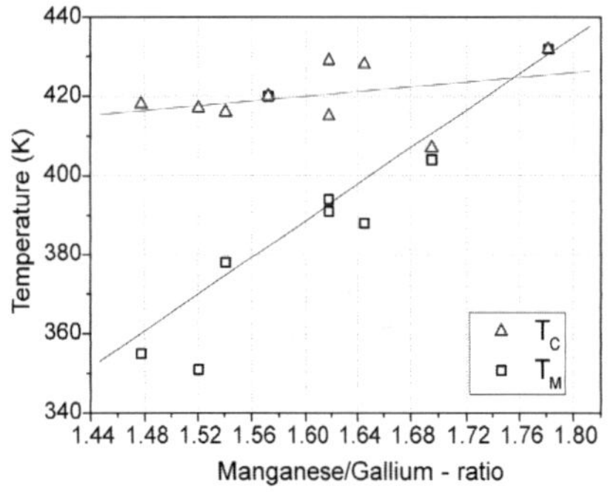

A

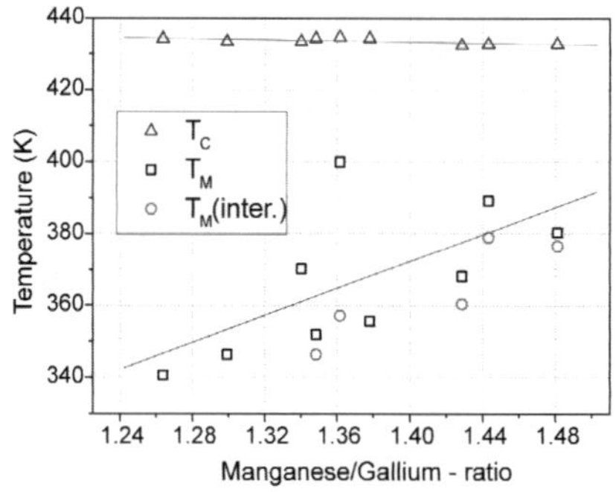

B

Fig 1: Phase trasition temperature of A: $Ni_{45}Mn_{31}Ga_{19.2}Co_{4.8}$ and B: $Ni_{44}Mn_{31}Ga_{19}Co_6$ determined by DSC versus the Mn/Ga ratio of the particular samples. The black squares indicate the Martensite temperature, the red circles the intermartensitic phase transition temperature and the blue triangles the Curie temperature.

The Martensite-temperature in both alloys is highly affected by the Mn/Ga-ratio. While the structural phase transition from the Martensite to the Austenite in samples with 4.8 at% Co is shifted from 350K in the Ga-richest sample to 435°C in the sample from the other side of the rod. (see Fig.1) In samples containing around 6 at% Co accompanied by a decreasing Ga-content an increasing Martensite temperature from 340 to 400K was verified.

Furthermore most of the determined samples with 6 at% Cobalt show a second intermartensitic phase transition, also increasing from 340K to 370K with the decreasing Ga-content.

3.2 *Structural properties of Co-alloyed Ni-Mn-Ga single crystals*

3.2.1 *Martensitic structure of samples with 4.8 at% Co*. The martensitic structures of sample 1-D and 1-E from the single crystalline rod with the weighted composition $Ni_{45}Mn_{31}Ga_{19.2}Co_{4.8}$ were determined by neutron diffraction. Both samples have exactly the same amount of cobalt, but with a different Mn/Ga – ratio.

Sample 1-D shows a tetragonal martensitic structure with the lattice parameters a = b = 6.56A and c = 5.49A, which results in a cell volume of $236.25A^3$. For the second sample 1-E with a higher Mn/Ga – ratio an orthorhombic unit cell was verified. The experiment showed the three different cell-parameters a = 6.62A, b = 5.58A and c = 5.48A and therefore a slightly smaller unit cell volume of $212.22A^3$.

3.2.2 *Martensitic structure of samples with 6 at% Co*. Two samples of the single crystalline rod with the weighted composition $Ni_{44}Mn_{31}Ga_{19}Co_6$ were chosen for the neutron diffraction experiments. The martensitic structure of sample 2-A and 2-D was determined at room temperature. Both samples show an orthorhombic nonmodulated structure with similar cell parameters. The longest axis of sample 2-A is c = 6.53A and for the shorter axis the values b = 5.57A and c = 5.40A were verified. The measurements on sample 2-D resulted in the cell parameters a = 6.58A, b = 5.52A and c = 5.51A.

Since this sample shows an intermartensitic phase transition at 379K, temperature dependent diffraction pattern showed a change in the lattice constants (see Fig. 2). With increasing temperature, the c – axis becomes smaller, while the length of a and b increases. Above the phase transition temperature a tetragonal nonmodulated structure with a = 6.54A and b = c = 5.54A was determined. By increasing the temperature up to 408K, again a decrease of the lattice constant c and an increase of a and b was verified.

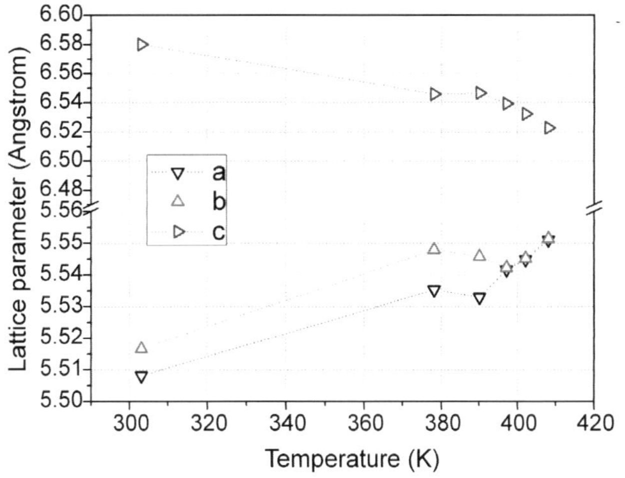

Fig. 2 Lattice constants of Sample 2-D dependent on the temperature. Above the intermartensitic phase transition the structure changes from an orthorhombic nonmodulated structure to a tetragonal nonmodulated structure. The diffraction experiments were done at the HZB Instrument E3 with a wavelength of 1.486A.

4. Conclusion

Replacing nickel with cobalt in Ni-Mn-Ga strongly influences both phase transition temperatures. The Curie temperature is shifted from approximately 380K [7] in Ni-Mn-Ga alloys with 50 at% Ni up to 435K in Ni-Mn-Ga alloyed with 6 at% Co. Comparing the Martensite temperature of Ni-Mn-Ga alloys with similar Mn/Ga-ratios and the Co-alloyed samples, an increase of around 20K is found [1].
Besides the nonmodulated tetragonal structure also found in Ni-Mn-Ga alloys, a nonmodulated orthorhombic structure was verified. The cell volume of Ni-Mn-Ga alloyed with 6 at% Co increases with the Mn/Ga ratio, which can be explained due to the higher amount of Mn, which has a bigger radius than Ga.

References

[1] K Rolfs, A Mecklenburg, J-M Guldbakke, R C Wimpory, A Raatz, J Hesselbach and R Schneider 2009 Mag. Mag. Mater. **321** 1063
[2] S H Guo, Y H Zhang, Z Q Zhao, J L Li, and X.L. Wang 2004 J. Rare Earths **22** 632
[3] G H Wu, W H Wang, J L Chen, L Ao, Z H Liu, W S Zhan, T Liang and H B Xu 2002 Appl. Phys. Lett. **80** 4 634
[4] N Lanska, O Söderberg, A Sozinov, Y Ge, K Ullakko and V K Lindroos 2004 J. Appl. Phys. **95** 8074
[5] A Mecklenburg, S Fiechter, H-P Nabein, R Schneider 2005 DE102004018664A1
[6] R C Wimpory, P Mikula, J Saroun, T Poeste, J Li, M Hofmann and R Schneider 2008 Neutron News 19 (1) 16
[7] V. V. Khovailo, V. Novosad, T. Takagi, D. A. Filippov, R. Z. Levitin and A. N. Vasil'ev, Phys.Rev. B 70, 174413 (2004)

Neutron diffraction study of Levantine Middle Bronze Age cast axes

El'ad N. Caspi[1], Sariel Shalev[2,3], Sana Shilstein[3], Anna M. Paradowska[4], Winfried Kockelmann[4] , Yossi Levy[5]

[1]Physics Department, Nuclear Research Centre – Negev, POBox 9001, 84190 Beer-Sheva, Israel

[2]Leon Recanati Institute for Maritime Studies & Institute of Archaeology, University of Haifa, Mount Carmel, 31905 Haifa, Israel

[3]Kimmel Center for Archaeological Sciences, Weizmann Institute of Science, 76100, Rehovot, Israel

[4]Rutherford Appleton Laboratory, ISIS Facility, Chilton Didcot, Oxfordshire OX11 0QX, UK

[5]Israel Antiquity Authority, PO Box 586, Jerusalem 91004, Israel.

E-mail: caspie@nrcn.org.il

Abstract. A neutron diffraction study on 6 Middle Bronze Age axes, cast from tin bronze or from arsenical copper, has been carried out using the ENGIN-X beamline at ISIS. The gauge volumes dimensions were $4x4x10mm^3$; data were collected along the lengths of the objects in their central parts, as well as on the blades, in order to establish the spatial phase contents. Average phase fractions were determined by Rietveld analysis. The main phases identified were solid solutions, corrosion phases and metallic Pb inclusions. We have observed distributions of lattice constants of the solid solutions Cu–Sn, and Cu-As inside each gauge volume in the central parts of the axes due to segregation, or liquation effects. However, the Cu-Sn variations were significantly less pronounced in comparison with typical inhomogeneity effects in as cast objects. The results indicate that the studied Middle Bronze Age axes were probably treated at high temperatures for homogenization necessary for generating sufficient hardness, especially on the blades.

1. Introduction

Socket axes in duckbill- or chisel-shapes were found in "Warrior tombs" of the Middle Bronze Age (about 2000 years b.c.) in Rishon LeZion, Israel. These axes were produced by casting of tin bronze or arsenical copper containing lead. The Sn content (up to about 14%) or As content (up to 5%) is important for improving the mechanical properties of an axe in comparison with a pure Cu axe. Some amount of Pb (up to 2–3%) is applied to improve liquidity of a molten metal during the casting process. Usually, some spatial variation of the doping metals (Sn, As, and Pb) has been observed in such cast tools. In many ancient objects variation in spatial distribution of both doping metals (Sn, and Pb, or As, and Pb) had been established, using mainly XRF. In some cases, the local Pb concentration was found to be 20% [1]. Since the effective depth of an XRF analysis in copper alloys is in the order of tens of micrometers, it is important to verify that the above mentioned spatial distribution is in fact a bulk property. Additionally, investigation of the existence (or non-existence) of Sn, or As inhomogeneities, typical for cast objects is of great interest as indication of the metal working. The importance of studying archaeological artefacts using bulk methods was demonstrated in our previous work, where neutron diffraction (ND) method for ancient artifacts had been used [2,3]. In this research, a pilot ND study was carried out on 6 Middle Bronze Age axes using ENGIN-X at ISIS.

2. Samples and experimental set up

ND analyses were conducted on six artefacts obtained from the Antiquity Authority of Israel (Fig. 1): one tin bronze duckbill-shaped axe BA8 (A-2731/97, L.743, B.7327); three tin bronze chisel-shaped axes BA3 (A-2502/96, L.506, B.5077), BA6 (A-2731/97, L.654, B.6362), and BA10 (A-2731/98, L.1075, B.6727); two arsenical copper axes BA12 (A-2731/98, L.1118, B.9418), and BA16 (A-2731/98, L.755, B.7844). Only the BA3 axe was in a virgin corroded state. All other axes were cleaned mechanically and treated for conservation.

Diffraction measurements on ENGIN-X [4] were carried out with gauge volumes of 4x4x10mm^3. Measurements were performed along the lengths of the objects in their central parts, and also on the blades. The analyses were carried out to establish the phase contents (Cu–Sn or Cu–As solid solutions, intermetallic compounds of Cu, and Sn or As, Pb metal, corrosion products) and inhomogeneities (segregation) of Cu–Sn or Cu–As solid solutions.

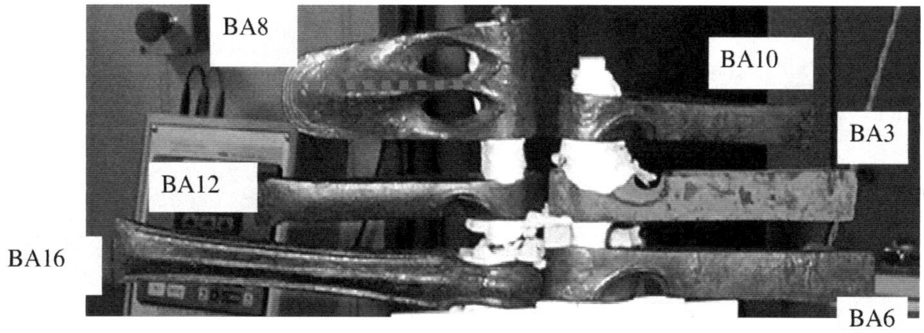

Fig. 1 – The axes on the ENGIN-X sample holder. Left (from top to bottom): BA8, BA12, BA16, right (from top to bottom): BA10, BA3, BA6. Note for scaling that the chisel-shaped axe BA16 is 165 mm long and that the chisel-shaped axe BA10 is 102 mm long. The dotted line on BA8 exemplifies the ND line scan.

Data (d-range between 1.25 Å – 2.9 Å) were analyzed in detail using the Rietveld refinement method with the GSAS (EXPGUI) code [5,6]. The refined model contained between 16 and 22 variables including phases weight fraction values, cell parameters (with starting values taken from the corresponding literature [7-10]), background, and scale factors. For peak-shape modeling we used the time-of-flight function #3 available in the GSAS code ([5], and references therein), usually used for ENGIN-X data analysis, by refining only σ_1 for the main solid solution phases. The resulted weighted profile agreement factor values, R_{wp}, [5] varied in the range 7-12%.

3. Results and Analysis

The ND patterns of all studied axes show three strong lines associate with the {111}, {200}, and {220} family of reflections (c.f. Fig. 2), belonging to a face centered cubic type lattice. The lattice parameters estimated from these reflections are in the range of 3.66 Å to 3.68 Å in the Sn bronze axes, and in the range of 3.62 Å to 3.63 Å in the As bronze axes. These values are in good agreement with cell parameters values of the α-phases, or solid solutions: Cu-Sn (tin bronze) and Cu-As (arsenical copper), respectively [11,12]. In most of the diffraction patterns, additional small reflections lines are observed associated with impurity phases and corrosion products. These lines are strongest on both edges of the studied axes (close to the edge of the blade, and in the handle). The intensities and positions of the impurity lines are in agreement with the CuCl [7], Cu_2O [10], SnO_2 [8], and Pb [9] phases for the tin bronze axes. Except for SnO_2 the same impurity lines appear in the diffraction patterns of the arsenical copper axes. Neither in the tin bronze axes nor in the arsenical copper axes were secondary intermetallic phases found (e.g. δ-phase in Cu-Sn [11]; or Cu_8As, and Cu_3As in Cu-As [12].

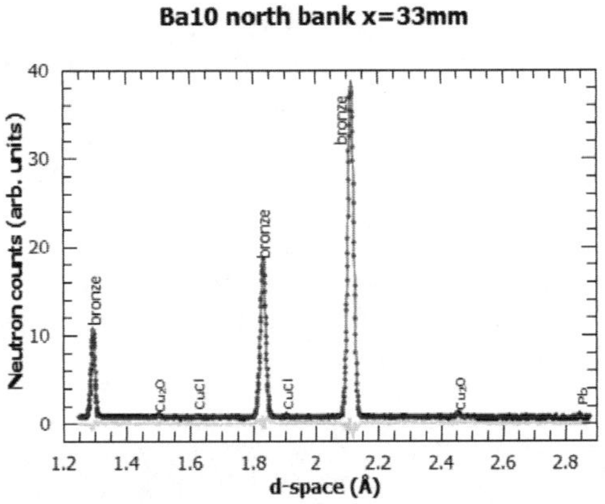

Fig. 2 – Rietveld refinement analysis of the neutron powder diffraction data taken at 33 mm from the blade edge of the tin-bronze BA10 axe, measured on ENGIN-X. Solid circles, continuous line and bottom line are measured data, refined model and the difference between them, respectively. The refined model includes two tin-bronze solid solutions along with three additional phases Cu_2O, CuCl, and Pb.

Close inspection of the diffraction patterns taken from all tin-bronze axes shows significant broadening, or strong peak asymmetry of the copper alloy reflections in the bulk data, collected at a distance from the edges (~20-30 mm from the edges of the blades; henceforth "bulk") compared to lines in patterns obtained right on the edges. This broadening, or peak asymmetry either increases or retains the same value as a function of the studied gauge volume distance from the blade edge of each of the three chisel shaped axes (BA3, BA6, and BA10). In contrast, for the handle part of the duck-bill shaped axe BA8 (approximately 63 mm from the blade edge, close to "eyes"), the observed line width is similar to the one observed on the edge of the axe.

A detailed Rietveld analysis of all measured diffraction patterns was undertaken. The refined model used consisted of two main phases (solid solutions with different cell parameters), corrosion phases (i.e. Cu_2O, CuCl for all tin-bronze and arsenical copper axes, and an additional SnO_2 phase for the tin-bronze axes), and a metallic Pb phase. Good agreement between refined model and observed data is achieved for all diffraction patterns (R_{wp} between 7-12%). As an example, the cell parameter values of the two main tin-bronze solid solutions as a function of spatial position, x, along the BA6 axe, are depicted in Fig. 3, alongside with a picture of the studied axe. It is important to emphasize here that the use of the two solid solutions model for the analysis of the ND data is undertaken in order to simplify the data analysis process. The existence of a continuous compositional shift of solid solubility could not be ruled out by the observed data. Nevertheless, the use of this model does not affect our main conclusions.

For BA6 the Rietveld refinement of patterns collected at distances of up to ~30 mm from the blade edge gives a good fit with only one tin-bronze solid solution in the model. At a distance from the blade edge, i.e. in the bulk of the axe, a good refinement requires two different solid solutions (about 60% and 40% in weight), in agreement with the observed lines shape. Moreover, the difference in the cell parameters of the two solid solutions increases as a function of distance (Fig. 3). Similar behavior is also observed in BA3, and BA10, which are all chisel-shaped tin-bronze axes. In the duckbill-shaped axe, BA8, the refinement indicates one solid solution phase close to the blade, with two different solid solutions in the bulk part, and, again, with one solid solution close to the "eyes". Following the calibration curve published in [11], and using the refined values of the cell parameters associated with the tin-bronze solid solutions we are able to estimate the tin content in the different parts of the studied axes. The average tin content for all tin-bronze axes is in the range of 7 – 10 wt% for the chisel-shaped axes, and ~ 11 wt% for the duck-bill-shaped axe. In all tin-bronze axes, the average tin content is constant along the length of the axe. However, the width of the tin distribution, as evidenced by the necessity to use two copper alloy phases, increases from edge to bulk. As an example, the width of the tin distribution of the axe BA6 (Fig. 3) is 5-11 wt% in the bulk, indicating a considerable degree of tin segregation. On the other hand, the sharp tin distribution on the edge indicates that the axe had undergone considerable heat treatment for hardening of blade edge.

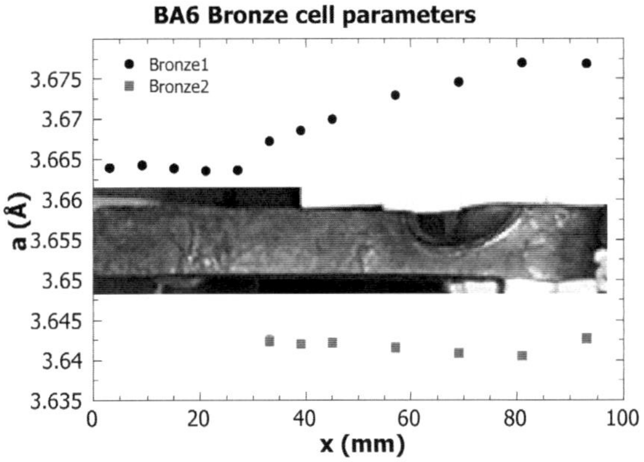

Fig. 3 – BA6 tin-bronze refined cell parameters values as a function of distance from the blade edge, as deduced by Rietveld refinement of the "two phases' model" (see text) to the ENGIN-X data. Error bars values, deduced from the refinement process, are smaller than symbols sizes.

In the diffraction patterns of the arsenical copper BA12 axe no change of the line widths of the main phase is observed as a function of distance from the blade edge. However, the diffraction lines measured for this axe are broader than those from BA16 lines. Moreover, the BA12 peak broadening is even more pronounced than from the edges of the Cu-Sn axes, indicating considerable segregation effects in this arsenic copper axe, even on the edges.

The refined Pb weight fractions show maximal values on the blade edges, but in no case do they exceed ~3 wt% (optimal for casting). It means that the earlier observed high Pb local concentrations (~20 wt%) by XRF [1] are definitely connected with surface effects, probably related to conservation treatments.

In conclusion, we found that neutron diffraction provides additional unique non-destructive information on the internal structure of Bronze Age axes, especially regarding the technology of their production. In particular, the indications in the observed microstructure of heat treatment for mechanical hardening demonstrate rather high qualification of the metal experts nearly 4000 years ago. Moreover, such post-manufacture treatment is a demonstration that the presently studied Middle Bronze Age axes were probably destined for use as weapons, rather than for use as ritual objects.

References

[1] Shalev S, Shilstein S, Levy Y 2008 *Abstratcs of 37th International Symposium on Archaeometry* (Siena, Italy) 406.
[2] Caspi E N, Ettedgui H, Rivin O, Peilstöcker M, Breitman B, Hershko I, Shilstein S, Shalev S 2009 Journal of Archaeological Science, doi:10.1016/j.jas.2009.09.011.
[3] Kockelmann W, Siano S, Bartoli L, Visser D, Hallebeek P, Traum R, Linke R, Schreiner M, Kirfel A 2006 *Appl. Phys. A* **83** 175.
[4] Daymond M R, Priesmeyer H G 2002 *Acta Mater.* **50** 1613.
[5] Larson A C, von Dreele R B 2004 *Los Alamos National Laboraory Report* LAUR 86-748.
[6] Toby B H 2001 *J. Appl. Cryst.* **34** 210.
[7] Pfitzner A, Lutz H D 1993 *Z. Krist.* **205** 165.
[8] De Meyer C, Bourée F, Evans J S, De Buysser K, Bruneel E, Van Driessche I, Hoste S 2004 *J.*

Mater. Chem. **14** 2988.
[9] Klug H P 1946 *J. Amer. Chem. Soc.* **68** 1493.
[10] Machado K D, de Lima J C, Grandi T A, Campos C E, Maurmann C E, Gasperini A A M, Souza S M, Pimenta A F 2004 *Acta Cryst. B* **60** 282.
[11] Siano S, Bartoli L, Zoppi M, Kockelmann W, Daymond M., Dann J A, Gargnani M G, Miccio M 2003 *Proc. Archaeometallurgy in Europe* **2** 319.
[12] Subramanian P R, Laughlin D E 1988 *Bull. Alloy Phase Diagrams* **9** 605.

Study of phase formation in metal injection moulding through real time neutron diffraction

R E Whitfield[1] **D J Goossens**[2] **A J Studer**[3]

[1] Research School of Physics and Engineering, Australian National University, Canberra 0200, Australia

[2] Research School of Chemistry, Australian National University, Canberra 0200, Australia

[3] Bragg Institute, Australian Nuclear Science and Technology Organization, Lucas Heights, Sydney, Australia

E-mail: `ross.whitfield@anu.edu.au`

Abstract. The sintering of metal injection moulded stainless steel was investigated using *in situ* neutron diffraction with different sintering temperatures, from 1270°C up to 1390°C, with sintering profiles that were based on those used in industry. The production of an unwanted high temperature phase, δ-ferrite, was observed during sintering and is seen to be retained in the final product after sintering. Ferrite production during sintering acts to speed up the sintering process by forming in the grain pores but is unwanted in the final product as it is a soft and malleable phase. The ferrite that was formed at high temperature was observed to not completely disappear during cooling as a result of the coexistence of dual high temperature phases delta-ferrite and gamma-austenite during the high temperature soak. This suggests the segregation of the alloying elements between the two phase which changes the composition of the phase grains and allows the ferrite to exist during cooling, resulting in the unwanted phase in the final product.

1. Introduction

Metal injection moulding (MIM) is the process of producing small and delicate castings, as small as 300 microns across and 30 mg in weight, without the need for melting but with near liquid cast strengths. The process is to make a mould which is 10 to 20% oversize and inject this with a mix of metal powders and an organic binder. The binder is then cooked away at around 600°C leaving a porous material and then the casting is heated to around 1400°C, to allow densification to occur, shrinking the sample to the final size with a near to liquid casting strength.

17-4PH stainless steel was looked at as it is the most common precipitation hardened stainless steel and contains around 17 wt.% chromium and 4 wt.% nickel along with other alloying elements which gives it high strength and good corrosion resistance [1]. Chromium is a ferrite-stabilising element while nickel is an austenite-stabilising element [2]. The high temperature δ-ferrite phase is wanted during sintering as it acts to speed up the sintering by forming in the pores between the grains but is unwanted in the final sample. Previous investigations of MIM 17-4PH have focused on the shrinkage rates during the sintering [1]. It was found that 17-4PH can only be sintered to near-full density in pure hydrogen or a vacuum [3, 4, 5]. This investigation looks at the phase composition during the sintering process too better understand the densification process which leads to insights into the sintering of metal injection moulded

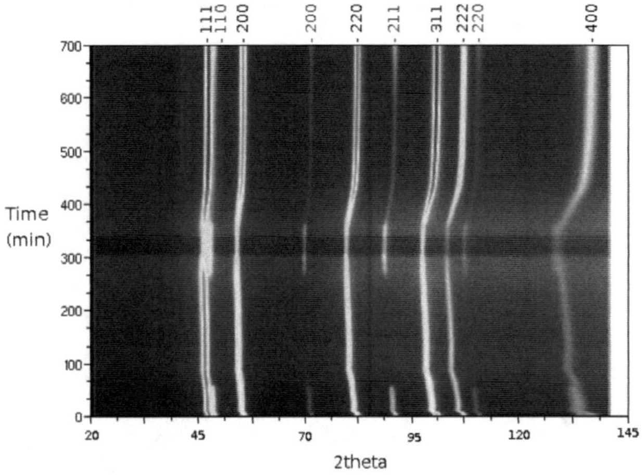

Figure 1: The diffraction pattern intensity as a function of 2θ, with intensity represent by colour, blue low intensity, red high intensity as it varies with time. The sample is heated to a maximum of 1390°C. The reflections are labeled with austenite black and ferrite in blue. Wavelength of the neutrons used was 1.664 Å.

stainless steel that can be used to optimise sintering profiles. The relationship between the shrinkage rate, shown in Figure 1 in ref [1] and phase composition is explored and the source of the production of ferrite in the final sample suggested.

2. Experimental

The samples were produced for the neutron diffraction experiments by a MIM manufacturer [6]. The samples were first debindered before being sintered in the beam. The sintering process as used by ref [6] involves a fast heating ramp of 10°C/min up to 50°C below the maximum temperature then 3°C/min up to the maximum. The sample is then held at that temperature for an hour before cooling at 10°C/min down too 1000°C where the furnace is turned off and the sample allowed to cool to room temperature.

The sintering was done in a vacuum furnace while on WOMBAT, the High Intensity Powder Diffractometer at ANSTO [7] using the 115 reflection on the Ge monochromator with a wavelength of 1.664 Å. This allowed diffraction patterns to be collected in real time and the phase transformations to be observed. Upon heating the steel went through two phase changes first from the α-ferrite to γ-austenite then to a δ-ferrite phase. A time resolution of about 1 min for each diffraction pattern was used as this is quick enough to see the phase changes as they happen and also long enough to get a large number of counts in the diffraction pattern. One of the evolution of diffraction patterns shown in Figure 1. The sintering was done with a range of maximum temperature ranging from 1270°C to 1390°C. Sintering temperature are listed in Table 1.

3. Results and Discussion

Rietveld analysis was performed using MAUD[8, 9] on every diffraction pattern collected for each of the sintering profiles. This allowed quantitative phase analysis as a function of time and temperature to be performed. The results of the weight fraction for austenite and ferrite for one of the sintering profiles with a maximum temperature of 1390°C is shown in Figure 2. The temperature profile used for the 1390°C sinter was reproduced from ref [1] too allow easier comparison with their shrinking rate data.

The sample that was sintered at 1270°C didn't reach a temperature high enough to enter the δ-ferrite phase. The amount of ferrite remaining in the sample depended on the maximum temperature reached, with the highest amount found, around 10.6% Wt, with the sample sintered at 1390°C and around 1.2% and 1.5% found with the samples sintered at 1330°C and 1360°C

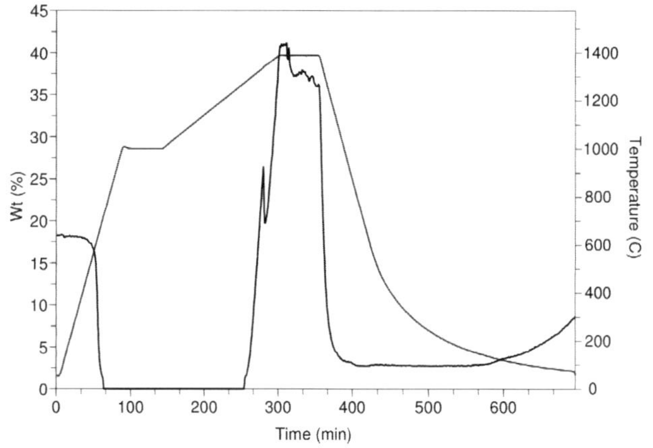

Figure 2: Weight percent vs time for α or δ-ferrite in black. Temperature vs time in blue. The sample is heated to a maximum of 1390°C.

Sintering temperature	δ-ferrite max Wt %	α-ferrite final Wt %	Final volume (mm^3)	% Shrinkage
1270	0	0	3094	9.6
1330	4.8	1.2	3093	10.0
1360	22.4	1.5	2692	12.8
1390	42	10.6	2406	

Table 1: The debinding temperature relating the maximum ferrite found at high temperature and the final ferrite in the sample. The final volume is given to represent the amount of shrinkage, initial volume was 4236 mm^3.

respectively. Looking at Table 1 shows the relationship between sintering temperature and shrinkage of the sample with relation to the amount of ferrite in the sample. There is little difference between the shrinkage when sintered at 1270°C and 1330°C but sintering at 1360°C results in much more shrinkage. The production of δ-ferrite acts to increase the rate of sintering of the sample where this is seen in our samples with little or no δ-ferrite not shrinking as much as sample that entered further into a pure δ-ferrite phase.

It can be seen that 1360°C versus 1330°C gives almost the same ferrite phase fraction after cooling, but the much higher δ-ferrite at high temperature gives better shrinking at higher temperature, Table 1, which means better densification and stronger component. This is only true if the sample is not quenched from high temperature. This explains why the experimentally derived profile finds it best to slowly cool to 1000 C before quenching as this reduces the ferrite that remains in the sample.

It is suggested that the high temperature dual phase allows the segregation of the stabilising compounds into their preferred phases due to their different solubility in different phases. The result is a δ-ferrite with a higher amount of chromium and γ-austenite with a higher amount of nickel in their local grains. So the δ-ferrite attains more chromium and less nickel which shifts its position in the phase diagram and allows the ferrite to exist during cooling and in the final product. It was also observed that the higher temperature which results in the higher amount of δ-ferrite caused more ferrite to remain in the sample which suggests a greater amount of segregation occurred between the two high temperature phases.

The relationship between the shrinkage of the sample, Figure 1 in ref [1] and the phase composition, Figure 2, to be seen with the major shrinkage occurring with the production of a δ-ferrite phase which acts to close up the pores in the sample. While the phase change into the high temperature δ-ferrite acted to increase the shrinkage, which can be seen by the different final densities of the 1330°C and 1360°C sintered steel, the segregation of the alloying elements has to be inhibited to prevent the δ-ferrite being stabilized to lower temperatures: this maintains the desired mechanical properties of the final product.

To prevent the segregation of the alloying elements the two phase system at high temperature has to be avoided. The steel has to be either sintered at a lower temperature in a pure austenite phase but for a much longer time as the δ-ferrite acts as a catalyst for shrinkage, or heated up higher to a pure δ-ferrite phase. This will prevent segregation because there is only one phase but large grain growth is likely to occur in the δ-ferrite. Grain growth may have negative effects on the physical properties especially when dealing with very small and delicate objects since δ-ferrite grains can grow to 2-3 mm in size. Large grains also result in bad mechanical properties.

Sintering with an oscillating temperature around the δ-ferrite phase change temperature could allow for a faster sinter because δ-ferrite will form in the grain pores, with the additional benefit of having a shorter time in a dual δ-ferrite/γ-austenite phase region which could decrease the segregation of alloying elements and in turn decrease the amount of ferrite in the final product. Alternatively because little shrinking happens while at the maximum temperature, Figure 1 in ref [1], the hold at that point could be removed. So the sample is heated to the maximum temperature then is immediately cooled. This should have no effect on the densification of the sample but will reduce the time needed for sintering and also reducing the ferrite in the final sample.

Further work could involve the use of a scanning electron microscope (SEM) to observe the final samples microstructure which will give the ability to see which alloying elements are undergoing segregation in different phases. Also the effects of varying the heating and cooling rates the final phase composition could be investigated along with removing the high temperature soak in the sintering cycle.

4. Conclusion

This research has been able to show like never before the phase transitions happening during the sintering of the metal injected moulded stainless steel. It leads to insights into the sintering of metal injection moulded stainless steel which cannot be done using any other technique. The phase composition during the sintering has been shown and it has been suggested that the production of δ-ferrite in the final product can be reduced by minimising the time in the high temperature dual phase while still allowing the complete densification of the sample. This means a lot less additional processing after the sintering is require to remove the ferrite in the sample. This can be used to optimize sintering profiles used in industry.

References

[1] Blaine D, Wu Y, Schlaefer C, Marx B and German R 2003 Sintering Shrinkage and Microstructure Evolution during Densification of a Martensitic Stainless Steel, *In Proceedings Sintering*

[2] Kennedy R, Roberts G and Krauss G 1998 *Tool Steels.* ASM International, 5th edition

[3] Zhang H and German R 1992 *Proc. 1992 Powder Injection Molding Symp.* Metal Injection Industries Federation, pages 219-27

[4] Banerjee S, Booker P, Gaspervich J and German R 1992 *Proc. 1992 Powder Injection Molding Symp.* Metal Powder Industries Federation, pages 181-92

[5] Baba T, Miura H, Honda T, Tokuyama Y, Philips M and Porter J 1995 *Advances in powder metallurgy and particulate materials.* Metal Powder Industries Federation, **2**:6.271-78

[6] Teo J Ceramet technologies 2008 Private Communication

[7] Noakes T, Studer A and Hagen M 2006 *Physica* B 385-386:1013-15

[8] Wenk H, Lutterotti L and Matthies S 1999 *Proceeding of the Twelfth International Conference on Textures of Materials* **1**:1599

[9] Wenk H, Lutterotti L and Matthies S 1999 *IUCr: Newsletter of the CPD* **21**:1415

International Conference on Neutron Scattering 2009 IOP Publishing
Journal of Physics: Conference Series **251** (2010) 012049 doi:10.1088/1742-6596/251/1/012049

Neutron Diffraction Residual Stress Measurements in Key-Hole Laser Formed Weldments

A.M. Paradowska[1], W. Suder[2] and S. Williams[2]

[1] ISIS, Rutherford Appleton Laboratory, Didcot, OX11 0QX, Oxfordshire, UK

[2] Welding Engineering Research Centre, Cranfield University, Cranfield, MK43 0AL, Bedfordshire, UK

E-mail: anna.paradowska@stfc.ac.uk

Abstract

There is a new generation of lasers available for materials processing. The new lasers include fibre and disc laser are characterized by very high beam quality at very high power levels providing extremely high energy density. In general laser welding is selected over other welding processes because of the minimisation of the heat input, residual stresses and distortion, for this reason they are increasingly popular over a wide range of industries. The narrowness of the welds makes it challenging to resolve the details of the residual stresses produced by the welding process. Residual strain/stress measurements of two specimens produced using the fibre laser were performed on Engin-X. Across the weld and through thickness distributions are reported in this paper.

1. Introduction

Laser welding has many advantages compared with traditional electric arc-based methods. The most important is the high energy density available which allows welds with high depth to width ratio. This means that the same penetration depth can be obtained using less energy or with a faster processing speed compared with traditional welding methods [1]. This leads to a reduction of heat input into the material and results in high process efficiency. This is beneficial as heat input is the main cause of residual stresses and distortion of joined components.

New types of lasers such as fibre and disc are now available commercially and these allow even better control of the heat input through their inherent high stability. Also the high beam quality of these lasers enables further increases in energy density and therefore further reductions in heat input [2,3].

Residual stresses are formed in weld structures primarily as the result of differential plastic and elastic strain during the heating and cooling cycles. During heating a large amount of plastic strain occurs whilst on cooling most of the strain is accommodated elastically. This leads to large residual stresses developing in the region of the weld zone. These stresses can have important consequences on the performance of engineering components [4]. There are several ways of measuring residual stresses in small volumes. The most common ones involve mechanical invasive methods (e.g., hole drilling or cutting [5]) and non-destructive methods using radiation such as x-ray (laboratory or synchrotron) or neutron diffraction (ND) [6,7,8]. ND is outstanding in its ability to obtain residual stresses non-destructively within the subsurface and deep within the bulk of the components. Neutrons can evaluate the three principal directions of residual stress in the weld structures, with a spatial resolution of 1 mm (or less) and to a depth of many millimetres below the surface of the weld (up to 50 mm for steel). In order to investigate the relation between heat input and residual stresses two bead-on-plate welds were produced with widely different process conditions. This approach enables comparison of welds obtained with relatively high and low heat inputs. The detailed residual stress distributions across the weld zone and through the plate thickness were measured and discussed in relation to the different heat inputs.

2. Materials and welding procedure.

The welds were made in 60mm by 200mm and 12mm thick S355 mild steel plates. A YLR 8000 IPG fibre laser with a maximum power of 8kW and a 630μm beam diameter was used to make the welds.. Pure shield argon with a flow rate of 15 l/min was used as a shielding gas. All welds were sectioned, polished and investigated under the optical microscope. Two bead-on-plate welds were manufactured, one with a high heat input and one with a low heat input. Welding parameters for both samples are shown in Table 1. The macrographs are presented in Fig.1 (note the different magnification for the two pictures).

Table 1. Welding parameters used in the experimental work

Sample	Laser Power [kW]	Travel speed [m/min]	Penetration depth [mm]	Weld width [mm]	Heat input (kJ/m)
A1	6	5	3.24	1.4	72
A2	6	0.5	9.15	6.3	720

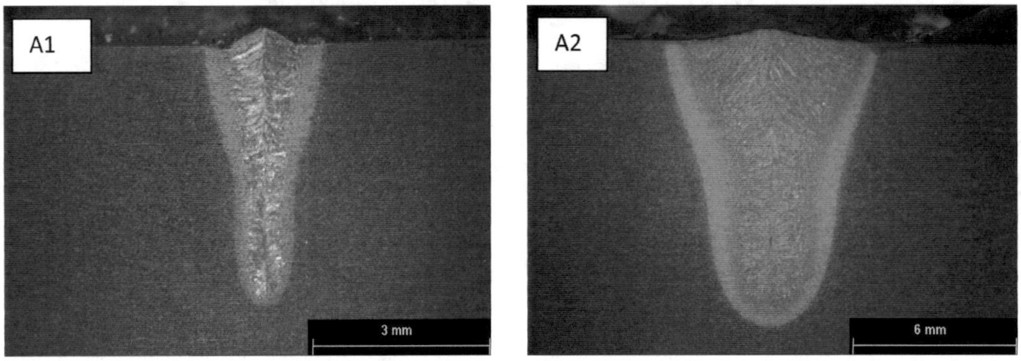

Figure 1. Macrograph of (a) Sample A1, and (b) Sample A2 (note the change of scale).

3. Residual stress measurements

The residual strain/stress measurements were performed on the ENGIN-X beam-line at ISIS [9]. For the longitudinal measurements the gauge volume was 2×2×2 mm^3 but for the transverse, normal and reference sample measurements this gauge volume could be relaxed in one direction to 2×2×10 mm^3. The gauge volume could be considered as being relatively large, especially for the Sample A1, However by using small steps across the weld zone (with the gauge volume overlapping at each step) we were able to extract the full detailed strain/stress distribution by deconvoluting the effect of the gauge size. The reference "stress free" value was measured at the corner of the reference sample cut from the parent metal (10x12x20 mm^3). The longitudinal, transverse and normal components of the residual stress field were calculated for each strain-measurement position in the weldment from the relevant three strain components. This was done using standard elasticity theory assuming an elastically isotropic material and that the measurement directions correspond to the principal axes of the stress tensor [10]. For this conversion a Young's modulus of 207 GPa and Poisson's ratio of 0.3 were used. The focus of this research was on the detailed line scans across the weld 2mm below the surface and through the thickness of the plate (Fig. 4) in the middle of the weldments.

Figure 2a show the residual strain distribution across the weld for Sample A2. Figure 2b shows the comparison of longitudinal residual stress for Sample A1 and A2. Figure 3a shows the residual strain distribution through the thickness for Sample A2. Figure 3b shows through thickness comparison of longitudinal residual stress for Sample A1 and A2.

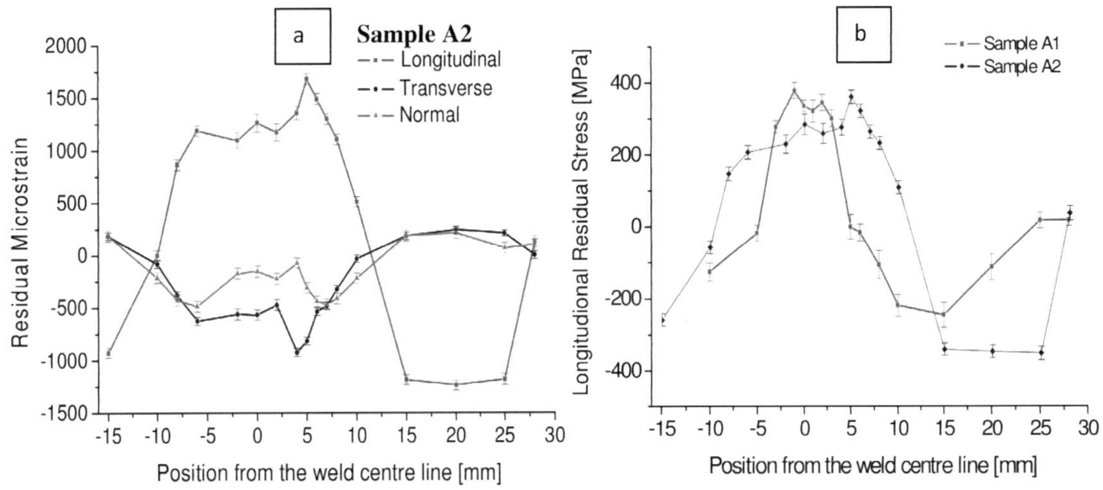

Figure 2. Across the weld distributions of (a) residual strain for Sample A2, and (b) longitudinal residual stress of Sample A1 and A2 (2mm below the surface of the plate).

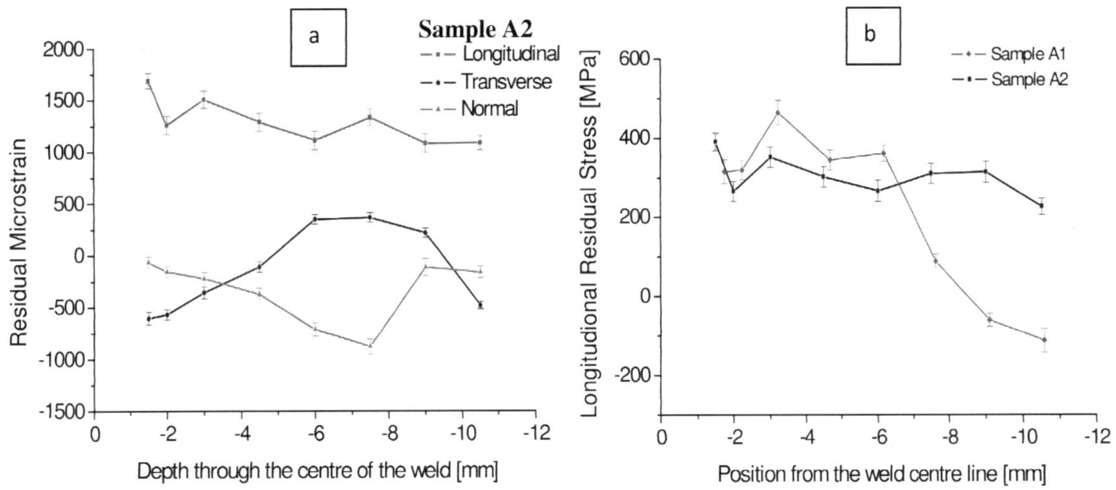

Figure 3. Through thickness distribution of (a) residual strain for Sample A2, and (b) longitudinal residual stress of Sample A1 and A2.

4. Discussion

The use of a neutron beam as a non-destructive method of measuring residual stress produced by laser welding has been explored. The narrowness of the welds make the measurements more challenging as the spatial resolution of the neutron measurements is in range of 2 mm. However we have shown that it is still possible to resolve the residual stress distribution in all three directions of principal stress with a high level of detail by using small incremental steps across the weld zone.

The highest residual strain/stress was found in the longitudinal direction near the middle of the weld and heat affected zones (Figures 2 and 3) for both samples. The tensile residual strain/stress is than balanced by compressive stress away from the weld zone. Towards the edges (x=28mm) of the sample the residual stress was approximately zero in all three directions. The residual strain in normal and transverse directions for sample A2 was compressive in the area of the weld and nearby. As is shown in Figure 2b the width of the longitudinal tensile peak corresponds to the size of the weld. The larger the weld (i.e. higher heat input) the larger the tensile peak width (Figure2b). However the peak height of the longitudinal stress was independent of the heat input.

The largest residual strain through the thickness in the middle of the weld was observed in the longitudinal direction (Figure 3a). The normal residual strain was compressive throughout the thickness. The transverse strain on the other hand is compressive towards the both surfaces of the plate but tensile in the mid-thickness. Sample A1 has uniform tensile longitudinal stress throughout the depth (Figure 3b). In contrast Sample A2 varies from tension on the topside, where welding was performed, to compression on the bottom side. The difference is because Sample A2 is nearly full penetration (9mm deep in 12mm) whereas Sample A1 is only partial penetration (3.2mm deep in 12mm).

5. Acknowledgements

This work was conducted as a part of NEGLAP project funded by Corus and EPSRC through Cranfield IMRC.

6. References

[1] Ion, J.C., *Laser processing of engineering materials : principles, procedure and industrial application*. Amsterdam ; Oxford: Elsevier/Butterworth-Heinemann. (2005) xviii, 556 p.: ill.; 26 cm.
[2] Verhaeghe G.., Welding Journal, (2005) 84 (8) 56
[3] Ozden H., Welding Journal, (2007) 86 (5) 26
[4] Webster G.A., Wimpory R.C, Journal of Neutron Research, (2001) 9 281
[5] Pang J. W. L., Preuss M., Withers P. J., Baxter G. J., Small C., Materials Science and Engineering A, (2003), 356 (1-2) 405
[6] R. A. Owen, R. V. Preston, P. J. Withers, H. R. Shercliff and P. J. Webster, Materials Science and Engineering A, (2003), 346 (1-2) 159
[7] Webster J., Ananthaviravakumar N., Hughes D.J., Mills G., Preston R.V., Shercliff H.R., Withers P.J., Applied Physics A (2002), 74, S1421
[8] Paradowska A M, Price J W H, Finlayson T R, Lienert U, Walls P and Ibrahim R, *J. Physics: Condensed Matter*.(2009) 21 124213
[9] Daymond M.R.,. Priesmeyer H.G, Acta Mater. (2002). 50 1613
[10] Lodini A., Calculation of residual stress from measured strain in *Analysis of residual stress by diffraction using neutron and synchrotron radiation*, eds. Fitzpatrick M E and Lodini A (Taylor & Francis Inc., London) (2003) 4

International Conference on Neutron Scattering 2009 IOP Publishing
Journal of Physics: Conference Series **251** (2010) 012050 doi:10.1088/1742-6596/251/1/012050

Evaluation of Residual Stress Measurements Before and After Post-Weld Heat Treatment in the Weld Repairs

Anna M. Pardowska[1], John W.H. Price[2], Trevor R. Finlayson[3], R. Ibrahim[2]

[1] ISIS, Rutherford Appleton Laboratory, Didcot, OX11 0QX, Oxfordshire, UK
[2] Department of Mechanical Engineering, Monash University, Clayton, VIC 3800, Australia
[3] School of Physics, The University of Melbourne, Melbourne, VIC 3010, Australia

E-mail: anna.paradowska@stfc.ac.uk

Abstract

Welding repairs are increasingly a structural integrity concern for aging pressure vessel and piping components. It has been demonstrated that the residual stress distribution near repair welds can be drastically different from that of the original weld. Residual stresses have a significant effect on the lifetime performance of a weld, and a reduction of these stresses is normally desirable. The aim of this paper is to investigate residual stresses in various weld repair arrangements using the non-destructive neutron diffraction technique. This research is focused on characterization of the residual stress distribution: (i) in the original weld; (ii) in a shallow toe weld repair; and (iii) after conventional post-weld heat treatment. The focus of the measurements is on the values of the sub-surface strain/stress variations across the weld.

1. Introduction

In welded structures residual stresses (RS) are formed primarily as the result of differential contractions which occur as the weld metal solidifies and cools to ambient temperature. RS have a significant effect on corrosion, fracture resistance, creep and corrosion/fatigue performance and a full understanding of these stresses is desirable. Therefore, experimental measurements are essential to establish a quantitative understanding of the sign, magnitude and distribution of the RS around the repair, within acceptable limits and to validate increasingly demanding finite-element models. Many non-destructive and destructive techniques are employed for detecting RS in welded components [1], such as hole drilling, conventional X-ray, neutron (ND), synchrotron diffraction (SD) or ultrasonic examination. Non-destructive measurement is a key issue in the confirmation of the theoretical work. ND is outstanding in its ability to obtain RS non-destructively within the subsurface and interior of components [2,3].

RS in welded joints and repairs are sometimes reduced by post-weld heat treatment (PWHT) [4] or by mechanical stress relieving. The main aim of PWHT is to reduce or completely relax the RS induced by the welding process. It is an important, but costly step in the manufacture or repair of pressure vessels and pipe components. Stress relieving heat treatments are generally avoided unless specified as mandatory by Codes and/or Standards due to the high cost involved and the potential consequence of an incorrect post-weld heat treatment procedure.

In the work reported in this paper an ordinary quality weld and weld repair in carbon steel (AS 1548-7) were investigated. Detailed residual stress distributions were measured using ND at The Australian Nuclear Science and Technology Organization (ANSTO), Lucas Heights, Australia, The focus of these measurements is on detailed, surface, line scans across the weld, with particular interest in the weld and heat affected zone (HAZ). The residual stress distributions were compared to establish the influence of the shallow repair on the original weld, as well as to investigate the residual stress relaxation after conventional PWHT.

2. Materials and Welding Procedure

Two types of specimens were manufactured, an unrepaired butt weld (Sample I) and the same butt weld with partial penetration repair at the toe of the weld (Sample II). The dimensions of the parent metal plates were the length of 300 mm, the width of 150 mm and the thickness of 12 mm. Both samples were restrained during the welding process. Macrographs of the specimens are shown in Figure 1. Sample I was a 60° preparation butt weld with a 2 mm gap between the plates. Four beads were deposited to fill the weld as it is shown in Figure 1(a). The second weld, Sample II, was produced in the same manner but after the welding, attempts were made to remove a hypothetical crack (4 mm in depth and 60 mm in length) from the HAZ with a 100 mm long repair.

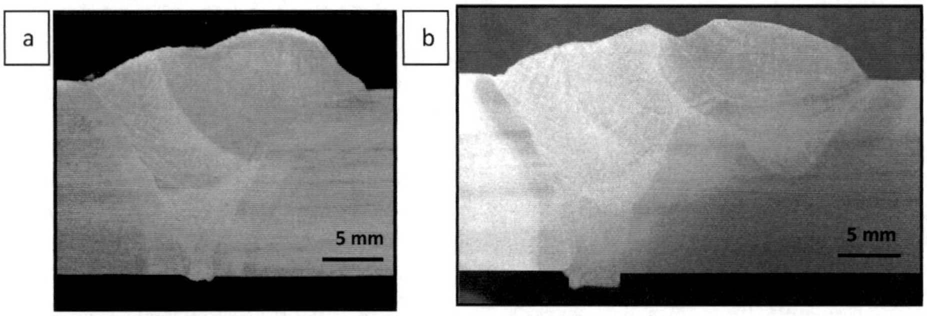

Figure 1. Macrographs of (a) original weld - Sample I, and (b) repair weldment - Sample II.

Table 1. Welding parameters used in the experimental work

Parameters	Root run (1)	Runs (2-6)
Electrode diameter [mm]	1.6	1.6
Current range [A]	260-280	260-280
Voltage range [V]	28-30	28-30
Traverse speed [mm/min]	480	360
Wire feeding speed [mm/min]	3600	3600
Electrode stick-out distance [mm]	20	20
Gas flow rate [l/min]	20	20

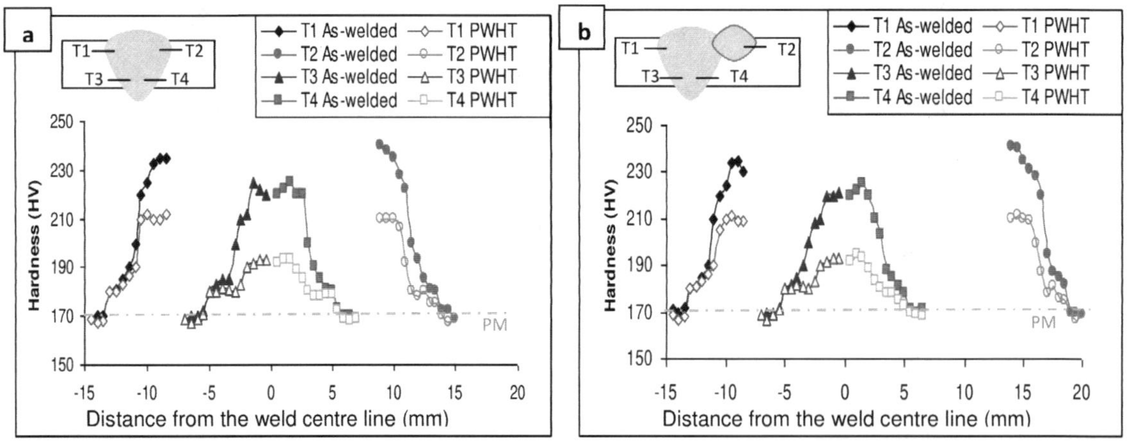

Figure 2. Vickers hardness traverses (T) measured before and after post-weld heat treatment (PWHT) for (a) Sample I and (b) Sample II.

The repair comprised two stages, the first being the grinding of the grove to remove the "crack", followed then by the deposition of the fifth and sixth beads. The welds were produced using a semi–automatic flux core arc welding process. The welding parameters are shown in Table 1. The conventional post-weld heat treatment (PWHT) (at the furnace where the holding temperature was 600°C for 1hour) was applied to both samples after they were measured in the as-welded condition. Hardness profiles were measured on both samples before and after PWHT (Figure 2). This information was obtained to ascertain the effect of post-weld heat treatment on hardness. No degradation of the parent metal was observed in any sample.

3. Neutron Diffraction Procedure

ND measurements of RS were undertaken on The Australian Strain Scanner (TASS) at ANSTO, Australia. The neutron wavelength used was 1.40 Å. Measurements were made using the αFe (211) reflection, at the detector angle, 2θ, of approximately 73.5° with the scattering vector parallel to each of the three axes identified as longitudinal (parallel to the length of the weld), transverse (perpendicular to the length of the weld and parallel to the plate) and normal (perpendicular to the length of the weld and perpendicular to the plate). The measurements were made in the middle of the weldments across the weld and 2 mm below the surface. The gauge volume for the longitudinal and d_0 measurements was approximately 1.5×1.5×2 mm^3 but for the transverse and normal measurements this gauge volume could be relaxed in one direction to 1×1×20 mm^3. To establish the "stress-free" lattice parameter, d_0, a "comb" reference sample was produced from an identical weldment. (For more details see [5].) No significant variations were found therefore the average of all measurements across the "comb" reference sample was used to determine the strains at all the points in the weldment. The longitudinal, transverse and normal components of the residual stress field were calculated for each strain-measurement position in the weldment from the relevant three strain components. For this conversion, a Young's modulus of 207 GPa and Poisson's ratio of 0.3 were used.

4. Results and Discussion

Figure 3(a) shows the residual stress across Sample I measured by ND. The RS in the longitudinal direction within the weld and at the weld toe are very high, up to the yield strength of the weld material. The transverse stresses are lower, up to 110 MPa at the toe, on the side of the last deposited bead and to -60 MPa on the opposite side. This may be due to the larger size of the weldment which results in a high constraint condition.

After the successful repair (Sample II) a broadening of the peak of the longitudinal residual stress distribution was observed and high RS were found within the weld metal and HAZ, as shown in Fig. 3(b).

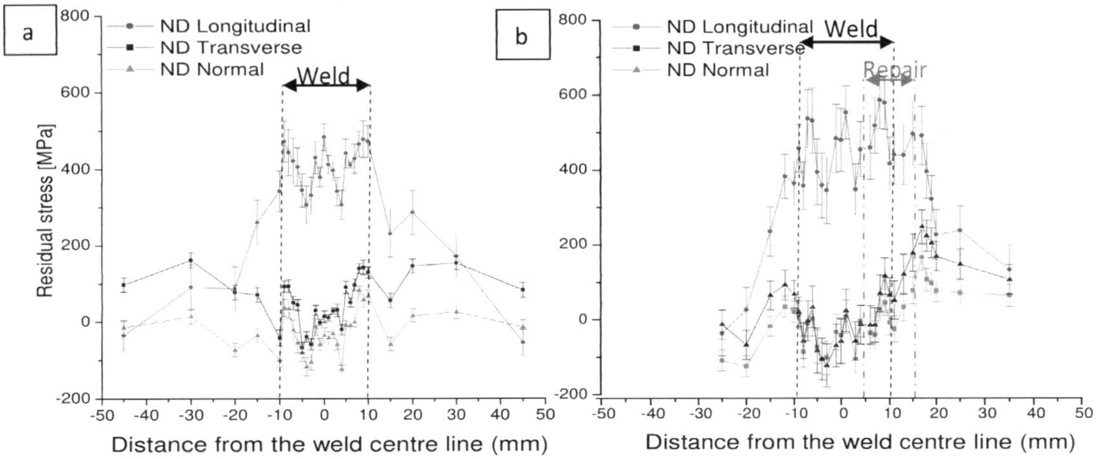

Figure 3. Residual stress distribution across (a) the original butt weld Sample I and (b) repair weld Sample II measured by neutron diffraction (1.5 mm below the surface).

Additionally, following the repair an increase in the transverse stresses was found on the opposite side of the repair, up to 280MPa, at the weld toe.

After the residual stress measurements in the as-welded condition were completed, the PWHT was applied to both samples and the residual stress measurements were repeated. The residual stress distribution across the weld repair, Sample II, is shown in Figure 4. The residual stress significantly decreases in all directions in the weld and parent metal, which proves that conventional post-weld heat treatment is a very effective stress relaxation technique.

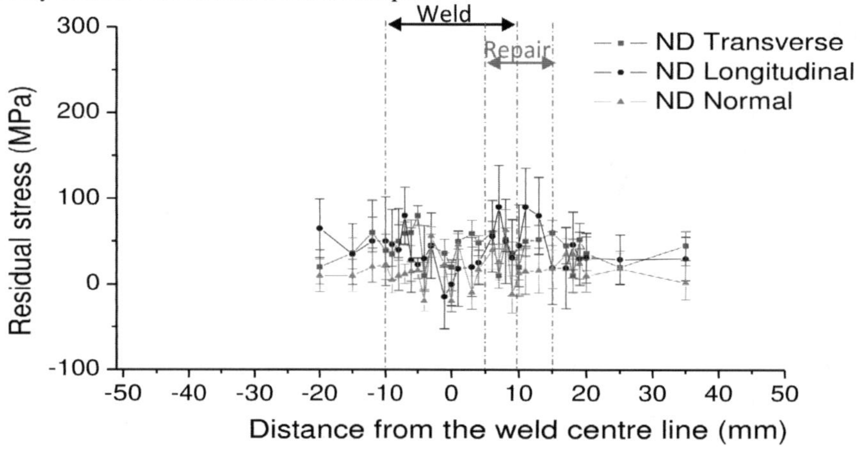

Figure 4. Residual stress across repair weldment (Sample II) after PWHT (1.5 mm below the surface).

5. Conclusions

High longitudinal RS were found within the weld as well as at the weld toe, in the as-welded condition, exceeding the uniaxial yield value of the parent metal. Additionally, in the region of the HAZ the longitudinal residual stress varied rapidly from highly tensile in the weld to a balancing compressive value in the PM. Transverse RS are much smaller in the original weld but they increased significantly after the deposition of the weld repair. Tensile transverse residual stress may be particularly detrimental to the weld life performance, as in many cases in-service stresses will superimpose in that direction. PWHT was very effective in reduction of high RS around the weld and weld repair. Conventional PWHT is strongly recommended for original and repair welds, especially where RS play a role in driving in-service cracking.

6. Acknowledgements

This work was conducted with the assistance of an Australian Research Council Grant supported by the Welding Technology Institute of Australia (WTIA). Other assistance has been received from the Monash University Research Fund and the Australian Nuclear Science and Technology Organisation (ANSTO). The Australian Institute of Nuclear Science and Engineering (AINSE) is acknowledged for financial assistance (Award No. AINSTU1604) to enable measurements on TASS to be conducted.

7. References

[1] Withers P J and Bhadeshia H K, 2001, *Mater. Sci. & Technol.* **17** 355
[2] Owen R A, Preston R V, Withers P J, Shercliff H R and Webster P J, 2003, *Mater. Sci. & Eng. A* **346** 159
[3] Webster G A and Wimpory R C, 2001, *J. Neut. Res.* **9** 281
[4] Sedek P, Brozda J, Wang L and Withers P J, 2003, *Int. J. Pressure Vessels and Piping* **80** 705
[5] Paradowska A M, Price J W H, Finlayson T R, Lienert U, Walls P and Ibrahim R, 2009, *J. Physics: Condensed Matter.* **21** 124213.

Neutron diffraction residual strain measurements in alumina coatings deposited via APS and HVOF techniques

R Ahmed[1*], N H Faisal[1], R L Reuben[1], A M Paradowska[2], M Fitzpatrick[3], J Kitamura[4], S Osawa[4]

[1] School of EPS, Heriot-Watt University, Edinburgh, UK
[2] Rutherford Appleton Laboratory, ISIS, Didcot, UK
[3] Department of Materials Engineering, The Open University, Milton Keynes, UK
[4] Thermal Spray Materials Department, Fujimi Incorporated, Gifu, Japan

*Corresponding author E-mail: R.Ahmed@hw.ac.uk

Abstract. Residual strains in air plasma sprayed (APS) and high velocity oxy-fuel (HVOF) alumina (Al_2O_3) coatings were investigated by means of neutron diffraction technique. Results of this investigation conclude that through thickness residual strain in the APS coating was mainly tensile whereas the HVOF coating had both compressive and tensile residual strain. Further analysis of Vickers indentation fracture behaviour using acoustic emission (AE) technique concluded that the nature and magnitude of these residual strain fields had a direct effect on the fracture response of two coatings during the indentation process.

1. Introduction

Thermally sprayed alumina (Al_2O_3) coatings are used in industry for thermal and electrical resistance applications. Measurement of residual strains in these coatings is critical for modeling and designing improved components. Non-destructive through thickness residual strain measurements in thermal spray coatings is possible via the high penetration depth achieved by the neutron diffraction technique [1-2]. The influence of coating process parameters on the residual stress fields of Al_2O_3 coatings has been a topic of research of previous investigations [3]. Recent advancements in high velocity oxy-fuel (HVOF) technique has made it possible to deposit much finer alumina powders at relatively lower temperatures than those achieved by air plasma spraying (APS) process [4]. This has the effect of limiting phase transformations during coating deposition, which along with the relatively higher velocity and lower temperature of HVOF process can influence residual stress behavior. This paper therefore aims to investigate the through thickness changes in the residual strain profile of finer powder and conventional Al_2O_3 coatings deposited by the HVOF and APS processes, respectively. Further analysis interlinking fracture behavior and residual strain is also presented.

2. Experimental procedures

Industrially optimized coating process parameters [5] were used to deposit 250 μm thick HVOF and APS coatings on the surface of 31 mm diameter and 8 mm thick AISI 440 steel substrate discs. HVOF (θ-gun) coatings were deposited using finer powder (1-5 μm). APS (9 MB) coatings were deposited using conventional coating powder in the size range of 10-45 μm. Both coating powders (98% purity) had angular and irregular morphology. Coating microstructure was investigated via cryogenic fracture in liquid nitrogen to reveal coating porosity and splat morphology. X-ray diffraction (XRD) measurements were made to reveal phase composition using Cu-Kα radiation (λ = 0.1542 nm). Neutron diffraction measurements (ENGIN-X) [1-2] were conducted in vertical scan mode to measure through thickness residual strain profile of the coating substrate system. To achieve a through thickness residual strain profile a partial submerged beam was used for measurements near the coating surface, and a beam submerged in the coating and substrate materials near the coating substrate interface. By using the fast vertical scan and careful leveling of specimen using theodolites, the coating surface and the coating substrate interface was located up to a resolution of about 50 μm. Residual strain was calculated via the shift in individual neutron diffraction peaks. Vickers indentation fracture tests and its AE measurement [6] were conducted at nine different loads between 98 and 490 N (five indentations at each load) with an AE sensor on the surface. The data were sampled at 2.5 MHz so that the full band-width of the AE (0.1-1 MHz) could be captured.

3. Results and Discussion

3.1. Coating microstructure

The cryogenic fracture images (Figure 1) reveal the splats morphology, porosity and linear cracks in both coatings. Unmolten Al_2O_3 particles can also be observed in HVOF coating. This figure also shows relatively higher porosity for APS coatings compared to the HVOF coatings. The coating surface was dense and exhibited a similar structure throughout with no surface connected porosity.

Figure 1. Coatings cross-sections: (a) APS Al_2O_3 with thick large splats, and (b) HVOF Al_2O_3 revealing relatively thin small splats and unmolten particles (indicated by arrows).

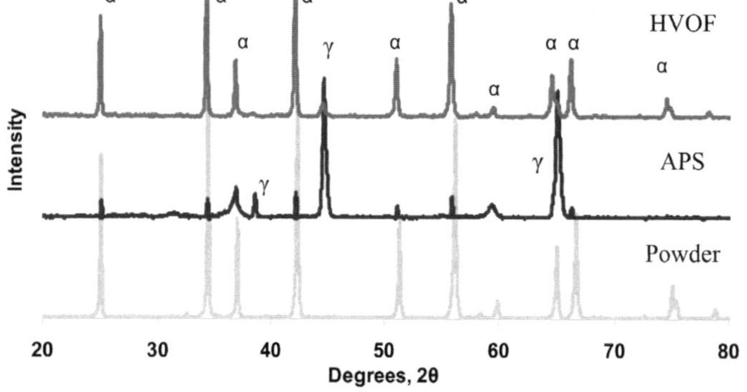

Figure 2. XRD pattern of Al_2O_3 powder, APS and HVOF coatings.

The XRD spectrum (Figure 2) of APS Al_2O_3 coatings showed relatively higher $\alpha \rightarrow \gamma$ transformation when compared to the HVOF process, due to rapid solidification and relatively higher temperature of the APS process. It can be seen that the HVOF coating predominantly contained the primary alumina phase (α-Al_2O_3) with very little γ-Al_2O_3 transformation.

3.2. Neutron diffraction residual strain and indentation fracture behaviour

Figure 3 summarizes the residual strain results for both coatings, plotted on log-scale to highlight the differences in coating strain. These strain values illustrate the combined effect of phase transformations, thermal mismatch, peening effect, quenching of lamellas and differences in the coefficient of thermal expansion (CTE) of the coating substrate system [1-3]. The difference in CTE of Al_2O_3 (7×10^{-6} °C^{-1}) and steel (12.2×10^{-6} °C^{-1}) had a significant influence on the net residual strain in the coating layer. The average residual strain in the HVOF coating changes from tensile to compressive, whereas the average strain in the APS coating is tensile balanced by a corresponding opposite strain in the substrate. In APS coating, the effect of higher temperature leading to tensile strain is more dominant due to the CTE mismatch between the coating and substrate materials. Similarly, within the substrate material, the strain at the coating-substrate interface in HVOF coating is close to zero, indicative of low substrate heating during HVOF spraying, whereas the strain at the APS coating-substrate interface is compressive indicative of relatively higher substrate heating. Apart from the temperature difference of APS and HVOF processes, there were three other factors responsible for the relative differences in the residual strain of the two coatings i.e. i) peening effect caused by the relatively higher velocity of the partially melted, though relatively smaller (1-5 µm) Al_2O_3 particles in HVOF spraying, ii) influence of finer powder particle size on splat morphology, porosity and hence residual stress and iii) relatively higher $\alpha\rightarrow\gamma$ phase transformation in the APS coatings. The peening effect in HVOF process is understood to cause compressive residual stress in thermal spray coatings. Similarly, investigations relating to the influence of powder particle size on the residual stress of thermal spray coatings have indicated that the decrease in powder particle size increases the residual stress in thermal spray deposits [7-8]. This effect is attributed to the higher surface area to particle volume ratio of smaller particles, resulting in changes in the pore and intersplat morphology, which resists stress relaxation in the deposited layer. In the investigation by Coats et al. [7] on WC-Co coatings, it was concluded that with the increase in powder particle size the tensile stress in the Co and the compressive stress in the WC decreased. In the current investigation, a similar trend can be observed for the compressive residual strain in the HVOF coating, although the tensile residual strain is lower for the finer powder HVOF coating. The shrinkage in APS coating deposit caused by the $\alpha\rightarrow\gamma$ phase transformation [5] also contributes to the residual differences between the two coatings. These results indicate that the tensile residual strain in coatings can be substantially reduced using the fine powders deposited by the HVOF process, which can improve the fracture and delamination resistance of Al_2O_3 coatings.

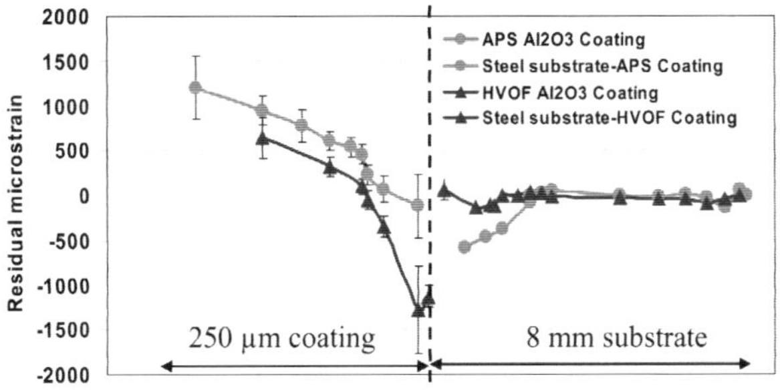

Figure 3. Neutron diffraction residual strain profile of Al_2O_3 coatings and steel substrate.

Figure 4 shows influence of this residual stress profile and microstructural morphology on the indentation fracture behavior of both coatings. Figure 4a,b shows typical fracture morphology of both coatings after Vickers indentation. The fracture morphology for APS coating shows spallation with no fully developed cracks from indent corners. This behavior was different to the HVOF coatings. Figure 4c shows the total AE energy of Vickers indentation fracture of both coatings tested under a variety of loads. It can be observed from this figure that the AE energy increases with load until 245 N, followed by a sudden decrease, after which the AE energy broadly remains constant. This indicates that from lower to moderate loads the effect of indentation surface tensile stress dominates the surface cracks and eventually at moderate to higher loads the indentation compressive stress dominates the subsurface cracks. Also in most of the cases the AE energy for APS coatings is higher than that for HVOF coatings, which is due to the combined effect of residual tensile stress and microstructural morphology.

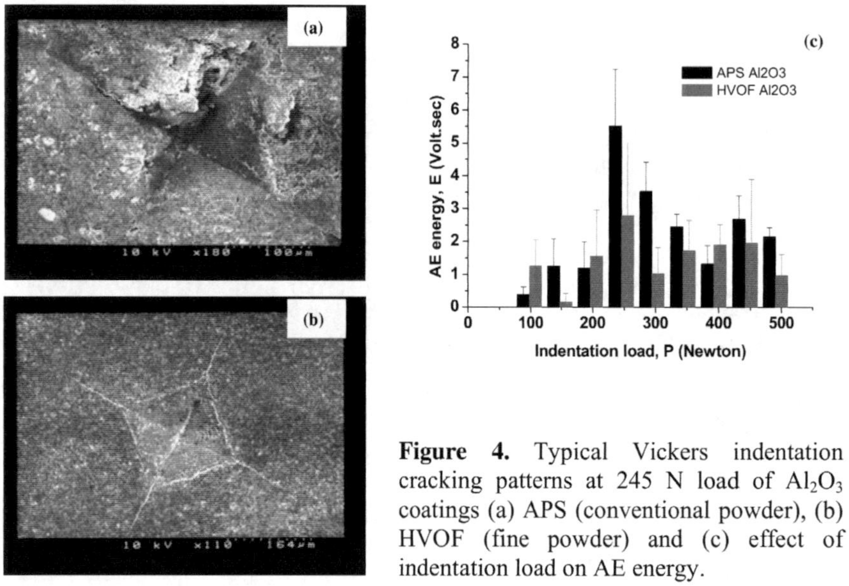

Figure 4. Typical Vickers indentation cracking patterns at 245 N load of Al_2O_3 coatings (a) APS (conventional powder), (b) HVOF (fine powder) and (c) effect of indentation load on AE energy.

4. Conclusions

It is concluded that the through thickness residual strain profile in the APS coating was mainly tensile whereas the HVOF coating had a both compressive and tensile residual strain. The nature and magnitude of these residual strain fields and coating morphology had a direct effect on fracture response of two coatings during the indentation process.

Acknowledgements

Authors acknowledge the support of Rutherford laboratory UK (RB810413) for neutron diffraction measurements.

References

[1] Ahmed R, Yu H, Stoica V, Edwards L, Santisteban J R 2008 *Mater. Sci. Eng. A* **498** 191–202.

[2] Ahmed R, Yu H, Stewart S, Edwards L and Santisteban J 2007 *ASME J. Tribol.* **129** 411-418.

[3] Kulkarni A, Gutleber J, Sampath S, Goland A, Lindquist W B, Herman H, Allen A J and Dowd B 2004 *Mater. Sci. Eng. A* **369** 124-137.

[4] Itsukaichi T, Osawa S and Ahmed R 2003 *Proc. Int. Therm. Spray Conf.* (Florida), ed C Moreau and B Marple (Ohio: ASM Internaltional) pp 819-824.

[5] Morishita T, Osawa S and Itsukaichi T 2004 *Proc. Int. Therm. Spray Conf.* (Germany), Proc. CD Section: HVOF-Processes and Materials I, (Ohio: ASM Internaltional) pp 1–4.

[6] Faisal N H, Steel J A, Ahmed R and Reuben R L 2009 *J. Therm. Spray Technol.* (Preprint DOI: 10.1007/s11666-009-9334-1).

[7] Coats D L, Krawitz A D 2003 *Mater. Sci. Eng. A* **359** 338-342.

[8] Herold TG, Prask H J, Biancaniello F S 2004 *Recent Advances in Experimental Mechanics* (DOI 10.1007/0-306-48410-2) 507-514.

SANS study of microstructural evolution in oxide strengthened EUROFER steel after heat treatments

R Coppola[1] , R Lindau[2] , R P May[3] , A Möslang[2] , M Valli[4] , P Vladimirov[2] , A Wiedenmann[3]

[1] ENEA-Casaccia, FPN, CP 2400, 00100 Roma, Italy

[2] Forschungszentrum Karlsruhe, IMF, PO Box 3640, D-76021 Karlsruhe

[3] ILL, 6, rue Jules Horowitz, 38042 Grenoble, France

[4] ENEA-"Clementel", V. M. di Montesole, 40129 Bologna, Italy

roberto.coppola@enea.it

Abstract. Oxide dispersion strengthened (ODS) reduced activation ferritic/martensitic steels are being developed for structural applications in future fusion reactors since they would allow consistently increasing the operation temperature The 9%CrWVTa Eurofer-97 steel has been selected as a base material strengthened by Y_2O_3 contents of 0.3 and 0.5 wt% and produced in a powdermetallurgical route by mechanical alloying and hot isostatic pressing (HIP). The mechanical properties are strongly correlated with Y_2O_3 particle average size and spatial distribution, therefore small-angle neutron scattering (SANS) is quite useful to investigate how these evolve in the course of the production steps. Solid polycrystalline samples as well as mechanically alloyed powders have been investigated after thermal treatments thus simulating the HIP conditions. The obtained SANS results show that annealing at high temperature results in a redistribution of scattering centres sizes, which might be attributed to the growth of small Yttria particles and possibly of larger carbides.

1. Introduction

Oxide dispersion strengthened (ODS) reduced activation ferritic/martensitic steels are attractive candidates for structural applications in future fusion reactors since they would allow increasing the operation temperature by approximately 100°C, with a consequent improvement of the reactor efficiency [1]. The 9%CrWVTa Eurofer-97 steel (8.9 Cr, 1.1 W, 0.2 V, 0.14 Ta, 0.42 Mn, 0,11 C wt%) has been selected as a base material strengthened by Y_2O_3 contents of 0.3 and 0.5 wt% and produced in a powdermetallurgical route by mechanical alloying and hot isostatic pressing. ODS Eurofer-97 steel was produced by inert gas atomisation of the base material and subsequent mechanical alloying in industrial ball mills; hot isostatic pressing was then chosen as the appropriate consolidation process [2]. The internal structure of the ODS particles in consolidated materials has been studied by analytical transmission electron microscopy (TEM) applying microanalytical techniques (EDS and EELS, Ref. [3]) but a direct observation of these nano-particles by TEM in mechanically alloyed powders has not been possible until now because of difficulties in preparing the specimens for such observations. Small-angle neutron scattering (SANS) has therefore been utilized to investigate how these nano-sized particles evolve under isochronal annealing thus simulating the thermal conditions during the HIP process. Previous SANS measurements [4] carried out on polycrystalline ODS Eurofer-97 with the same chemical composition as the one investigated in this study, have in fact shown the usefulness of this technique to characterize Y_2O_3 particle average size and distribution, obtaining results in good agreement with TEM and relevant to understand the changes in the mechanical properties, which are strongly correlated with such a distribution.

2. Experimental technique and results

The investigated samples were mechanically alloyed powders of ODS Eurofer-97 steel and Yttria (0.5 wt%) in the as-received state, annealed 2 h at 750°C and annealed 2 h at 1100°C in vacuum. It is supposed that in this temperature range the Y_2O_3 particles will evolve and may change their size distribution and new phases, such as carbides, may nucleate. The SANS measurements were carried out at the D22 instrument at the High Flux Reactor of the Institut Max von Laue - Paul Langevin, Grenoble. A sample-to-detector distance of 2.00 m with a wavelength, λ, of 6 Å were used. Defining the modulus of the scattering vector $Q = \frac{4\pi \sin\theta}{\lambda}$ (where 2θ is the full scattering angle), these experimental conditions gave a Q interval ranging from 0.02 to 0.3 $Å^{-1}$, which corresponds to particle sizes ranging from 10 to 300 Å approximately.

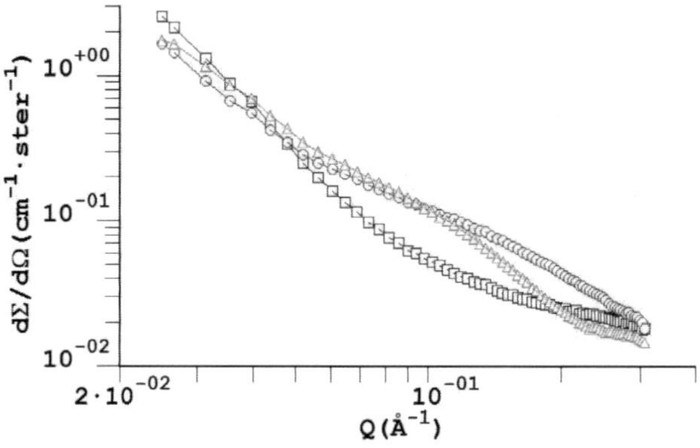

Figure 1. Nuclear SANS cross-sections of Eurofer-97 ODS powders reference mechanically alloyed (squares), mechanically alloyed plus annealed at 750°C (circles) and at 1100°C (triangles); the experimental errors range from 5 to 10%.

The powders were placed in a quartz cell 1 mm thick and the empty quartz cell was used as a reference for the background. Calibration to absolute SANS cross-section (expressed in cm^{-1} $sterad^{-1}$) was obtained by measurement of water in a quartz cell; the data were treated by the ILL standard programs [5]. A horizontal magnetic field was applied perpendicular to the incoming neutron beam in order to fully align the magnetic moments in the sample. Thus only nuclear scattering occurs in the horizontal plane, while nuclear and magnetic scattering occur in the vertical one. The purely magnetic scattering is obtained as the difference between the vertical and horizontal SANS cross-sections. In fact, in the case of magnetic samples, the total SANS cross-section $\frac{d\Sigma(Q)}{d\Omega}$ (where Ω stands for the solid angle) can be written as the sum of two terms

$$\frac{d\Sigma(Q)}{d\Omega} = \left(\frac{d\Sigma(Q)}{d\Omega}\right)_{nucl} + \left(\frac{d\Sigma(Q)}{d\Omega}\right)_{mag} \sin^2\alpha \qquad (1)$$

where α is the azimutal angle on the detector plane. The ratio of the "vertical" to the "horizontal" SANS components:

$$R(Q) = \frac{\left(\frac{d\Sigma(Q)}{d\Omega}\right)_{nucl} + \left(\frac{d\Sigma(Q)}{d\Omega}\right)_{mag}}{\left(\frac{d\Sigma(Q)}{d\Omega}\right)} = 1 + \frac{(\Delta\rho)^2_{mag}}{(\Delta\rho)^2_{nucl}} \qquad (2)$$

is related to the composition of the microstructural inhomogeneities and its dependence on Q implies that defects of different size or composition are present in the investigated sample, $(\Delta\rho)^2$ being the "contrast" or square difference in neutron scattering length density (nuclear and magnetic respectively) between the observed nuclear and magnetic inhomogeneities and the matrix. However, as in the previous study of ref. [4], no difference was found in the Q-dependence between nuclear and magnetic scattering, i. e. nuclear and magnetic inhomogeneities are of the same sizes.

The SANS nuclear and magnetic cross-sections can each one be written as

$$\frac{d\Sigma(Q)}{d\Omega} = (\Delta\rho)^2 \int_0^\infty dR\, N(R) V^2(R) \left| F(Q,R) \right|^2 \qquad (3)$$

where $N(R)$ is the number per unit volume of centers with a typical size between R and $R+dR$, V their volume and $\left| F(Q,R) \right|^2$ their form factor (assumed to be spherical in the present case) and $(\Delta\rho)^2$ is the nuclear or magnetic "contrast". The volume distribution function is defined as:

$$D(R) = N(R)R^3 \qquad (4).$$

$N(R)$ was determined, by transformation of eq. (3), using the method described by in [6] and more recently discussed in [7]. This code assumes that the size distribution function can be described by a set of cubic B-spline functions, with equispaced knots in log$R$ scale. The number of splines is determined by the R-range where the size distribution is to be explored (always larger than the range where different sizes can be effectively resolved) and by the required degree of detail. The logarithmic representation of $N(R)$ is quite suited for the case of technical alloys, such as Eurofer-97, where different kinds of microstructural inhomogeneities with sizes differing in order of magnitude may be simultaneously present.

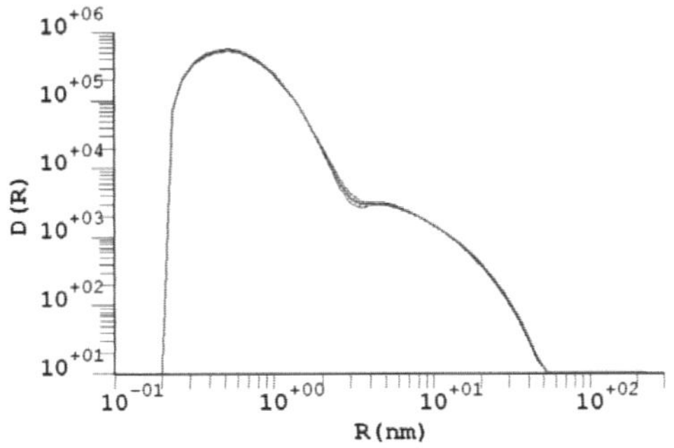

Figure 2. Volume distribution in A. U. ($D(R)$ volume per unit volume of Y_2O_3 particles with a radius between R and $R+dR$) for Eurofer ODS mechanically alloyed plus annealed at 750°C

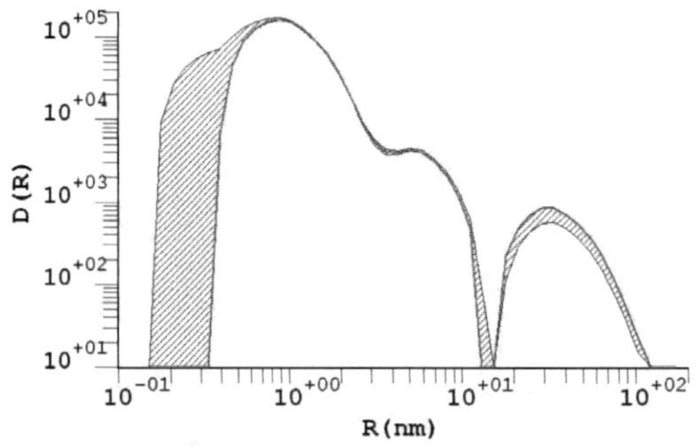

Figure 3. Volume distribution in A. U. (*D(R)* volume per unit volume of Y_2O_3 particles with a radius between *R* and *R+dR)* for Eurofer ODS mechanically alloyed plus annealed at 1100°C, the hatched area represents the 80% confidence band.

Figure 1 shows the nuclear SANS components of the as-milled powder and of the powders annealed at 750°C and 1100°C. The SANS spectrum of the mechanically alloyed powder serves only qualitatively as a reference and cannot be subtracted from the annealed ones. It is evident that, compared to the as-milled sample, small particles are present at 750°C. Their size slightly increases with increasing the temperature up to 1100°C. The corresponding size distributions, have been determined on a relative scale since the chemical composition of the scattering inhomogeneities is not precisely known and therefore the neutron "contrast" cannot be determined. These distributions, reported in figures 2 and 3 for 750°C and 1100°C annealing temperatures respectively, show a distribution of particles which are tentatively identified as Yttria ones. Furthermore, a growth of larger particles at 1100°C is observed, as also indicated by the small *Q* values of the corresponding SANS data, which might be attributed to nucleation of carbides. This is consistent with the observations made in [4]. It is evident that annealing at high temperature results in a redistribution of scattering centers sizes, but additional microstructural information and measurements extended to smaller Q values are needed to clarify these findings.

3. Conclusion - SANS is quite sensitive to microstructural changes produced under high temperature annealing in nano-structured Eurofer-97 ODS material, which turns out to be quite useful since these powders are difficult to observe by TEM. Micro-analytical information is needed to confirm the chemical composition of these scattering particles. Additional information is expected by planned high-resolution neutron diffraction measurements aiming to identify the crystallographic phases present in these powders. Once this experimental procedure will be fully developed it is planned to investigate by this technique the effect of annealing in a wider temperature range, including possibly *in-situ* heating, in order to follow as closely as possible the onset and the evolution of the precipitate phases detected by these measurements.

References

[1] Schaaf B v d, Gelles D S, Jitsukawa S, Kimura A, Möslang A, Odette G R 2000 *J. Nucl. Mat.* **283-287** 52
[2] Lindau R, Möslang A, Schirra M, Schlossmacher P, Klimenkov M 2002 *J. Nucl. Mat.* **307-311** 769
[3] Klimenkov M, Lindau R, Möslang A 2009 *J. Nucl. Mat.* **386-388** 557
[4] Coppola R, Klimiankou M, Lindau R, May R P, Valli M 2004 *Physica B* **350** e545
[5] Ghosh R E, Egelhaaf S U, Rennie A R 1998 *ILL report ILL98GH14T*
[6] Magnani M, Puliti P, Stefanon M 1988 *Nucl. Inst & Meth.* **A271** 611
[7] Coppola R, Kampmann R, Magnani M, Staron P 1998 *Acta Mat.* **46** 5447

The Future of Neutron Scattering in China: Meeting Increasing Demand with New Facilities?

Chun-Keung Loong[a], Jie Wei[b], Chuanxiang Tang[b], Hesheng Chen[c], Dongfeng Chen[d], and Yuntao Liu[d]

[a]School of Physics and Engineering, Sun Yat-Sen University, Guangzhou, 510275, China
[b]Engineering Physics Department, Tsinghua University, Beijing, 100084, China
[c]Institute of High Energy Physics, Chinese Academy of Sciences, Beijing, 100039, China
[d]Neutron Scattering Laboratory, China Institute of Atomic Energy, Beijing, 102413, China

ckloong@gmail.com

Abstract. The China Advanced Research Reactor (CARR) and the China Spallation Neutron Source (CSNS), like other big national facilities for basic research in China, are viewed as vehicles for engaging in international R&D efforts, furthering the nation's technological advancement, and facing arduous challenges such as energy and materials needs. In 2009, CARR—a 60MW steady-state reactor in Beijing—is expected to reach criticality while the CSNS—an 120 kW pulsed source in Dongguan—is to begin construction. In spite of the differing schedule, the road leading to full operation of the sources and productive utilization of the neutron instruments will be a long one. From the outset both projects benefit a great deal through interactions with the worldwide neutron communities. An even more trenchant task is to build a domestic user base, presumably from China's numerous universities and research laboratories, to imbue new scientific ideas into using and exploiting of the neutron methodology. Arguably, a smaller neutron facility yet apt at flexibility and optimization for education and academic R&D is needed to bridge 'big' facilities and 'small' sciences. Here, we introduce the plan of a compact pulsed neutron source (driven by a 13MeV-proton accelerator system with a Be target and room-temperature and cryogenic moderators), to be built in about 3 years as a part of a Hadron Application Research Center at Tsinghua University. We discuss the mission of these neutron sources, the complementarity therein, and their synergetic relation with the academia and other facilities, and possible cooperation with international counterparts.

1. Neutron-source projects in China

During the last 10 years efforts to modernize China's neutron-scattering capability have persisted, resulted in the undertaking of new neutron-source projects. The outstanding cases are the construction of the China Advanced Research Reactor (CARR) and the development of the China Spallation Neutron Source (CSNS). The basic design goals and the interim progress of CARR and CSNS have been previously reported elsewhere [1]. Very recently, a newly approved project—the Compact Pulsed Hadron (i.e., neutron and proton) Source (CPHS) as the Phase 1 of the eventual Tsinghua Hadron Accelerator Research Center (THARC) has been launched at Tsinghua University in Beijing, China.

The rationale behind these developments reflect the increasing demand of particle-beam experimental platforms for expansive applications. Here, we first update the recent activities at CARR and CSNS and introduce the CPHS project, then describe the inter-relation between these three neutron sources.

1.1. The China Advanced Research Reactor

The CARR of the China Institute of Atomic Energy (CIAE), located in the southwest suburb of Beijing, is to succeed the Heavy Water Research Reactor (HWRR) that retired in 2007 after 50 years of operation. Currently with all major construction completed, CARR is expected to reach criticality in 2009. It is a 60 MW tank-in-pool type reactor using a D_2O reflector for inverse neutron trap, designed to yield an optimal thermal neutron flux of $8x10^{14}$ n/cm^2/s. The Neutron Scattering Laboratory (NSL)—a user facility—under CARR is to conduct neutron-scattering for fundamental research utilizing 7 horizontal beam tubes, each accommodating dual ports, of the reactor. Two of these beam tubes will transmit neutrons from a liquid-hydrogen cold source to the guide hall via four neutron guides that were purchased abroad. Installation of all major reactor components except the cold source has been completed. CARR currently carries out the final safety testing and is scheduled to reach criticality in 2009. Afterward, the thermal-neutron instruments (see below) will be commissioned while the reactor refines its operation. However, operation of the reactor has to be re-adjusted in conjunction with the installation of the cold source in ~2 years later.

From the outset NSL has promoted and maintained cooperation with the international neutron-scattering community especially in personnel training and instrumentation development. Through collaborative agreements a number of instrumental components have been relocated to CARR: a triple-axis spectrometer, a four-circle diffractometers and a texture diffractometer from Jülich Center for Neutron Science, Germany and a residual stress diffractometer from Studsvik Neutron Research Laboratory, Sweden. Participation research teams (PRT) also play a key role in development of additional beamlines and scientific programs. For example, a PRT from the Institute of Chemistry of the Chinese Academy of Sciences (CAS), through collaboration with the National Institute of Standards and Technology (NIST) of USA, is building a small-angle scattering instrument and a reflectometer. These beamlines, together with two (a high-resolution and a high-intensity) powder diffractometers and a radiography/imaging station under construction by NSL, will constitute a suite of instruments offering a broad range of capabilities to users starting in 2010. The layout of the instrument is shown in Figure 1.

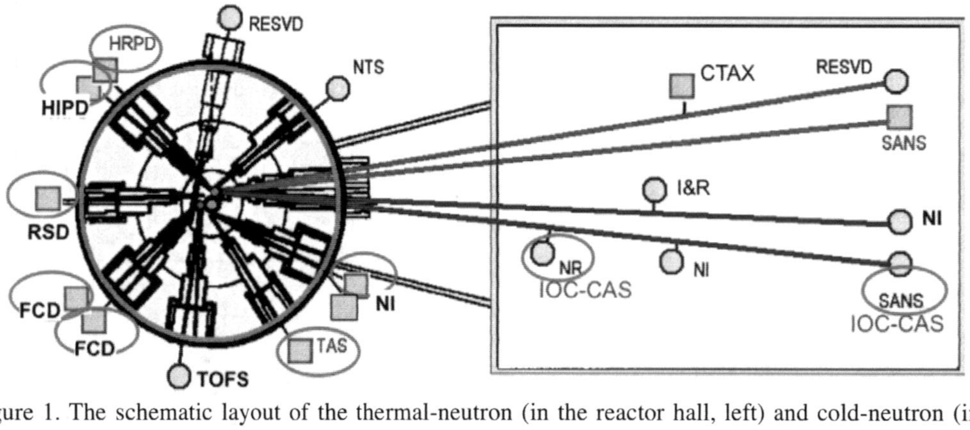

Figure 1. The schematic layout of the thermal-neutron (in the reactor hall, left) and cold-neutron (in the guide hall, right) instruments at CARR. The circled ones denote instruments to be commissioned at the initial stage by NSL.

NSL has a young staff to cover the installation, commissioning, and operation of the initial set of instruments. They work closely with the PRT scientists and the veteran researchers from the previous HWRR program. Furthermore, domestic and international advisory committees were set up to review NSL's progress in instrumentation and scientific-program development and to communicate the findings to CIAE upper management. Thus far NSL has received strenuous support from CARR-CIAE. For example, the plan of a new building to provide office and laboratory facilities to users has been approved by CIAE. Nevertheless, mechanisms have yet to be identified and implemented for securing a long-term budget to ensure steady capital investments and sustainable programs for the NSL. Perhaps heightened awareness of CARR's important role in neutron science R&D by the governing organization of CIAE—the China National Nuclear Corporation may improve the situation.

1.2. The China Spallation Neutron Source

The CSNS is a proton-accelerator driven short-pulsed neutron source (proton energy 1.6 GeV, repetition rate 25 Hz, beam average power 160 kW) to be built in Dongguan City of Guangdong Province in south China, see Figure 2. The proposed budget of 1.4B CNY (~$200M) becomes extremely tight after accounting for inflation. The strategy is to glean the knowledge and knowhow that are currently lacking through international cooperation so that the responsible organizations of the project, the Institute of High Energy Physics (IHEP) and the Institute of Physics (IoP) of CAS, can develop and test prototypes of all the major components. Eventually, the technology will be transferred to domestic industry for production at lower costs. Since 2005 active R&D and prototyping in key areas of accelerator system (ion source, RFQ, DTL, ceramic vacuum chambers, beam extraction, ring magnets and RF), neutron target station (target, reflector, and moderators), neutron instrumentation and beamline design have been carried out using discretionary funds provided by the Guangdong government and CAS. The project is benefited immensely by generous assistance from foreign neutron centers such as SNS, J-PARC, KEK, ISIS, and PSI and international advisory committees.

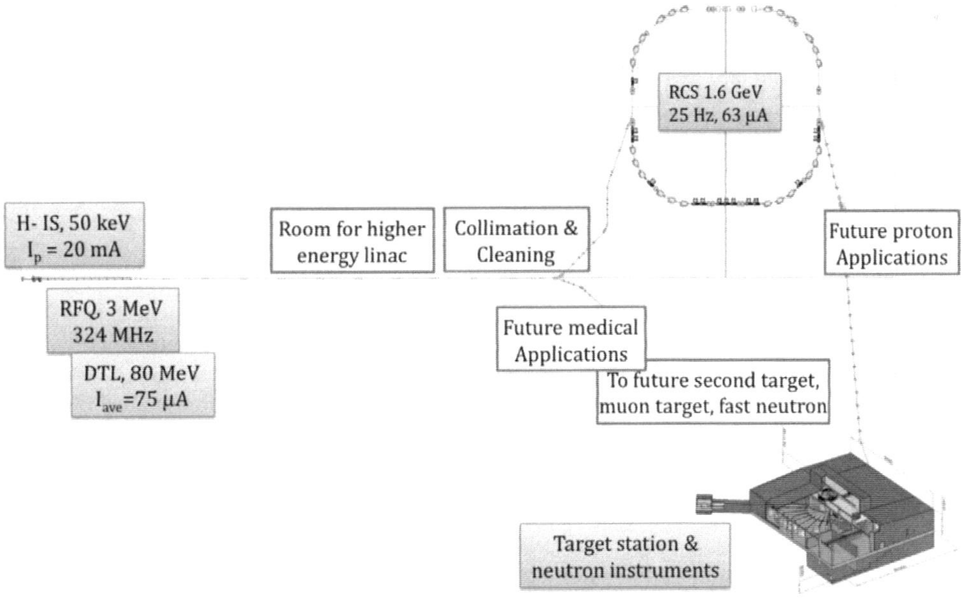

Figure 2. The schematic illustration of the linear accelerator system, the rapid cycling synchrotron and the experimental hall of the CSNS facility. Future upgrade and additional applications are also included.

In parallel to technical R&D, IHEP has conducted the environmental impact studies and responded to various reviews of the project mandated by the State Development and Reform Commission. In September of 2009, the Commission officially approved the CSNS site in Dongguan City. The remaining task is to finish the conceptual design and technical documentation. Final approval of the CSNS project is expected to materialize in the spring of 2010. Construction of the entire facility is to take 78 months. Currently, a major effort is underway to organize the IHEP personnel to be dispatched to Dongguan and to coordinate with the local government to prepare for construction. The most severe repercussion from the tight budget is the trimming the scattering instrument building down to 3 machines (a powder diffractometer, a small-angle scattering instrument and a reflectometer) out of the 18 available beamlines. It is hoped that as soon as CSNS construction begins, IHEP and IoP will establish a scheme to work with universities in seeking external funding for building additional world-class scattering instruments.

1.3. The Compact Pulsed Hadron Source of the Tsinghua Hadron Accelerator Research Center

The CPHS is a newly approved project led by the Department of Engineering Physics (DEP) of the Tsinghua University in Beijing, China. It is to be housed in an existing building on the Tsinghua campus, previously built and used for the now completed cargo-inspecting accelerator system. As illustrated in Figure 3, Phase I of the CPHS consists of a high-intensity proton linac (ion source, RFQ, and DTL), a neutron target station (Be target, reflector, and solid-methane and water moderators), and a small-angle neutron scattering instrument and an imaging/radiography station. Phase 2 will expand to include a proton research platform and additional neutron beamlines (for powder diffraction, reflectometry, irradiation/activation analysis, neutron therapy, and neutron optical devices R&D). The primary parameters of the CPHS proton-accelerator are: power on target 16 kW, energy 13 MeV, average and peak beam current 1.25 and 50 mA respectively, pulse repetition rate 50 Hz, and pulse width 0.5 ms. The rationale behind the CPHS-THARC is three-fold: 1) to establish a medium-flux, university-based neutron source for education of students and training of users, 2) to enable multi-disciplinary research and neutronics/device development, and 3) to pave the way for further R&D of innovative hadron technologies, from materials characterization to medical radiology to energy/environment applications. Since the CPHS adopts primarily mature accelerator and neutronic technology, requires a base cost much less than those of CSNS and CARR, and affords a fast-track construction, the facility even in the initial phase can advance scientific utilization and instrumental development of cold-to-epithermal neutrons that are much needed in China. CPHS will serve research partners and users from a large community, including but not limited to Tsinghua University.

After the launching of the CPHS-THARC project in February 2009, the university soon allotted the first-year budget of ~$3M. In April 2009 the Project contracted the construction of the ion source by the Institute of Modern Physics at Lanzhou, China. By virtue of the completed design of the RFQ and DTL, procurement processes for their fabrication and the associated power supply are now underway. The design and optimization of the neutron target station, on the other hand, requires considerable research, from understanding of the basic neutronics to Monte Carlo simulations. Not until an optimized design is achieved can engineering/construction of the target station commence. For this reason the DEP will convene the first international mini-workshop in June 2009 focusing on the neutron physics that governs the target-reflector-moderator performance and the processes towards the realization of an optimal configuration. Ad hoc meetings aiming at tackling problems in beamline optics and neutron-scattering instruments will be organized. The first goal is to complete Phase 1 of the project as well as to start R&D of Phase 2 in 3 years. Admittedly, such a timeline is rather ambitious. Therefore, the Project welcomes the participation of the international hadron-accelerator and neutron communities in advisory and collaborative activities with Tsinghua University and will actively take part in international collaborative such as the International Collaboration on Advanced Neutron Sources (ICANS) and symposia of accelerator and neutron moderation development. One of the top-priority items is assertive recruitment of young talented faculty, particularly in the area of neutronics and neutron science/instrumentation. Since experienced researchers in these fields are scarce in China, attracting talent from overseas is a reasonable alternative. Concurrently, Tsinghua University will seek cooperation with the CSNS and CARR projects and with interested academic institutes in order to build an alliance for the development of hadron science and applications.

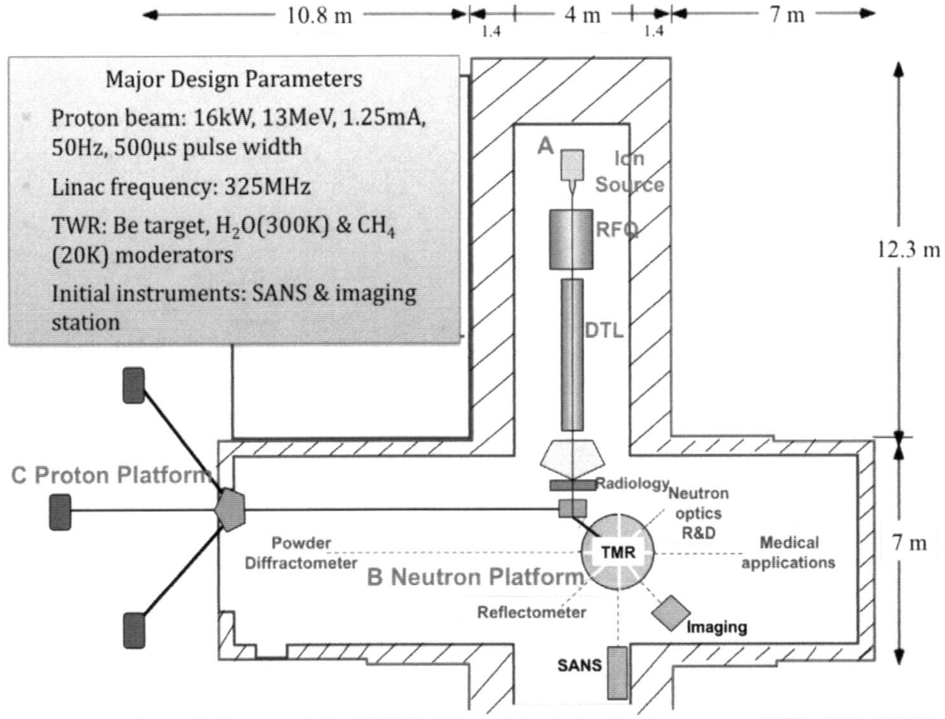

Figure 3. The schematic layout of the CPHS. Phase 1 includes the ion source, RFQ, DTL, TMR, and the SANS and Imaging beamlines. Phase 2 adds more neutron instruments and the proton platform.

2. Development of a domestic neutron user community and a network linking the international counterparts

Since 2005 four user meetings [2] and two US-China Workshops on neutron science and technology [3] were held. Attendants coming from 20-30 universities and ~10 laboratories nationwide listened to introductions of the CARR and CSNS and scientific applications of the neutron method from project scientists and foreign expert speakers. They also participated in discussions of possible outreach activities such as student training, joint research programs, and co-development of instruments. A nationwide survey of potential users has recently been conducted by CSNS. It identified the neutron-related research topics and priorities of desirable scattering instruments of ~30 research groups from universities and national laboratories. With continuing outreach efforts the assessment indicated that 30-40 user groups will conduct ~60 experiments per year using the 3 CSNS instruments. If all 18 instruments are available, the number of user groups will increase to ~100 with ~500 experiments performed per year. In the future the three neutron-source project will undoubtedly concert their efforts in user development, which will enhance the effectiveness of the scientific utilization of the facilities.

In the previous user meetings an informal alliance of neutron users was proposed but it came short of forming a national neutron science society. In contrary, neutron societies have been established in many countries in Asia-Oceania, Europe and America. For many years the neutron societies in each continent/region have organized their own regional activities for promoting cooperation in neutron science and instrumentation such as topical schools, student exchange, international conferences, and prizes and fellowships. The establishment of a Chinese neutron science society will enable China's full participation in the international events sponsored by the Asian-Oceania Neutron Scattering Association (AONSA) as well as effectively championing domestic support of neutron-related research through a coherent voice. The organization of a university-led neutron science society in China is currently under discussion.

3. Conclusions

The advent of three neutron-source projects in China accedes the fact of increasing demand for multi-disciplinary R&D using neutrons, fueled by the rapid growth of international efforts in neutron applications versus the lagged behind status in China. The high-flux national projects of the CARR and CSNS have undergone different stages of development for a decade. Significant progress has been made yet arduous challenges remain in the next 2-7 years before fruits of the hefty investments are to be harvested. A medium-flux university-based source, owing to its relatively low cost and fast-track development, may play a pivotal role in user outreach within the academia and promotion of multi-disciplinary applications of accelerator and neutronics technology. The CPHS of the THARC project may produce cold-to-epithermal neutrons for small-angle scattering and imaging in 3 years and expand neutron/proton R&D facilities in subsequent years.

Acknowledgements

We thank Profs. Dingsheng Wang and Fangwei Wang of the Institute of Physics for providing useful information regarding user statistics.

References

[1] D. F. Chen, Y. T. Liu, C. Gou, C. T. Ye, Development of neutron scattering on 60 MW research reactor in CIAE, Physica **B385-386** (2006) 966.
J. Wei *et al.*, China spallation neutron source: Design, R&D, and outlook, Nucl. Instr. Methods **A600** (2008) 10.
[2] Neuron News **16**(1) 2005 and *ibid* **17**(1) 2006.
[3] Neuron News **19**(1) 2008 and *ibid* **20**(2) 2009.

The Second Target Station at the ORNL Spallation Neutron Source

R. K. Crawford, R. Dean, P. Ferguson, J. Galambos, F. Gallmeier, T. McManamy, M. Rennich

Oak Ridge National Laboratory, Oak Ridge, TN

E-mail: crawfordrk@ornl.gov

Abstract. After the Spallation Neutron Source (SNS) construction project was completed in June, 2006, development of plans to construct a second target station (STS) at SNS began. These plans have evolved to the establishment of a reference concept for a STS and associated neutron beam instruments, and the evaluation of the expected performance for this station, all of which have been documented in a White Paper. Based on this White Paper, the Department of Energy has approved development of a detailed conceptual design leading to a construction project for the STS. The STS reference design is based on pulse stealing from the 60 Hz SNS accelerator system, with one pulse of every three going to the STS and the other two going to the first target station. The STS would operate in long-proton-pulse mode with no pulse compression in the accumulator ring, and would be optimized for production of intense beams of cold neutrons. The reference concept for the STS and the estimated performance for this concept will be discussed

1. Introduction

The Spallation Neutron Source (SNS) facility at Oak Ridge National Laboratory (ORNL) became operational in the spring of 2006. The SNS is a short-pulse spallation source with a linac accelerating a beam of H^- ions to 1 GeV. This beam has a macro-pulse length of 1 ms which is chopped into a number of mini-pulses and injected into an accumulator ring where the mini-pulses are accumulated on top of one another. After the ring is fully loaded, the beam is extracted in a single turn to produce a ~700 ns proton pulse that impinges on a mercury target where the spallation process produces very intense short pulses of neutrons. Several moderators slow these neutrons to thermal or sub-thermal energies for use in neutron scattering experiments. This process repeats at a 60 Hz rate.

The SNS was designed and implemented such that the power of the facility could be upgraded with minimal disturbance, and the facility design and site layout facilitate provision of an additional target station to accommodate an expanding experimental instrument suite. A Power Upgrade Project (PUP) for the SNS is already underway, and a "white paper" has developed and evaluated a preliminary concept for a second target station (STS) at SNS. This concept, referred to as the "reference concept" will be described in this paper. This reference concept was optimized for the production of intense beams of cold neutrons, thus emphasizing areas of science for which the first target station was not fully optimized.

2. Reference concept

2.1. Reference concept for the accelerator system

In the STS reference concept the accelerator system operates in a "pulse-stealing" mode at 60 Hz with every third pulse going to the second target station (20 Hz second target station operation) and the remainder of the pulses going to the first (present) SNS target station (40 pulses per second in a "pseudo-60 Hz mode"). For the 20 pulses per second going to the second target station, the long proton pulse (~1 ms long) from the linac is sent directly to the second target station with no accumulation in the ring. This "long-pulse" operating mode enables the delivery of about 50% more power from the linac for the same pulse duration because the chopping of the proton beam necessary for storage and extraction in the ring can be eliminated, although acceleration of this additional beam power requires additional RF power to the linac. The use of long proton pulses also allows the use of a mercury target for the reference concept for the second target station even at the higher power per pulse, because spreading the proton pulse out over ~1 ms significantly diminishes any mercury cavitation effects.[1] The anticipated energy and beam current provided by the accelerator systems will permit operation in this mode with at least 1.3 MW of short pulses directed to the first target station and at least 1 MW of long pulses directed to the STS, using accelerator RF equipment upgrades consistent with the PUP.

2.2. Reference concept for the target building and target station

The reference concept for the second target station is similar to that of all recent spallation facilities—a horizontal proton beam colliding with a target located in the center of an iron and high-density-concrete shielding monolith. In the reference concept neutron beamlines arrayed on each side of the target shielding monolith are illuminated by moderators above and below a mercury target. A shielded target service bay used to replace the targets and contain the mercury cooling loop is located downstream of the proton beam. The target and associated shielding monolith are placed asymmetrically within the target building, so that the instruments on one side have more room to fit within the building while instruments on the other side will mostly be on longer beamlines that would not fit within the target building in any case.

The optimization studies for the STS reference concept led to the selection of two large pre-moderated supercritical para-hydrogen moderators, closely coupled in wing geometry to a mercury target similar to that at the first SNS target station. The cylindrical moderators are each 22 cm diameter and 14 cm tall, surrounded by water pre-moderators on all sides except where neutron beams are to be extracted, and embedded inside a beryllium reflector assembly. Figure 1 shows the STS reference concept target, moderators, and neutron beamlines inside the shielding.

2.3. Reference instrument suite

In order to better assess the full potential of the facility concept defined in the preceding sections, a set of neutron beam instruments has been selected as a reference or "straw man" instrument suite for STS. The specific instruments selected are based on concepts proposed and evaluated as part of the SNS Second Target Station Instrumentation Workshop. In some cases, the instrument designs proposed at the workshop have been modified to fit circumstances (e.g., the maximum beam length that fits on the site is ~130 m) and the performance reevaluated for the new parameters. Instruments selected for the STS reference instrument suite are those instruments proposed during the workshop that provide the greatest scientific payoff and benefit the most from the intense cold beams and low operating frequency at the STS.

2.4. Conventional facilities

Figure 2 shows the layout of the STS reference concept target building and instruments on the SNS site. The location of the target building is optimized to accommodate the longer beam lines appropriate for this type of facility.

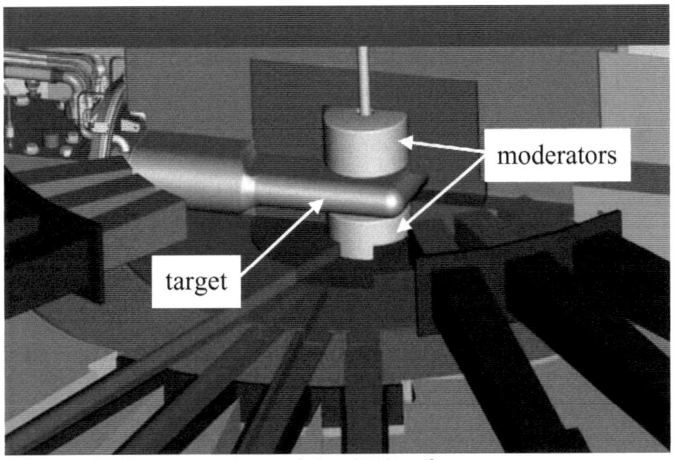

Figure 1. Cutaway view of the STS reference concept target, moderators, and neutron beamlines inside the shielding monolith.

Figure 2. Second target station building reference conceptual model shown in its approximate location on the SNS site. New infrastructure buildings required to support the STS will be in close proximity but are not shown in the figure.

3. Estimated performance

The neutronic performance of the STS reference target station and the associated performance of the reference suite of instruments were calculated for both the short-proton-pulse (~1 μs) and long-proton-pulse (~1 ms) operating modes. In the short-pulse option, the pulse shape is determined solely by the neutron moderation process. Shorter-wavelength neutrons are emitted in narrower and more peaked pulses, whereas long-wavelength neutrons need longer moderation times, resulting in wider pulses with less pronounced peaking. In the long-pulse option, the pulse shapes derive from a time-folding of the proton pulse shape and the short-pulse neutron pulse shapes.

Figure 3 shows the peak and time-averaged brightness calculated for the STS reference concept in both short-proton-pulse and long-proton-pulse modes. The time-averaged brightness is the important parameter for instruments that can use the full pulse width (i.e., the full pulse width provides adequate timing resolution), and the peak brightness is the most important parameter for instruments that need to chop the source neutron pulse to provide adequate resolution. One additional performance difference between STS and the first target station is that STS would operate at 20 Hz, leaving a much longer period between pulses. This longer period gives access to a larger bandwidth.

The estimates show that in long-pulse mode, instruments on the STS that can take full advantage of both of these factors can perform more than an order of magnitude better than they would at the first target station, and on average the STS in long-pulse mode provides about an order-of-magnitude gain over the first target station for a broad class of important scientific applications. This improved performance makes it practical to tighten angular and/or energy resolution or to use other techniques to probe previously inaccessible ranges of parameters. It also makes it practical to consider the use of neutron beams focused to well below 100 microns in size, since the neutron intensity that will be available in such focused beams at STS will be enough to measure the very weak absorption or scattering produced by the relatively small number of sample atoms illuminated by such a beam.

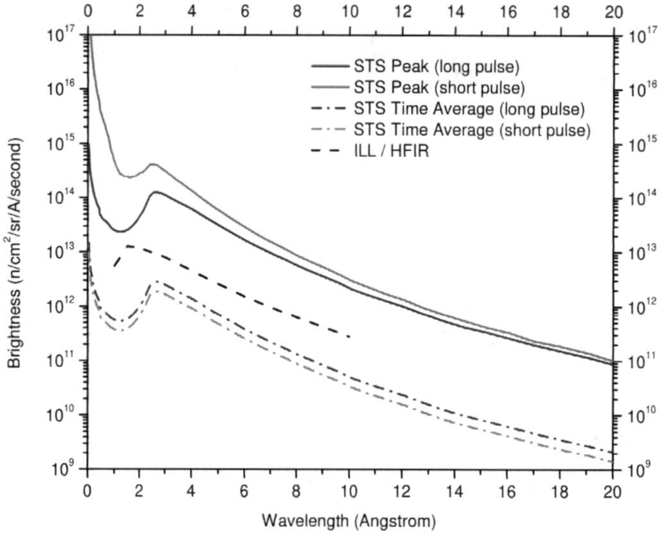

Figure 3. Peak and time-averaged moderator brightness calculated for the STS reference concept target-moderator system. For comparison, time-averaged cold-neutron brightness is also shown for the ILL cold source. Calculations are for 1 MW to the STS in long-pulse mode and 0.67 MW to the STS in short-pulse mode, reflecting the difference in the linac power in the two cases.

4. Summary

A reference concept has been developed and evaluated for a second target station at the SNS. By focusing on and optimizing for the production of cold neutrons this new facility will provide much higher cold-neutron intensities than heretofore available on any pulsed neutron source. The evaluation shows that such a second target station will provide major new capabilities that significantly extend the types of scientific problems that can be fruitfully addressed with neutron scattering.

Acknowledgement

ORNL/SNS is managed by UT-Battelle, LLC, for the U.S. Department of Energy under contract DE-AC05-00OR22725.

References

[1] B. W. Riemer, private communication (2009).

International Conference on Neutron Scattering 2009 IOP Publishing
Journal of Physics: Conference Series **251** (2010) 012055 doi:10.1088/1742-6596/251/1/012055

CryoEDM: a cryogenic experiment to measure the neutron Electric Dipole Moment

C A Baker, S N Balashov, V Francis, K Green, M G D van der Grinten, P S Iaydjiev[1], S N Ivanov[2], A Khazov[3], M A H Tucker and D L Wark
Science and Technology Facilities Council, Rutherford Appleton Laboratory, Chilton, Didcot, OX11 0QX UK.

A Davidson, J R Grozier, M Hardiman, P G Harris, J R Karamath, K Katsika, J M Pendlebury, S J M Peeters, D B Shiers, P N Smith, C M Townsley and I Wardell
University of Sussex, Falmer, Brighton BN1 9QH, UK.

C Clarke, S Henry, H Kraus and M McCann[4]
University of Oxford, Keble Road, Oxford, OX1 3RH, UK.

P Geltenbort
Institut Laue-Langevin, BP156, F-38042 Grenoble, France.

H Yoshiki
University of Kure, Japan

corresponding author: maurits.van-der-grinten@stfc.ac.uk

Abstract. We have constructed an instrument, CryoEDM, to measure the neutron electric dipole moment to a precision of 10^{-28} *e* cm at the Institut Laue-Langevin. The main characteristic is that it is operating entirely in a cryogenic environment, at temperatures of 0.7 K within superfluid helium. Ultracold neutrons are produced in a superthermal source and stored within the superfluid in a storage cell which is held in a magnetic and electric field. NMR measurements are carried out to look for any shifts in the neutron Larmor precession frequency associated with the electric field and the neutrons are detected *in-situ* in the superfluid. Low temperature SQUID magnetometry is used to monitor the magnetic field. We report on the current status of the project that is now being commissioned and give an outlook on the future exploitation of the instrument.

[1] on leave of absence from INRNE, Sofia, Bulgaria
[2] on leave of absence from PNPI, St Petersburg, Russia
[3] joint with the University of Sussex, Falmer, Brighton BN1 9QH, UK
[4] joint with Rutherford Appleton Laboratory, Chilton, Didcot, OXON OX11 0QX, UK

1. Introduction

Experimental searches for particle electric dipole moments (EDMs) form an important part of particle physics: these experiments provide a possibility to look at the breaking of fundamental symmetries of nature. For particles to have electric dipole moments, the forces concerned in their structure must be asymmetric with regard to space-parity (P) and time reversal (T). T violation also implies, through Lorentz invariance, that the combined charge-parity (CP) symmetry is violated. P violation is a well-known intrinsic feature of the weak interaction, but the combined charge-parity (CP) violation is much less understood and subject of intense research effort both at high energy physics facilities as well as in low energy physics experiments such as EDM searches.

CP violation is required to explain a number of observed phenomena, most importantly the baryon asymmetry [1]: our Universe is predominantly made out of matter despite the fact that the Big Bang should have produced equal amounts of matter and antimatter. CP violation is incorporated in the Standard Model of elctroweak interaction but is orders of magnitude too small to generate the baryon asymmetry as we see it today. The origin of CP violation we are searching for must thus lie beyond the Standard Model.

The neutron is a very good system to look for the required CP violation. The Standard Model gives a contribution to the neutron EDM of the order of 10^{-31} to 10^{-33} e cm which, because it is second order in the weak interaction coupling constant, is very small. Extensions to the Standard Model though, such as additional Higgs fields, right-handed currents or supersymmetric partners invariably give rise to dipole contributions which are of first order. These are necessarily much larger, and are typically of order 10^{-26} to 10^{-28} e cm [2]. The neutron thus allows to look for CP violation without a background signal that would originate from Standard Model interactions.

The EDM experiments so far have given upper limits on the value of the EDM and have placed ever tighter constraints on the models that go beyond the Standard Model. The current upper limit on the neutron EDM, $d_n < 2.9\ 10^{-26}$ e cm [3], is based on a previous measurement by some of the CryoEDM collaboration. The CryoEDM experiment is designed to improve the measurement by two orders of magnitude.

2. Experimental techniques & set-up

2.1. Neutron EDM measurement techniques

The experimental techniques used in the neutron EDM experiment are based on measuring the Larmor precession frequency ν_L of the neutron in a magnetic holding field B_0 to a very high precision. The Larmor frequency is given by $\nu_L = -2B_0\mu_n/h$, where μ_n is the neutron magnetic moment and h is the Planck constant. The presence of an EDM d_n will shift the Larmor frequency by $\pm 2Ed_n/h$ if an electric field E is applied, where the plus and minus signs correspond to the electric field being parallel and antiparallel to the magnetic field. The difference between the Larmor frequencies measured in opposite electric field directions associated with these frequency shifts is directly proportional to the neutron EDM:

$$\Delta\nu_L = -4Ed_n/h$$

The measurement of the precession frequency is performed using Ramsey's method of separated oscillatory fields. Neutrons, polarised parallel to the B_0 holding field, are submitted to two short $\pi/2$ rf pulses separated by a precession period. If the rf pulses are exactly on resonance with the Larmor precession frequency the neutrons will have a final polarisation directed antiparallel to the magnetic field. If the rf frequency is off resonance, the phase difference between the neutron signal and the rf signals will build up with each precession, and it accumulates to a sizeable phase difference at the end of the precession period. The final spin state of the neutrons is dependent on this phase difference. By counting the number of neutrons with polarisation parallel to B_0 as a function of the applied rf, the Larmor frequency can be determined to a high precision. The precession time between the two rf

pulses can be maximised using ultra-cold neutrons, neutrons of energy low enough (of the order of 10^{-7} eV), so that they can be stored in a bottle for long times.
The statistical sensitivity of an EDM measurement is given by

$$\sigma(d_n) \approx \frac{\hbar}{2ETP^2\sqrt{N}}$$

where E is the electric field strength, T is the neutron storage time, P is the neutron polarisation and N is the number of neutrons in a measurement. These are the parameters that need to be optimised to make a high precision EDM measurement.

2.2. Experimental set-up

The main characteristic of the CryoEDM set-up is that it is operating entirely in a cryogenic environment at temperatures of 0.7 K: ultra-cold neutrons (UCNs) are produced in a super-thermal source, stored within superfluid helium and afterwards detected also within the superfluid. Clearly this adds a degree of complexity to the design and operation of the experiment but the gains are compelling. The UCN densities produced are orders of magnitude higher than in previous experiments and the electric field sustainable in liquid helium is also considerably stronger.
We operate a large cryogenic system to reach the temperatures required for the UCN production and EDM measurements: a ^{3}He circulation cryostat cools the helium down to a base temperature of 420 mK in conjunction with two other cryostats that hold liquid helium for thermal and magnetic shields. The ^{3}He circulation is driven by a pump station consisting of four large roots pumps operating in series. A separate pumping station ensures that the general liquid He system, serving the cryogenics operation and supplying helium to the UCN source, is kept at low temperatures. This provides the cryogenic basis around which we have constructed the EDM experiment. Figure 1 shows a picture of the CryoEDM apparatus installed at the ILL showing the EDM measurement region and some of the cryostats. A drawing of a cross section of the apparatus is shown in Figure 2

Figure 1: The experimental apparatus installed on the H53 beam position at ILL. In the foreground the measurement region with its magnetic shields is seen. The superthermal UCN source coupled to the measurement region is in the background of the picture

The experimental set-up furthermore comprises the following parts:

2.2.1. The primary neutron beam. The experiment is running on the H53 cold neutron beam at ILL. Before the beam enters the source it is spin polarised by a 2.3 meter long neutron polariser developed for the experiment. The polariser also acts as a filter: the incoming neutron spectrum is cut off at a wavelength below 4.5 Å and the transmission linearly increases to 100% at 9 Å. Filtering the primary beam in this way reduces radiation background by a large factor and it protects the ^3He circulation and transfer cryostats from being activated. The capture flux is measured to be 3×10^9 cm^{-2} s^{-1} at the H53 guide exit. The capture flux of the polarised beam is 1.5×10^9 cm^{-2} s^{-1} with a polarisation measured to be over 90% at 9Å, the polariser has a slightly focusing arrangement. Measurements of the energy spectrum of the polarised neutron beam entering the experiment showed the flux to be 3×10^7 cm^{-2} s^{-1} Å^{-1} at 9 Å. These numbers are sufficiently high to perform an improved nEDM measurement using CryoEDM.

2.2.2. The superthermal UCN source. The UCN are produced by down scattering the 9 Å neutrons in superfluid helium at a temperature of about 0.65 K. In this process a phonon is created in the helium that balances the energy and momentum lost by the neutron. With the intensity of the primary beam at H53 the UCN production rate is 1 cc^{-1} s^{-1} [4] and can generate high UCN densities. There is a number of UCN loss channels that needs to be kept under control in order to retain the produced UCN. A possible loss is through upscattering: a phonon being absorbed by a neutron will raise the energy of the neutron and the UCN is lost for the experiment. The temperature of the helium must be kept well below 1K to minimise the phonon density and thus to suppress this process. Another loss channel would be through absorbtion by ^3He: ^3He has a very large absorption cross section for neutrons of UCN energy. The helium in which the UCN are created must be isotopically pure ^4He which does not absorb neutrons. In our experiment we produce our own isotopically pure ^4He by operating a series of so-called superleaks [5]. Liquid helium from the main tank in the cryostat is passed through these superleaks and the superfluid component only, the ^4He, will be passed on to the UCN source. The purity of the helium has been measured and showed a ^3He abundance better than 10^{-10} which is low enough to ensure long UCN storage lifetimes.

2.2.3. *UCN guide system and UCN valves.* The UCN guides need to be able to store and transport

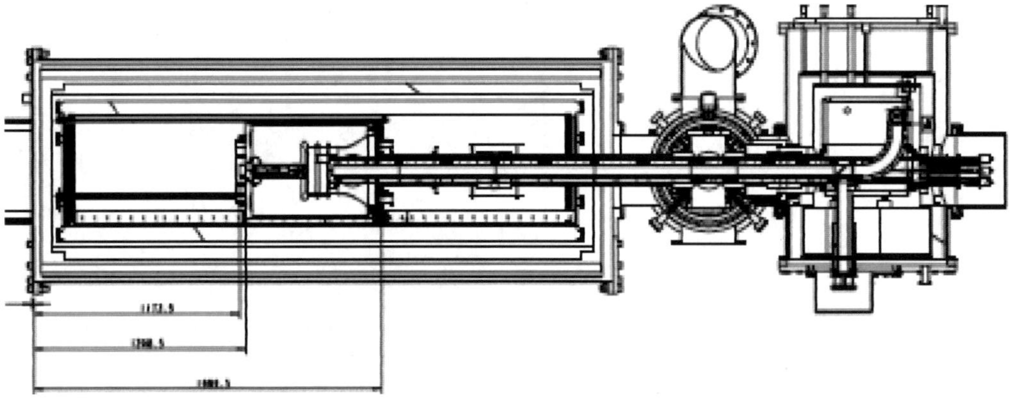

Figure 2: A cross section of part of the experimental apparatus. The storage cells are mounted at the centre of the magnetic shields and are coupled to the superthermal UCN source on the right with UCN guides. The detector chamber is situated under the transfer cryostat. The UCN source is at right angle to this cross section and is not shown.

UCN with minimal losses. In addition our UCN source also serves as a guide through which the 9 Å primary beam passes the superfluid helium so the guide needs to be reflective for cold neutrons. Furthermore, the guides need to be made and placed such that the polarisation of the UCN is not lost. To comply with these requirements the neutron guides have been made from copper coated with a layer of beryllium. Beryllium has a high Fermi potential of 250 neV and makes good UCN storage material. The UCN source is a 3m long tube tube of diameter 68 mm, providing a total volume of some 10 litres for UCN production. The source tube is connected to the transport guides that lead to the storage cells.

The experiment is fitted with four UCN valves that are operated in the superfluid helium. The valves are used to keep the UCN stored in a part of the experiment and to let the UCN move between the different chambers of the experiment. A defined storage time in the UCN source and in the storage cells can thus be set and the UCN can be directed towards the detector chamber at the end of a data cycle.

2.2.4. Magnetic field and shielding. The magnetic shielding consists of three layers of mu-metal and a superconducting lead screen that is kept in a cylinder filled with liquid helium at a temperature of 4K. A holding field of 50 mG is generated by a superconducting NbTi solenoid, this solenoid is wound on a former kept in the helium filled cylinder just inside superconducting lead screen. Various trim coils are mounted around the storage chambers that can optimise the holding field and increase its homogeneity further. The magnetic shields assembly has its own independent helium cryostat and cryocoolers.

2.2.5. UCN storage cells. The UCN are stored in a double cell arrangement where the actual Ramsey resonance operations take place. The cells consist of two BeO cylinders each capped with Ti electrodes coated with 1 μm of diamond like carbon [6]. The cells hold a volume of 2.2 litres each. The common electrode between the two cells is kept at ground potential as is the electrode through which the UCN enter the cells. On the third electrode a voltage is applied, one cell is thus holding an electric field and is neighboured by a cell being in a neutral electric field. The neutron precession measurements in the two cells are treated separately and give a handle on systematic effects that would change the neutron precession frequency and generate a false EDM signal. The storage cells are placed in a superfluid containment vessel made of a non magnetic titanium alloy at the centre of the magnetic shielding.

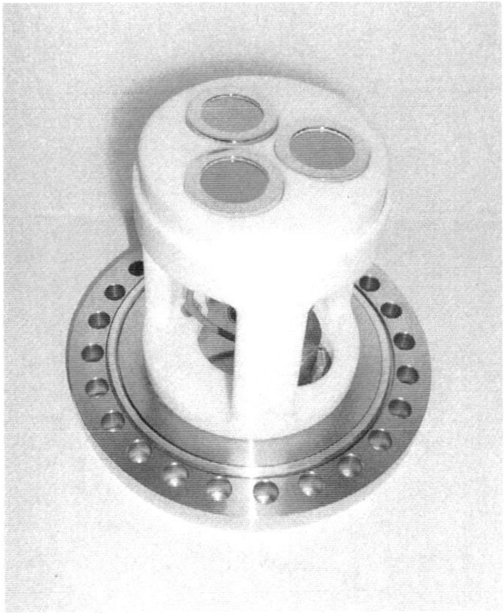

Figure 3: The UCN detector assembly. The surface barrier detectors are coated with ^{6}LiF for neutron to charged particle conversion and an Fe layer for UCN polarisation analysis purposes

2.2.6. UCN detectors and polarisation measurement. The detector system (Figure 3) consists of a set of solid state surface barrier detectors covered with ^{6}LiF. The signal in the detector is generated by the alpha and triton particles from the n(^{6}Li,^{3}H)^{4}He reaction [7]. This is a system that has been shown to work efficiently within superfluid helium [8]. The detectors are placed in a magnetic field that the UCN will have followed adiabatically. Two of the detectors have an additional Fe layer on top of the ^{6}Li coating that will reflect one spin state of the UCN. The ratio in neutron counts in the detectors with Fe layer to the counts in the detectors without Fe layer is a measure of the UCN polarisation.

2.2.7. Magnetometry. SQUID magnetometers are used to monitor the magnetic field changes inside the SCV. For this an assembly of superconducting pick-up loops are fed back to the SQUIDS. The SQUIDS are operating in a section between the UCN source and magnetic shields inside a cryogenic tube that is filled with the superfluid. The pick-up loops are situated under the neutral electrode and can monitor different components of the field changes in the SCV. In addition to the SQUIDS cryogenic fluxgates measure the magnetic field inside the shields to provide an absolute field measurement close to the storage cells.

3. Performance & Outlook

One of the major challenges in the experiment is cooling down apparatus of this size to the sub-Kelvin temperatures that is required for producing and storing UCN. We have run the cryogenic apparatus in various configurations in order to characterise the performance of the separate parts of the set-up and have cooled down the full apparatus as part of the commissioning process. We now have an operational superthermal UCN source with a measured UCN production rate sufficiently high for us to reach a sensitivity of the order of 10^{-27} *e* cm at the current H53 beam position.

UCN detection is operational to a high efficiency as is the polarisation analysis of the UCN. The magnetic shields and both the fluxgate and SQUID magnetometry is operational, additional improvements are envisaged to enhance the shielding properties further.

The experiment is now in its commissioning phase and we envisage to make a nEDM measurement of the order of 10^{-27} *e* cm at the H53 beam position at ILL. In a later stage (year 2012) the experiment will move to a higher intensity 9 Å beam at ILL that will allow for a 10^{-28} *e* cm measurement which is the design aim of the apparatus.

References

[1] M. Barr, Int. J. Mod. Phys. A**8** (1993) 209-236

[2] T. Ibrahim and P. Nath, arXiv:0705.2008v2 [hep-ph] (2008)

[3] C.A. Baker, D.D. Doyle, P. Geltenbort, K. Green, M.G.D. van der Grinten, P.G. Harris, P. Iaydjiev, S.N. Ivanov, D.J.R. May, J.M. Pendlebury, J.D. Richardson, D. Shiers, K.F. Smith *Improved Experimental Limit on the Electric Dipole Moment of the Neutron*, Physics Review Letters **97** (2006) 131801.

[4] C.A. Baker, S.N. Balashov, J. Butterworth, P. Geltenbort, K. Green, M.G.D. van der Grinten, P.G. Harris, P.S. Iaydjiev, S.N. Ivanov, J.M. Pendlebury, D.B. Shiers, M.A.H. Tucker and H. Yoshiki, *Experimental measurement of ultracold neutron production rate in superfluid ^{4}He*, Physics Letters A **308-1** (2003) 67-74.

[5] H. Yoshiki, H Nakai and E. Gutsmiedl, *A new superleak to remove He^3 for UCN experiments* Cryogenics, **45-6** (2005) pp. 399-403

[6] M.G.D. van der Grinten, J.M. Pendlebury, D. Shiers, C.A. Baker, K. Green, P.G. Harris, P.S. Iaydjiev, S. N. Ivanov and P. Geltenbort, *Characterization and Development of Diamond-Like Carbon Coatings for Storing Ultracold Neutrons*, Nuclear Instruments & Methods-A **423** (1999) pp. 421-427.

[7] C.A. Baker, K. Green, M.G.D. van der Grinten, P.S. Iaydjiev, S.N. Ivanov, S. Al-Ayoubi, P.G. Harris, J.M. Pendlebury, D.B. Shiers and P. Geltenbort, *Development of solid state silicon devices as ultra cold neutron detectors*, Nuclear Instruments & Methods-A **487-3** (2002) 511-520.

[8] C.A. Baker, S.N. Balashov, K. Green, M.G.D. van der Grinten, P.S. Iaydjiev, S.N. Ivanov, J.M. Pendlebury, D.B. Shiers, M.A.H. Tucker, H. Yoshiki and P. Geltenbort, *Development of low temperature solid state silicon ultra cold neutron detectors working within superfluid ^{4}He,* Nuclear Instruments & Methods-A **501-2** (2003) 517-523.

Journal of Physics: Conference Series **251** (2010) 012056 doi:10.1088/1742-6596/251/1/012056

Aiming for the theoretical limit of sensitivity of Bonse-Hart USANS instruments

John M. Carpenter [1,2] **and Michael Agamalian** [2]

[1] Argonne National Laboratory, Argonne, IL 60439

[2] Oak Ridge National Laboratory, Oak Rige, TN 37831

E-mail: jmcarpenter@anl.gov

Abstract. The basis of the Double-Crystal Diffractometer (DCD), used for ultra-small angle neutron scattering (USANS), is that the reflectivity function is very near 1.0 for $|y| < 1$, where $y = (\theta - \theta_B)/\ \delta\Theta_D$, falls off rapidly for $|y| > 1$, eventually decreasing as y^{-2}. In the Bonse-Hart multi-bounce crystal the reflectivity function R(y) transforms into $R^m(y)$ and in the wings, for large $|y|$, theoretically decreases as y^{-2m} after m consecutive Bragg reflections inside the channel-cut crystal. Here θ and θ_B are respectively the diffraction angle and the Bragg angle (for given wavelength) and $\delta\Theta_D$ is the width of the Darwin plateau. However, the experimental reflectivity $R^m{}_{exp}(y)$ obtained for m = 3 exceeds the theoretical prediction by over two orders of magnitude in the range of the far wings, which creates limitations for USANS studies of weakly scattering objects.

We used the pulsed-source neutron time-of-flight (TOF) technique to study this discrepancy in more detail. Two identical Si(111) crystals, a slab-shaped single-bounce and a channel-cut triple-bounce, were measured at the nominal Bragg angle $\theta_B = 24.4^o$ in the TOF powder diffractometer GPPD at IPNS, in the range $0.2 < \lambda < 4.0$ Å of the first seven Bragg reflections from Si(111) family. Cadmium shielding protected the detectors from view of the first-bounce crystal. The experimental data obtained from the single-bounce crystal shows thermal diffuse (phonon) scattering (TDS) filling the region between the Bragg reflections, and exhibiting the symmetry of the reciprocal lattice. With appropriate shielding installed, the triple-bounce Bragg reflections, in contrast, are TDS-free in the range $0.6 < \lambda < 3.0$ Å; however, the intensity of TDS grows in the range $\lambda < 0.5$ Å, reaching the level of TDS registered for the single-bounce reflections (777) and (888). The growth of TDS correlates with the increase of the Cd transmission $T(\lambda)$ in the range $0.2 < \lambda < 0.5$ Å, which in the vicinity of (888) reflection is $T(\lambda) \approx 0.9$. Therefore, the Cd shielding of the triple-bounce crystal becomes ineffective and the parasitic single-bounce back-face reflection and TDS, blocked for $0.5 < \lambda < 3.0$ Å, reappear for $\lambda < 0.5$ Å.

It is practically impossible to separate this parasitic scattering from the triple-bounce reflection at steady state neutron beam lines except by the use of highly curved neutron guides. However, the TOF-USANS instrument, which is currently under construction at SNS, allows separation of the orders of Bragg reflections and the residual TDS in time-of-flight and thus the discovered parasitic effect cannot compromise its sensitivity. Thus, we expect to approach the theoretical limit of sensitivity for the SNS TOF-USANS instrument.

1. Introduction. We studied Bragg diffraction and Thermal Diffuse Scattering (TDS) from a Si(111) channel-cut triple-bounce crystal using the time-of-flight technique at a pulsed neutron source. Cadmium shielding restricted the detector's direct view of the first bounce surface. The shielded channel-cut crystal dramatically suppresses TDS in the vicinity of the (111), (333) and (444) Bragg reflections; however, TDS appears and increases with the decrease of wavelength in the range of the (555), (777) and (888) orders where cadmium becomes transparent and the single-bounce reflections and TDS contaminate the triple-bounce (555), (777) and (888) reflections.

U. Bonse and M. Hart [1] introduced channel-cut multi-bounce crystals into diffraction physics more than four decades ago as new optical components of the X-ray Double Crystal Diffractometer (DCD). Theoretically the multi-bounce reflection reduces by many orders of magnitude the tails, or wings, of the Darwin reflectivity function while the resolution, that is, the width of the reflectivity plateau, and the peak reflectivity are preserved. Originally Si and Ge channel-cut crystals were tested with X-rays [1]. The experimental results appeared satisfactory only for the resolution and the peak reflectivity, but the reflectivity in the wings was many orders of magnitude higher than expected. The dramatic departure from theoretical prediction still is not understood completely and is known in dynamical diffraction physics as the long-standing "wings problem." This parasitic scattering significantly decreases the sensitivity of DCDs, creating limitations in structural investigations of weakly scattering objects. In spite of this limitation, the Bonse-Hart technique found broad application in small-angle neutron and x-ray scattering (SANS and SAXS) and DCDs on channel-cut crystals are nowadays routinely available at all major synchrotron and neutron laboratories worldwide. Therefore, solution of the wings problem is a subject of great interest for experts in diffraction physics as well as for the large community of scientists using Bonse-Hart DCDs for supra-molecular structural studies of condensed matter.

The Bonse-Hart concept of a "tailless" reflectivity function assumes that the classical Darwin function $R_D(y)$ transforms into $R_D^m(y)$ after m consecutive front-face Bragg reflections inside the channel, which is typically cut from a massive single crystal with diffractive faces parallel to a chosen crystallographic plane. The argument $y = (\theta - \theta_B)/\delta\theta_n$ is the dimensionless angular parameter; $(\theta - \theta_B)$ is the deviation from the Bragg angle θ_B, and $\delta\theta_n$ is the half-width of the Darwin plateau for each wavelength,

$$\delta\theta_n = (\pi V_0 \sin 2\theta_B)^{-1} \times [F(h, k, l)\exp\{-W\}\lambda_n^2]. \qquad (1)$$

In equation (1), n is the order of Bragg reflection, λ_n is the wavelength, V_0 is the unit cell volume, $\exp\{-W\}$ is the Debye-Waller factor and F(h, k, l) is the structure factor. In the range of total Bragg reflection $|y| \leq 1$, $R_D^m(y) = R_D(y) = 1$; however, $R_D^m(y)$ decays according the m power law in the range $|y| > 1$ because in this range $R_D(y) = [|y| - (y^2 - 1)^{0.5}]^2$. The Darwin model assumes that the crystal is transparent and infinitely thick; therefore, only the front-face reflection is taken into account. The Ewald function is identical to the Darwin function $R_E(y) = R_D(y) = 1$ when $|y| \leq 1$; however, it becomes different in the range $|y| > 1$ because the Ewald theory calculates the total reflection from a transparent, infinitely long crystal of finite thickness, $R_E(y) = 1 - (1 - y^{-2})^{0.5}$. The relationship between $R_D(y)$ and $R_E(y)$ was established in [2] and can be expressed in terms of the series of consecutive front- and back-face reflections

$$R_E(y) = R_D(y)+[1-R_D(y)]^2 \times R_D(y) \times [1+R_D^2(y)+R_D^4(y)+\ldots] = 2R_D(y)/[1+R_D(y)], \qquad (2)$$

where $R_D(y)$ and $[1-R_D(y)]^2 \times R_D(y)$ correspond to the first front- and back-face reflections respectively. In the far wings where $R_D(y) << 1$, formula (2) can be approximated as $R_E(y) \approx 2\ R_D(y)$, reflecting the fact that the first front- and back-face reflections give the same contribution to the wings. Experimentalists can study the $R_D^m(y)$ and $R_E^m(y)$ functions with the DCD by measuring the rocking curves of the analyzer crystal. This technique allows determining parasitic scattering after fitting the experimental results to the theoretical rocking curves, which are calculated as the convolution of the monochromator and analyzer reflectivity functions.

The parasitic scattering from a thick Si triple-bounce channel-cut crystal in the vicinity of Bragg reflections originates from various sources. Usually it is difficult to determine the source of parasitic scattering from only the rocking curve measurements, and thus additional experiments involving channel-cut crystal modifications are required. Figure 1 shows the current design of a Si triple-bounce channel-cut crystal, the most applicable in SANS studies, along with the cadmium shielding plates. The yellow triangles indicate the internal areas of diffractive walls blocked for incoming and outgoing neutrons. The neutron beam incident at the Bragg angle can pass the channel only by undergoing triple front- and back-face reflections inside the shaded gray zones. Because Si is nearly transparent for neutrons, the strongest mechanism of parasitic scattering is the admixture of single-bounce back-face reflections [3] passing through the long diffractive wall (see the red dashed arrows in Fig. 1). To eliminate this effect, the original Bonse-Hart channel-cut crystal was modified for application in neutron diffraction by cutting an additional groove in the middle of the long wall and inserting a cadmium plate blocking the parasitic beam (see Fig. 1). Also, the edges of the channel-cut crystal are covered by cadmium to eliminate contamination of the triple-bounce reflections by the edge-face diffraction [4]. The surface-induced scattering, as was found in [5], is another parasitic effect, which can be suppressed by polishing and etching the diffractive surfaces of the channel-cut crystal. Nevertheless, in spite of all the described improvements, the experimental reflectivity functions of the Si triple-bounce crystals in the range of $|y| > 4$ still show significant departure from the theoretical prediction, reaching several orders of magnitude at $|y| = 12$ [5]. In contrast, excellent agreement of the experimental data with the $R_D^3(y)$ function in the restricted angular range $0 \leq |y| \leq 4$ has been reported in [6].

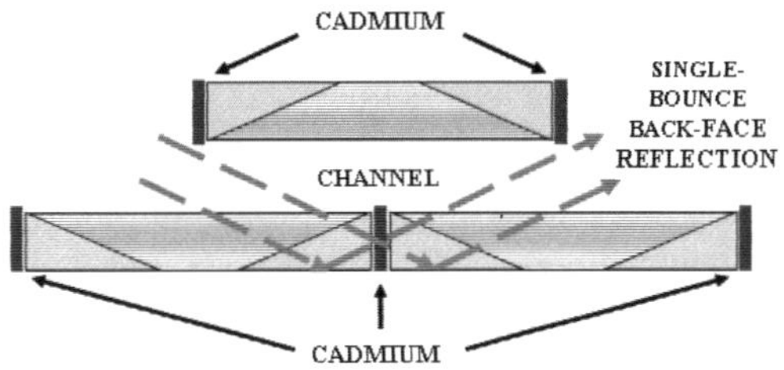

Figure 1. Top view of the channel-cut triple-bounce crystal protected with cadmium plates blocking parasitic single-bounce reflections and edge-face diffraction.

Figure 2 summarizes the introductory part of this paper showing the theoretical single- and triple-bounce reflectivity functions $R_D(y)$, $R_E(y)$, $R_D^3(y)$, and $R_E^3(y)$ in the range $0 \leq |y| \leq 12$. Here the departure from the Bonse-Hart model experimentally observed in many studies (see, e.g., in [5, 7]) in the range $|y| > 3$ is simulated with the admixture of 0.1% single-bounce reflection, $0.001\ R_D(y)$, to the cubed Ewald function $R_E^3(y)$. It should be underlined that we show this simple $[0.001\ R_D(y) + R_E^3(y)]$ model only to demonstrate the typical y-range and the magnitude of deviation from the theory observed at all the Bonse-Hart DCDs worldwide. This is quite the opposite of the unique experimental proof reported in [6]. That contradiction suggests the existence of additional parasitic effects not discussed yet in this paper: refraction from the edges of Cd plates studied in [8] and thermal diffuse scattering (TDS) from thick Si crystals [9]. Because refraction from the edges can be suppressed by changing the rectangular profiles of the Cd plates for curved ones, the most probable mechanism in our opinion is TDS. The hypothesis that TDS is responsible for the wings problem has not been supported by calculations or experimental results so far. The only attempts to prove that TDS is responsible for the departure from the Bonse-Hart theory were made in [10, 11]. In these studies the multi-bounce channel-cut crystals of a DCD were cooled down with cold ^{4}He to observe a reduction of the rocking curve wings. However, the technical difficulties related to this approach did not enable observation of the expected results. Therefore, we decided to use the pulsed-source neutron time-of-flight technique to investigate the connection between TDS and the wings problem.

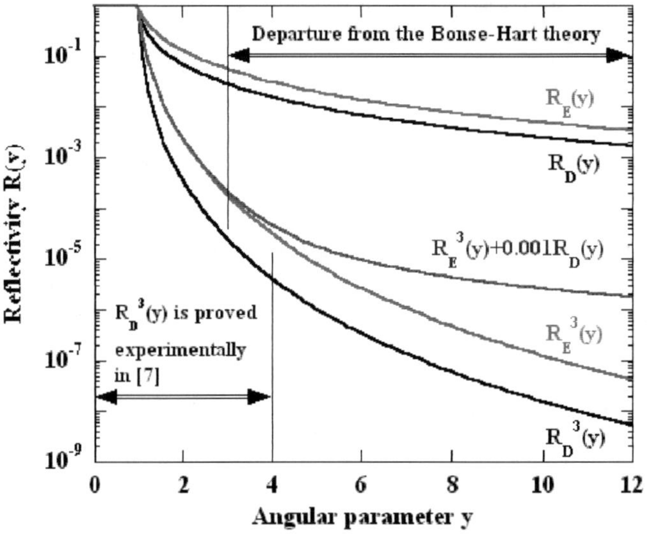

Figure 2. The reflectivity functions $R_D(y)$, $R_E(y)$, $R_D^3(y)$, $R_E^3(y)$ and $\{R_E^3(y)+0.001R_D(y)\}$.

It is well known from previous studies (see, e.g., [9]) that thick Si crystals generate TDS in the vicinity of Bragg reflections. The contribution of TDS to the integrated intensity of Bragg reflections increases as $I(\lambda)_{TDS} / I(\lambda)_B \sim \lambda^{-3}$ [12], suggesting that TDS originating in the vicinity of high-order Bragg reflections should be significantly stronger than that generated near the first Bragg reflection. The Oak Ridge National Laboratory (ORNL) DCD used in [5] was equipped with a Si(111) mosaic pre-monochromator installed at an unfiltered thermal neutron beam. Therefore, high-order Bragg reflections and TDS were present in the neutron beam entering the double-crystal arrangement. Because it is not possible to reveal this mechanism of parasitic scattering from rocking curve measurements at steady-state reactors, we addressed the problem using the time-of-flight General Purpose Powder Diffractometer (GPPD), at the Intense Pulsed Neutron Source, Argonne National Laboratory.

2. Experiment. The first experiment has been run on a slab-shaped single-bounce Si(111) crystal at $\theta_B = 45^\circ$ to test the technique. Figure 3 shows the diffraction data observed for all of the GPPD detectors in the angular range $18^\circ < |2\theta_B| < 156^\circ$ (there are gaps in the angle coverage) as a function of wavelength calculated from the time-of-flight as for elastic scattering. Several families of Bragg reflections appear as tiny bright spots, our particular attention being the (111) family. Surrounding is the scattering that we interpret as TDS, which increases at short wavelengths, as it should. The intensity distribution displays the periodicity of the Si reciprocal lattice and the phonon dispersion relations, convincingly supporting its origin as TDS.

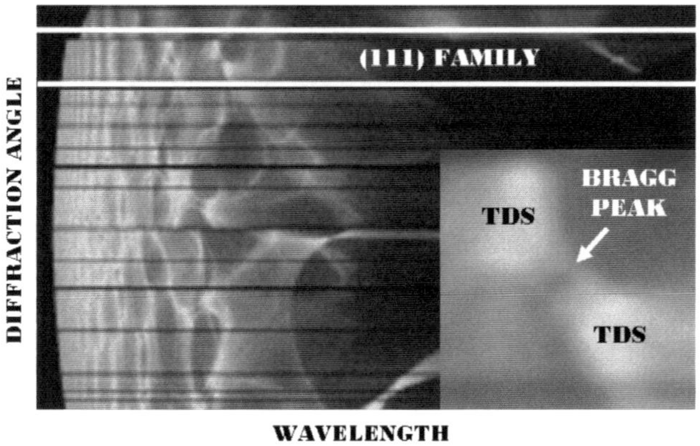

Figure 3. Two-dimensional representation of the GPPD data from a Si(111) slab-shaped single-bounce crystal at $\theta_B = 45^o$. The insert shows separately the Si(333) Bragg reflection and nearby TDS.

The measurements run on the Si(111) channel-cut triple-bounce and the identical single-bounce crystals spanned the wavelength range $0.2 < \lambda < 4.0$ Å. The data collected at three detectors in the angular range $2\theta_B = 24.4 \pm 0.75^o$ were summed and normalized by the monitor rate and the spectrum of the primary beam. Comparison of the two spectra (see Fig. 4) clearly shows strong suppression of TDS nearby the triple-bounce (111), (333), and (444) Bragg reflections; however, the reflections (555), (777), and (888) are contaminated with TDS.

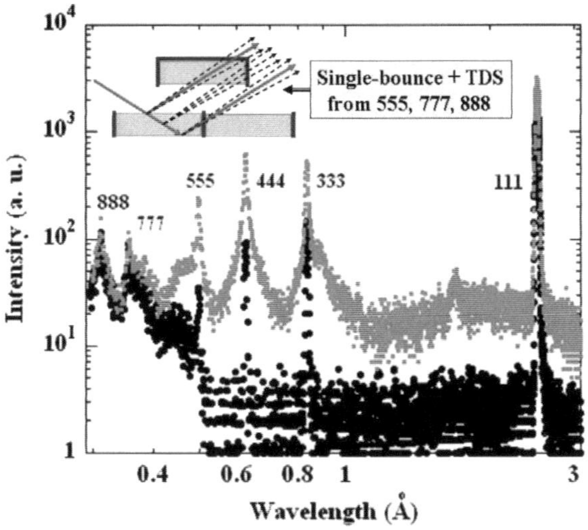

Figure 4. Single-bounce Bragg reflections and TDS (upper curve) from a slab-shaped Si(111) crystal compared with that from the triple-bounce channel-cut Si(111) crystal (lower curve). The insert shows front- and back-face Bragg reflections and TDS emerging from the scattering of non-Bragg-correlated neutrons in the volume.

3. Interpretation. Figure 5 separately presents the spectrum registered from the triple-bounce crystal, highlighting the fact that the increase of TDS in the range of wavelength $0.3 < \lambda < 0.6$ Å correlates with the transmission factor $T(\lambda)$ of 0.75-mm-thick Cd used as the shielding for the Si(111) channel-cut crystal. In the vicinity of the (888) reflection, $T(\lambda) \approx 0.9$, which explains the similarity of the single- and triple-bounce spectra in this range of wavelength. The insert in Fig. 4 demonstrates penetration of (555), (777), and (888) single-bounce front- and back-face reflections and TDS originating along the neutron path inside the crystal wall through the Cd shielding.

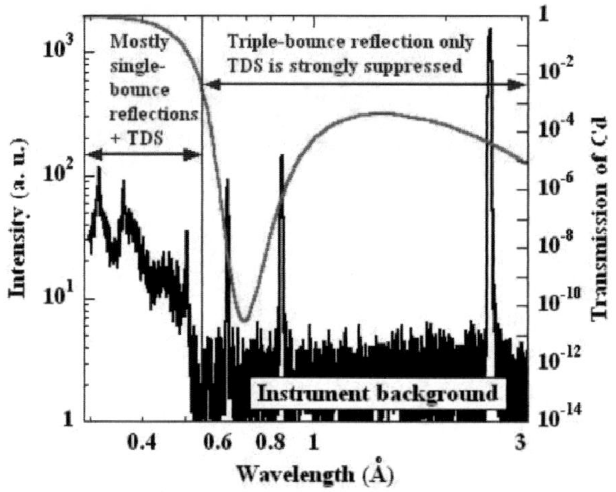

Figure 5. Bragg diffraction and TDS from the Si(111) triple-bounce crystal compared with the transmission of 0.75-mm-thick Cd plate (red line).

Figure 6 shows the region including the (333) and (444) reflections on an expanded wavelength scale. Here. TDS severely contaminates the single-bounce Bragg peaks (see the upper curve), while the triple-bounce peaks are free of TDS and what remains nearby is the general instrumental background. Popa and Willis [12] express the integrated intensity of TDS-contaminated Bragg reflections as $I_{Total} = I_B + I_{TDS} = I_B(1+ \alpha)$, where $\alpha = (I_{TDS} / I_B)$. We integrated our experimental single-bounce data in the range of wavelength $\lambda_B - 0.07$ Å $< \lambda < \lambda_B + 0.07$ Å and the triple-bounce data in $\lambda_B - 0.018$ Å $< \lambda < \lambda_B + 0.018$ Å, subtracting a linearly interpolated background. Subtracting the triple-bounce intensities from the single-bounce intensities and calculating α_{333} and α_{444}, we found that $\alpha_{333} = 7.0$ and $\alpha_{444} = 14.2$ (a rather large admixture of TDS). Popa and Willis [12, Eq. 3] indicate that α is proportional to λ^{-3}; in our case the experimental ratio $\alpha_{333} / \alpha_{444} = 0.493$, while the ratio of wavelengths $(\lambda_{333} / \lambda_{444})^{-3} = 0.422$. This agreement of the experimental and theoretical α values supports one more time that the observed scattering is indeed TDS.

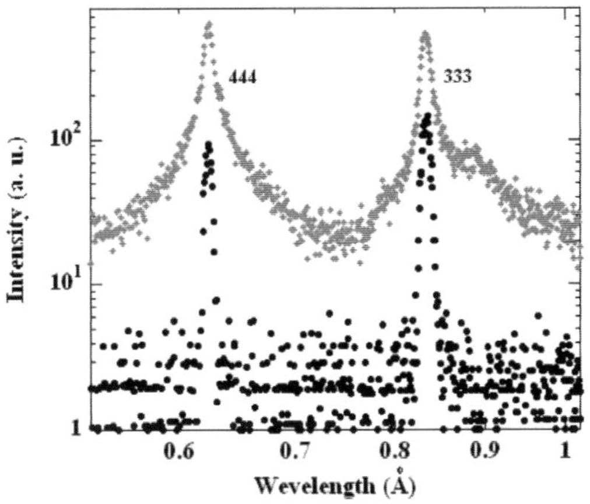

Figure 6. Zoomed view of Fig. 4 in the vicinity of the (333) and (444) Bragg reflections.

The time-of-flight experiments illuminate the TDS contribution to the wings of the reflectivity function from Si channel-cut triple-bounce crystals in the range $|y| < 3$ (see Fig. 2). This parasitic effect explains the discrepancy between the results obtained at ORNL [5] and those reported in [6]. The upstream neutron optics of the DCD used in [6] consist of a curved ^{58}Ni neutron guide with the radius R = 500 m having the characteristic wavelength $\lambda_c \sim 5.3$ Å and a PG(002) premonochromator Bragg reflecting $\lambda_1 = 5.24$-Å neutrons to the double-crystal arrangement. This configuration of the upstream neutron optics practically does not transmit the wavelength range $\lambda < 0.6$ Å to the Bonse-Hart DCD, and thus the effects described in the present paper could not in principle be observed in [6]. From this point of view the perfect fit to the Bonse-Hart theoretical model reported in [6] supports our conclusion about contamination of the reflectivity function wings with the TDS originating in the vicinity of high-order Bragg reflections with $\lambda < 0.6$ Å.

4. Conclusion. The Time-of-Flight Ultra-Small-Angle Neutron Scattering (TOF-USANS) instrument [13] currently under construction at ORNL's Spallation Neutron Source will incorporate this method of suppressing TDS. The instrument will be equipped with Si(220) channel-cut triple-bounce crystals designed for $\theta_B = 70^o$, and thus the first seven orders of Bragg reflection will be in the range $0.5 < \lambda \leq 3.6$ Å. Therefore, the first seven triple-bounce Bragg reflections from Si(220) crystals in the TOF-USANS instrument will be protected from contamination by TDS.

Acknowledgement. The authors are thankful to A. D. Stoica for the discussions and fruitful notes and to Ashfiya Huq and Evan R. Maxey for assistance in the measurements. The U. S. Department of Energy supported the work at Argonne National Laboratory under contract number DE-AC02-06CH11357 between Chicago Argonne, LLC and the Department of Energy. This Research for Oak Ridge National Laboratory's Spallation Neutron Source was sponsored by the Scientific User Facilities Division, Office of Basic Energy Sciences U. S. Department of Energy.

References

[1] U. Bonse and M. Hart, *Appl. Phys. Lett.* **7**, 238 (1965).
[2] T. Takahashi and M. Hashimoto, *Phys. Lett.* A **200**, 73 (1995).
[3] M. Agamalian, R. Triolo, and G. D. Wignall, *J. Appl. Crystallogr.* **30**, 345 (1997).
[4] C. G. Shull, *Appl. Crystallogr.* **6**, 257 (1973).
[5] M. Agamalian *et al.*, J. *Appl. Crystallogr.* **31**, 235 (1998).
[6] A. G. Wagh, V. C. Rakhecha and W. Treimer, *PRL.* **87**, 125504-1 (2001).

[7] J. G. Barker *et al.*, *J. Appl, Crystallogr.* **38**, 1004 (2005).
[8] W. Treimer *et al.*, *Phys. Lett.* A. **305**, 87 (2002)
[9] H. A. Graf *et al.*, *Acta Cryst.* A **37**, 863 (1981).
[10] K. Aizawa and H. Tomimitsu, *JAERI-Review* **2001-005**, 84 (2001).
[11] K. Aizawa and H. Tomimitsu, *JAERI-Review* **2002-006**, 107 (2002).
[12] N. C. Popa and B. T. M. Willis, *Acta Cryst.* **A53**, 537 (1997).
[13] J. M. Carpenter *et al.*, J. *Appl. Cryst*,, **36**, 763 (2003).

International Conference on Neutron Scattering 2009 IOP Publishing
Journal of Physics: Conference Series **251** (2010) 012057 doi:10.1088/1742-6596/251/1/012057

ZEEMANS - a new facility to probe matter at high magnetic field through neutron scattering

A. T. Savici[1,2], G. E. Granroth[2], C. Broholm[1], Y. S. Lee[3] and M. D. Bird[4]

[1]Department of Physics and Astronomy, Johns Hopkins University, Baltimore, MD, USA
[2]Neutron Scattering Science Division, Oak Ridge National Laboratory, Oak Ridge, TN, USA
[3]Department of Physics, MIT, Boston, MA, USA
[4]National High Magnetic Field Laboratory, Tallahassee, FL, USA

E-mail: savici@pha.jhu.edu

Abstract. We describe ZEEMANS, a new instrument proposed for the Spallation Neutron Source (SNS) to probe matter at extremely high magnetic fields. The complexity of the high field magnet demands a versatile neutron scattering instrument, capable of performing diffraction (powder and single crystal), SANS, reflectometry, and inelastic spectrometry, with minimal modifications between configurations. In this paper we present a conceptual design for neutron scattering instrumentation to be built around the horizontal conical high field magnet. Monte Carlo simulations and analytical calculations indicate performance on a par with other SNS instrumentation despite limited access to the sample.

1. Introduction

High magnetic fields can perturb condensed matter to reveal or alter properties while neutrons can provide detailed information about nano-scale structure and dynamics. As recommended in a recent NRC report,[1] a world class facility is under development to combine these techniques and create a powerful new tool for a wide range of materials science.

Two recent technological advances enable this new instrument that we call ZEEMANS, in honor of Pieter Zeeman. (1) The increased brightness of the Spallation Neutron Source enables experiments on small samples under extreme thermodynamic conditions and (2) NSF investment to develop a new hybrid high field technology at the National Magnetic field Laboratory makes it feasible to bring a neutron beam into a >25 Tesla DC magnet and operate it cost effectively at the duty cycle of a neutron source.

The ZEEMANS facility will provide unique materials research capabilities in areas including quantum magnetism, correlated metals, molecular magnetism, nano-structured magnets, superconductivity, metallurgy, macro molecular crystallography, hydride structure determination.[2, 3, 4] This basic research has the potential to impact technological areas such as high-density magnetic information storage, quantum computing, superconducting power transmission, steel processing, pharmacology, and hydrogen based energy distribution.

The central component of ZEEMANS is an all superconducting electro-magnet capable of reaching 25 T. Designed at the National High Magnetic Field Laboratory,[5] it is a horizontal solenoid with a conical bore to provide access for the sample environment system and neutron scattering. The full conical opening angle will be 30 degree total at either end.

To make best possible use of the unique high field conditions, the neutron scattering instrument will be capable of all major neutron scattering techniques: powder and single crystal diffraction, small angle scattering, reflectometry, and inelastic scattering.

For versatility and high intensity in the relevant energy range, the instrument will view the coupled liquid H_2 moderator, from beamline 14A at SNS. The magnet is large, has considerable utility requirements, and must be sufficiently removed from other instruments to avoid fringe field interference problems. Consequently ZEEMANS will be outside the target building, with a sample to moderator distance L_{m-s} of 70 m. The magnet will be able to rotate about a vertical axis, to vary the angle between the incident beam and the field axis from 0 degrees for SANS to 15 degrees for access to scattering angles of thirty degrees. Detectors at L_{s-d}=5 m will be placed in both the forward and backward scattering directions. Fixed detectors, a rotating magnet, and a common detector-magnet vacuum space is the chosen configuration for flexibility in scattering, operational convenience, and engineering feasibility. A conceptual drawing of the instrument is shown in Fig1(a).

2. Projected performance for Neutron Scattering

ZEEMANS will accommodate up to a 2 cm by 2 cm sample located 70 m from the 10 cm wide, 12 cm tall liquid H_2 moderator. To compensate for the limited angular access, a wide incident energy (wavelength) range E_i = 1-150 meV (λ = 0.8-10 Å) will be employed.

Monte Carlo optimizations for different guide configurations were performed using McStas.[6] Independent optimizations were performed for the horizontal and vertical profiles. An m=3.6 guide coating was used in the simulations. The horizontal profile of the guide will consist of a curved section, 15 m in length, with a 8200 m radius of curvature, followed by an 8 channel bender, with the same length, and a 726 m radius of curvature. Close to the sample position, the end of the guide has a parabolic profile. The elliptical ballistic guide concept[7] was used for the vertical profile, and placement of the focal points was optimized for the finite sized source and sample. The maximum height of the guide is 15 cm.

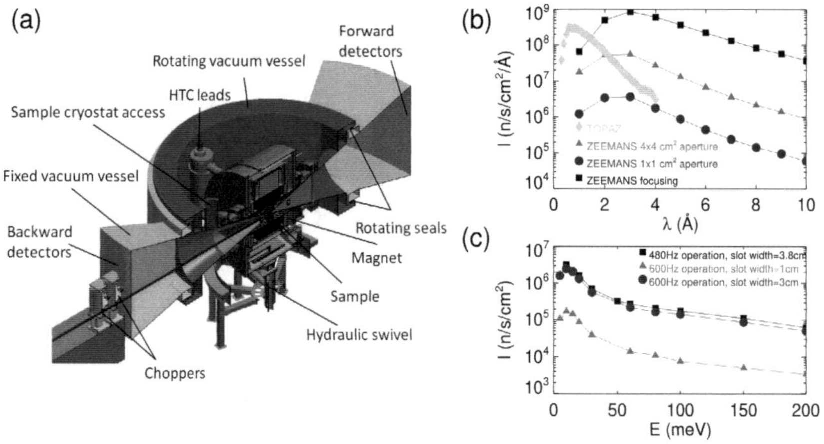

Figure 1. (a) Conceptual design for ZEEMANS. Neutron flux on sample for ZEEMANS in different experimental configurations: (b) diffractometer mode with and without apertures 5 m before the sample, and comparison to TOPAZ; (c) spectrometer mode. Note it will also be possible to run the instrument in rep-rate multiplication mode where several discrete incident energies appear at the sample during a single frame.

Four bandwidth choppers and at least two high speed choppers will be required for the various experimental configurations. For the elastic scattering configurations, the maximum bandwidth without frame overlap is $\Delta\lambda = 0.876$ Å. An aperture can be introduced 5 m before the sample, to control beam divergence. Alternatively, to maximize flux, a focusing guide section can be used. The wave length dependent flux on sample is shown in Fig. 1(b), for various configurations. Also shown is a simulation for Topaz,[8] a single crystal diffractometer at SNS. The higher flux at shorter wave lengths on Topaz is due to a bender configuration optimized for diffraction alone. For the direct geometry spectroscopy, high speed choppers are used. The energy dependent flux is shown in Fig. 1(c) for typical experimental setups, corresponding to different chopper transmission times, and therefore different energy resolutions.

The conical bore allows scattered neutrons to emerge from the magnet in both the forward and backward directions. To increase the maximum accessible momentum transfer, **Q**, for a fixed E_i, the magnet can rotate as described earlier. Rotation of the sample around the magnet axis will further extend momentum space coverage.

For elastic scattering, the **Q** space available is shown in Fig. 2(a). Forward scattering (detailed in Fig. 2(b)) can be used for both nuclear and magnetic scattering, and will probe scattering with wave vector transfer roughly perpendicular to the field direction. Due to the form factor, magnetic scattering in the backward direction will be reduced in strength but importantly will probe scattering with wave vector transfer approximately parallel to the field. The energy-**Q** range of ZEEMANS is shown in Fig. 2(c)-(e) for several values of E_i, and for magnet orientations at 0° and 15° with respect to the incident beam.The corresponding range of energy transfer is indicated by the color scheme in Fig. 2(f). The lack of overlap for accessible $\mathbf{Q}_\perp$ regions at the highest energy transfers in Fig. 2 (c)-(e) indicates that it is necessary to be able to rotate the magnet to non-extremal positions.

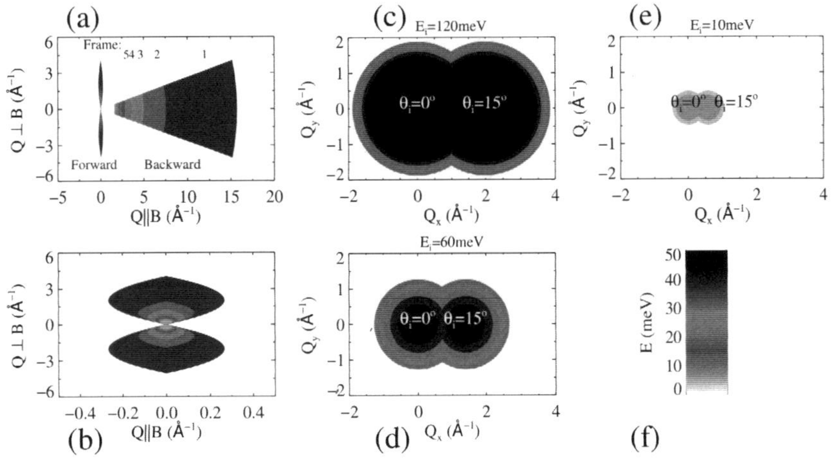

Figure 2. Momentum range for elastic scattering (a) with detail for the forward detectors (b). E-$\mathbf{Q}_\perp$ ranges for different incident energies (c)-(e), at the values of energy transfer indicated by (f). As only $\mathbf{Q}_\parallel$ depends on forward or backward scattering, these figures apply for both cases.

For momentum resolution calculations, we considered geometrical factors, beam divergencies, and the neutron emission time uncertainty from the moderator.[9, 10] Results are shown in Fig. 3(a)-(b). The resolution in the forward direction is dominated by geometrical factors, while in the backscattering detectors moderator contributions are most important. For energy resolution, McStas simulations for different running conditions are shown in Fig. 3(c).

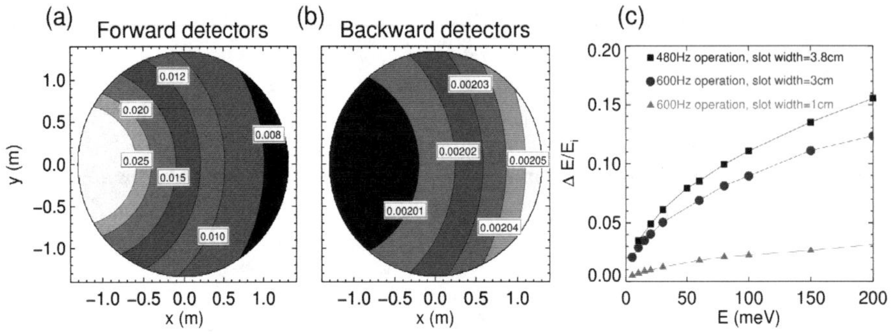

Figure 3. Projected resolution for ZEEMANS. Relative momentum $\Delta Q/Q$ or lattice spacings $\Delta d/d$ resolution for (a)forward detectors and (b) backward detectors, with the magnet rotated 15° with respect to the incident beam. (c) McStas simulation of elastic energy resolution, under different running conditions.

3. Conclusions

Built around a unique all superconducting magnet that will be able to reach at least 25 Tesla, the ZEEMANS instrument will be extremely versatile. It will be capable of performing diffraction, reflectometry, and spectroscopy, with minimal configurational changes. The expected flux and resolution of the instrument are comparable to state of the art instrumentation at the SNS. ZEEMANS will thus provide a first view of structure from nano-meters to angstrom and dynamics from nano to pico-seconds in materials perturbed by fields beyond 25 Tesla.

Acknowledgments

We acknowledge fruitful discussions with F. Mezei, P. Smeibidl, M. Steiner, and others from Helmholtz Zentrum Berlin, where a similar instrument is under development. We thank SNS instrument scientists, for providing information on their instruments. The National Science Foundation funded this work under DMR-0603126. GEG is funded by the US DOE contract DE-AC05-00OR22725.

References

[1] Committee on Opportunities in High Magnetic Field Science 2005 *Opportunities in High Magnetic Field Science* (The national Academies Press), *http://www.nap.edu/openbook.php?isbn=0309095824.*
[2] Yoshii S, Yamamoto T, Hagiwara M, Takeuchi T, Shigekawa A, Michimura S, Iga F, Takabatake T and Kindo K 2007 J. Magn. Magn. Mater. **310** 1282.
[3] Jo Y J, Balicas L, Capan C, Behnia K, Lejay P, Flouquet J, Mydosh J A and Schlottmann P 2007 Phys. Rev. Lett. **98** 166404.
[4] Zhao J K 2005 Physica B **356** 168.
[5] Bird M D, Bole S, Gundlach S and Toth J 2006 IEEE Trans. Appl. Supercond., **16** 957.
[6] Willendrup P, Farhi E and Lefmann K 2004 Physica B **350** 735.
[7] Schanzer C, Boni P, Filges U and Hils T 2004 *Nucl. Instr. Meth. A* **529** 63
[8] Hoffmann C, Stoica A, Lee H, Wang X -L, Schultz A, Piccoli P B, Koetzle T, Bau R, Thomison J and Davis L *http://neutrons.ornl.gov/instrument_systems/beamline_12_topaz/pdf/poster-1105-advances-1.pdf*
[9] Windsor C G 1981 *Pulsed Neutron Scattering* (London: Taylor and Frances); Crawford R K 1978 in *IPNS a National Facility for Condensed Matter Research Argonne National Laboratory Report ANL 78-88* 206; Carlile C J, Taylor A D, and Williams W G 1985 *Neutron Scattering in the 90s* IAEA-CN_46/8P 421.
[10] Iverson E B, Ferguson P D, Gallmeier F X, and Popova I I 2002 *Detailed SNS Neutronics Calculations for Scattering Instrument Design: SCT Configuration, Spallation Neutron Source Report SNS 110040300-DA0001-R00, http://neutrons.ornl.gov/instrument_systems/components/moderator/110040300-DA0001-R00.pdf*

International Conference on Neutron Scattering 2009 IOP Publishing
Journal of Physics: Conference Series **251** (2010) 012058 doi:10.1088/1742-6596/251/1/012058

SEQUOIA: A Newly Operating Chopper Spectrometer at the SNS

G. E. Granroth[1],A. I. Kolesnikov[1], T. E. Sherline[1], J. P. Clancy[2], K. A. Ross[2], J. P. C. Ruff[2], B. D. Gaulin[2], S. E. Nagler[1]

[1] Neutron Scattering Sciences Division, Oak Ridge National Laboratory, Oak Ridge, TN USA
[2] Department of Physics and Astronomy, McMaster University, Hamilton, Ontario, Canada

E-mail: granrothge@ornl.gov

Abstract. A fine resolution chopper spectrometer (SEQUOIA) recently received first neutrons at the SNS. The commissioning phase of the instrument is underway. SEQUOIA is designed to utilize neutrons of an incident energy (E_i) between 10-2000 meV. A monochromatic beam is provided on a sample, 20 m from the decoupled ambient temperature H_2O moderator, by filtering the white beam with a Fermi chopper located 18 m from the source. After interacting with the sample, neutrons are detected by an array of ^3He linear position sensitive tubes located on a vertical cylinder with a radius of 5.5 m. This contribution presents current results from the commissioning experiments and compares SEQUOIA's actual and predicted performance. These commissioning experiments include characterization of the beam by monitors, determination of the chopper phase offsets, and runs with V and $C_4H_2I_2S$. The predicted performance is provided by analytical calculations and Monte Carlo simulations.

1. Introduction

The SEQUOIA spectrometer at the Spallation Neutron Source, is a fine resolution, Fermi chopper instrument that can utilize E_i between 10 meV and 2 eV. The instrument received first neutrons on October 7, 2008. SEQUOIA is currently in its commissioning phase. This contribution describes its current status in commissioning.

Figure 1 shows a schematic view of the completed instrument with the major components labeled. The neutron beam originates from the decoupled ambient temperature H_2O moderator and travels 20 m to the sample position. A Fermi chopper, located 18 m (L_1) downstream of the moderator and 2 m upstream of the sample (L_2), monochromates the beam. Two Fermi chopper choices are available on a translation table for ease of changing energy range and resolution conditions. Fermi chopper 1 has 3.6 mm slits and a channel curvature of 1.53 m. Fermi chopper 2 has 2 mm slits and a channel curvature of 0.58 m. A T_0 chopper is located 9.8 m from the moderator. It blocks the highest energy neutrons generated when the proton pulse hits the target and neutrons that would be transmitted through additional openings of the Fermi chopper that occur during the 60 Hz time frame of the source. Neutron guide is utilized to provide a high flux of neutrons on a 5 cm by 5 cm sample. Details of the guide and T_0 chopper design are described elsewhere[1]. A cylindrical array of 1.2 m long by 2.5 cm wide detectors is located 5.5 m (L_3) downstream of the sample position. This detector array currently spans angular ranges of $-30°$ to $60°$ in the horizontal and $\pm18°$ in the vertical. The cylindrical arrangement provides more continuous detector coverage than a spherical geometry. The effect

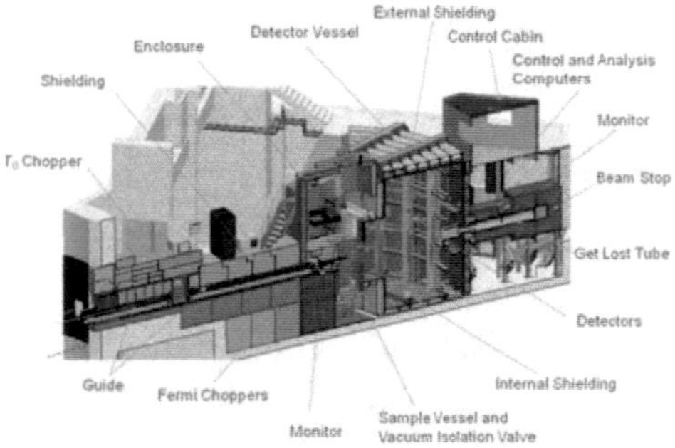

Figure 1. A Schematic view of the completed SEQUOIA

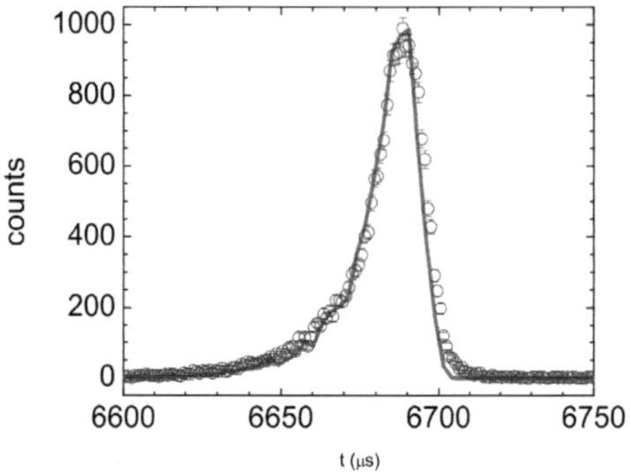

Figure 2. The signal observed in monitor 2 for E_i = 98 meV neutrons. The solid line is the results of a Monte Carlo simulation of the instrument in the same conditions.

of the cylindrical detector geometry on the resultant data was considered in detail. Lechner [3] showed that a cylindrical, pixelated detector array does not adversely effect the instrumental resolution for isotropic scatterers. Furthermore we used a McStas [4, 5] Monte Carlo simulation of SEQUOIA to examine the resolution effects of the cylindrical detector array. These studies showed that when pixel sizes in energy (ω) and momentum ($\mathbf{Q}$) transfer are much smaller than the instrumental resolution, the cylindrical geometry has a negligible effect on subsequent data processing. The 128 position pixels along the detector tubes and the $1\mu s$ timing resolution of the detector electronics provide more than sufficient pixelization. Longer detector tubes were considered for the detector array, but the chosen length preserves the ~1 cm pixel resolution and minimizes dead time effects in data collection.

To reduce background by removing Al windows, the detectors are located inside a detector chamber that can be evacuated. This chamber is contiguous with the evacuated sample vessel. A large gate valve connects the two chambers to isolate them during sample change out. This valve expedites sample changes as the largest evacuated volume remains under vacuum. To further reduce background, the inside of the detector chamber is lined with 2.5 cm of ZHIP (Zero Hydrogen in Product) mix, boron carbide powder bound together with a non hydrogenous glue [2] and the outside of the detector vessel is covered with 10 cm of 5% borated polyethylene.

There are two ^{3}He beamline monitors in the beam path; they are located immediately downstream of the Fermi chopper (L_{m1} = 18.2 m from the moderator) and in the get lost tube (L_{m2} = 29 m from the moderator). Each monitor has an efficiency of 1×10^{-4} for 1 Å neutrons. These monitors were used to characterize the incident beam. The points in Figure 2 shows a typical pulse observed in the downstream monitor. This data was taken when the source was running at 700 kW and the detector and sample chambers contained air at atmospheric pressure. Fermi Chopper 1 was spinning at 600 Hz and phased to provide E_i=98.0 meV neutrons with $\frac{d\omega(\omega=0.0)}{E_i} = 2.0\%$ McStas Monte Carlo simulations were run, using the same simulation code as was used in reference [1], to compare the expected peak shape and intensity to the observations. The solid curve in Figure 2 is the results of this simulation once they are scaled for beam power, Al in the beam, air in the beam, and the energy response of the monitor. No arbitrary scale factor is included in the scaling. The peak position is adjusted in time to compensate for an electronic offset in the data acquisition chain. Similar response was also seen for the first monitor. This Monte Carlo simulation also monitors the flux at the sample position. Therefore by comparison

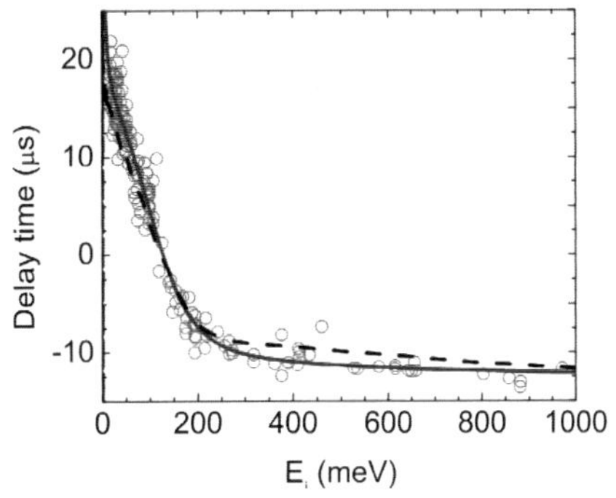

Figure 3. The points are the τ as described in the text. The solid line is a fit to Equation 1. The dashed line shows the prediction for the moderator as given in Ref. [6]

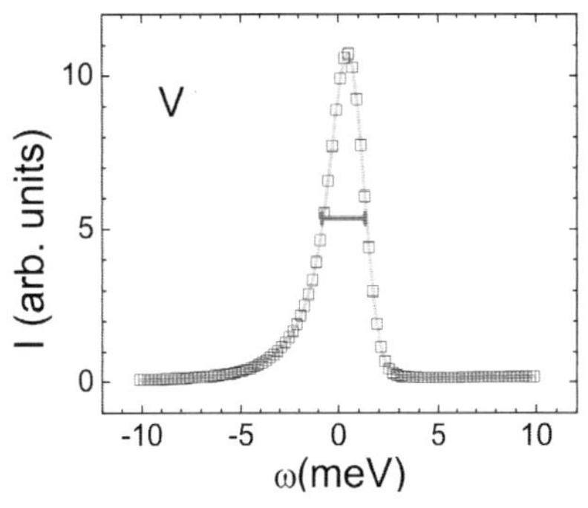

Figure 4. The incoherent elastic line in the neutron scattering spectrum from V for $E_i = 91.3$ meV. The horizontal errorbar gives the FWHM as predicted by equation 2

with the simulation, the time averaged flux on the sample is inferred to be $1 \times 10^5 n/cm^2/s$.

Measurements with the two monitors as a function of Fermi chopper phase, were used to determine the combination of the delay in emission of the neutrons after the protons hit the target plus the electronic offset (τ). Before the start of each measurement, the phase was set for a given E_i assuming $\tau = 0$ and L_1. The center of gravity of the neutron pulse in each monitor, corresponding to E_i, along with L_{m1} and L_{m2} provide the information to determine the actual E_i and τ. The results are shown as the circles in 3. Electronic offsets cause the observed negative τ for the highest values of E_i. The solid line is a fit of the empirically determined function,

$$\tau = \frac{1 + tanh(\frac{E_i - a}{b})}{2cE_i} + \frac{1 - tanh(\frac{E_i - a}{b})}{2dE_i} + f + gtanh(\frac{E_i - a}{b}) \tag{1}$$

to the data. The resultant parameters are: $a = 92.0$, $b = 81.3$, $c = 1.3 \times 10^{-3}$, $d = -75.1$, and $g = -14.5$ The dashed curve is the expected τ from the MCNPX calculations shifted only by a constant to account for the electronic offset[6]. The expected and observed τ values agree.

Several standard samples were used in the initial characterization of the scattered beam. A vanadium rod, 1.12 cm in diameter, was used to test the energy resolution ($d\omega$) at zero energy transfer($\omega = 0$). A typical measurement with Fermi chopper 2 spinning at 600 Hz and phased for $E_i = 91.3$ meV is shown in Figure 4. The FWHM of this peak corresponds to $d\omega/E_i = 2.28\%$ which compares favorably to the predicted value of 2.35% given by the analytical function,

$$d\omega = m_n \sqrt{\left(\frac{v_i^3}{L_1} + v_f^3 \frac{L_2}{L_1 L_3}\right)^2 dt_m^2 + \left(\frac{v_i^3}{L_1} + v_f^3 \frac{(L_2 + L_1)}{L_1 L_3}\right)^2 dt_c^2 + \left(\frac{v_f^3}{L_3}\right)^2 dt_d^2}, \tag{2}$$

where v_i and v_f are the initial and final velocities of the neutrons, respectively, dt_m is the time spread of the moderator [6], dt_c is the width of the time pulse through the Fermi chopper, and dt_d is the time spread from the detector and the sample widths given in quadrature[7, 8, 9].

The material 2,5-diiodothiophene ($C_4H_2I_2S$) has many rotational and vibration rotational energy modes. These sharp peaks in ω were used to characterize $d\omega(\omega)/E_i$. The data shown in Figure 5a were taken using similar instrument conditions as the V run; differing only by the sample size (5 cm x 5 cm x 2 mm flat plate) and orientation of 15° turned about a vertical axis. For a preliminary analysis, the FWHM/E_i of several of the peaks is plotted in Figure 5b. The solid curve shows the expected values as given by equation 2. The predictions and the results agree for most of the points. The exceptions are the 0 meV and 68 meV peaks. The measured $\omega = 0$ peak is expected to be narrowed because this analysis does not distinguish coherent and incoherent scattering. Measurements of the 68 meV peak on TOSCA also showed its width as greater than the instrumental resolution[10]; implying that the added width is from the sample and not the instrument.

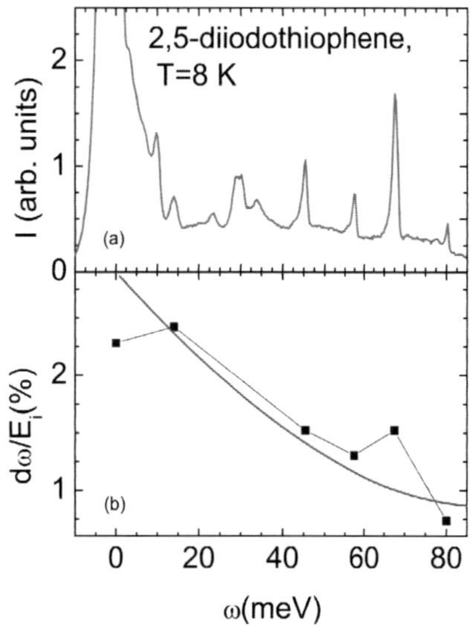

Figure 5. (a) The inelastic neutrons scattering spectrum for 2,5-diiodothiophene ($C_4H_2I_2S$) for $E_i = 91.3$ meV (b) The points are the measured FWHM/E_i and the solid line is the analytical prediction as described in the text.

A new Fermi chopper spectrometer is operational at the SNS. Experiments with the monitors have verified that the flux on sample is as expected. Experiments with standard samples have verified the instrumental resolution is as expected.

Acknowledgments

We acknowledge fruitful discussions with E. Iverson, C. Wildgruber, D. Abernathy, and M. Hagen. We thank D. Mikkelson for assistance with the ISAW portion of the data analysis. This work was sponsored by a contractor of the United States Government under contract DE-AC05-00OR22725 with the United States Department of Energy.

References

[1] Granroth G E, Vandergriff D H, Nagler S E 2006 *Physica B* **385-386** 1104
[2] ZHIP mix is a product of Dielectric Sciences incorporated.
[3] Lechner R E, private communications.
[4] Lefmann K and Nielsen K 1999 *Neutron News* **10** 20
[5] Willendrup P, Farhi E, and Lefmann K 2004 *Physica B* **350**, 735
[6] Iverson E B, Ferguson P D, Gallmeier F X, and Popova I I 2002 *Detailed SNS Neutronics Calculations for Scattering Instrument Design: SCT Configuration, Spallation Neutron Source Report SNS 110040300-DA0001-R00*
[7] Crawford R K 1978 in *IPNS a National Facility for Condensed Matter Research Argonne National Laboratory Report ANL 78-88* 206
[8] Windsor C G 1981 *Pulsed Neutron Scattering* (London: Taylor and Frances)
[9] Carlile C J, Taylor A D, and Williams W G 1985 *Neutron Scattering in the 90s* IAEA-CN 46/8P 421
[10] Michell P C H 2003 *Hydrogen in Hydrotreating and Other Catalysts Studied by Inelastic Neutrons Scattering -Spillover Hydrogen and H-containing Compounds, Proc. of 18th North American Catalysis Society Meeting* http://www.nacatsoc.org/18nam/Orals/168-Mitchell-Hydrogen%20in%20Hydrotreating%20and%20Other%20Catalysts.pdf

BioRef – a time-of-flight neutron reflectometer combined with in-situ infrared spectroscopy at the Helmholtz Centre Berlin

M Strobl[1,2], R Steitz[2], M Kreuzer[1,2], A Nawara[3], F Mezei[4], M Rose[2], M Grunze[1] and R Dahint[1]

[1]University of Heidelberg, Im Neuenheimer Feld 229, 69120 Heidelberg, Germany
[2]Helmholtz Centre Berlin, Glienickerstr. 100, 14109 Berlin, Germany
[3]Slovak Academy of Sciences, Dubavska cesta 9, 84538 Bratislava, Slovakia
[4] Los Alamos National Laboratory, Los Alamos, NM 87545, USA

Abstract

The novel BioRef reflectometer currently under construction in neutron guidehall-I of the BER-2 reactor at the Helmholtz Centre Berlin is designed for applications in soft matter science at solid-liquid interfaces especially under dynamic conditions. In addition to a time-of-flight (TOF) approach to be realised with three choppers in order to adapt resolutions and wavelength bands to the needs of specific applications, the instrument will be equipped with an in-situ infrared (IR) spectrometer operating in its surface sensitive ATR-geometry for simultaneous spectroscopic and conformational studies. The instrument will become operational in 2010.

1. Introduction

Neutron reflectivity measurements are a well established tool for studies of surfaces and interfaces in soft matter [1] and hard matter physics [2]. Specular reflectivity profiles provide information on the (scattering length) density profile perpendicular to the surface and hence constituted layers and layer thicknesses on a surface or interface. In contrast to monochromator based reflectometers the TOF method enables, as well as other recently proposed but not yet realised approaches [3,4], the simultaneous coverage of a defined q-range (where q denotes the scattering vector component perpendicular to the surface) without changing the angular alignment of the sample in the beam, i.e. for a fixed incidence angle θ. This makes the technique superior to monochromatic instruments in cases in which the surface structure might be time-dependent, instable or sensitive to movements. However, most importantly, the TOF approach enables to match the instrument resolution to the requirements brought forward by the system under study. This can be particularly beneficial for biological samples, where for instance ultrathin lipid membranes adsorbed to solid walls are considered.

The BioRef instrument has been designed with a strong focus on time-dependent studies under conditions like shear and flow at solid-liquid interfaces. Such dynamic conditions are important for investigations of biological model systems under more realistic physiological conditions. Corresponding sample cells and environment are developed along with the instrument taking into account the horizontal scattering geometry of BioRef. Such geometry has been chosen to allow for wide-angle scattering studies, and with respect to potential future neutron optical and spin defining and manipulating implementations in the primary and secondary reflectometer arm. The foreseen availability of simultaneous investigations of the sample by attenuated total reflection - Fourier transform infrared spectroscopy (ATR-FTIR) with a beam geometry perpendicular to the neutron beam is especially useful for applications under dynamic conditions, i.e. when changes in thickness or scattering length density of interfacial layers in an applied external field (shear or pressure) are to be linked with concurrent conformational changes of the constituting molecules.

2. Instrument design

The instrument under construction has a total length of 8.22 m and is located at the end of the 25 m long cold neutron guide NL3b in neutron guidehall-I of the BER-2 reactor at the Lise Meitner Campus of the Helmholtz Centre Berlin (former Hahn-Meitner Institute). The guide, which has a radius of curvature of 500 m, a cross section of 3 cm width and 5 cm height and which is currently coated with ^{58}Ni provides neutrons with a minimum wavelength of approximately 3 Å and a peak flux of about 1.4 10^7 n cm^{-2}s^{-1}Å^{-1} at 5.2 Å. An upgrade of the guide system with supermirror coatings has been envisaged and will change this situation significantly. It therefore had to be considered in the instrument layout. In this paper we focus on the current conditions under which the instrument will become operational. Upstream of the instrument only the perfect crystal monochromator of the V12a diffractometer [5] intersects the neutron guide with negligible influence on the neutron spectrum.

The design of the instrument is based on a three chopper system and is optimised with respect to the available space which limits the instrument length (figure 1). A first pair of choppers with variable distance defines the initial neutron pulse, while the third chopper located 3.22 m downstream of the first chopper is used to limit the utilised wavelength band. The pulse shaping choppers 1 and 2 have identical Al discs with a diameter of 0.6 m, windows of 26 deg and $^{10}B_4C$ coating. In the range of desired wavelength resolutions $\Delta\lambda/\lambda$ from 1% to 5% they can be operated in a mode described by A. van Well [6], where the second chopper opens when the first chopper closes, i.e. with a dephasing of 26°. Consequently the wavelength resolution $\Delta\lambda/\lambda$ depends exclusively on the distance z_0 between the choppers (for a fixed length of the instrument) and is constant up to a certain wavelength λ_0 where the transmission has its maximum and which is defined by z_0, the window width D and the frequency ω.

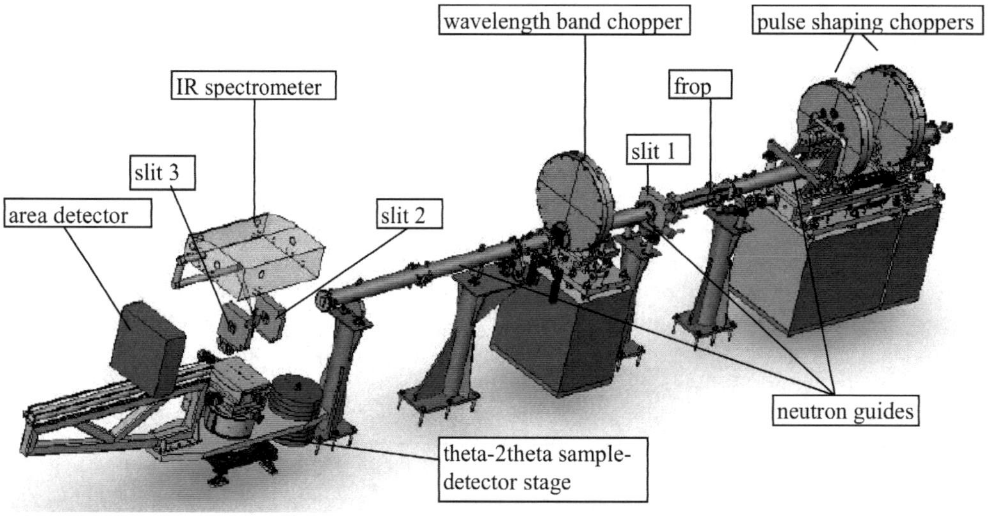

Figure 1 Schematic view of BioRef (without shielding)

The wavelength band chopper 3 with a window of 126° limits the utilized wavelength band at a maximum wavelength of $\lambda_{max} \leq \lambda_0$ and a minimum wavelength, λ_{min}, well above the cut-off of the guide system (currently at 3Å). The wavelength band chopper enables to tune the instrument resolution without changing the window size of the choppers (e.g. by double disk approaches [7]) or the pulse frequency. All choppers are operated at the same frequency. Additionally, a variation of the chopper frequency enables a focusing of the instrument onto a certain wavelength band width and hence onto a certain q-range, which is recorded simultaneously (figure 2). This allows to optimize a specific measurement for a distinguished q-range of interest especially in cases when time resolution is required.

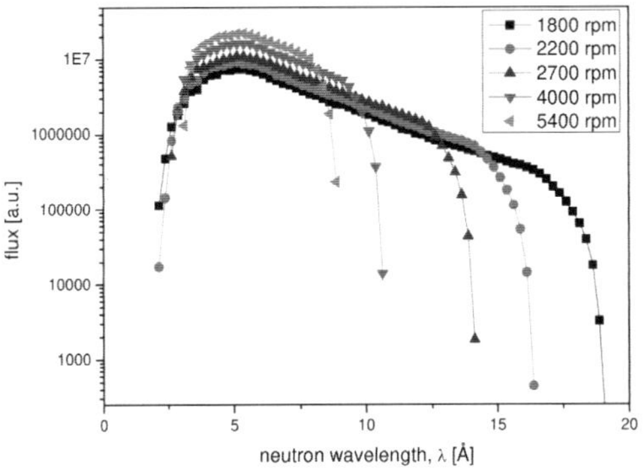

Figure 2 Simulated neutron spectra at sample position as a function of chopper speed.

In cases where a more relaxed resolution ($\Delta\lambda/\lambda > 5\%$) is required in favour of higher intensity, the approach of a constant resolution over the wavelength band can be abandoned easily (figure 3). A frame overlap mirror between the second pulse shaping and the wavelength band chopper hinders parasitic cross talk between pulses at wavelengths longer than 16 Å. Two slits define the beam collimation and hence the angular resolution. The first slit about 4 m upstream the sample position is located behind the frame overlap mirror, while the second slit, defining the footprint of the beam on the sample, is located directly in front of the sample position (distances of a few cm can be set). A third slit behind the sample limits the neutron entrance window of the shielding mounted on the secondary reflectometer arm towards the area detector of 30 cm x 30 cm. The spatial resolution is approximately 2 mm and the maximum distance between sample and detector is 2 m. A very first slit right upstream of chopper 1 is used to limit the beam width to enable high wavelength resolution and to reduce background.

The sample position as well as sample holders and substrates (Si-crystals) and a specific sample environment are being designed such that neutron reflectivity measurements can be combined with in-situ IR spectroscopy in its surface sensitive ATR-mode (Fig. 1). The last section of the neutron guide, mounted between chopper 3 and sample, is supermirror coated (m3) only on top and bottom faces like all the guide parts after the first collimation slit and is removable in order to allow for future installation of various neutron optics, polarizers or else. The whole set-up is compatible with the future upgrade of the guide system at BER-2, which includes m3 supermirror coatings at NL3b and will be discussed elsewhere.

3. Expected performance

Various simulations have been performed using the software package VITESS [8] in order to predict the performance that is expected from the novel instrument. To achieve reliable results, respectively in order to verify those, simulations were performed on the source and the guide system only. Hence, a neutron spectrum at the guide exit is provided, which has been compared to TOF measurements of the spectrum at the given position. For this purpose the latter data have been scaled by flux measurements utilizing a gold foil. Discrepancies in the flux density of more than half an order of magnitude were found. Consequently, all simulation results were downscaled accordingly. Slight differences between measurements and simulations concerning the spectral distributions have been concluded to be most likely due to misalignments of the current guide system and do not play a major role for the results. Selected results are displayed in figure 2 and 3. In particular the expected flux values given in figure 3a are promising with respect to the design goal to achieve competitive reflectivity measurements down to a level of about 10^{-7} with a still reasonable q-resolution of e.g. 7% (black curve) when a background $\leq 10^{-2}$ $cm^{-2}s^{-1}$ is considered. The highest flux is achieved without operating the second chopper. Consequently the resolution is wavelength dependent and ranges from approximately 6 % to 14%. The angular resolution for this simulation was set to a mean value of 10%. Fig 3b shows the results achieved in the simulated measurement at a 50 nm Au layer (on a Si-substrate). The insert compares the expected resolution effects for the three cases shown in Fig. 3a.

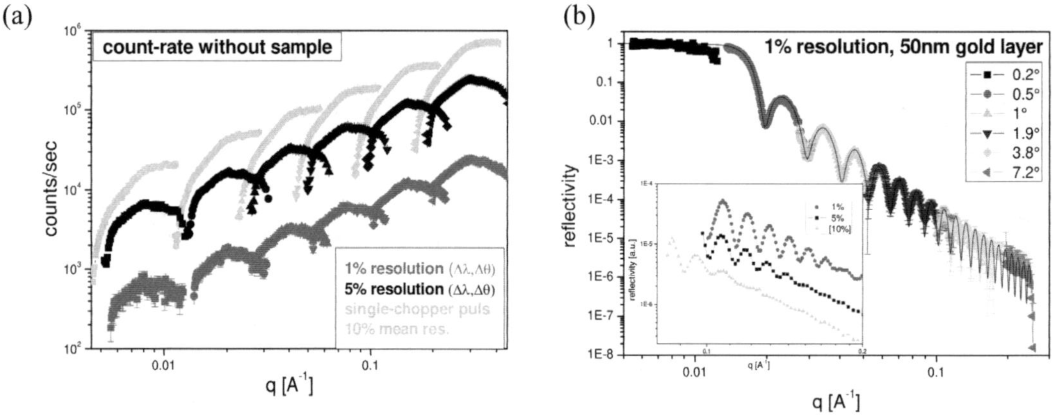

Figure 3 (a) Neutron flux at detector without sample for different instrumental resolutions (left side); (b) Simulated reflectivity curve at BioRef for a 50 nm Au layer measured with 1% angular and wavelength resolution and a chopper frequency of 90Hz; the insert compares a certain q-range simulated for different resolution as shown in (a); for the [10%]-curve the chopper speed is 4600 rpm.

4. Summary

The new TOF reflectometer BioRef in its current design is dedicated to soft matter investigations at liquid-solid interfaces, especially under dynamic conditions. In addition, the set-up is highly flexible with respect to different resolution requirements as well as future upgrades and implementations with regard to a broader range of applications. The instrument will offer the unique option for complementary in-situ ATR-FTIR measurements and will become operational in early 2010.

Acknowledgement

The presented work has been supported by the Federal Ministry of Education, Research and Technology (BMBF) of Germany (Contract No. 05KN7VH1).

References

[1] Steitz R et al., *Langmuir* **19** (2003) 2409-18.
[2] Gierlings M, Prandolini M J , Fritzsche H, Gruyters M and Riegel D 2002 *Phys. Rev.* B **65** 092407.
[3] Cubitt R, Shimizu H M, Ikeda K, Torikai N 2006 *Nucl. Instr. and Meth.* A **558** 547-50.
[4] Ott F and Menelle A 2006 *Physica* B **385–386** 985–88.
[5] Strobl M, Treimer W, Ritzoulis C, Wagh A G, Abbas S, Manke I 2007 *J. Appl. Cryst.* **40.**
[6] van Well A A 1992 *Physica* B **180 & 181** 959-61
[7] Kampmann R et al. 2006 *Physica* B **385–386** 1161–63
[8] http://www.hmi.de/projects/ess/vitess/

International Conference on Neutron Scattering 2009 IOP Publishing
Journal of Physics: Conference Series **251** (2010) 012060 doi:10.1088/1742-6596/251/1/012060

Project of the new multifunctional reflectometer GRAINS with horizontal sample plane at the IBR-2M pulsed reactor in Dubna

M.V. Avdeev[1], V.I. Bodnarchuk[1], V.V. Lauter-Pasyuk[2], H. Lauter[3], V.L. Aksenov[1], S.P. Yaradaikin[1], V.A. Ulyanov[5], V.A. Trounov[5], S.I. Kalinin[5]

[1]Frank Laboratory of Neutron Physics, Joint Institute for Nuclear Research, 141980 Dubna, Moscow Reg., Russia
[2]Oak Ridge National Laboratory, P.O. Box 2008 MS-6477, USA
[3]Institute Laue-Langevin, Avenue des Martyrs, BP 156, 38042 Grenoble, Cedex 9, France
[4]Russian Research Center "Kurchatov Institute", 123182 Moscow, Kurchatov sq. 1, Russia
[5]Petersburg Nuclear Physics Institute, Orlova Roscha, Gatchina, Leningrad Reg., 188350 St. Petersburg, Russia

Corresponding author's e-mail: bodnarch@nf.jinr.ru

Abstract. The new multifunctional reflectometer GRAINS is under construction at the modernized high flux IBR-2 pulsed reactor (IBR-2M) of the FLNP JINR (Dubna, Russia). The principal feature of this reflectometer, the horizontal sample plane (or vertical scattering plane), enables the study of liquid-containing interfaces. The reflectometer will operate in the time-of-flight regime with constant sample illumination during measurements. The important advantage is the constant angular resolution with unvaried background. The additional modes of the GRAINS reflectometer comprise (1) off-specular scattering and GISANS, which are measured simultaneously in the TOF regime at a 2D position-sensitive detector; (2) angular encoding in the horizontal plane, which is provided by a Larmor precession region limited by current sheets in front of the sample; (3) 3D polarimetry in reflection, which is provided by a Larmor precession region around the sample position. The design of the reflectometer is optimised to take better advantage of an exceptionally broad wavelength band of the new cold moderator at the IBR-2M. The set-up will open up principally new possibilities for investigations in the field of interface nano-science at the IBR-2M reactor.

1. Introdution

A present-day tendency in the neutron reflectometry instrumentation is the development of set-ups with horizontal sample plane. The combination of such geometry with the possibility of neutron polarization covers the whole range of research problems connected with the structure of non-magnetic and magnetic nanosystems on interfaces. The scientific program at the modern reflectometer with horizontal sample plane includes the topics such as biological complexes and membranes [1,2]; magnetic colloidal systems [3,4]; polymers and surfactants [6]; layered nanostructures [6-8]. Also, one can mention the activity in the development of new research methods based on the Larmor precession of neutron spin [9-11].

In 2007 at the Frank Laboratory of Neutron Physics, JINR (Dubna, Russia) the project on the new multifunctional reflectometer GRAINS (GRAzing Insidence Neutron Scattering) was started. The ideology of this set-up is based on the realization of last achievements in neutron reflectometry with the use of the pulsed regime at the neutron source—the IBR-2M reactor (modernized version of the previously successful pulsed reactor IBR-2). The main features of the new reflectometer are (1) the horizontal sample plane, which makes it possible to study liquid-containing interfaces; (2) polarization of the neutron beam; (3) full 3D reflectometry including analysis of specular, off-specular and grazing incidence small-angle scattering components; (4) new methods based on Larmor spin precession.

The IBR-2M reactor is a high flux long-pulse neutron source with 5 Hz repetition rate. Such a source provides ideal conditions for the reflectometry. The start-up of IBR-2M for experiments is scheduled for 2011.

2. The principle scheme and parameters of the reflectometer

The principle scheme of the GRAINS reflectometer is presented in Figure 1. The very important new element (in particular, for reflectometry) is the cold mesitylene moderators on the IBR-2M reactor [12]. The new sources will produce a broad wavelength band combined with the gain for cold neutrons. So, a unique momentum transfer band with a dynamic range of about 25 will be available at the GRAINS reflectometer. The immense advantage is that the experiments can be performed in time-of-flight (TOF) regime in one shot without changing the incoming scattering angle. This removes difficulties in fitting the reflected and off-specular scattered intensity measured in several steps in Q-space with different resolution in the overlap regions. The GRAINS set-up is constructed at beam 10 of the IBR-2M reactor, which has an advantage in the wide area of the cold moderator surface (20×20 cm^2) faced to the beamline. The available flight-pass of more than 20 m allows to optimize the intensity yield for the various focusing possibilities. All elements of the set-up have been optimized for the horizontal geometry.

The head part of the reflectometer containing the vacuum system starts after the cold moderator (2.5 m) and extends towards the collimation system (collimators). The aim of the background chopper after the head part is to cut out the subsidiary power peaks of the reactor. It opens only during the neutron pulse. After the chopper the vacuum collimating system continues up to the sample position and pass through the first resonant spin flipper. Flexible slit-collimators serve to tailor the beam shape for different applications. The deflector and the polarizing mirror are exchangeable units composed of a stack of adjustable mirrors. With deflector or polarizing mirror the direct view from the sample position to the cold moderator is prohibited. The reflection angle on the sample is changed by the regulation of the mirror inclination.

The two-coordinate position-sensitive detector (PSD) for the GRAINS reflectometer is constructed at FLNP JINR. Its design is based on a multiwire proportional chamber with delay line readout. Signals from both delay lines and an anode signal come through preamplifiers to a constant fraction timing five-channel discriminator and are registered in data acquisition and accumulation block (DAAB), which contains of signal level converter NIM-TTL, field programmable gate array (FPGA) and high-speed interface with optical communication link to a personal computer. The data acquisition and accumulation system can operate in two modes: in the mode of histogram accumulation in internal memory of the DAAB block and in the list mode, when "raw" data are accumulated directly on a PC hard disk. All functions of the block are programmable. Logic and time simulation of operation of the block was carried out using Quartus II package from ALTERA.

The expected parameters of the GRAINS reflectometer are presented in Table 1.

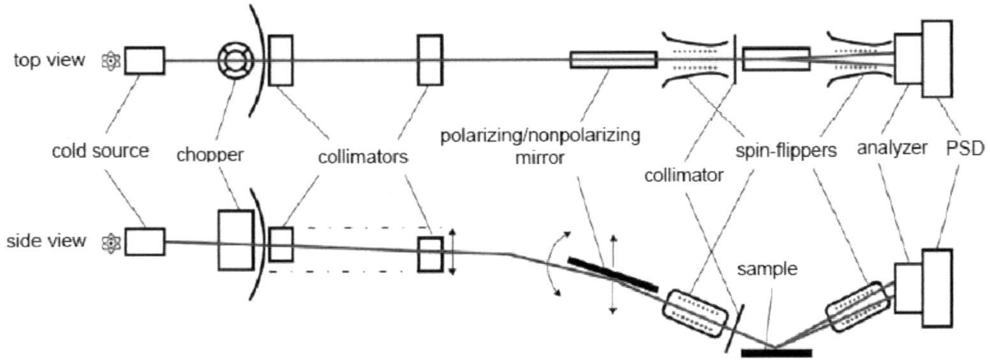

Figure 1. Principle scheme of the GRAINS horizontal reflectometer in top view (upper scheme) and side view (lower scheme). The upper length limit (moderator-detector) is 30 m. The Larmor-precession device is not shown for simplicity; it will have two positions, before and around the sample, respectively. "Polarizing mirror" stands for a stack of adjustable mirrors for polarizing and non-polarizing modes.

Table 1. Main expected parameters of reflectometer GRAINS

Beam size	100×10 mm^2
Wavelength range	0.5 – 15 Å
Incident angle range	3 – 25 mrad
Q-range	0.001 – 0.6 $Å^{-1}$
Angular resolution	3 - 10 %
Sample size	50×50 mm^2
Neutron flux on the sample	3.0×10^6 cm^{-2} c^{-1}
Deflecting mirrors	Supermirror, m = 2
Analyzer	Fan type, m = 2
Detector	PSD, ^3He, 200×200 mm^2, resolution 2×2 mm^2 count rate 10^6 s^{-1}

3. Main options of the reflectometer

At present, the majority of published data in reflectometry concerns specular reflection, from which the structural information perpendicular to the surface of a sample is obtained. For this purpose, the $Q_\perp$ vector perpendicular to the surface is scanned as a function of the incoming scattering angle in the monochromatic mode or the wavelength in time-of-flight. A further step in the sample characterisation is to obtain information about the lateral structure of surfaces and interfaces from off-specular scattering and GISANS. A combined measurement of specular reflection, off-specular and GISANS, called complete reflectometry [13], probes lateral correlations in a depth sensitive way providing a 3-dimensional structural analysis. In complete reflectometry the detailed neutron wave field along the surface normal created by dynamical scattering is taken into account. These recent developments are taken into account in the lay-out of the new horizontal reflectometer. A flexible collimation system allows to install a slit-shaped incoming beam for the set-up of specular reflection with off-specular scattering and a pencil-shaped incoming beam for the set-up of complete reflectometry. In particular a good resolution along $Q_\perp$, which is the necessary base for the depth sensitivity of off-specular scattering and GISANS, is guarantied.

Magnetic GISANS measurements including the polarisation analysis of specular together with off-specular scattering following [13] is taken into account in the lay-out of the new horizontal reflectometer.

Magnetic and non magnetic GISANS implies, as already mentioned, a pencil-shaped collimation of the incident beam. A considerable decrease of the flux on the sample is the consequence. This decrease of flux will be counterbalanced by the application of angular encoding with Larmor precession. If the incoming scattering angle is encoded its angular divergence can be enlarged and thus flux on the sample is increased. A pilot experiment on the TOF-reflectometer REMUR (IBR-2) has been performed [14]. A second important application of Larmor precession is 3-dimensional analysis of magnetisation distribution in layered magnetic samples. In standard experiments the external magnetic field and the neutron magnetic moment are oriented parallel to the sample surface. From such experiment one gets information about the depth distribution of the mean value of in-plane magnetic moments. The off-specular magnetic scattering probes the magnetic fluctuations along the surface plane and gives information about their magnetisation state through spin-flip or non-spin flip scattering. However, a full 3-dimensional analysis is only obtained if the polarisation of incoming neutron can be adjusted with respect to the magnetisation directions in the sample. In reflectometry this effect can be obtained if the neutron performs Larmor precession with the plane of its magnetisation rotation perpendicular to the external field [15]. Like this missing matrix elements in the magnetic interactions are measured in an elegant way in TOF. For this purpose a weak magnetic Larmor-precession field around the sample position is contained in the lay-out of the new horizontal reflectometer.

References

[1] Heinrich, T. Ng, D. J. Vanderah, P. Shekhar, M. Mihailescu, H. Nanda, M. Loesche, Langmuir 25 (2009) 4219
[2] T.Gutberlet, B.Kloesgen, R.Krastev, R.Steits, Adv. Eng. Mater. 6 (2004) 832
[3] V. Lauter-Pasyuk, H. Lauter, G. Gordeev, P. Müller-Buschbaum, B. P. Toperverg, W. Petry, M. Jernenkov, A. Petrenko, V. Aksenov, Physica B 350 (2004) E939
[4] A.Vorobiev, J.Major, H.Dosch, G.Gordeev, D.Orlova, Phys. Rev. Lett. 93 (2004) 267203
[5] T.L. Kuhl, J. Majewski, J.Y. Wong, S. Steinberg, D.E. Leckband, J.N. Israelachvili, G.S. Smith, Biophys. J. 75 (1998) 2352
[6] V.L.Aksenov, K.N.Jernenkov, Yu.N.Khaidukov, Yu.V. Nikitenko, A.V. Petrenko, V.V. Proglyado, G. Andersson, R. Waeppling, Phisica B 356 (2005) 9
[7] V.L.Aksenov, Yu.V.Nikitenko, A.V.Petrenko, V.M.Uzdin, Yu.N.Khaidukov, H.Zabel, Cryst. Rep. 52 (2007) 403
[8] V.F.Kozhevnikov, C.V.Giuraniuc,M.J.Van Bael, K.Temst, C.Van Haesendonck, T.M.Mishonov, T.Charlton, R.M.Dalgliesh, Yu.N.Khaidukov, Yu.V.Nikitenko, V.L.Aksenov, V.N.Gladilin, V.M.Fomin, J.T.Devreese, J.O.Indekeu, Phys. Rev. B **78** (2008) 012502
[9] R. Andersson, L. F. van Heijkamp, I. M. de Schepper, W. G. Bouwman, J. Appl. Cryst. 41 (2008) 868
[10] W.G. Bouwman, J.Plomp, V.O. de Haan, W.H. Kraan, Ad A. van Well, K. Habicht, T. Keller, M.T. Rekveldt, Nucl. Instrum. Methods 586 (2008) 9
[11] M. Jernenkov, S. Klimko, V. Lauter-Pasyuk, B.P. Toperverg, M. Milyaev, L. Romashev, V. Ustinov, H. Lauter, V. Aksenov, Nucl. Instrum. Methods 586 (2008) 116
[12] S.Kulikov, E.Shabalin, In Proc. of the 17th Meeting of the International Collaboration of Advanced Neutron Sources, Santa Fe, NM, USA, Apr. 24-29, 2005, Eds. G.J.Russel, J.J.Rhyne, B.V.Maes, Los-Alamos: LANL, 2006,Vol.2, p.341.
[13] Lauter-Pasyuk V., Lauter H., Jernenkov M., Toperverg B., In ILL Annual Report 2004, ILL, Grenoble, 2005.
[14] Lauter H., Toperverg B., Lauter-Pasyuk V., Petrenko A, Aksenov V., Physica B, 350 (2004) E759
[15] Lauter H., Lauter-Pasyuk V., Toperverg B., Romashev L., Milyaev M., Krinitsina T., Kravtsov E., Ustinov V., Petrenko A., Aksenov V., Journ. Mag. Mag. Mat., 258-259, 338-341, (2003)

Journal of Physics: Conference Series **251** (2010) 012061 doi:10.1088/1742-6596/251/1/012061

Adapting a triple-axis spectrometer for small angle neutron scattering measurements

M P Nieh, Z Yamani, N Kučerka and J. Katsaras

Canadian Neutron Beam Centre, Steacie Institute for Molecular Sciences, National Research Council, Chalk River, ON K0J 1J0, Canada

Mu-ping.nieh@nrc.gc.ca

Abstract. Small angle neutron scattering (SANS) instruments typically require a cold source and a large spread of wavelengths in order to enhance incident flux. However, as a direct result of the unavailability of a cold source at the Canadian Neutron Beam Centre (CNBC), we have resorted to adapting a triple-axis spectrometer for SANS measurements. This is achieved through the use of multiple converging incident beams which enhance the neutron flux on the sample by a factor of 20, compared to a single beam of same spot size. Furthermore, smearing effects due to vertical divergence from the slit geometry are reduced through the use of horizontal Soller collimators. As a result, this modified triple-axis spectrometer enables SANS measurements to a minimum q value (q_{min}) of 0.006 $Å^{-1}$ [1]. Data obtained from the modified triple-axis spectrometer are in good agreement with those data from the 30 m NG3-SANS instrument located at the National Institute of Standards and Technology (Gaithersburg, MD, USA).

1. Introduction

Small angle neutron scattering (SANS) has proven a powerful technique for the study of molecular structures and morphologies with length scales ranging from 10 to 1000 Å. Dedicated SANS instruments normally cover a scattering vector range (q range) from 0.001 to 0.6 $Å^{-1}$, where q is defined as $(4\pi/\lambda)\cdot\sin(\theta/2)$, where λ and θ are the neutron wavelength and the scattering angle, respectively. Measuring the lowest possible q value, q_{min}, usually requires long wavelength neutrons and a small incident beam. Long wavelength neutrons are usually produced from a cold neutron source, shifting the thermal neutron energy distribution toward lower energies. A small incident beam is generally produced using either a highly collimated or focused [2] neutron beam. Using a velocity selector, which increases the incident neutron bandwidth ($\Delta\lambda/\lambda$), can greatly compensate the reduced incident neutron flux caused by the restricted collimation.

Presently, the Canadian Neutron Beam Centre (CNBC) does not have a cold source to produce long wavelength neutrons. Therefore, we resorted to adapting a triple-axis spectrometer capable of carrying out SANS measurements. However, in order to so a number of technical obstacles had to be overcome. Incident neutron wavelength is selected through the use of a single crystal monochromator ($\Delta\lambda/\lambda \sim 1\%$), instead of a velocity selector ($\Delta\lambda/\lambda \geq 10\%$). One of the methods used to increase incident neutron flux, while not increasing the beam size, is to employ multiple incident beams that converge at a single spot on the detector. This concept of one-dimensional confocal beams was first proposed and tested by Nunes [3] and was later developed for two-dimensions (2D) by Glinka, *et al.*[4]. At the CNBC we have designed and implemented a confocal Soller collimator (CSC) for use at the N5 triple-axis spectrometer, making it suitable for SANS measurements, where $q_{\min}$ of 0.006 $Å^{-1}$ is achievable. This development allows for the ubiquitous triple-axis spectrometer to double, with minimal cost and effort, as a capable SANS instrument. Due to the slit geometry on the scattered side, additional horizontal Soller collimators (HSCs) are sometimes required to reduce the smearing due to vertical divergence. The comparison between the N5 data and those collected at the 30 m NG3-SANS instrument located at the National Institute of Standards and Technology (NIST) shows excellent agreement.

2. Instrument Configuration

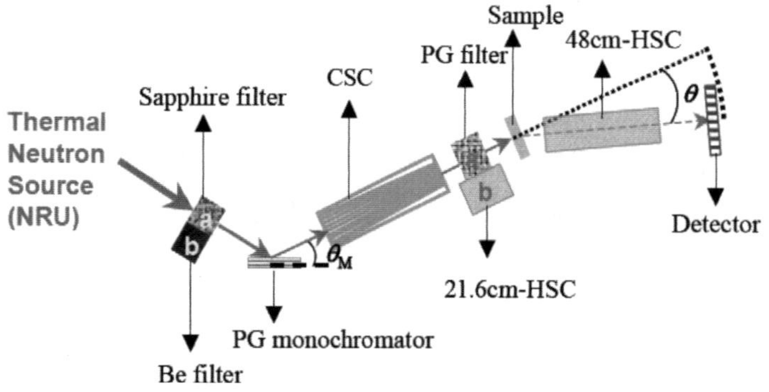

Figure 1. Schematic of the N5-SANS. The widths of the CSC are 0.13 cm and 0.1 cm at the monochromator and sample ends, respectively. The height of individual channels of both HSCs is 0.25cm.

Figure 1 shows a schematic of the N5-SANS configuration. λ was selected using the (002) reflection of a pyrolytic graphite (PG) monochromator, from which three λs (i.e., 2.37, 4, and 5.23 Å) were selected in the current setup. The higher order harmonics of the fundamental neutron wavelength (i.e., λ/2, λ/3, etc.) are reduced through the use of either a beryllium (Be) (for $\lambda > 3.99$ Å) or a PG filter (for $\lambda = 2.37$ Å) cooled by liquid nitrogen. A sapphire filter is optionally used for reducing the presence of "fast neutrons" [5]. Depending on the desired wavelength, either a sapphire (for $\lambda < 3.99$ Å), or Be (for $\lambda > 3.99$ Å) filter is placed upstream of the PG monochromator (Fig. 1). Selected neutrons then go through the CSC (length L_{CSC} of 66 cm) which is made up of 23 channels, whereby each channel is separated by steel blades coated with Gd_2O_3. All channels converge at the same spot on the detector. After the CSC, neutrons go through either a PG filter, or in the case of 4 and 5.23 Å neutrons, a 21.6 cm long HSC. The neutrons then interact with the sample, and the scattered neutrons then go through another 48 cm long HSC. A 32-wire ^{3}He position-sensitive detector is placed after the second HSC with an effective sample-to-detector distance, D_{SD}, of 1.43 m. Each wire is capable of detecting scattered neutrons at the corresponding θ.

3. Data Reduction

The reduced scattered intensity $I_{RED}(q)$ is obtained through a standard data reduction procedure applied to the raw scattered intensity, $I_{RAW}(q)$ through the following equation.

$$I_{RED}(q) = \frac{[I_{RAW}(q) - I_{BGD}(q)]}{T_{SAM}} - \frac{[I_{EMP}(q) - I_{BGD}(q)]}{T_{EMP}}, \quad (1)$$

Journal of Physics: Conference Series **251** (2010) 012061 doi:10.1088/1742-6596/251/1/012061

where $I_{BGD}(q)$, $I_{EMP}(q)$ are the background (blocked beam) and empty cell scattering, respectively, and T_{SAM} and T_{EMP} are the sample and empty cell transmitted intensities, respectively.

4. Results

4.1. Characterization the CSC

Figure 2 shows a comparison of the total intensity (all channels open) with that from individual channels. When fitted with a Gaussian function, the centers of the peak positions vary within 0.03° from each other, an indicator of each channel's alignment. The total incident flux from the 23 channels is approximately 20 times that from an individual channel (ideally 23 times).

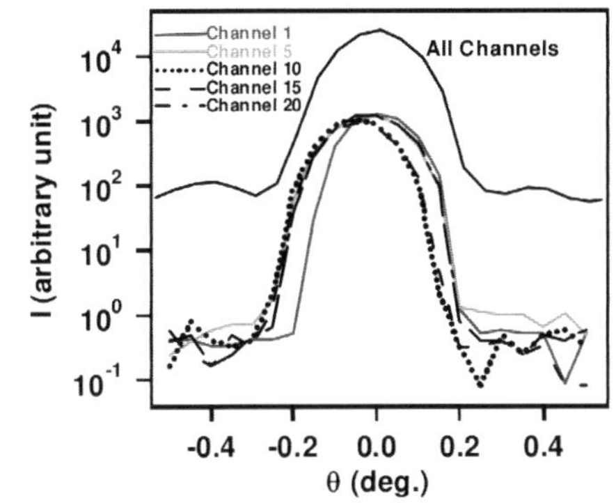

Figure 2. Comparison of the incident beam profile for all channels open compared to individual channels.

4.2. The resolution function

A SANS instrument's resolution function depends on its geometry (e.g., pinhole, slit, etc.). In slit geometry, scattering from isotropic samples is smeared due to vertical divergence. Therefore, the use of HSCs is sometimes needed.

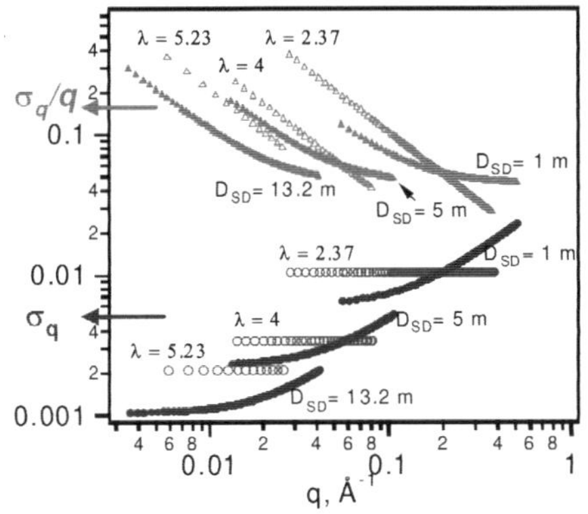

Figure 3. The instrumental resolution of the N5-SANS and NG3-SANS instruments. Both $\sqrt{\sigma_q^2}/q$ (red triangles) and $\sqrt{\sigma_q^2}$ (blue circles) are plotted as a function of q. Hollow and solid symbols represent N5-SANS and NG3-SANS data, respectively.

The resolution function σ_q for the slit scattering geometry was derived by Mildner and Carpenter [6,7]. Figure 3 shows a comparison of the calculated σ_q for the N5-SANS and a pinhole geometry instrument (i.e., the NG3-SANS located at the NIST Center for Neutron Research) [8]. NG3-SANS' σ_q is calculated based on a neutron wavelength of 6 Å, a $\Delta\lambda/\lambda$ of 15% and three $D_{SD}s$ (1, 5 and 13.2 m). For NG3-SANS, σ_q (Fig. 4) increases at higher q values as a result of a poor $\Delta\lambda/\lambda$, eventually surpassing the gains from distance collimation. However, this increase in σ_q does not affect the smearing effect as σ_q /q continues to decrease with increasing q. In the case of the N5-SANS, σ_q plateaus for a given λ because the main contribution is from distance collimation. At low q (q <0.05 $Å^{-1}$) the instrumental resolution, σ_q /q of the N5-SANS is worse than the NG3-SANS, but comparable and sometimes even better at $q > 0.1$ $Å^{-1}$.

4.3. Experimental Results

Two standard polystyrene microsphere (diameters of 24 nm and 50 nm from Bangs Laboratories) solutions of concentration 1 wt.% in D_2O were examined. The absolute intensities for both samples are using the N5-SANS and NG3-SANS are shown in Fig. 4(a) and (b). The N5-SANS data are rescaled for each q-range and then re-plotted on the same figure. The q_{min} of N5-SANS data is found to be ~0.006 $Å^{-1}$. Slight discontinuities in the data are observed in the overlap regions, presumably due to different resolution functions associated with the different q-ranges. We find excellent agreement between the NG3-SANS and N5-SANS data proving the successful implementation of this SANS design.

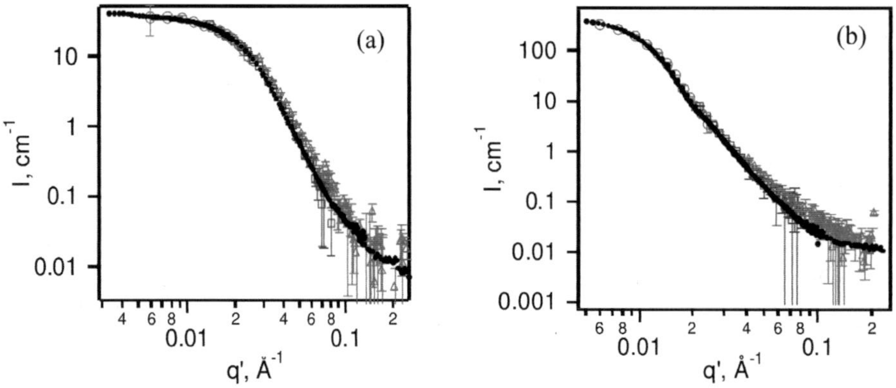

Figure 4. SANS data from 1 wt.% microspheres of diameter (a) 24 nm and (b) 50 nm obtained from the N5-SANS (open symbols) using $\lambda = 5.23$ (circles), 4 (squares), and 2.37 (triangles) Å and the NG3-SANS (solid circles).

5. Future Work

As a result of this experience, CNBC is now constructing a dedicated SANS instrument using a 2D image plate detector. The success of the N5-SANS has validated the concept of the CSC, which will now be applied to the design of a composite 2D-CSC, effectively two orthogonal 1D-CSCs. As a result, the HSCs will not be required.

References

[1] Nieh M-P *et al* 2008 *Rev. Sci. Instrum.* **79** 095102
[2] Choi S M *et al* 2000 *J. Appl. Cryst.* **33** 793
[3] Nunes A. C 1974 *Nuclear. Instr. Methods* **119** 291
[4] Glinka C J *et al* 1986 *J. Appl. Cryst.* **19** 427
[5] Nieman H F *et al* 1980 *Rev. Sci. Instrum.* **51** 1299
[6] Mildner D F R and Carpenter J M 1984 *J. Appl. Cryst.* **17** 249
[7] Mildner D F R and Carpenter J M 1987 *J. Appl. Cryst.* **20** 419
[8] Glinka C J *et al* 1998 *J. Appl. Cryst.* **31** 430

Neutron Spectral Brightness of Cold Guide 4 at the High Flux Isotope Reactor

B.L. Winn[1,2], J.L. Robertson[2], E.B. Iverson[2], D.L. Selby[2]

[1] Neutron Scattering Group, Condensed Matter Physics and Materials Science Department, Brookhaven National Laboratory, Upton, NY 11973-5000, USA
[2] Neutron Facilities Development Division, Oak Ridge National Laboratory, Oak Ridge, TN 37831, USA

E-mail: winnbl@ornl.gov

Abstract. The High Flux Isotope Reactor resumed operation in June of 2007 with a super-critical hydrogen cold source in horizontal beam tube 4. Cold guide 4 is a guide system designed to deliver neutrons from this source with a reasonable flux at wavelengths greater than 4 Å to several instruments, and includes a 15-m, 96-section, 4-channel bender. A time-of-flight spectrum with calibrated detector was recorded at port C of cold guide 4, and compared to McStas simulations, to generate a brightness spectrum.

1. Introduction

The High Flux Isotope Reactor (HFIR) at Oak Ridge National Laboratory resumed operation in June of 2007 with a super-critical hydrogen cold source in horizontal beam tube 4. Neutron scattering instruments that view the cold source receive an enhanced neutron spectrum for wavelengths greater than 4 Å. A time of flight measurement that viewed the moderator directly, when nearest guide sections were removed,[1] compares fairly well with previous MCNP model results, as shown in Figure 1. They disagreed above 3 Å, where brightness was fortunately better than expected, with a 60% increase at 5 Å before the guide.

Cold Guide 4 is a guide system designed to deliver neutrons with wavelengths between 2 Å and 6.4 Å from this source at reasonable flux to several instruments, and includes a 15-m, 96-section, 4-channel bender with total deflection of 8°, with a vertical trumpet, as shown in Figure 2a. A typical neutron will bounce about 12 times in getting through the 8° bend, so the transmission through the bender depends strongly on the quality of the mirror coatings.

In this paper we report the results of a new time of flight measurement after the guide system, at a position where a cold neutron triple axis spectrometer will be installed.

Table 1. Fit parameters for a 3 Maxwellian function manual fit to measured spectral brightness as shown in figure 3A.

Maxwellian	I_0	T
	(n/s/cm²/ster/Å)	(K)
1	7.5E12	300
2	3.3E13	64
3	7.0E12	22

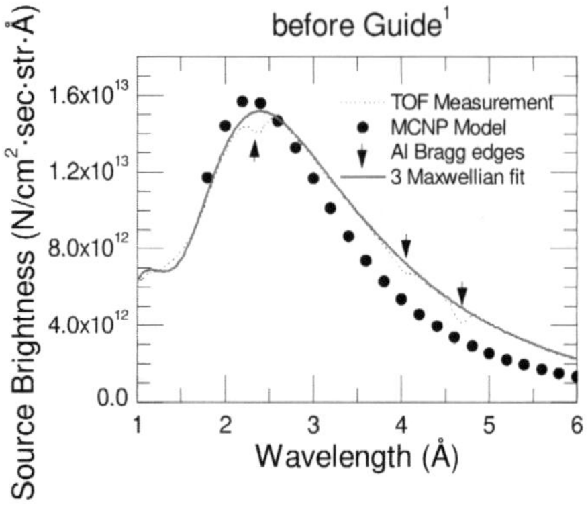

Figure 1. The spectral brightness of the HB-4 cold source at reactor power level of 85 MW and moderator temperature of 22.5 K as measured, modeled and fit using parameters in Table 1. The arrows indicate aluminum Bragg edges.

2. Experiment and Simulation

A time of flight spectrum with a calibrated detector was recorded at port C of cold guide 4, using an apparatus described elsewhere[1]. It consisted of three components: a disk chopper, a neutron detector, and a data acquisition system for processing the detector signals. The guide, chopper and detector are shown in Figure 2. The disk chopper is a rotating disk with the rotation axis parallel to the neutron beam and is used to cut the continuous neutron beam into short pulses. To produce the pulses, the disk has a slot through which the neutrons can pass as the slot rotates through the beam and the neutron 'window' opens. The disk is 1 cm of boron nitride backed by 0.7 mm of cadmium, which together block the neutron beam while the neutron window is closed. The chopper disk is preceded by an entrance aperture consisting of a 2.54 cm thick plate of boron nitride with an opening sized, shaped and aligned to match the slot in the chopper disk. The angular width of the neutron window was set at 1.8°, with a radial extent between 11.0 and 12.7 cm. The detector is a Reuter Stokes parallel plate fission counter with the anode located 2.451 m from the center of the chopper disk. The detector was shielded on all sides by 1 cm of boron nitride and 0.7 mm of cadmium. A circular entrance aperture with a diameter of 9.5 mm was machined into the detector shield. The data acquisition system is a Peripheral Component Interconnect (PCI)-based multichannel scaler triggered by an optical pickup on the chopper disk, with 10 μs bins, corresponding to 0.016 Å, and controlled by in-house LabView software. The DAQ system integrated the spectrum for 18,000 pulses then saved the data, cleared, and began a new histogram. In this way we could monitor any changes in the spectral brightness over time with an approximately fifteen-minute granularity.

Differences between the current and previous setup are the use of a 20 Hz chopper frequency instead of 60 Hz, a distance between the middle of the chopper to the detector of 2.451 m, instead of 2.473 m, and a detector aperture of diameter 9.5 mm instead of 10 mm.

In our McStas[2] simulation, we employed a source term that was a sum of three Maxwellians, fit to the measured brightness determined previously[1], as shown in Figure 1 with manually determined fit parameters listed in Table 1. In this way we can evaluate the performance of the guide system alone.

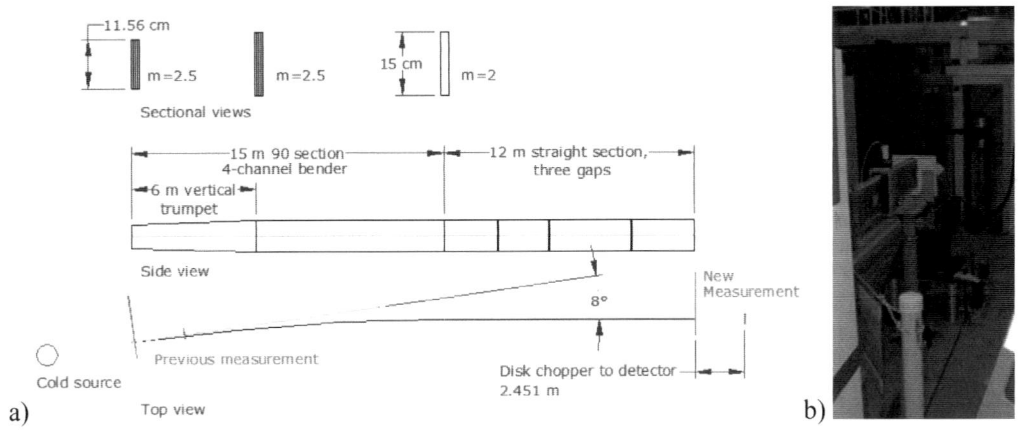

Figure 2. (a) Layout of guide and Time of Flight experimental apparatus. Views are not to scale. (b) Apparatus is shown at the end of the guide.

3. Results

Figure 3 shows measured neutrons per second at the detector, compared to the McStas simulation. This plot accounts for detector efficiency in the measured data. For the modelled data it accounts for (incoherent) attenuation in aluminium windows and in air, and beam reactor power. No attempt was made to account for the effects of Bragg scattering or multiple scattering by the air or aluminium. The wavelength offset of the measured data was roughly set using the dip in intensity at the (2 0 0) and (1 1 1) aluminium Bragg peaks, but with such a broad convolution in wavelength, the offset could easily be off by ~0.1 Å.

We measured an unexpected background signal at ~8% of the measured peak flux, roughly independent of time. This background was not observed in the previous study at the HB4 beam tube. For the plot, it was subtracted from the data prior to scaling by detector efficiency for comparison, but it will be further analyzed. The measurements will not be repeated however, because a monochromator shield has been installed at the port C for the cold neutron triple axis spectrometer.

To obtain brightness at the end of the guide from measured neutrons per second at detector as shown in Figure 3b, we first account, on both model and measurement, for the wavelength-dependent chopper transmission f_t due to finite disk thickness (0.01 m). Here $f_t = 1-\lambda/\lambda_m$, where for 20 Hz λ_m=92.5 Å. An analytical form of the acceptance in the detector is difficult due to the presence of both rectangular and circular apertures, but an approximation leads to an acceptance of 6.8 x 10^{-6} cm^2·str, the same acceptance used in the previous study[1]. We assume 100% power instead of 10% power, and correct for transmission through air and aluminium after exiting the guide. We account for the chopper duty factor of 0.005. For the measured data we again account for detector efficiency. Finally, no attempt was made to deconvolve either the measured data or the McStas results with respect to the chopper slot time window, which corresponds to a width of 0.404 Å.

4. Discussion and Summary

In Figure 3, we note that the measured and modelled transmission of the guide match very well for wavelengths above 3.5 Å, and the measured transmission is slightly better than that modelled between 2 Å and 3.5 Å. Below 1.5 Å, the measured count rate is slightly less than that modelled, but this is in part due to a slight overestimate of the source term as shown in Figure 1. Comparing Figures 1 and 3b, the aluminium Bragg edges are broader post-guide due to the use of a 20 Hz chopper frequency instead of 60 Hz for the disk chopper frequency. It is unclear how much of the flux loss near these edges is due to the aluminium scattering. The spectra at short wavelengths is more difficult to compare due to the uncertainty in the wavelength offset of the measured data, the steeper slope below 2.5 Å, and aluminium Bragg edges between 2 Å and 3 Å.

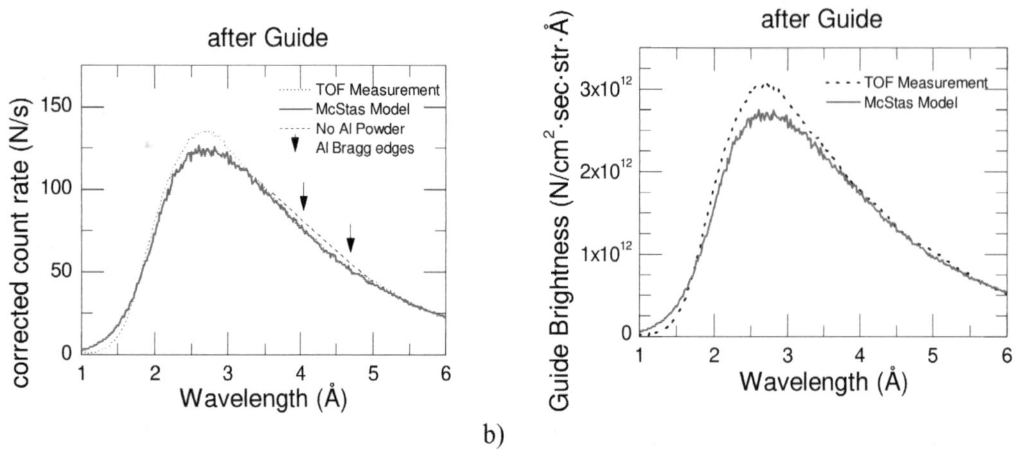

Figure 3. (a) Neutrons per second at the detector as measured and modeled, for a reactor power level of 8.5 MW. The arrows indicate aluminum Bragg edges and the 'No Al Powder' line is a guide to the eye for locating these edges in the measured data. (b) The spectral brightness at port C of cold guide 4, at reactor power level of 85 MW, as measured and modeled.

We were pleasantly surprised that a known septa misalignment problem did not significantly impact the guide system transmission. We knew that the bender septa were misaligned, due to thesepta of varying thickness lying in oversized slots. On one assembly of 6 sections, we measured at each end a typical gap of ~40 µm or 8% of the width of the Si plates, and found the plates lying loosely in the slots. We therefore performed McStas studies to estimate flux loss from misaligned septa occlusion, and found that for this typical gap attenuation may be up to 50%, independent of wavelength. The model results shown assume no misalignment.

Based on these results, we can proceed with installation of several new instruments, including a new cold neutron triple axis spectrometer, and a quasi-Laue diffractometer, confident that we can predict the performance of these instruments.

We wish to acknowledge the efforts of Ralph Moon, who developed the design concept of, and developed an earlier model for, all four guide systems[3] for horizontal beam tube 4 at HFIR. This experiment was supported by the Department of Energy's Office of Science.

References

[1] Robertson J L and Iverson E B 2008 Measured Spectral Brightness of the HFIR Supercritical Hydrogen Cold Source *Neutron News* vol 19 pp 17-19

[2] Lefmann K and Nielsen K 1999 McStas, a General Software Package for Neutron Ray-tracing Simulations *Neutron News* vol 10 p 20

Willendrup P, Farhi E and Lefmann K 2005 McStas 1.7 - a new version of the flexible Monte Carlo neutron scattering package *Physica B* vol 350 pp 735-737

[3] Moon R 2001 Final Design and Performance of Cold Guide Systems *internal HFIR memorandum*

Guide design study for the high-resolution backscattering spectrometer FIRES

C. Pelley[1], F. Kargl[1], V. Garcîa Sakai[2], M.T.F Telling[2], F. Fernandez-Alonso[2], F. Demmel[2]

[1]Institute of Physics and Mathematics, Aberystwyth, SY23 3BZ, UK
[2]ISIS Facility, Rutherford Appleton Laboratory, Didcot, OX11 0QX, UK

E-mail: franz.demmel@stfc.ac.uk

Abstract. Different options are considered to transport cold neutrons along 90 m for the proposed new spectrometer FIRES at the ISIS facility. Monte Carlo simulations using the McStas programme package are used to assess the performance of various guide designs from the biological shield to the sample position. By employing a curved geometry, to avoid the direct line of sight, a hybrid design which combines a curved ballistic guide and an elliptic focusing section appears to be the best solution.

1. Introduction

FIRES will be a new high-resolution backscattering spectrometer, proposed for the ISIS facility [1]. It aims to achieve an order of magnitude higher energy resolution compared to the existing near-backscattering spectrometers IRIS and OSIRIS, enabling new measurement possibilities in diverse fields ranging from biopolymers and viscous liquids to hydrogen storage materials and quantum magnets.

To obtain such a high resolution short-time incoming pulses have to be provided and a long flight path is advantageous. One possible approach is to position the instrument on a decoupled or even poisoned moderator. Such a configuration is used at the SNS backscattering spectrometer BASIS [2]. A different solution is to install a chopper to produce short pulses [3]. Furthermore, a pulse-shaping chopper allows changes to the resolution width by altering the incoming pulse width, hence trading resolution against intensity. Such a design has been proposed for the DNA spectrometer at JPARC in Japan [4]. The secondary spectrometer of these high resolution instruments consists of large-area silicon analysers in near backscattering geometry and position-sensitive detectors (psd).

To transport neutrons a long way the ballistic guide idea was proposed [5]. Later nonlinear tapering geometries have been suggested, like parabolic and elliptic geometries [6], which can be combined in various forms [7]. These new geometries combined with new technological developments, which promise supermirror coatings up to $m = 6$ values [8], provide numerous options to design sophisticated guide geometries. The m-value defines the factor of critical angle of a supermirror compared to the critical angle of a natural nickel guide.

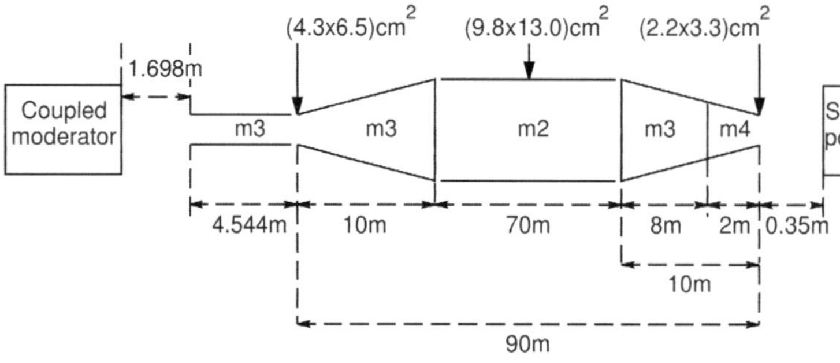

Figure 1. Schematic layout for the optimized ballistic guide configuration. Vertical and horizontal dimensions are depicted.

2. Simulations

A suitable compromise between high intensity and dynamic range proposes a flight path length of about 90 m for the FIRES instrument [1]. We are focusing on the specific layout at ISIS which consists of a distance of 1.7 m from the moderator to the guide entrance and a length inside the biological shield of 4.54 m (see figure 1). Both lengths will not be changed and the guide dimensions (4.3 cm × 6.5 cm) inside the biological shield will not be varied, only a change of the coating will be considered. To allow for large sample environments, like a high field magnet, at the sample position, a necessary distance of 0.35 m between the end of the guide and sample position was chosen. The FIRES instrument will use a silicon (111) analyser system in backscattering and the final wavelength will be 6.28 Å. The simulations are focused around this wavelength range by restricting the wavelengths to $5 - 7$ Å. For the simulations we used the McStas package [9]. The main components utilized were the *guide* and the *guide_curved* components with the standard reflectivity profile, using $\alpha = 5.07$ for the linear decrease in reflectivity beyond $m = 1$. This corresponds to a reflectivity of 70% for the $m = 4$ edge. We optimized the integrated intensity at the sample position which was defined as a rectangle of 2 cm × 3 cm and a square of 1 cm^2. The larger one is a standard size for quasielastic scattering experiments and the latter one resembles typical single crystal sizes or small amount biological samples. FIRES will be a high energy resolution instrument therefore larger divergence will be accepted to raise the intensity. Divergence was monitored to achieve a reasonably homogenous distribution. An inhomogenous divergence may influence the lineshape, in particular for single crystal experiments. We concentrated on curved guide designs to avoid direct line of sight. A direct line of sight will permit the transmission of high energetic neutrons from the proton pulse hitting the target and could create background peaks. The curvature radius was chosen to be $R = 5000$ m. This radius of curvature corresponds to a characteristic wavelength of $\lambda \approx 2\pi\sqrt{(2a/R}/k_\perp = 1.9$ Å with $k_\perp = 2.14 \cdot 10^{-2}\text{Å}^{-1}$ the maximum vertical wavevector for a $m = 2$ guide and a guide width of $a = 10$ cm. This large radius will allow to access large energy transfers with the silicon (111) reflection and the efficient use of the silicon (333) reflection.

3. Results

As a reference we took a guide similar to the existing one for the IRIS spectrometer [10], but now 90 m long. The resulting integrated intensity over the two sample sizes served as references and gain factors are quoted with respect to those intensities. The IRIS guide is a curved $m = 1$ guide with dimensions 4.3 cm × 6.5 cm, which is focused to end dimensions of 2.2 cm × 3.3 cm

by a 1.76 m long $m = 2$ linear focusing section.
As a first configuration we investigated a ballistic type of guide. The principle of such a guide is to widen the beam so that the number of reflections is reduced. A proper defocusing section will reduce the divergence according to the Liouville theorem and therefore the reflection losses. To obtain an optimized design, an iterative procedure was adopted. At first, the width and height of the ballistic section was varied, then the length and coating of the focusing section. The vertical and the horizontal lengths have been constrained to the same focusing length. After a further iteration now including the length and coating of the diverging part, an optimized set-up was found. The parameters for this set-up are indicated in fig. 1. The optimized value for the ballistic guide width is about a factor 4 larger than the width of the intensity distribution at the sample position, which corresponds to the achieved increase in accepted divergence with the $m = 4$ supermirrors in the focusing section. The gain factor is 2.3. The full widths at half maximum (FWHM) of the intensity distribution at the sample position are 2.4 cm horizontally and 3.3 cm vertically. The corresponding intensity distribution from a psd is depicted in fig. 2. A change of the coating in the long curved section to $m = 3$ did not provide a higher transmission, even by using larger m-values in the focusing and defocusing section. The next configuration

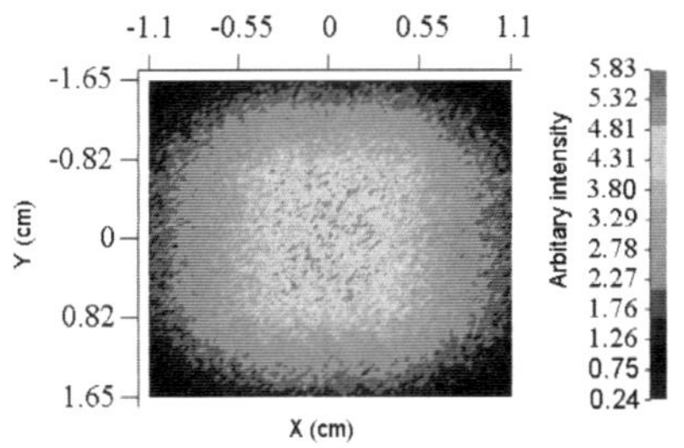

Figure 2. The intensity distribution at the sample position is shown for the ballistic guide set-up.

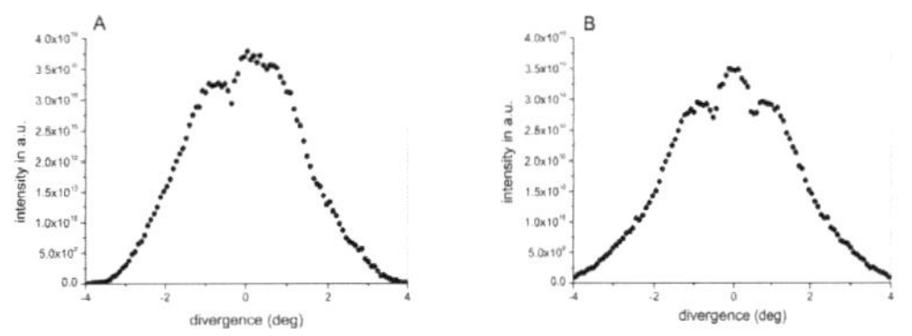

Figure 3. The horizontal (A) and vertical divergences (B) are depicted for the set-up with the elliptic focusing section. Due to the curvature of the guide the horizontal distribution is asymmetric.

studied was a combination of a curved guide with a nonlinear focusing section, an elliptic one. The component *guide_tapering* from the McStas-package allows elliptic and other designs, but is restrictive in some options. The two focal lengths have to be provided and with the entrance dimensions of the guide, the ellipse geometry is fully determined. We created an elliptic shaped guide by calculating all guide segments individually using a Fortran program. For the elliptic focusing section at the end of the curved guide we added an ellipse which starts with the end dimensions of the optimized curved guide 9.8 cm × 13.0 cm. These values are the minor axes of the ellipses in this case. The second focal point of the ellipse is fixed to the sample position, 0.35 m from the end of the guide. The major axis was then obtained by varying its length, for which we finally chose 8 m. Due to the high eccentricity of the ellipse the length of the segments can be chosen according to an exponential distribution [11], with longer segments at the center of the ellipse. For this focusing section we utilized 50 linearly tapered guide segments which resulted in a smallest length of 4.5 cm at the end of the ellipse. The coating was optimized to a distribution of m-values starting from $m = 3$ to $m = 5$ for the last 3 pieces. The guide-end dimensions are 2.2 cm×3.3 cm and the FWHMs of the intensity distributions are 2.4 cm×2.5 cm, resembling a more circular spot. The optimized design delivered a gain factor of 3.9 for the larger sample area and 4.0 for the smaller one, a factor 2 improvement over the ballistic design. Due to the guide curvature the horizontal divergence distribution is not symmetric as shown in fig. 3, but still more homogenous compared to the straight focusing guide.
The next natural step is to investigate a vertical elliptical guide over the whole length from

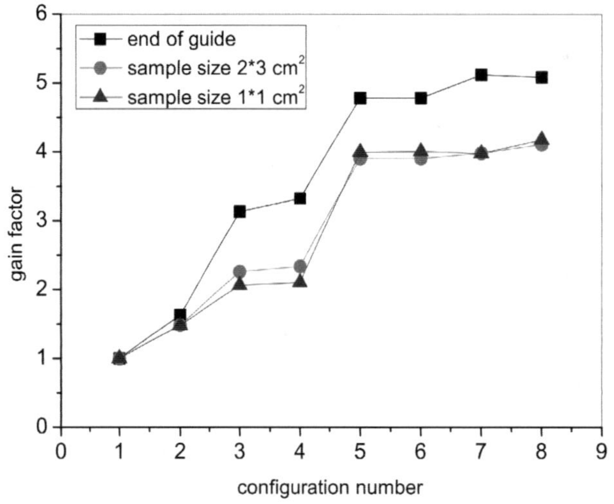

Figure 4. Gain factors are depicted for the different configurations in respect to the IRIS-type instrument. Configuration 1 denotes the IRIS-type instrument, 2 the OSIRIS-type instrument, 3 the m2 ballistic guide, 4 the m3 ballistic guide, 5 the m2 curved guide combined with an elliptically focusing section, 6 the m3 curved guide combined with an elliptically focusing section, 7 the full elliptic geometry in the vertical direction and 8 the height restricted nonsymmetric bi-elliptic geometry.

the biological shield combined with a curved guide in the horizontal dimension. The horizontal guide geometry used the ballistic curved guide combined with the elliptic focusing section. This design resulted in a maximum guide height of 32 cm, but only negligible gains in intensity, which might be related to a too long distance between the focal point and the source and therefore an under illumination of the elliptic guide. Interestingly, a design which restricted the guide height to 10 cm and comprised of a combination of two elliptical sections delivered similar gain factors (configuration 8). The elliptical sections are not symmetrical, but are divided up in a quite long first one (70 m) and then a shorter second part. The partial illumination of the symmetric ellipse might explain the similar performance of this bi-elliptical set-up with reduced heights.
Fig. 4 depicts a comparison of all of the obtained gain factors. It shows a two-step increase first by the ballistic guide geometry and second by the elliptic focusing designs. The clear increase for the elliptic designs combined with the more homogenous divergence distributions demonstrates a distinct advantage of these nonlinear geometries. Finally, we used the full elliptic guide component to get an idea what can be gained by a new guide design starting from the moderator and modifying the biological shield insert. This full elliptic design yielded ellipses with minor axes of 24.6 cm and 30.5 cm. The entrance dimensions are 1.6 cm × 2.8 cm and the guide end dimensions are 3.0 cm × 3.7 cm. With a $m = 3$ coating we got a gain factor of 12. This large gain factor might be understood due to a much reduced distance between the focal point of the ellipse and the moderator. For the used 6 Å neutrons the acceptance angle towards the source is a factor two larger with the full elliptic design compared to the $m = 3$ in-pile part. This factor 2 in one dimension might explain the overall factor 3 increase in gain with the full elliptic design.

4. Conclusions

From these results we conclude that a new guide design must consider also a change of the biological shield insert geometry. Even so larger gain factors for the curved design with modified inserts can be expected, we suppose that none will reach the high gain factor of the full elliptic design.

References

[1] Demmel F and Andersen K H 2008 *Meas Sci and Tech* **19** 034021
[2] Herwig K W and Keener W S 2002 *Appl Phys A* **74** *Materials Science* S1592
[3] Zsigmond G, Mezei F, Wechsler D and Streffer F 2001 *Nucl Inst Meth A* **457** 299
[4] Takahashi N, Shibata K, Sato TJ and Arai M 2007 *J. of Neutron Research* **15** 61
[5] Mezei F 1997 *J of Neutron Res* **6** 3
[6] Schanzer C, Böni P, Filges U and Hils T 2004 *Nucl Instr Meth A* **529** 63
[7] Böni P 2008 *Nucl Instr Meth A* **586** 1
[8] Maruyama R, Yamazaki D, Ebisawa T and Soyama K 2009 *Nucl Instr Meth A* **600** 68
[9] http://neutron.risoe.dk/mcstas
[10] Carlile C J and Adams M A 1992 *Physica B* **182** 431
[11] Kleno K http://www.fys.ku.dk/~kaspar/speciale
[12] Telling M and Andersen K H 2005 *Phys Chem Chem Phys* **7** 1255

International Conference on Neutron Scattering 2009 IOP Publishing
Journal of Physics: Conference Series **251** (2010) 012064 doi:10.1088/1742-6596/251/1/012064

Virtual design of the neutron guide for the TOF spectrometer NEAT

Zunbeltz Izaola[1] **and Margarita Russina**[1]
[1]Methods and instruments of neutron scattering, Helmholtz-Zentrum Berlin GmbH, Hahn-Meitner Platz 1, Berlin, Germany
E-mail: zunbeltz.izaola@helmholtz-berlin.de

Abstract. We present the results of a virtual design study based on Monte-Carlo neutron ray tracing techniques for the neutron guide of the time of flight (TOF) spectrometer NEAT. We studied several configurations with linearly or elliptically tapered compressors with different degrees of focusing and different guide coatings. The calculations were performed and cross-checked using two software packages which produced similar results. The geometrical arrangement of selected guide components was optimised with the Particle Swarm Optimisation algorithm. The results of the Monte Carlo simulations confirm an expected intensity gain factor of approximately 5, that can be achieved by the optimal configuration.

1. Introduction

NEAT is a direct time-of-flight cold neutron spectrometer designed to study the dynamics and structure in the time domain 10^{-13} s–10^{-10} s and on the length scale ranging from microscopic (0.5 Å) to nanoscale dimensions (up to about 50 nm). To transport neutrons from the neutron source to the sample NEAT uses a neutron guide. The current guide of NEAT starts 1.5 m behind the cold source with a cross section 3 cm wide and 12.5 cm high (Figure 1). At a position 30.8 m behind the cold source the guide is horizontally split in half and only the upper half of the guide with a vertical dimension of 5.5 cm is used for the NEAT spectrometer. To monochromatise neutrons NEAT uses disc chopper devices which are installed in four chopper housings, located 35.1 m, 37.5 m, 42.6 m and 47.2 m from the cold source. The last section of the guide employs a so-called "eye-of-the-needle" section to optimise the intensity for the short chopper pulses. This device was invented in the course of the design process of NEAT [1]. In front of the last chopper the guide is horizontally focused to a width of 15 mm by a converging section and then decompressed towards the sample by using the diverging guide section. The neutron guide is coated with ^{58}Ni, except for the converging-diverging sections which are made of supermirrors.

2. Simulations for the new configurations

Thanks to the number of important innovations at its conception, NEAT had equivalent flux, in spite of the medium flux of the reactor, and somewhat better ultimate resolution than the other best in class instrument worldwide, IN5 at ILL, for the first 10 years of operation. The fast evolution of TOF neutron spectroscopy in the last years resulted in an increase of the data rate of more than an order of magnitude at the world-wide leading instrument, and new developments at pulsed spallation sources promise a tremendous progress in performance. To maintain the

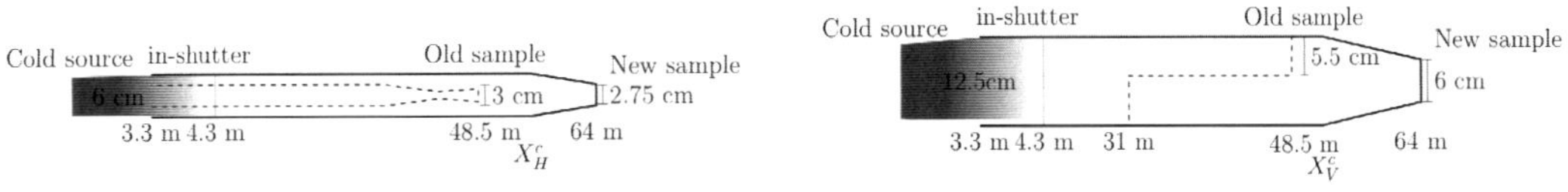

Figure 1. Horizontal (left) and vertical (right) views of the present guide (dashed line) and upgraded guide (solid line) configuration. X_H^c and X_V^c are the starting positions of the compressor in the horizontal and in the vertical configurations.

competitive edge of NEAT, a full upgrade of the instrument has recently been proposed [2]. The proposal includes the replacement of the existing neutron guide by modern supermirror neutron optics and a ballistic guide concept [3] to gain a factor of 5 in incident neutron flux compared to the present instrument. Here we present the results of the computer simulation study of the upgrade of the NEAT neutron guide. The new guide (Figure 1) is 12.5 cm high and 6 cm wide. The sample position has been moved 15.5 m downstream. To couple the neutron guide to the sample dimensions the neutron beam is compressed horizontal and vertically.

We conducted a computer simulation study of the new guide configurations using the VITESS software package [4]. To simplify the optimisation of the guide design, we decoupled the vertical and the horizontal dimensions in the calculations. For this purpose the neutron guide was simulated with perfectly reflecting mirrors in the dimensions of no interest, and the gaps between the different components, namely the guide exit and the sample position, were simulated with a perfectly absorbing mirror in the dimension of interest. Thus the total transmission is a product of the transmissions in both dimensions. For the verification we independently ran complete 2D simulations with final optimised parameters. Two detection windows were assumed in the calculations: one at the position of the guide exit and another one at the position of the sample. The intensity of the neutrons arriving at the sample was used as a final parameter for the neutron guide optimisation. VITESS calculations were combined with the Particle Swarm Optimisation (PSO) algorithm used to optimise the geometrical parameters of the guide and to maximise the total flux at the sample. PSO is a stochastic, population-based evolutionary algorithm used to find the optimum solution of a problem [5, 6]. As a final step, the calculated wavelength-dependent flux was divided by the flux delivered by the present NEAT guide.

2.1. Calculations of the horizontal configuration

As mentioned above the horizontal dimension of the guide equals 60 mm. It starts with a straight section 3.34 m long coated with supermirrors of $m = 3\theta_c$. To filter fast neutrons the guide includes a curved 40 m length long section coated with supermirrors with $m = 3\theta_c$ for the outer side and $m = 1.5\theta_c$ for the inner side. It is followed by a straight section and a compressor section. The length of the compressed section in the simulation was varied in the range of 2.01 m–7.34 m depending on the coating of the guide. For the optimisation of the guide configuration, we studied two options: (1) The neutron beam was compressed by a single linear part to deliver the beam to the sample. (2) This configuration included two parts, i.e., first a linear converging section that compresses the beam to 18 mm in front of the last chopper, which is placed 1.3 m from the sample, and then transports the neutrons by linear or diverging part to the sample. We simulated several configurations with different coatings and different compressor-decompressor distances (Table 1)

Figure 2 shows the gain factor and Table 1 shows the optimised geometrical parameters.

Table 1. Optimised exit width and converging length of the compressor part in the horizontal dimension for the cases of focusing on the sample and focusing on the last chopper, with samples of size 2.2/2.75/3.0 cm.

	Focus on the sample		Focus on the last chopper	
m	Width (cm)	Length (cm)	Width (cm)	Length (cm)
3.5	2.90/3.20/3.47	209.1/201.2/273.0	2.30/2.36/2.36	618.0/618.0/618.5
3.0	2.91/3.24/3.47	208.4/234.9/266.9	2.57/2.35/2.57	604.2/620.6/604.4
2.5	3.10/3.39/3.47	211.7/264.6/270.9	2.43/2.34/2.42	611.6/612.5/611.8

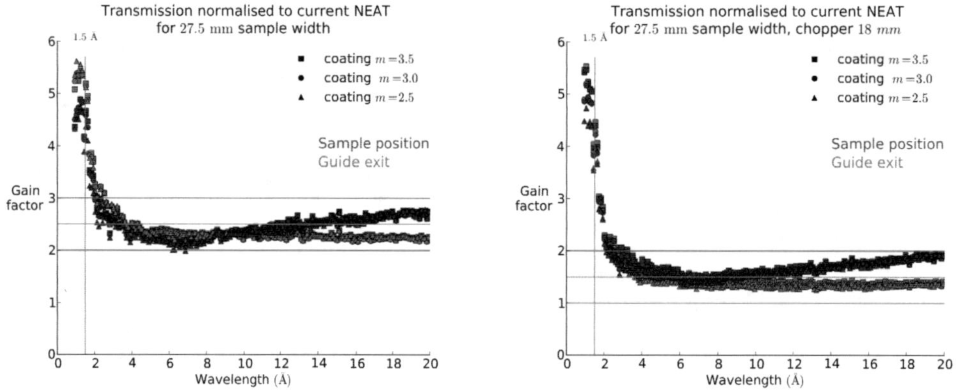

Figure 2. Gain factor for horizontal configuration with focusing on the sample (left) and focusing on the last chopper (right), for a sample width of 3 cm.

3. Discussion

Our calculations confirm an efficiency gain of a factor of even more then 5 of the new improved guide with increased dimensions for the neutron transport, as it was indicated by analytical approximations (Figure 3). The larger gain is observed in the short wavelengths range which is particularly important for low temperature and magnetism experiments, areas which are of strategic importance at Helmholtz-Zentrum Berlin.

The main problem with using a wider or higher neutron guide is the increased divergence of the beam when the neutrons are compressed towards the sample area with a smaller cross section and higher supermirror coatings. Because of the higher divergence fewer neutrons reach the sample, if they are not guided by the guide. Thus we identified the gap between the guide exit and the sample position as one of the main sources of the neutron losses. For the horizontal configuration with the guide–sample distance of 70 cm a gain factor of no more than 1.5 could be achieved. The gain increases up to the factor of 2.5 when the distance between the guide exit and the sample is decreased to 30 cm. The divergence can be decreased by increasing the sample width and height, which on the other hand spoils the resolution of the secondary spectrometer. Thus, we have two contradictory conditions which influence the guide configuration: On one hand the guide dimensions on the extraction part should be as large as possible to extract the largest possible number of neutrons, on the other hand the divergence should be reduced to avoid the losses. These conditions and the performance requirements of the secondary spectrometer motivated our choice for sample dimensions limited to a width of 27.5 mm and a height of 60 mm.

Furthermore, we observed that the intensity and the homogeneity of the beam on the sample improves if the dimensions of the guide exit are slightly larger than the dimension of the sample

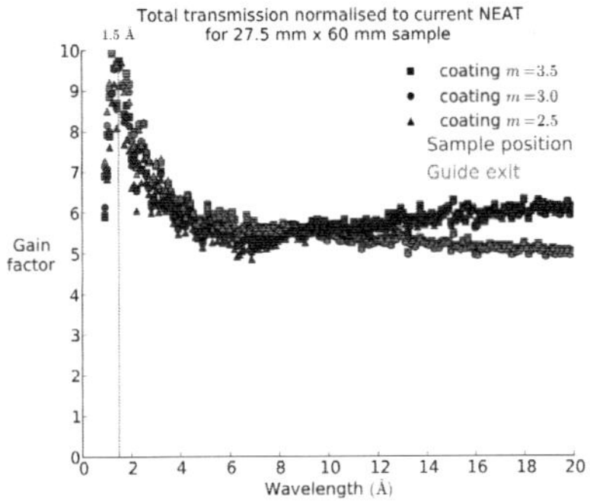

Figure 3. Total gain factor for $2.75 \times 6.0\,\mathrm{cm}^2$ sample.

Table 2. Total intensity (T) and intensity at a 1 cm wide region of the centre (D) in the horizontal dimension for the cases of focusing on the 3 cm wide sample and $m = 3\theta_c$, for different guide exit sizes.

Guide exit (cm)	T (10^{12} n)	D (10^{12} n)	D/T
2.5	5.7	3.2	0.56
2.75	5.8	3.1	0.53
3.0	5.8	2.9	0.59
3.25	5.9	2.8	0.47
3.5	5.7	2.7	0.47
3.75	5.4	2.5	0.46

(Table 2). We observed this trend for all configurations in both dimensions. In the horizontal dimensions the configuration with the neutron beam focused on the sample (a) proved to be more advantageous compared to the configuration with the compressor focused in front of the chopper (b). Thus, the gain factor of about 2.5 was observed for the configuration (a) while the configuration (b) only allows for a gain factor of 1.2–1.5. The configuration (a) is best optimised for the medium resolution of the primary spectrometer, when the chopper slit matches the guide width of 40 mm. However, it still allows the high-resolution experiments with chopper slits of 18 mm and delivers nearly the same intensity as in the case of configuration (b). Experience with NEAT user operation shows that about 80 % of the experiments use the medium range resolution configurations which will optimally be realised by configuration (a).

4. Conclusions

Detailed simulations of the different guide configurations using an approach combining VITESS and Swarm Particle Optimisation were carried out for configurations possible for the time-of-flight spectrometer NEAT. The horizontal and the vertical configurations were split and analysed separately. Our calculations show that utilisation the neutron guide of $12.5 \times 6\,\mathrm{cm}^2$ cross section coated with supermirrors of $m = 1.5\theta_c$ coupled to the sample of $2.75 \times 6\,\mathrm{cm}^2$ dimensions by supermirror compressor part of $m = 3\theta_c$ results in substantial intensity increase. The gain factor of 5 and higher was calculated for the broad wavelength range of 2 Å–15 Å.

Acknowledgments

The authors would like to acknowledge F. Mezei and K. Lieutenant for they comments and help with the simulations, and K. Diederichsen for linguistic assistance with the preparation of the manuscript.

References

[1] Lechner R E and Mezei F 1990 Optimization of the transmission of a double-trumpet neutron beam compressor ICANS-XI ed Misawa M, Furusaka M, Ikeda H and Watanabe N (Tsukuba-shi, Japan) pp 919–929
[2] Russina M 2009 PoF proposal for the Helmholtz-Zentrum Berlin: Upgrade of the Neutron Spectrometer NEAT
[3] Mezei F and Russina M 1999 Multiplexing chopper systems for pulsed neutron source instruments Advances in Neutron Scattering Instrumentation vol 4785 (SPIE) pp 24–33
[4] Wechsler D, Zsigmond G, Streffer F and Mezei F 2000 Neutron News **11** 25–28
[5] Kennedy J and Eberhart R 1995 1995 IEEE International Conference on Neural Networks Proceedings, Vols 1-6 1942–1948
[6] Bentley P M and Andersen K H 2009 Journal of Applied Crystallography **42** 217–224

Upgrade of the neutron guide system at the OPAL Neutron Source

D. Martin Rodriguez, S. J. Kennedy and F. Klose

Bragg Institute, ANSTO, PMB 1, Menai NSW 2234 (Australia)

dmr@ansto.gov.au

Abstract. The new research reactor at ANSTO (OPAL) is operating with seven neutron beam instruments in the user programme and three more under construction. The reactor design provides for expansion of the facility to eighteen instruments, and much of the basic infrastructure is already in place. However, an expansion of the neutron guide system is needed for further beam instruments. For this purpose, several possibilities are under consideration, such as insertion of multi-channel neutron benders in the existing cold guides or the construction of a new elliptic cold guide. In this work Monte Carlo (MC) simulations have been used to evaluate performance of these guide configurations. Results show that these configurations can be competitive with the best instruments in the world.

1. Introduction

The new research reactor at ANSTO (OPAL) is operating and seven of the first-wave neutron beam instruments are complete [1]. Three instruments, currently in design phase, are also expected to begin operation in the next three to four years. These instruments are powerful resources for condensed matter and materials physics research, and will provide high quality research outputs for Australia over the next decade. As the requirements of neutron beam science grows and diversifies there will be a demand for expansion of the research capabilities at OPAL. This will require new instruments as well as periodic improvements to existing instruments. Further expansion of the instrument suite beyond those under construction will require upgrade of the neutron guide system to provide additional neutron beams matched to the performance requirements of new instruments.

The design process for the original neutron guide system began in 1998, relying mainly on Monte-Carlo (MC) simulation methods [2]. The system was final built and commissioned by 2007; a period of around nine years. In recognition of the long lead time for upgrade of the guide system, we are now evaluating a range of designs for splitting of neutron beams and for creating more end-guide positions.

One possibility is insertion of multi-channel benders to fully exploit the potential of the existing cold neutron guides [3]. Different neutron bender designs have been evaluated in the present cold guides and their performance assessed for some potential new instruments.

Another possibility is insertion of new guides. The simplest approach would be insertion of large curved parallel sided supermirror guides between the existing guides (and similar in design to the existing guides). However, in recent years non-linear tapered guides have been built in some neutron laboratories with impressive performance [4]. In particular elliptical guides have been shown to deliver outstanding performance in transporting and focusing neutrons. For this reason, the design of elliptical guides is being considered as well.

2. Description of the simulated systems

MC simulations of possible modifications to the OPAL cold beam transport system were performed using VITESS program [5]. The source term is based on measurements of the characteristics of the OPAL cold neutron source (a 20 litre re-entrant liquid Deuterium moderator) [6]. The source is modeled as a flat rectangle (at the beam tube entrance), which has a cold spectrum in the central part (of the same area as the beam tube) and a thermal spectrum on either side. The sides account for neutrons that are either underthermalized in the cold source, or that travel from the cold source through the D_2O reflector and are partially re-thermalized on the way to the guide entrance. Therefore, the neutron spectrum is a sum of two Maxwellian distributions at 45K and 300K.

In the existing cold neutron guides, the beam tube is ~1.35 metre long and is followed by a 3 metre long supermirror guide up to the reactor face, and then by a 20.5 m. long curved guide section until the secondary shutter.

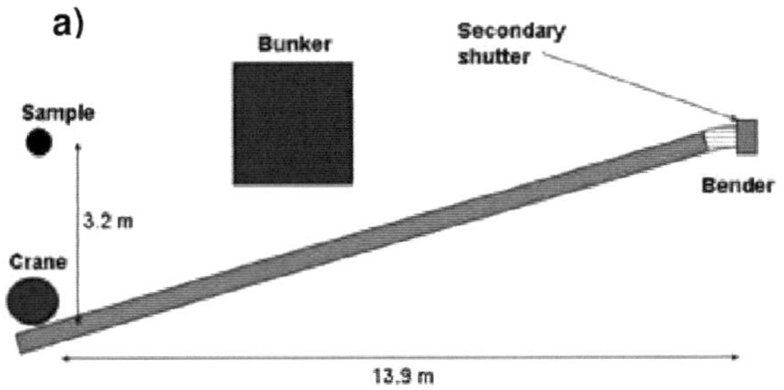

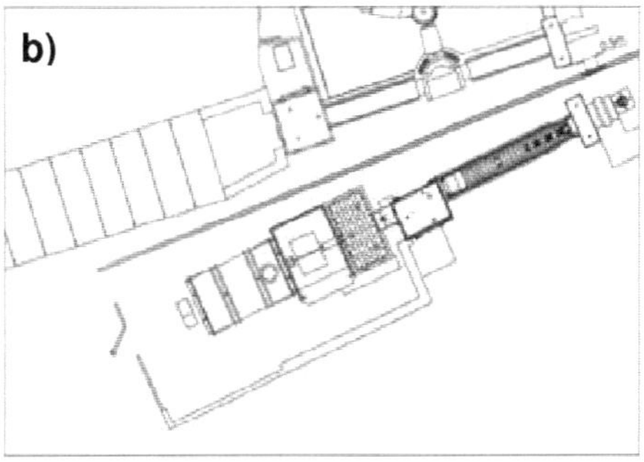

Fig. 1 schematics of a) a multichannel bender on CG3 and b) an elliptical cold guide between CG1 and CG3.

2.1. Creation of more guide end positions in existing guides with benders

The most obvious choice for insertion of a multichannel bender is CG3 (a cold guide), upstream of the PLATYPUS (the neutron reflectometer at OPAL). The cross-section of CG3 from source to secondary shutter is 20 x 5 cm, but PLATYPUS only uses the top 2 x 5 cm, leaving most of the beam free. The supermirror coating of the guides in the straight sections has m=3 for top and bottom parts and m=2 on the sides, while in the curved guide section m=2.5 in the outer side of the guide and in the rest is the same as the straight sections [2]. To minimize disturbance the bender must bypass the Platypus crane. This can be achieved by a combination of bender and straight guide (see Fig. 1 a). The bender must have short radius of curvature (ρ), and a large number of channels. Here we consider 16 or 32 channel benders and substrate thickness of 0.25 or 0.50 mm. (typical thickness of Si or glass substrate respectively). Several combinations with differing relative lengths of bender (L) and straight guide (L_{st}) sections have been simulated (see table 1 for details).

L (m)	ρ (m)	L_{st} (m)
1.0	4.1	13.16
2.0	8.0	12.33
3.0	11.5	11.29
4.0	14.7	10.27
5.0	17.7	9.32
6.0	20.3	8.31

Table 1: Parameters of the simulated multichannel benders where L is the bender length, ρ is the radius of curvature and L_{st} is the length of the straight guide placed after the bender and up to a position adjacent to PLATYPUS crane.

2.2. Insertion of an elliptic cold guide

There is space for insertion of another guide between the two existing cold guides. Here we consider elliptical guide geometry. Fig. 1 b) shows the sketch of the simulated elliptical guide. The entrance of the guide is at the end of the cold neutron beam tube and is elliptic in two dimensions. The minimum guide length required to access free floor space beyond PLATYPUS is 41.5 m. Various lengths have been simulated between 41.5 m and 1000 m. Although this range greatly exceeds the length of the OPAL neutron guide hall, analysis of longer lengths will show the limits of performance. In order to filter fast neutrons and gamma radiation from the reactor core two beam stops (ideal absorbers) with the same area as the guide entrance and exit are placed half way along the guide, separated by 50 cm. We have evaluated guide performance at the sample position (of area 3x3 cm) for two possible future instruments; a Small Angle Neutron Scattering (SANS) instrument and a Neutron Spin Echo (NSE) spectrometer. In these simulations the distance from guide end to sample are 3 m and 6 m, for NSE and SANS respectively.

Guide length	Focus at 3 m. (NSE)		Focus at 6 m. (SANS)	
(m)	Height (cm)	Width (cm)	Height (cm)	Width (cm)
41.5	41.09	10.28	31.69	7.92
61.5	48.50	12.13	36.52	9.13
81.5	54.93	13.74	40.80	10.20
101.5	60.68	15.18	44.68	11.17
251.5	93.14	23.30	67.03	16.76
501.5	130.38	32.61	93.04	23.26
751.5	159.13	39.80	113.24	28.31
1001.5	183.75	45.88	130.34	32.59

Table 2: Height and width at the guide centre of the simulated elliptical guides

3. Results of simulations of multi-channel benders

The dependence of total flux on bender length is plotted in Fig. 2a. We see that flux peaks at bender length of 5-6 m. or 3 m. for 16 channel or 32 channel benders respectively. We also see the strong dependence of flux on substrate thickness. Wavelength dependence of bender transmission (Fig. 2b) shows that long benders (and 32 channels) favour short wavelengths over longer wavelengths, but a short 16 channel bender is far superior for very cold neutrons. We can qualitatively explain these trends by recognising the shift from garland to zig-zag reflections with increasing wavelength.

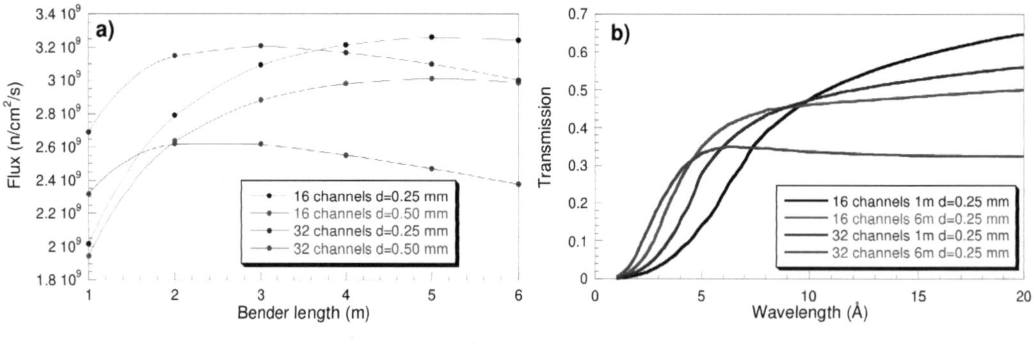

Fig. 2: a) Flux at the end of the straight guide for selected simulated benders. b) Wavelength dependence of transmission for selected benders (d=0.25 mm).

In order to benchmark potential instrument performance after the bender arrangement (Fig. 1a), we calculated flux at the sample position in either a medium resolution SANS instrument or a NSE spectrometer . Results indicate that a 32 channel bender in the length range 4 - 6 m., delivers highest usable flux at $\lambda \sim 6$ Å. Maximum calculated fluxes for a SANS instrument are 4.2 x 10^7 or 1.5 x 10^7 n/cm^2/s for 2.5 m. or 5 m. collimation length, respectively. By comparison, calculated sample flux, at $\lambda = 6$ Å, on QUOKKA (the SANS at OPAL) with 1 m. or 8m. collimation length is 1.8 x 10^8 or 2.6 x 10^7 n/cm^2/s, respectively[7]. For a NSE spectrometer, our calculated flux is 1.7 x 10^7 n/cm^2/s at $\lambda = 6$ Å, around half that of the IN11 spectrometer at ILL (~3.2 x 10^7 n/cm^2/s [8]). These comparisons indicate that bender options really are competitive for such neutron beam instruments.

4. Results of simulations of a cold elliptic guide

Comparison of the dependence of total flux on elliptic guide length with that of OPAL's existing cold guides (Fig. 3 a & b) shows radically different trends. Whereas flux delivered to sample decreases with increasing length in the existing guides, the elliptic geometry shows the opposite trend, with increasing flux up to guide lengths of order ~ 800 m. This surprising trend illustrates the main advantage of elliptic geometry; that the number of reflections of transmitted neutrons decreases markedly with increasing guide dimensions.

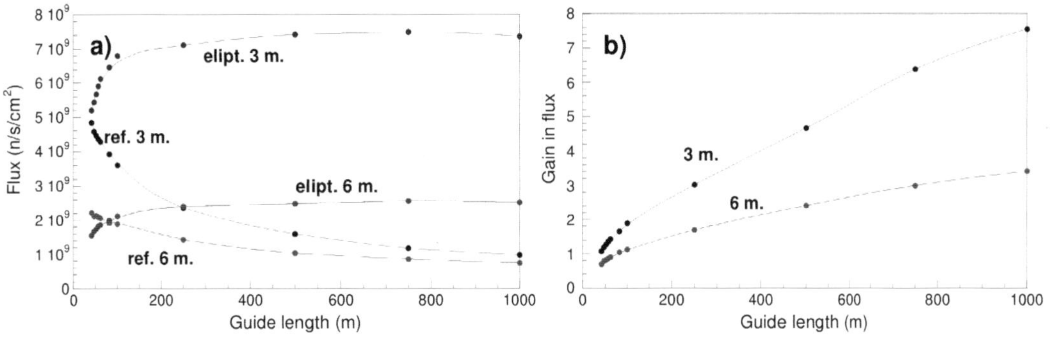

Fig. 3 a) Comparison of total flux at sample position for elliptic (elipt.) and existing (ref.) guides, and b) gain in flux of an elliptic guide over OPAL existing guides.

5. Conclusions

The neutron optical concepts considered here show promise for construction of additional competitive cold neutron instruments in the OPAL neutron guide hall. It is possible with multi-channel benders to provide high flux cold beams to instruments placed at large angles to the existing beam trajectories. Preliminary simulations of elliptical guides indicate the possibility of outstanding performance in the transport of neutrons over very long distances. A more thorough study is being made about the reliability of this kind of guides to the OPAL Neutron Guide Hall.

References

[1] http://www.ansto.gov.au/bragg/facilities/instruments
[2] S. J. Kennedy, B. A. Hunter, F. Mezei, L. Rosta, *Appl. Phys. A* **74**, S40-42 (2002)
[3] A. Schirmer and D. F. R. Mildner, *Meas. Sci. Technol.* **2**, 1059 (1991).
[4] P. Boeni, *Nucl. Inst. Meth. A* **586**, 1-8 (2008).
[5] G. Zsigmond, K. Lieutenant and F. Mezei, Neutron News **13**, 11 (2002).
[6] S. J. Kennedy and G. R. Davidson, to be published (2009)
[7] http://www.ansto.gov.au/bragg/facilities/instruments/quokka/specifications
[8] B. Farago, private communication

Spin-Echo Resolved Grazing Incidence Scattering (SERGIS) at Pulsed and CW Neutron Sources

Rana Ashkar[1], P Stonaha[1], A Washington[1], V R Shah[1], M R Fitzsimmons[2], B Maranville[3], C F Majkrzak[3], W T Lee[4], Roger Pynn[1, 4]
[1]Indiana University Cyclotron Facility, Bloomington, IN 47408, USA
[2]Los Alamos Neutron Scattering Center, Los Alamos, NM 87545, USA
[3]NIST Center for Neutron Research, Gaithersburg, MD 20899, USA
[4]Neutron Science Directorate, Oak Ridge National Lab, Oak Ridge, TN 37831, USA

e-mail: rashkar@indiana.edu

Abstract. We have used SERGIS to probe the surface structure of a silicon diffraction grating of period 140 nm. Experiments were performed at: the Los Alamos Neutron Science Center (LANSCE) pulsed neutron source and the National Institute of Standards and Technology (NIST) continuous wave (CW) reactor neutron source. Although both sets of data show peaks of the spin echo polarization at integer multiples of the grating period, as expected, the results differ in detail. We have developed a dynamical theory, based on a Bloch wave expansion, to describe neutron diffraction from a grating. The theory explains the differences between the two sets of data without any adjustable parameters.

1. Introduction

Physical constraints, such as grooves of a diffraction grating or steps on vicinal surfaces, have been recently used to direct the self-assembly of macromolecules [1, 2]. Neutron scattering is a useful technique for the characterization of the structures of such self-assembled films. Specular reflectivity [3] and a more recently developed technique, rotational SANS [4], have both been used to study the self-assembly of block copolymers (BCP) on surfaces. To study in-plane BCP ordering as a function of depth, grazing-incidence small angle scattering is needed and some experiments of this type have been performed with x-rays [5]. The SERGIS technique is being developed so that neutron scattering can be applied to such problems. As a first step in the process of developing SERGIS, we studied a bare diffraction grating of period $d = 140$ nm and groove depth $h = 65$ nm as indicated by the manufacturer and confirmed by AFM (Figure 1). Experiments on this grating were performed with the same SERGIS hardware at the LANSCE pulsed neutron source and the CW neutron source at the NIST Center for Neutron Research (NCNR). Subtle differences between the two sets of data were observed and a dynamical theory has been developed to explain the results.

2. Dynamical Theory

Neutrons scattered at grazing incidence by a periodic grating are analogous to electrons propagating in a crystal lattice in that they experience a periodic interaction potential. For neutrons the scattering potential, $V(\vec{r})$, is proportional to the scattering length density $\rho(\vec{r})$. A dynamical calculation based on a Bloch wave expansion has been used in the past [6, 7, 8] to describe the interaction of neutrons, x-rays and electrons with such periodic potentials, and we follow the same approach. The scattering geometry is such that neutrons are incident almost parallel to the lines of the grating at a grazing angle $\theta = 0.2^\circ$ to the surface (Figures 2, 3).

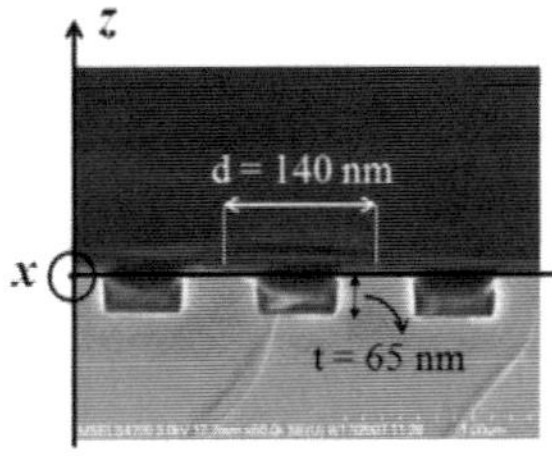

Figure 1. SEM image of the grating

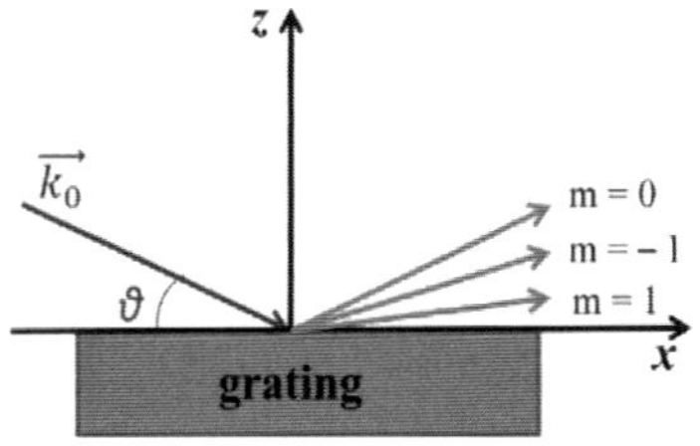

Figure 2. Side view of the scattering geometry

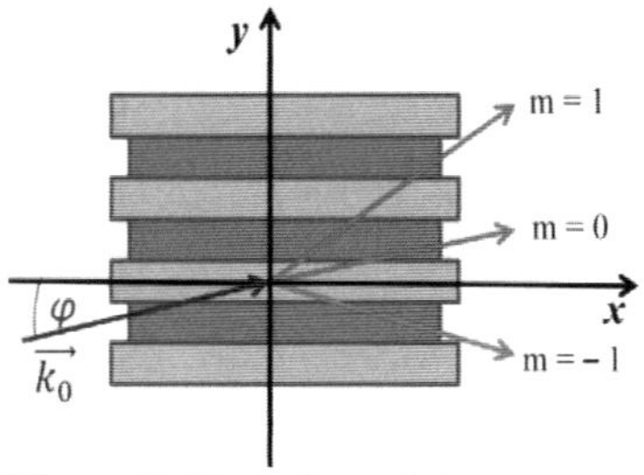

Figure 3. Top view of the scattering geometry

Neutrons incident on the grating with a wavevector $\vec{k_0}$ are described by a wavefunction, ψ, that obeys the Schrödinger equation:

$$\nabla^2\psi(\vec{r}) + \left(k_0^2 - 4\pi\rho(\vec{r})\right)\psi(\vec{r}) = 0 \qquad (1.1)$$

where $\rho(\vec{r})$ is the coherent scattering length density. We divided the grating into three layers: air with a zero scattering potential, a modulated Si-air layer with a periodic potential and a silicon substrate with a constant scattering potential. Within the modulated layer, the neutron wavefunction, ψ, can be expanded as a series of Bloch waves [6]:

$$\psi_{layer} = \sum_{n,m}[R_{1,n}e^{ik_{z,n}z} + T_{1,n}e^{-ik_{z,n}z}]b_{n,m}e^{im\left(\frac{2\pi}{d}\right)y}e^{ik_{0y}y} \qquad (1.2)$$

Similar expressions of the wavefunction are used in air and the substrate:

$$\psi_{air} = [e^{-ik_{0z}z} + \sum_n R_{0,n}e^{ik'_{z,n}z}e^{in\left(\frac{2\pi}{d}\right)y}]e^{ik_{0y}y} \;;\; \psi_{sub} = \sum_n T_{2,n}e^{-ip_{z,n}z}e^{in\left(\frac{2\pi}{d}\right)y}e^{ik_{0y}y} \qquad (1.3)$$

When these expressions are substituted in the Schrödinger equation, a secular equation is obtained [7]. The z-components of the wavevectors, $k_{z,n}$, are the square roots of the eigenvalues of this equation in the modulated layer and the coefficients $b_{n,m}$ are the elements of the associated eigenvectors. The wavevectors $k'_{z,n}$ and $p_{z,n}$ are the solutions of the secular equation in air and the substrate respectively. Note that both the y and z components of the diffracted wavevectors depend on the Bloch-wave index, m, as indicated in Figures 2 and 3. The amplitudes $R_{0,n}$, $R_{1,n}$, $T_{1,n}$ and $T_{2,n}$ of the Bloch waves in the different layers can be determined by equating the wavefunctions and their derivatives at the layer interfaces [8].

Using the T-matrix formalism, the scattering cross-section, $d\sigma/d\Omega$, can be expressed in terms of the matrix elements of the scattering potential operator between the state ψ and a plane wave [9]:

$$\frac{d\sigma}{d\Omega} \propto \left| < e^{i\vec{k}.\vec{r}} |V| \psi > \right|^2 \qquad (1.4)$$

where $\vec{k}$ is the outgoing wavevector. The normalized SERGIS spin-echo polarization, P/P_0, is given as the cosine Fourier transform of the scattering cross-section:

$$\frac{P(y_{se})}{P_0(y_{se})} = \frac{\int_{-\infty}^{\infty} \frac{d\sigma}{d\Omega}(q_y)\cos(q_y y_{se})dq_y}{\int_{-\infty}^{\infty} \frac{d\sigma}{d\Omega}(q_y)dq_y} \qquad (1.5)$$

where P_0 is the empty-beam polarization and y_{se} is the spin-echo length (Figure 4) defined as:

$$y_{se} = (1.476\times10^{14}T^{-1}m^{-2})LB\lambda^2\cot(\alpha) \qquad (1.6)$$

where B is the magnitude of the magnetic field at the interface of the triangular solenoids [10].

3. Experimental setup

The SERGIS setup consists of a neutron polarizer (placed before the upstream $\pi/2$ flipper to the left in Figure 4), a series of magnetic flippers and triangular solenoids that control the neutron spin precession, a polarization analyzer and a detector (set in reflection geometry after the downstream $\pi/2$ flipper to the right in Figure 4) that accepts all neutrons reflected from the surface of the grating. The polarizer polarizes the neutron beam in the $|z\rangle$ state. The first $\pi/2$ flipper creates an in-phase linear superposition of the $|z\rangle$ and $|-z\rangle$ states (corresponding to a classical spin in the y-direction). The triangular solenoids serve as birefringent media for neutrons, forcing the up and down spin states to follow different paths, as indicated in Figure 4 [10]. The separation between the paths of the two spin states as they leave the second triangular pair can be shown to be equal to the spin-echo length [10]. In our experiments, the grooves of the grating were aligned almost parallel to the x-axis ($-0.4° < \varphi < 0.4°$) thus allowing the neutron beam to probe the surface structure of the grating along the y-axis (Figure 4). The Larmor phase of each spin state depends on the time spent in each magnetic field region and consequently on the neutron trajectory. In the absence of a sample, the difference in the Larmor phase of the two spin eigenstates before the sample is cancelled in the second half of the instrument by the symmetry of the setup, and the initial neutron beam polarization is recovered as a spin echo. With a sample in place, the neutron is scattered to a different trajectory through the second half of the setup and a net Larmor phase difference between the two spin states is acquired, potentially changing the polarization of the outgoing beam. The normalized spin echo polarization is a weighted average of the cosine of the net Larmor phase difference over all detected momentum transfer i.e. the cosine Fourier transform of the scattering cross section along the y-direction, as expressed in equation (1.5).

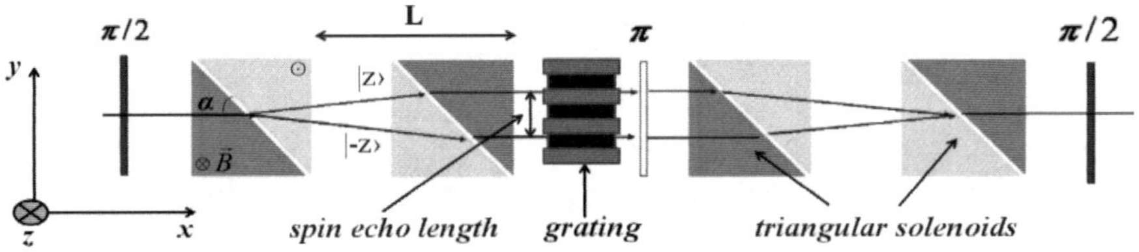

Figure 4. Top view of the experimental setup in the same configuration as Figure 3. Different shades of the triangles represent regions of opposite magnetic fields along the z-direction

4. Measurements

4.1. Data from a fixed wavelength source

Our experiment was performed at the AND/R beamline at NCNR with a beam of wavelength $\lambda = 0.5$ nm and a beam divergence in φ of 0.2° FWHM (Figure 3). The spin-echo length was scanned by varying the current in the triangular solenoids, providing a variable magnetic field strength at the interface between triangular solenoids. At 0.5 nm neutron wavelength and a current between 1A and 10A, the spin echo length ranged from ~23 nm to ~230 nm. The measured echo polarization shows peaks at integer multiples of the grating period (Figure 5).

4.2. Data from a pulsed source

At the ASTERIX beamline at LANSCE, the spin echo length varied as a function of the variable wavelength of the beam (0.4 nm $< \lambda <$ 0.9 nm) at a fixed current in the solenoids. With a current of 12A we obtained a range of the spin echo length between ~90 nm and ~450 nm. Again, experimental results for the echo polarization (with a beam of 0.1° FWHM divergence in φ) show peaks at integer multiples of d (Figure 6).

4.3. Analytical calculations

Calculations using the dynamical formalism introduced above showed that the spin echo polarization is very sensitive to the sample alignment (i.e. the angle φ) and the beam divergence. The first order Bloch expansion (which is a three-beam approximation) gives a good account of the results we obtained at both CW and pulsed sources as shown in Figures 5, 6. There are no adjustable parameters in the theory: all quantities (such as beam divergence, neutron wavelength, magnetic field in the triangular solenoids, etc) were set to their experimental values.

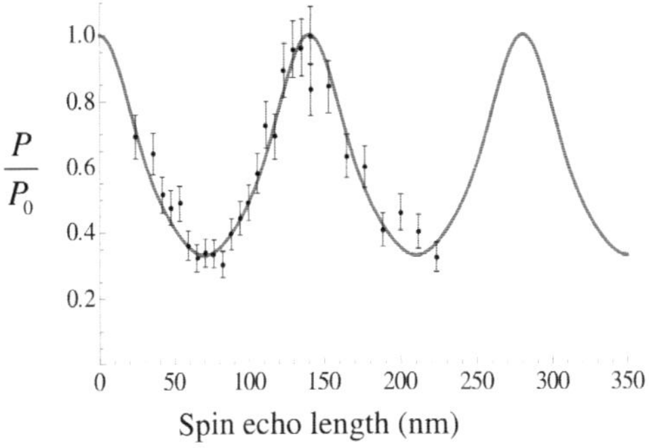

Figure 5. Plot of P/P_0 at NCNR

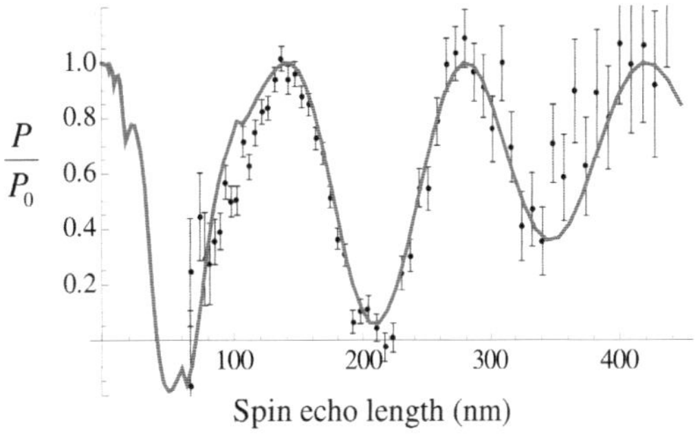

Figure 6. Plot of P/P_0 at LANSCE

4.4. Interpretation of results

Within a simple approximation such as the Distorted-Wave Born Approximation, the neutron scattering cross-section is proportional to the Fourier transform of the height-height correlation function of the reflecting surface, albeit modified by strongly varying optical factors [9]. Thus, we might expect the echo polarization to have a shape similar to the height-height correlation function of our grating. Since the grating profile is rectangular, the height autocorrelation function is a periodic array of isosceles triangles. Indeed, such a form with triangular peaks at integer multiples of the grating period is similar to the measured echo polarization although there are quantitative deviations from this simple form. For example, the amplitude of the oscillations is different for data at NIST and LANSCE and the second minimum in Figure 6 is not as deep as the first. The dynamical theory shows that P/P_0 measured in reflection from a periodic diffraction grating at a constant wavelength source such as NIST are expected to be periodic and the amplitude of the oscillations is determined by the beam divergence in φ. On Asterix, the fact that different wavelengths are used to access different spin echo lengths leads to a spin echo signal that is not periodic in y_{se}, i.e. the dips change in amplitude. At large values of y_{se} on Asterix, the spin echo polarization eventually approaches unity because the long wavelength neutrons used can no longer be Bragg reflected by the grating periodicity so only specular reflection is present. In addition, neutrons with large wavelengths cannot penetrate into the grooves of the grating.

5. Acknowledgements

This work was supported by Department of Energy through its Office of Basic Energy Sciences, Division of Material Science and Engineering (Grant No. ER46279). The NIST Center for Neutron Research, where one of the experiments described here was performed, is funded by the U.S. Department of Commerce. The Los Alamos Neutron Science Center, where the other experiment was carried out, is operated by the U.S. Department of Energy. The authors thank the staff of both facilities for their assistance.

References

[1] Kim S, Solak H, Stoykovich M , Ferrier N, de Pablo J and Nealey P *2003 Nature* **424** 411-14
[2] Park S, Lee D, Xu J, Kim B, Hong S, Jeong U, Xu T and Russell T P *2009 Science* **323** 1030-33
[3] Anastasiadis S H, Russell T P, Satija S K and Majkrzak C F *1989 Phys Rev Lett* **62** 1852-55
[4] Zhang X, Berry B, Yager K, Kim S, Jones R, Satija S, Pickel D, Douglas J and Karim A *2008* ACS Nano **2** 2331-41

[5] Rockford T, Mochrie S G J and Russell T P *2001 Macromolecules* **34** 1487-92
[6] Berrouane H, André J M, Barchewitz R, Moreno T, Sammar A, Khan Malek C, Pardo B and Rivoira R *1992 Nucl Inst and Meth in Phys Res* A **312** 521-530
[7] Humphreys C J *1979 Rep Prog Phys* **42** 1825-87
[8] Ott F, Humbert P, Fermon C and Menelle A *2001 Physica* B **297** 189-93
[9] Sinha S K, Sirota E B, Garoff S and Stanley H B *1988 Phys Rev* B **38** 2297-311
[10] Pynn Roger, Fitzsimmons M, Lee W T, Stonaha P, Shah V, Washington A, Kirby B, Majkrzak and Maranville B *2009 Physica* B **404** 2582-84

CASCADE with NRSE: Fast Intensity Modulation Techniques used in Quasielastic Neutron Scattering

C.J. Schmidt[1], F. Groitl[2], M. Klein[2], U. Schmidt[2], W. Häussler[3,4]

[1] GSI Detector Laboratory, Darmstadt, Germany
[2] Physikalisches Institut, Universität Heidelberg, Germany
[3] Forschungsneutronenquelle Heinz Maier-Leibnitz (FRM II), Technische Universität München, D-85747 Garching, Germany
[4] Physik-Department E21, Technische Universität München, D-85747 Garching, Germany

c.j.schmidt@gsi.de

Abstract. Fast intensity modulation techniques require neutron detectors with a response time below 1μs, a demand which can not be fulfilled by standard neutron detector technology. Here we present the CASCADE neutron detector developed at the University of Heidelberg, which is based on stacked, thin ^{10}B layers as neutron converters and particularly addresses these needs. After explaining the CASCADE concept in context with fast intensity modulation techniques we report on two test runs performed at the RESEDA spectrometer at FRM II, Munich, which indeed prove MHz time resolution in this spectrometer system with a reasonably high detection efficiency.

1. Introduction

Neutron **S**pin **E**cho [1] (NSE) is a well established high-resolution inelastic neutron scattering technique. In contrast to other inelastic neutron scattering techniques, which measure the scattering function as a function of energy transfer between sample and scattered neutrons, NSE measures the intermediate scattering function as a function of Fourier time. Increasing resolution, i.e. increasing the Fourier time does not decrease the intensity of the incoming neutron beam, because the achievable Fourier time does not depend to first order upon the energy spread of the incoming neutron beam, an intriguing feature of the NSE technique.

All NSE methods base on polarized neutrons and the polarization of the scattered neutrons is the quantity of interest, rather than intensity. Beside the classical NSE technique, which relies upon the Lamor precession the neutron spin undergoes, while flying through static magnetic fields, several advanced schemes exist, which employ resonant spin flips in a magnetic rf-field [2]. Here we focus on one special technique out of several existing **N**eutron **R**esonance **S**pin **E**cho (NRSE) schemes. This technique is called **M**odulation of **I**nt**E**nsity by **Z**ero **E**ffort (MIEZE) [3, 6, 7]. Within the MIEZE schemes, the rotating polarization of the neutron beam is transformed to a fast modulation of intensity by the help of a resonant spin flipper in combination with a polarizer. Thus, in contrast to all other spin echo methods, the change of intensity on time scales from 100μs down to below 1μs needs to be recorded, a demanding task for a neutron detector. While standard ^{3}He-Counters are by far too slow for this purpose, other detection concepts, like the CASCADE scheme developed at the Physikalische

Institut at Heidelberg [4], are well suited for this task. Before we present the CASCADE scheme and its adaptation to MIEZE measurements, we have to discuss the MIEZE principle and the implication for a detector based on it.

2. The MIEZE principle

The MIEZE scheme exists in two variants I and II. For our test experiments we used RESEDA at FRM II, Munich, which is a spectrometer in a standard NRSE configuration. Because modifying a NRSE setup to become a MIEZE II setup requires much less effort compared to becoming a MIEZE I setup, our preferred setup schematically shown in figure 1 was a MIEZE II setup. In addition switching between NRSE and MIEZE II mode allows us to separately determine the performance of the detector and the performance of the whole spectrometer.

The upper part of figure 1 shows the location and polarity of the resonance spin flip coils, while the lower part shows the kinetic energy levels of partial spin-up and spin-down wave functions assumed to form a coherent mixture. A detailed discussion of this full quantum mechanical plane wave approach to NRSE techniques of R. Golub et al. is given in [3].

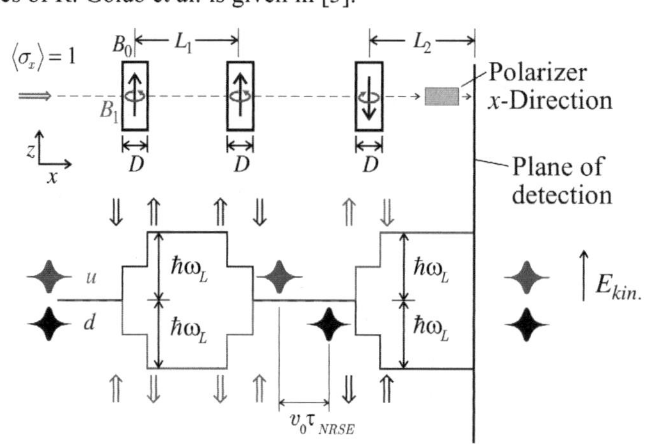

Figure 1. Scheme of the MIEZE II setup.

Neutrons, polarized with respect to the x-direction enter the setup from the left side. In the frame of the static magnetic field B_0, pointing along the z-direction, a neutron is seen as a coherent mixture of a partial spin up and a partial spin down wave packet. While entering the static B_0-field of the first spin flip coil, the up component is decelerated due to the field gradient, in contrast to the down component, which is accelerated. This effect is known as the longitudinal Stern-Gerlach effect. After a resonance spin flip, the same happens while leaving the first spin flip coil. Therefore the two partial wave packets spin-up and spin-down leave the coil at slightly different kinetic energy, drifting away form each other during the flight through the zero field region between the spin flip coils. At the end of the first spectrometer arm a second spin flip coil undoes the kinetic energy change, but leaves the two partial wave packets spatially separated. Therefore these wave packets arrive at different times at the sample region and thus probe the sample at different times. This time difference is precisely the Fourier time τ_{NRSE} . In a standard NRSE setup, the two spin flip coils of the second spectrometer arm serve to undo the spatial separation of the partial wave packets and brings them back to overlap, resulting in the same coherent mixture of the partial spin-up and spin-down wave packet as at the entrance of the first spectrometer arm, i.e. the neutrons are polarized with respect to the x-direction. The polarization measures the time correlation of the sample at the different scattering times. In a MIEZE II setup the second spin flip coil of the second spectrometer arm is replaced by a neutron detector with a polarizer in front of it. At the location of the detector, the spatial separation of the two partial wave packets, which can pass the second polarizer, is undone, but the kinetic energy is still different. So interference between the two partial wave packets surviving the second polarizer leads to an intensity distribution, fast modulated in time with the frequency $f_{MIEZE} = 2f_{rf}$ (f_{rf}: frequency of the rf in the second spectrometer arm) and in space. The wave length of the spatial variation of the intensity along the beam for a given time is to first order approximation given by $\lambda = v_0/f_{MIEZE}$, where v_0 is the mean velocity of the neutron beam. Therefore the spatial extension of the active detection material along the beam has to be small compared to this wave length λ, in order to avoid damping of the measured modulation through averaging. This is an important issue for the selection of a suitable detection mechanism for high resolution MIEZE applications. As the values of MIEZE frequencies reaches the MHz region, the value of the corresponding length λ shrinks below 1mm for cold neutrons, requiring the thickness of the neutron converter inside a detector to be below 100μm.

3. The CASCADE detector

In general, solid converting materials have the advantage that high absorption can be realized over very short distances. Moreover, the thin conversion layers intrinsically provide the potential for sub microsecond time resolution, being necessary for neutron intensity modulation techniques. The CASCADE detector combines and exploits the advantages of gas detectors and the well-defined locus of conversion within solid neutron converting materials. In our case we use isotopically pure ^{10}B with the dominant absorption reaction: $n + {}^{10}B \rightarrow \alpha + {}^{7}Li + 2.3MeV$. For a single ^{10}B layer however, the maximum detection efficiency for thermal neutrons is only 5%, caused by the maximum penetration depth of 3.5μm of the α and ^{7}Li in boron. Therefore adding more Boron layers is imperative to reach a higher efficiency. For a position sensitive area detector, however, such additional layers must be transparent for charge while opaque for neutrons. We resolved this apparent contradiction by employing gas electron multiplier (GEM) foils invented and developed by Fabio Sauli at CERN in 1997. GEMs are micro-structured, flexible foils of a 50μm thick Kapton substrate, sandwiched between 5μm thick copper claddings on either side. The foil is structured with a regular hexagonal grid of 50μm diameter holes at a spacing of 140μm. Depending upon the potentials applied to the two copper claddings on top and bottom, every hole serves as an electrostatic lens, which allows to guide and image charge from one side of the foil through the holes to the other side without loss of in-plane position information. With larger potential differences, the field strength in the holes can be made strong enough to cause charge amplification.

In our test experiments at RESEDA, the CASCADE-**M**IEZE detector was equipped with these GEM-foils as charge transparent substrates partly carrying boron layers. The GEMs respective the boron layers are stacked one behind the other to cumulate the detection efficiency of every layer of

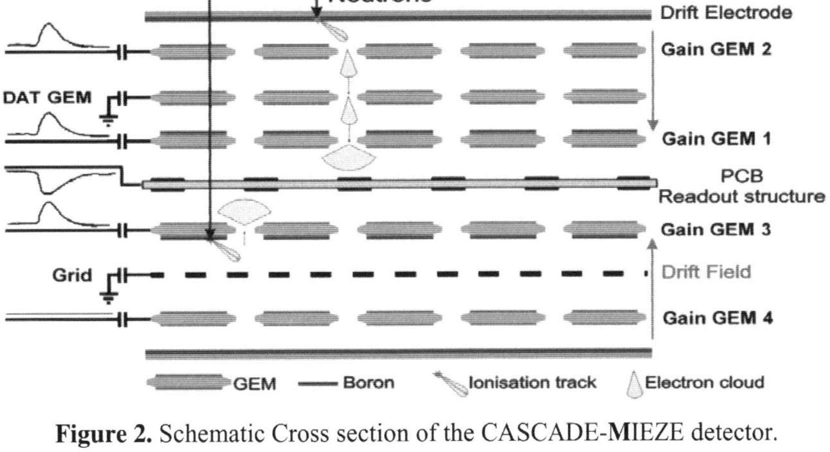

Figure 2. Schematic Cross section of the CASCADE-**M**IEZE detector.

1μm boron up to 20% for 5Å at RESEDA. The lateral position resolution is 2.5mm resulting from the ambient pressure of the counting gas. The charges generated by the fragmentation products in the gas spacing of 2mm between successive GEMs are channelled through the GEMs to the readout structure (see figure 2). The overall signal height is set to a convenient level by the potential difference on the Gain-GEMs. The two-dimensional signal readout is realized through a charge-sharing scheme. A cloud of charges when drifting onto the readout structure is collected by 128 x 128 electrode stripes corresponding to both dimensions x and y. Thus, simultaneous registration of signals on x- and y-readout-channels reveal the detection of a neutron at the corresponding (x, y) coordinates.

For MIEZE experiments it is essential to measure the fast oscillating neutron intensity. This requires a high time of flight resolution better than 1μs in case of several 100kHz MIEZE frequency. In addition, the neutron intensity is not only modulated in time but also in space along the propagation direction of the neutron beam. Therefore the thickness of the neutron converter has to be small compared to the longitudinal modulation period in order to avoid contrast damping due to averaging. For example in case of cold neutrons (λ = 5Å, v ≈ 800m/s) the modulation period is only 1.2mm in case of a MIEZE frequency of 654kHz. As a consequence for the CASCADE detector it is necessary to determine the particular boron layer in which the neutron was absorbed. For this purpose the detector is equipped with an additional readout channels sensing the mirror charge induced on each GEM when the cloud of charges passes through its holes. Because of the large capacity of more than 20nF between the copper claddings of a GEM, it is impractical to read the differential signal of one GEM. But also reading out the common mode signal between two GEMs was not successful because of strong AC-coupling. As a solution we introduced an AC-grounded **D**ecoupling-**At**tenuation-GEM (DAT-GEM) in the detector stack to decouple the different GEMs and their boron layers (see fig. 2). As an alternative concept an AC-grounded grid with 28% optical transmission was successfully tested, where we could show that the passage of the electric field lines through the grid does not influence the decoupling properties. The CASCADE detector system is conceived as a standalone, position sensitive neutron detection device with high time of flight resolution. The sensitive area of 200mm x 200mm is read out on 128 independent channels for each coordinate, resulting in an overall image of 2^7 x 2^7 = 16384 points. The entire electronic readout is directly mounted onto the rear side of the device. To take usage of the high rates detector front, the bandwidth of readout electronics immediately comes into focus. In response to this problem, we decided to pioneer the transfer of highly integrated ASIC-based readout technology from high-energy physics into the field of neutron instrumentation. For the CASCADE detector, the CIPix ASIC, originally developed by the Heidelberg ASIC lab in 2000 for the DESY H1 experiment, is employed as readout device. A single chip provides 64 independent analogue channels comprising pre-amp, shaper and programmable discriminator. Each channel can accept statistical data at a rate of 330kHz at 10% dead time, resulting in an immediate specification for the individual pixel rate capability of the same size, no matter how the pixel is read out. Two-dimensional readout on 128 x 128 pixels is realized through a correlation in time of signals from x- and y-coordinates, using four CIPix chips. The overall detector readout bandwidth is in this case 2MHz at 10% dead time.

Four CIPix chips produce a data output of 3,2GB/sec. Such a challenging data load can only be handled in real time by massive parallel processing. This can be managed nowadays with a one-chip solution using a powerful **F**ield-**P**rogrammable-**G**ate-**A**rray. This FPGA is located in our case directly on the rear side of the CASCADE detector. It collects all the data from all CIPix chips simultaneously, manages the onboard SRAM (8MB) and communicates over an optical link to any PC. As shown in figure 3, inside the FPGA a real time event reconstruction algorithm evaluates the incoming data from the readout structure together with the time stamp information t to one neutron event (x, y, t). The time stamp is derived from a phase-locked-loop (PLL) circuit controlled by RESEDA's MIEZE frequency f_{MIEZE}. The MIEZE frequency f_{MIEZE} triggers inside the FPGA the internal sweep refresh similar to a chopper pulse in TOF. With the 16-times higher frequency $16\times f_{MIEZE}$ of the PLL the internal time bin counter of the FPGA is triggered. Thus the entire detector electronic runs phase synchronous to the

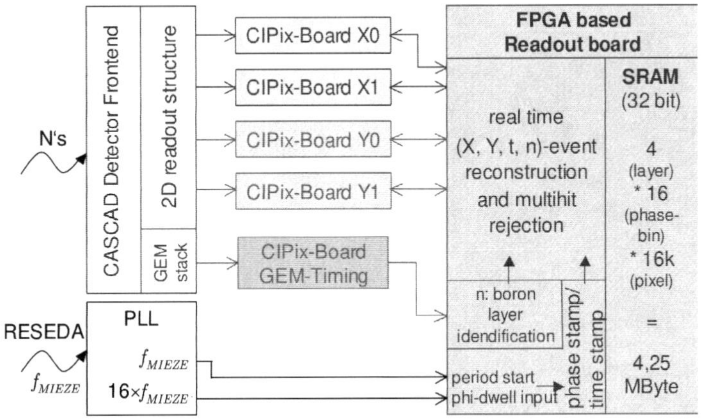

Figure 3. Schematic of data flow and processing of the CASCADE-MIEZE detector.

RESEDA rf-frequency and therefore synchronous to the oscillating neutron intensity. A fifth CIPix chip is used for coincident readout of all GEMs of the stack. These data are evaluated in parallel to identify, in which boron layer n the absorption of the neutron took place. So finally a complete TOF-spectra for each boron layer and each pixel of the readout structure is generated and stored in the SRAM. Communication with the detector module is realized through a fiber optical link, which decouples the system galvanically. It provides a high bandwidth data link (1GBit/sec), which serves to download the histograms of measurements, the module has acquired.

4. Experimental results

Two test runs were performed at FRM II, Munich using the RESEDA spectrometer. The mayor change was to replace the ^{3}He-detector of RESEDA by the CASACDE detector. Depending on the use of the fourth spin flip coil, we could operate in a standard NRSE or. alternatively, in a MIEZE configuration.

In the first test run we used a CASCADE setup, with only one Boron layer. Thus we could verify the proper functioning of the complete setup beside the readout of the GEM stack and the boron layer identification. Figure 4 shows the results obtained at the highest MIEZE frequency, we could obtain at RESEDA. Because the analyzer restricts the neutron beam to a 32mm × 32mm window, only 20 × 20 pixels of the CASCADE detector could be used. Figure 4.b shows the histograms for two pixels. By fitting a sin-function to the phase histogram of each pixel, we could extract the contrast and phase of the intensity modulation for each pixel from the data, shown in figure 4.a and 4.c. The contrast of a MIEZE measurement corresponds exactly to the polarization of a NRSE measurement and can be directly compared. We used this comparison to extract an upper limit of 2% for contrast losses of our

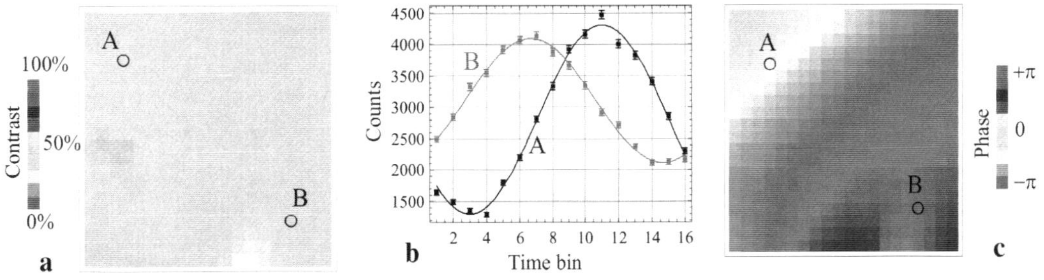

Figure 4. Results form Data taken at MIEZE frequency 645kHz (only one boron layer).
a: Contrast, **b**: Intensity modulation of Pixel A and B (marked in part **a** and **c**), **c**: Phase.

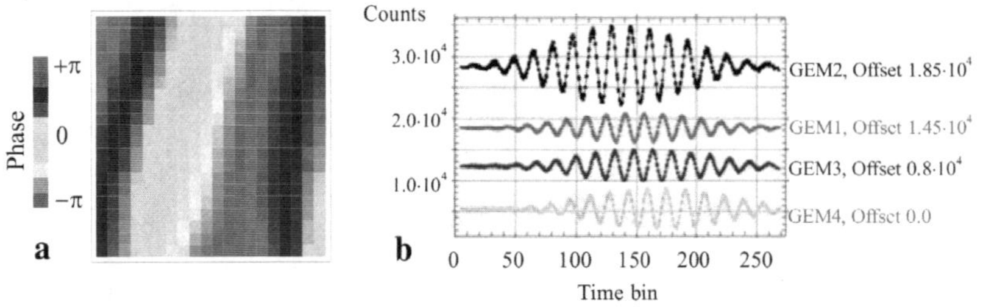

Figure 5 **a:** strong phase shifts caused by a slight turn of flipper coil (only one boron layer)
b: measurement of full spin echo group for each of the 4 boron layer.

MIEZE setup due to imperfections of the CASCADE detector. The major part of losses of polarization respectively contrast is caused by magnetic stray fields inside the spectrometer. There are reports on other experiments testing the feasibility of MIEZE using scintillation detectors [6, 7]. Unfortunately the information provided by these reports is not sufficient to distinguish between contras losses due to the detection system and losses caused by other components.

Figure 5.a shows the phase pattern, while we turned the third spin flip coil by an amount of 3.1°. The purpose for these measurements was to demonstrate the capability of a position sensitive detector to measure the contrast in case of strong phase shift across the beam i.e. strongly distorted phase fronts. From this measurement we could extract a mean contrast over the total beam 36.9%. Without a position sensitive detector the contrast would amount to as little as 0.4%.

In the second test run the CASCADE detector was equipped with 4 boron layers as shown in the scheme figure 2. The Measurement of the complete spin echo group averaged over the full beam cross section for each boron layer is shown in figure 5.b. One measurement with the MIEZE setup results just in one Period of the spin echo group. By carefully tuning the first spectrometer arm, we managed to shift the spin echo group period by period through the CASCADE detector, thus by combining all the data we obtained an image of the complete spin echo group shown in figure 5.b. These measurements were performed at a MIEZE frequency of 72kHz. From detailed analysis of the data we could extract an upper limit of 1% for false identification of the converting boron layer, which would result in a loss of contrast.

In addition to results shown in this paper, we compared data taken with the NRSE and the MIEZE setup. We used a standard sample for elastic and one for quasi elastic scattering. The measured polarization, respectively contrast, agrees within a few percent.

5. Conclusion

Our results show clearly that a MIEZE setup with a CASCADE detector has a competitive performance with respect to standard NRSE techniques. The contrast of a intensity modulated signal up to a modulation frequency of 645kHz, was measured, with less than 2% contrast loss attributed to imperfections of the CASCADE detector. A detection efficiency of 20% for the neutrons of 5.3 Angstrom wavelength was observed and is consistent with the particular choice of boron layers and may be optimized.

References

[1] F. Mezei, Z. Phys. **255** (1972) 146, F. Mezei (ed.), Neutron Spin Echo (Springer, Heidelberg, 1980)
[2] R. Gähler et al. Z. Phys. B **65** (1987) 269, R. Golub et al. Phys. Lett. A **123** (1987) 43
[3] R. Golub et al. Am. J. Phys. **62** (9) (1994) 779
[4] C. Schmidt et al, neutron News **17** (2006) 12-15
[5] W. Häussler et al. Physica B **397** (2007) 112
[6] R. Gähler et al. Physica B **180** & **181** (1992) 899
[7] M. Bleuel et al. Nucl Instr. and Meth. A 600 (2009) 220

International Conference on Neutron Scattering 2009 IOP Publishing
Journal of Physics: Conference Series **251** (2010) 012068 doi:10.1088/1742-6596/251/1/012068

Comparison of Polarizers for Neutron Radiography

M. Schulz[1,2], P. Böni[1], C. Franz[1], A. Neubauer[1], E. Calzada[2], M. Mühlbauer[1,2], B. Schillinger[1,2], C. Pfleiderer[1], A. Hilger[3], N. Kardjilov[3]

[1]Technische Universität München, Physik Department, E21, Garching, Germany
[2]Technische Universität München, Forschungsneutronenquelle Heinz Maier Leibnitz (FRM II), Garching, Germany
[3]Helmholtz Zentrum Berlin, Berlin, Germany

E-mail: michael.schulz@frm2.tum.de

Abstract. Imaging with polarized neutrons is a new method to measure the spatially resolved magnetisation. It is based on the interaction of a polarized neutron beam with magnetic fields. In this paper we present an overview of different polarizing techniques and their application for neutron imaging. The performance of various setups is discussed. It is demonstrated that radiography with polarized neutrons may evolve into an important technique for determining the homogeneity of ferromagnetic samples.

1. Introduction

In conventional neutron radiography the attenuation of the intensity of a neutron beam transmitted through a sample is measured. By means of longitudinal polarization analysis the contrast obtained in neutron imaging for ferromagnetic samples may be varied by making use of the interaction between the nuclear magnetic moment μ of the neutron and a magnetic field inside and/or around the sample. In particular, the depolarization of the neutron beam may be used to map out variations of the ferromagnetic properties and their dependence on external parameters such as stress, pressure, magnetic field or temperature as well as internal characteristics such as chemical composition, defects and stress. In fact, for sufficient resolution it may even become possible to determine the domain distribution inside samples [1].

Imaging with polarized neutrons is a new field of great current interest. Various techniques have been developed such as phase contrast imaging as well as the depolarization of a neutron beam on transmission through ferromagnetic samples [2–4]. In a series of exploratory first studies, conventional polarizing equipment for neutron scattering was used. In the following we compare various concepts for imaging with polarized neutrons paying special attention to the two major technical requirements: (i) how to maintain high spatial resolution, and (ii) to how to illuminate the sample homogeneously while maintaining a high beam polarization.

2. Experimental Technique

Neutron radiography is traditionally performed with a pinhole setup, where the resolution is determined by a beam collimation given as the ratio L/D of the distance L between pinhole and sample and the diameter D of a circular pinhole [5]. Typical L/D-ratios range from 300 to 1000 [6–8], where the collimation improves as the ratio increases. Together with the distance

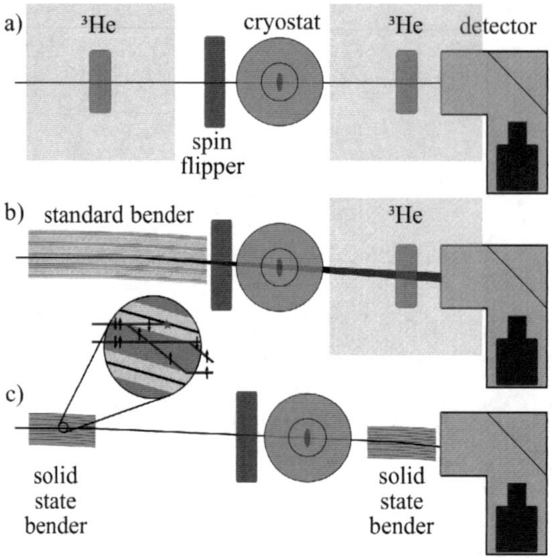

Figure 1. Various setups for neutron imaging with polarized neutrons. The setups (a) and (b) involving ^{3}He filters and setup (c) using solid state benders have been used at the ANTARES (FRM II) and CONRAD (HZB) beam lines, respectively.

d_{SD} between sample and detector (SDD) the ratio L/D determines the maximum geometrical resolution possible, given as $r = d_{SD}/(L/D)$. Clearly, as a prerequisite for radiography with polarized neutrons the equipment used must not degrade the L/D-ratio or d_{SD}. We have therefore tested various combinations of conventional polarizers and analyzers as shown in Fig. 1 to determine the ideal conditions, without special efforts to adapt these techniques to the special requirements of the pinhole setup used in radiography. In the following we compare the properties of these setups for neutron imaging as tested on various topical materials.

The basic setup used for all measurements consisted of a neutron polarizer, followed by a spin flipper, a polarization analyzer, and a position sensitive CCD detector that records the image of a neutron sensitive scintillator, which is placed behind the analyzer. The sample was attached to a GF-coldhead with a base temperature of 5 K. The depolarization of the transmitted neutrons was measured for various inhomogeneous, ferromagnetic samples. The measurements using a ^{3}He spin filter were carried out at the ANTARES beam line at FRM II, Munich, and those with solid state benders were made at the CONRAD beam line at HZB, Berlin.

3. Results

Fig. 2 shows typical data of the spatially resolved beam polarization for different samples as measured with the setups depicted in Fig. 1. White areas indicate areas of high polarization, i.e., parts of the sample with unchanged polarization. Black areas designate maximum depolarization. It is immediately evident that the depolarization of the beam for the samples studied is highly inhomogeneous.

Using ^{3}He as a polarizer has the advantage of leaving the beam collimation as well as the beam homogeneity unaffected, making it ideally suited for neutron imaging. However, the bulky magnetostatic box needed to create a homogeneous field surrounding the ^{3}He cells leads to a large d_{SD}. Therefore, the spatial resolution is seriously reduced. To improve the setup, the non-magnetic scintillator can be placed inside the box directly after the ^{3}He cell. This decreases

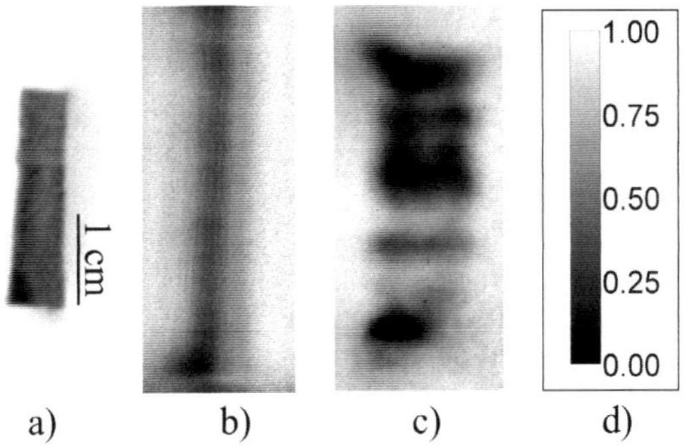

Figure 2. a) – c) Images of the beam polarization after transmission of ferromagnetic samples taken with the setups (a) through (c) shown in Fig. 1. The differences in image quality for the different methods are clearly visible. The smaples are: (a) a Fe alloy, (b) a Ni alloy and (c) a Pd alloy; (d) intensity code for the polarization shown in the images.

d_{SD} in our case to $\approx 30\,\mathrm{cm}$.

At FRM II, the ^3He gas is polarized at low pressure using the optical pumping facility HELIOS [9, 10] and filled into pyrex cells. These cells are transported to the beam line and installed in the magnetostatic boxes. Because of inhomogeneities of the magnetic field and collision with the wall the polarization of the ^3He gas relaxes, leading to a decrease of the polarization of the neutrons. Therefore, a regular exchange of the ^3He cells is required. Using ^3He as a polarizer and analyzer (see Fig. 1a) leads to a complicated time dependence of the polarization, which can in principle be corrected. Moreover, the transmission of the cells is also time-dependent. Currently, at the beam line ANTARES at FRM II the gas has to be exchanged every 24 hours. For typical applications one generally maximizes the term P^2T, which yields a beam polarization P of up to 80% and a transmission T of $\approx 50\%$ for 3.2 Å neutrons.

For most recent experiments, a RF spin flipper was integrated into the magnetostatic box that allowed to select the polarization of the cell thus allowing to use a broad wavelength band of neutrons, yielding a high neutron flux. For the polychromatic measurements, a Be-filter was inserted to remove neutrons with a wavelength $\lambda < 4\,\text{Å}$ [11]. This way, the polarization of the neutrons could be improved, i.e. for a ^3He polarization of 70%, a flipping ratio of 7 was obtained. The image shown in Fig. 2(a) has been obtained using this setup yielding a resolution of $\approx 0.4\,\mathrm{mm}$. Of course, for a quantitative determination of the field integral as described in [2, 3], a polychromatic beam cannot be used. By reducing the size of the magnetostatic box, the resolution could be improved further.

For the setup using a standard bender[12] as standard polarizer as shown Fig. 1(b) a poor spatial resolution is observed in the reflection plane, because the bender increases the beam divergence. Moreover, the beam becomes inhomogeneous. These effects show up due to the curved channels. Hence, the images are smeared in the horizontal direction. With $d_{SD} \approx 30\,\mathrm{cm}$, which was required by the ^3He analyzer, no satisfactory images could be obtained (see Fig. 2(b), i.e., the sample was completely smeared out in the horizontal direction. Even with a shorter d_{SD}, the image would still be severely blurred. A more favorable position for the bender would be to place it closely behind or before the pinhole. Such a setup was not feasible at the ANTARES beam line due to the geometrical constraints mentioned above. The divergence of the beam will still be increased but the bender acts as a pinhole source itself, thereby defining the collimation of the beam.

The length of a bender can be significantly reduced by using thin Si wafers coated with a polarizing supermirror. The neutrons are guided through the Si. A setup using two solid state benders was investigated at the radiography facility CONRAD at HZB, Berlin (see Fig. 1c). The performance of the polarizers has been evaluated as described in [13]. A high beam polarization of $\approx 95\%$ was achieved, however, the beam profile was inhomogeneous exhibiting a pronounced stripe structure caused by the wafers of the benders. Because these stripes are fixed for fixed detector and polarizer position, they cancel when the polarization is evaluated from the raw images. For the benders used in this experiment, the wafer thickness was $250\,\mu$m and the length was $\approx$ 80 mm, yielding a resolution of $\approx$ 0.5 mm [2] (see Fig. 2(c). The resolution can be improved by using thinner Si wafers, which would also allow to build a shorter polarizer, i.e. d_{SD} can be decreased too. Of course, the optimum position of a solid state polarizer would be in front of the pinhole of the beam line.

4. Conclusions and Outlook

The results of our experiments demonstrate that the combination of two ^{3}He cells, where the scintillator is placed directly behind the analyzer cell inside the magnetostatic box (see Fig. 1a) yields a resolution of ≈ 0.4 mm, which is the highest resolution of the setups tested.

Our results indicate that the highest resolution may be obtained if a ^{3}He spin filter as a polarizer is combined with an optimized solid state bender as an analyzer, if the setup has to be installed close to the sample area of an existing beam line for radiography. In this case the polarizer would not affect the beam collimation and homogeneity at all and also the size of the magnetostatic box would not be crucial for this application. A solid state bender analyzer could be manufactured with a Si wafer thickness of $100\,\mu$m leading to an overall length of the analyzer of only ≈ 5 cm. At a collimation of $L/D = 800$ as available at FRM II, Munich this would result in a geometrical blurring of $< 100\,\mu$m in the vertical direction and $100\,\mu$m in the horizontal direction. This would be a great step towards high resolution imaging with polarized neutrons.

If a dedicated beam line for polarized neutron imaging were to be built, the polarizer should be installed before or shortly after the pinhole. This setup would relax the restriction for the polarizer to conserve the beam collimation and thus would largely increase the choice of polarizing devices. Moreover, the background would be reduced significantly.

Acknowledgments

Financial support through DFG Forschergruppe FOR 960 on quantum phase transitions is gratefully acknowledged. We also wish to thank the staff of FRM II and HZB for their support.

References

[1] Badurek G, Hochhold M and Leeb H 1997 *Physica B: Physics of Condensed Matter* **234** 1171–1173
[2] Kardjilov N, Manke I, Strobl M, Hilger A, Treimer W, Meissner M, Krist T and Banhart J 2008 *Nature Physics* **4** 399
[3] Piegsa F, van den Brandt B, Hautle P, Kohlbrecher J and Konter J 2009 *Physical Review Letters* **102** 145501
[4] Schulz M 2009 *Nuclear Inst. and Methods in Physics Research, A* **in press**
[5] Schillinger B 1996 *Journal of Neutron Research* **4** 57–63
[6] Grünauer F, Schillinger B and Steichele E 2004 *Applied Radiation and Isotopes* **61** 479–485
[7] Hilger A, Kardjilov N, Strobl M, Treimer W and Banhart J 2006 *Physica B: Physics of Condensed Matter* **385** 1213–1215
[8] Materna T, Baechler S, Jolie J, Masschaele B, Dierick M and Kardjilov N 2004 *Nuclear Inst. and Methods in Physics Research, A* **525** 69–73
[9] Hutanu V, Masalovich S, Meven M, Lykhvar O, Borchert G and Heger G 2007 *Neutron News* **18** 14 – 16
[10] Andersen K, Chung R, Guillard V, Humblot H, Jullien D, Lelièvre-Berna E, Petoukhov A and Tasset F 2005 *Physica B: Physics of Condensed Matter* **356** 103–108
[11] Lorenz K 2007 *Proceedings of the 8th World Conference on Neutron Radiography* **in press**
[12] Schaerpf O 1989 *Physica B: Condensed Matter* **156-157** 639 – 646 ISSN 0921-4526
[13] Krist T, Kennedy S, Hicks T and Mezei F 1997 *Physica B: Physics of Condensed Matter* **241** 82–85

Transmission Bragg edge spectroscopy measurements at ORNL Spallation Neutron Source

A S Tremsin[1], J B McPhate[1], J V Vallerga[1], O H W Siegmund[1], W B Feller[2], H Z Bilheux[3], J J Molaison[3], C A Tulk[3], L Crow[3], R G Cooper[3], D Penumadu[4]

[1]Space Sciences Laboratory, University of California at Berkeley, Berkeley, CA 94720, USA
[2]Nova Scientific, Inc, Sturbridge, MA 01566, USA
[3]Spallation Neutron Source, Oak Ridge National Laboratory, Oak Ridge, TN 37831 USA
[4]Professor and JIAM Chair of Excellence, Department of Civil and Environmental Engineering, University of Tennessee, Knoxville, TN 37996, USA

E-mail: ast@ssl.berkeley.edu

Abstract. Results of neutron transmission Bragg edge spectroscopic experiments performed at the SNAP beamline of the Spallation Neutron Source are presented. A high resolution neutron counting detector with a neutron sensitive microchannel plate and Timepix ASIC readout is capable of energy resolved two dimensional mapping of neutron transmission with spatial accuracy of ~55 μm, limited by the readout pixel size, and energy resolution limited by the duration of the initial neutron pulse. A two dimensional map of the Fe 110 Bragg edge position was obtained for a bent steel screw sample. Although the neutron pulse duration corresponded to ~30 mÅ energy resolution for 15.3 m flight path, the accuracy of the Bragg edge position in our measurements was improved by analytical fitting to a few mÅ level. A two dimensional strain map was calculated from measured Bragg edge values with an accuracy of ~few hundreds μstrain for 300s of data acquisition time.

1. Introduction

Modern pulsed neutron sources that provide sub-100 μs bright neutron pulses uniquely enable non-destructive studies of texture, phase and residual strain within metal samples with high resolution [1-5]. Presently there are a number of instruments used for strain mapping from neutron diffraction with time of flight (TOF) measurement for each detected neutron [2-5]. The measured time of flight can be converted into neutron energy with the resolution determined by the width of the initial pulse, length of the flight path and the timing resolution of the neutron detection system. The conventional TOF neutron diffraction techniques measure the scattered neutrons. This method allows the measurement of the lattice parameter within the volume corresponding to the intersection of incident (constrained by slits) and diffracted beams (defined by detector collimators). Scanning this volume through the sample allows non-destructive measurement of the wavelengths of Bragg edges and as a result spatially correlated lattice parameter variations within the sample along the incident beam direction.

An alternative approach is to measure the spectrum of neutrons transmitted through the sample [1-3,5]. In this case, the lattice parameter is averaged over the neutron path length within the sample. However, in this mode the entire 2-dimensional map of averaged lattice parameter variation can be reconstructed if transmission spectra are measured in each pixel of the detection system. Thus a high resolution position sensitive detector in the transmission mode allows accurate two-dimensional mapping of combined effects from all the scattering and reaction processes occurring in the sample. The spatial and temporal resolutions of the detection system are the crucial parameters defining the accuracy of neutron transmission spectroscopic measurements.

In this paper we present the results of proof of principle transmission Bragg edge spectroscopic experiments performed at the SNAP beamline of the Spallation Neutron Source of Oak Ridge National Laboratory. A two dimensional map of the Fe 110 Bragg edge positions was measured for a pre-stressed steel screw sample. The map of residual strain was then calculated from Bragg edge positions with accuracy of ~few hundred μstrain.

Figure 1. Photograph of bent steel screw used in the measurements. Dashed square shows 14x14 mm^2 area imaged by the detector.

Figure 2. Neutron transmission image of the screw. Full beam spectrum is used. 100 sec acquisition. Squares indicate areas where strain histograms of Figure 6 are measured.

2. Experimental setup

A high resolution neutron counting detector with a neutron sensitive microchannel plate and a Timepix ASIC readout [6] was used in our measurements. The active area of the detector was limited by the TimePix readout, which is 14x14 mm^2, with a 256x256 matrix of 55 μm square pixels, while the neutron sensitive microchannel plate stack has ~25 mm active area in diameter. The high detection efficiency (~40% for cold and ~20% for thermal neutrons [7]), spatial (55 μm [8,9]) and temporal (~μs [10,11]) resolution proven by our recent experiments enabled high resolution measurement of two dimensional neutron transmission spectrum in time of flight mode. The distance between the source of neutrons and the detector was only 15.3 m, compared to a 50 m flight path at ENGIN-X facility [3,11]. The neutron transmission spectra in our measurements in principle could have been acquired in each 55 μm pixel of our detector, providing there are enough statistics. The energy resolution of measured raw data was limited by the duration of the initial neutron pulse (~30 to 130 μs, depending upon the neutron wavelength), corresponding to ~7-33 mÅ. However, the position of Bragg edge could be determined with much better accuracy (down to ~1 mÅ or ΔE/E~0.025% at 4.08 Å) after the analytical fit to the measured edge data.

A pre-bent steel screw shown in Figure 1 was installed ~14 mm from the active area of our detector. The SNS neutron source operated at 600 μA proton beam current at 60 Hz frequency. The beamline has a decoupled poisoned supercritical hydrogen moderator. The full spectrum transmission image of the screw acquired over 100s is shown in Figure 2. The open beam area was used for flux normalization in our measurements. Our current serial detector readout electronics does not allow simultaneous measurements of neutrons across a wide range of energies, and only a narrow range around a particular Bragg edge was measurable in one acquisition run. The 110 Bragg edge data shown below was acquired in 300 seconds. Position (with ~55 μm accuracy) and timing (~1 μs) of each detected neutron was recorded into raw data files, which then could be rebinned into spectral data with a given spatial and spectral resolution.

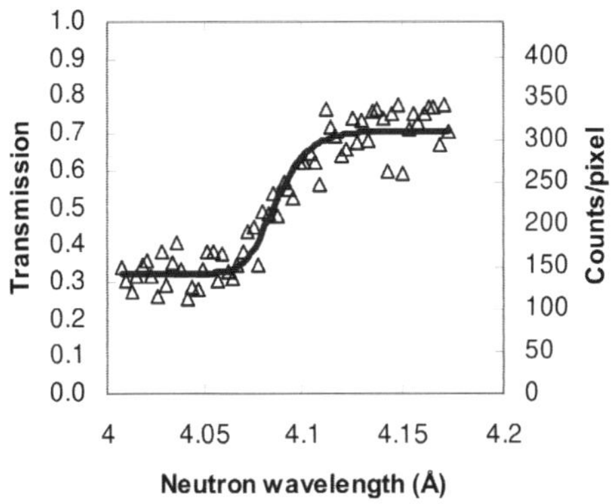

Figure 3. Experimental 110 Bragg edge of steel screw measured in the area 2 of Figure 2. The width of the edge is defined by the width of the initial neutron pulse. Analytical fit improves the accuracy of λ_0 position to ~few mÅ level. 300 sec data acquisition.

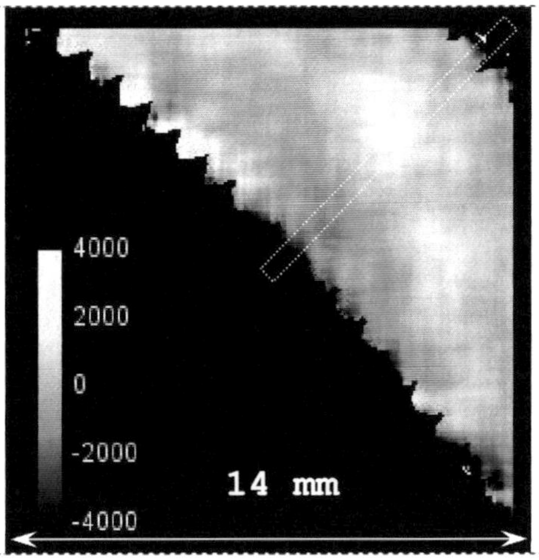

Figure 4. Strain map image of the steel screw. The strain is calculated from 110 Bragg edge positions λ_0 mapped across the sample. Spectral data is calculated for a 20x20 pixel area around each pixel. The colour bar indicates strain values in μstrain

3. Results and discussion

Due to limited neutron statistics (only 37 counts per 55x55 μm^2 pixel within 170 mÅ range around 110 edge) our measurement did not provide spectral information in each pixel. Therefore we have rebinned data into spectra calculated within 20x20 pixel area with 10 μs timing (2.59 mÅ energy) resolution. Thus each pixel in our 256x256 2-dimensional map of Bragg edge spectrum contains data from a 20x20 pixel area. Figure 3 shows a typical transmission curve calculated for the box N2 of Figure 2. The width of the wedge is determined by the width of the initial beam pulse and by the flight path of 15.3 m. Despite relatively poor statistics and large edge width, the position of Bragg edge can be calculated quite accurately with the help of analytical function fitting [1,2]. In each pixel, the wavelength of Bragg edge centre λ_0, was calculated from the spectral data integrated over the 20x20 pixel area, and the resulting map of λ_0 values was converted into the residual strain map. Figure 4 shows the map of calculated residual strain in the sample. The strain in that image varies within ±4 mstrain. The diagonal cut through the bent screw area in that image is shown in Figure 5. The noise on that curve indicates that the resolution of a few hundred μstrain in each pixel of strain map was achieved in our measurement. The histogram of strain values in boxes shown by dotted lines in Figure 2 are shown in Figure 6, together with the histogram for the entire strain map image. It is clearly seen that compression in box 1 and stretching in box 2 are on the scale of ~1.5 mstrain, while areas 3 and 4 do not have much of the strain, as expected in that sample.

4. Conclusion

The transmission Bragg edge spectroscopy measurements with MCP-Timepix detector installed at a pulsed neutron beamline indicate that 2-dimensional maps of strain can be measured with accuracy of ~few hundred μstrain and spatial resolution as good as the pixel size of the detector, providing enough statistics is acquired. Five minute acquisition experiments at SNAP beamline and SNS operating at 50% power are clearly not enough to resolve strain within 55 μm pixels. However, increased neutron flux, dedicated beamline optimized for this type of measurements and optimized detector electronics should enable very high resolution strain mapping in the near future.

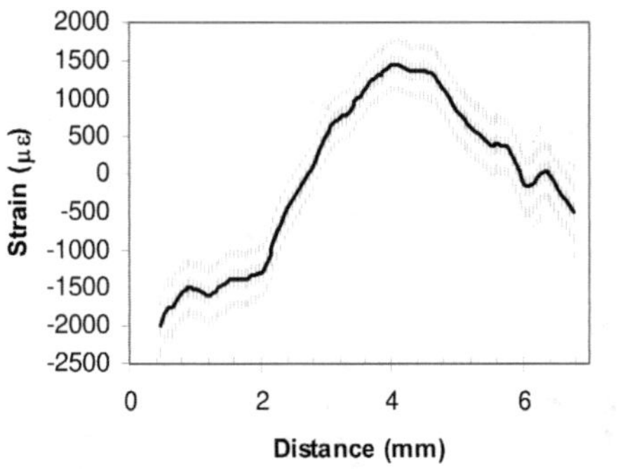

Figure 5. Cross section through the image of Figure 4 showing the strain variation across the bent area of the screw.

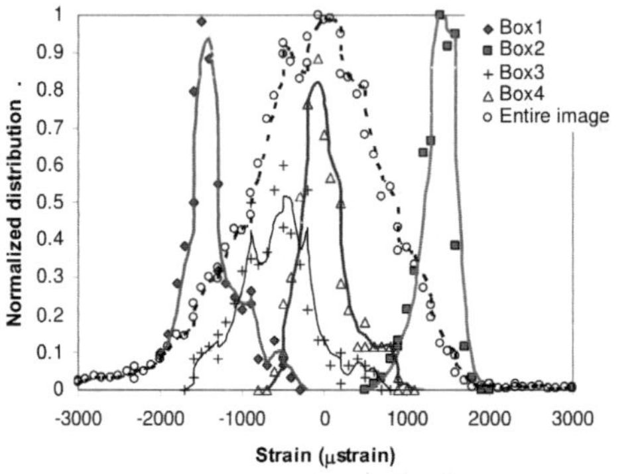

Figure 6. Histograms of strain values from Figure 4 measured in the areas shown in by boxes in Figure 2 as well as that of the entire stain map image.

Acknowledgments

We would like to thank Dr. A. Stewer from ESS for his very helpful discussions and Antonio Des Santos of ORNL for the help with the measurements. This work was supported in part by the U.S. Department of Energy under Grant No. DE-FG02-07ER86322, DE-FG02-08ER86353 and National Science Foundation Grant No. DMR-0753599.

References

[1] Steuwer A, Santisteban J R, Withers P J, Edwards L, Fitzpatrick M E 2003 *J. Appl. Crystallography* **36**, 1159
[2] Steuwer A, Withers P J, Santisteban J R Edwards L 2005 *J. Appl. Phys.* **97**, 074903
[3] Santisteban J R, Daymond M R, James J A, Edwards L 2006 *J. Appl. Cryst.* **39**, 812
[4] Wang X -L et al. 2006 *Physica B* **385–386**,673
[5] Kiyanagi Y et al. 2009 *Nucl. Instr. Meth. A* **600**, 167
[6] Llopart X, Ballabriga R, Campbell M, et al. 2008 *Nucl. Instr. Meth. A* **585**, 106
[7] Tremsin A S, McPhate J B, Vallerga J V, Siegmund O H W, Hull J S, Feller W B, Lehmann E 2009 *Nucl. Instr. Meth. A*, in press, doi:10.1016/j.nima.2009.01.041
[8] Siegmund O H W, Vallerga J V, Tremsin A S, McPhate J B, Feller W B 2007 *Nucl. Instr. Meth. A* **576**, 178
[9] Tremsin A S, Vallerga J V, McPhate J B, Siegmund O H W, Feller W B, Crow L, Cooper R G 2008 *Nucl. Instr. Meth. A* **592**, 374
[10] Tremsin A S, Feller W B, Siegmund O H W 2008 *IEEE Trans. Nucl. Sci.* **55**, 1664
[11] Tremsin A S, McPhate J B, Kockelmann W A, Vallerga J V, Siegmund O H W, Feller W B 2008 *IEEE Nucl. Sci. Symposium* N42-2, subm. *IEEE Trans. Nucl. Sci*

Imaging of a spatial distribution of preferred orientation of crystallites by pulsed neutron Bragg edge transmission

H Sato[1], O Takada[1], K Iwase[2], T Kamiyama[1] and Y Kiyanagi[1]

[1]Graduate School of Engineering, Hokkaido University, Kita-13 Nishi-8, Kita-ku, Sapporo 060-8628, Japan

[2]Frontier Research Center for Applied Atomic Sciences, Ibaraki University, 162-1 Shirakata-shirane, Tokai, Ibaraki 319-1106, Japan

E-mail: hakuryu@eng.hokudai.ac.jp

Abstract. A pulsed neutron transmission coupled with a two-dimensional position sensitive neutron detector gives a time-of-flight spectrum at each pixel of the detector, which depends on the total cross-sections of materials. In order to extract quantitative information of the preferred orientation included in the Bragg scattering total cross-section data, a spectral analysis software for the 2D imaging has been developed, and the transmission data of an unbent iron plate were analyzed. The 2D images with respect to the preferred orientation were successfully obtained, and the effectiveness of spectroscopic neutron transmission imaging was indicated.

1. Introduction

It is very important for improving the strength, the rolling processability and the electromagnetic characteristic depending on the preferred orientation inside a steel sheet to investigate and control the texture. An engineering neutron diffractometer is generally utilized to analyze such crystalline anisotropy inside a material. Moreover, a mapping of such information is performed by the real-space scanning using a pin-point beam, combined with the neutron detectors coupled with a radial collimator.

On the other hand, we have been developing the spectroscopic neutron transmission imaging technique by a time-of-flight (TOF) method at a pulsed source [1], including also energy-selective neutron radiography [2]. For example, there are the spectroscopic Bragg edge radiography for crystallographic imaging [3] and the spectroscopic resonance-absorption tomography for nuclide-selective density imaging and internal thermography [4]. The Bragg edge transmission spectroscopy can give a d-spacing map only with one measurement because it can use a two-dimensional position sensitive neutron detector (2D-PSND) for the imaging of the real-space distribution. This technique has capability to give a 2D real-space distribution of crystal structure information, as requested from material sciences and industrial fields.

So far, the Bragg edge imaging has been applied to a strain imaging [3,5], a crystalline phase imaging [6] and a texture imaging [2,7]. In particular, the quantitative analysis of Bragg edge shapes is indispensable to extract texture information concerning the preferred orientation. Therefore, we present the novel procedure for the imaging of quantitative information of the preferred orientation in a textured material, as well as a Rietveld refinement software development in order to analyze position-dependent pulsed neutron Bragg edge transmission spectra.

2. Experimental

We prepared an experimental sample for which the spatial distribution of the preferred orientation became inhomogeneous due to the plastic deformation. This was a flat iron plate (30 mm height, 105 mm width and 5 mm thickness) stressed at two lines along the height-direction which is parallel to the rolling-direction. This sample was made by bending and unbending at each line. It has been known that this sample has large residual-strain [3], by using the Bragg edge transmission measurement at the spallation neutron source KENS at High Energy Accelerator Research Organization (KEK) in Japan.

The 2D-PSND was the 256 (16×16) pixel ^{6}Li-glass-scintillation detector [8]. The detection efficiency was more than 95 % for cold neutrons. The maximum counting rate was 8.3 MHz/cm^2 which was sufficient for intense beams. The time resolution was set into 10 μs in this experiment. The spatial resolution was 3 mm which was the order of the gauge volume in diffractometry.

The pulsed neutron radiography experiment was carried out at Neutron Beam-line for Observation and Research Use (NOBORU) [9] at Materials and Life Science Experimental Facility (MLF) at Japan Proton Accelerator Research Complex (J-PARC) in Japan. NOBORU directly viewed the decoupled 18 K supercritical para-H_2 moderator at MLF. The MLF accelerator beam power was 4 kW during this experiment. The neutron flight distance from the source to the sample was 14 m. The collimation ratio L / D was 140. The neutron wavelength resolution $\Delta\lambda / \lambda$ at $\lambda = 0.4$ nm was 0.35 %.

The unbent iron sample was placed just in front of the 2D imaging detector, and the incident beam and the transmitted beam were measured with the 2D-PSND by using a TOF method simultaneously.

3. Spectral analysis of position-dependent Bragg edge transmission spectra

In order to extract crystallographic information such as preferred orientation from the Bragg edge transmission spectrum obtained at each pixel, we have been developing a new whole-pattern analysis software based on the Rietveld refinement technique [10]. The algorithm was based on the total cross-section simulation code CRIPO [11] and the Bragg edge Rietveld analysis code BETMAn [12].

The neutron transmission spectrum $Tra(\lambda)$ depending on the neutron wavelength λ can be given by

$$Tra(\lambda) = \exp\left[-\sum_{p} \sigma_{tot,p}(\lambda)\rho_p t_p \right]. \tag{1}$$

Here, σ_{tot} is the total cross-section, ρ is the atomic number density and t is the thickness, summed over the number of the crystalline phases p. σ_{tot} of a certain phase consists of elastic coherent, elastic incoherent, inelastic coherent, inelastic incoherent scattering and absorption parts as shown below.

$$\sigma_{tot}(\lambda) = \sigma_{coh}^{ela}(\lambda) + \sigma_{incoh}^{ela}(\lambda) + \sigma_{coh}^{inela}(\lambda) + \sigma_{incoh}^{inela}(\lambda) + \sigma_{abs}(\lambda) \tag{2}$$

The elastic coherent (Bragg) scattering cross-section can be given by

$$\sigma_{coh}^{ela}(\lambda) = \frac{\lambda^2}{2V_0} \sum_{d_{hkl}=0}^{2d_{hkl}<\lambda} |F_{hkl}|^2 d_{hkl} B_{hkl}(\lambda, d_{hkl}) P_{hkl}(\alpha_{hkl}(\lambda)) Ext(\lambda, S). \tag{3}$$

Here, V_0 is the unit cell volume of a crystal. d_{hkl} is the crystal-lattice-plane spacing with respect to the reflection Miller index *hkl*. The Bragg edge of *hkl* diffraction appears at $\lambda = 2d_{hkl}$ indicating the back scattering on the Bragg's law. F_{hkl} is the crystal structure factor including the Debye-Waller factor. B_{hkl} is the resolution function describing the Bragg edge broadening [5] due to neutron pulse width. *Ext* represents the reduction term describing the extinction effect depending on the crystallite size S [13].

P_{hkl} is the probability function of the crystallite orientation depending on α_{hkl} (the acute angle between the incident beam and the orientation of equivalent crystal-lattice-planes $\{hkl\}$) described by

$$\alpha_{hkl}(\lambda) = \frac{\pi}{2} - \arcsin\left(\frac{\lambda}{2d_{hkl}} \right) = \frac{\pi}{2} - \theta_{hkl} \tag{4}$$

where θ_{hkl} is a half of the angle between the incident beam and the diffracted beam. In order to describe the preferred orientation in a Bragg edge transmission spectrum, we used the March-Dollase function [14] as P_{hkl}. This function is used in many Rietveld codes for neutron and X-ray powder diffraction. The March-Dollase function for the transmission geometry [12] is

$$P_{hkl}(\alpha_{hkl}(\lambda)) = \frac{1}{\pi}\int_{-\frac{\pi}{2}}^{\frac{\pi}{2}}\left[\left(r_{hkl}^2 - \frac{1}{r_{hkl}}\right)(\cos\alpha_{hkl}(\lambda)\cos\beta_{hkl} - \sin\alpha_{hkl}(\lambda)\sin\beta_{hkl}\sin\phi)^2 + \frac{1}{r_{hkl}}\right]^{-\frac{3}{2}} d\phi \quad (5)$$

where r_{hkl} indicates the degree of crystalline anisotropy and β_{hkl} represents the most probable angle of preferred orientation in this study. Since Eq. (5) is described by averaging over φ on the Debye-Scherrer cone, we can investigate orientation information only about θ. $r = 1$ means the isotropic distribution of crystallites (no preferred orientation). In practice, however, r is not unity because we frequently measure more or less a textured material. The most probable α corresponds to $\pi / 2 - \beta$ in case of $r > 1$, and to β in case of $r < 1$. These relations are applicable to fiber-shaped crystallites or plate-shaped crystallites. Fig. 1 shows the calculation results of Eq. (2) depending on β_{110} in case of $r_{110} > 1$, considering only the preferred orientation of {110} crystal-lattice-planes. In this study, we corrected only the longest wavelength Bragg edge shape although a typical Rietveld code considers about the all peaks. This is because it was experimentally examined that the longest wavelength Bragg edge shape can drastically change due to the preferred orientation effect [3], and the correction method of the effect of forward-scattered neutrons detected by 2D detector pixels has not been developed yet. Such forward-scattered neutrons from Fe atoms may increase at $\lambda < 2d_{110}\sin 45°$ (about 0.29 nm).

Next, we tried to extract preferred orientation information β_{110} and r_{110} from the measured pulsed neutron transmission spectra by the Rietveld refinement using a spectral fitting technique. The fitting algorithm was the Levenberg-Marquardt method [15] for a non-linear least-squares estimation. Fig. 2 shows some results of the measured transmission spectra fitted with the Rietveld simulations using Eq. (1). The fitting was improved by introducing the March-Dollase function. Thus, we succeeded in extracting the refined parameters of not only d_{110} but also $\rho \times t$, β_{110} and r_{110}.

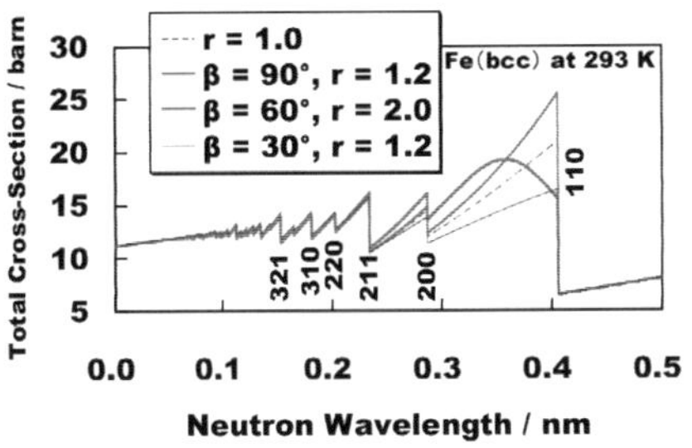

Figure 1. Results of the simulation calculation of the total cross-section depending on the March-Dollase model considering the preferred orientation of a textured polycrystalline material, obtained the developed Rietveld analysis code system for a pulsed neutron Bragg edge transmission spectroscopy.

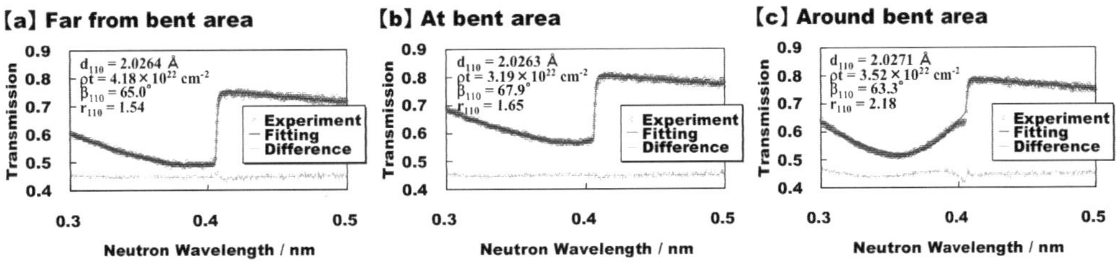

Figure 2. Results of the Rietveld fitting analysis around {110} Bragg edge of the transmission data (a) far from the bent position, (b) at the bent position and (c) around the bent position, measured by the 2D imaging detector. The anisotropic trend becomes strong around the bent area because r is large.

4. Results and discussion

Fig. 3 shows the quantitative images of the projected density of Fe atoms ($\rho \times t$), the most probable angle of the preferred orientation (β_{110}) and the degree of crystalline anisotropy (r_{110}) of the stressed iron sample. Fig. 3 (a) indicates that the density or the thickness decreases only at the bent positions since the $\rho \times t$ values around x = - 13.5 and + 18 mm positions without the plastic deformation are close to 4.25×10^{22} cm^{-2} (the theoretical value). Fig. 3 (b) indicates that the β_{110} values change to 68° at the bent positions, and to 63° around those. Fig. 3 (c) indicates that the r_{110} values increase above 2.0 only around the bent positions. Thus, the spectroscopic neutron imaging coupled with spectral analyses can non-destructively give the images concerning the physical quantity of a material but not the transmitted neutron intensities. Essentially, these are significant images for crystallography, and are different from the transmission images obtained by an energy-selective neutron radiography.

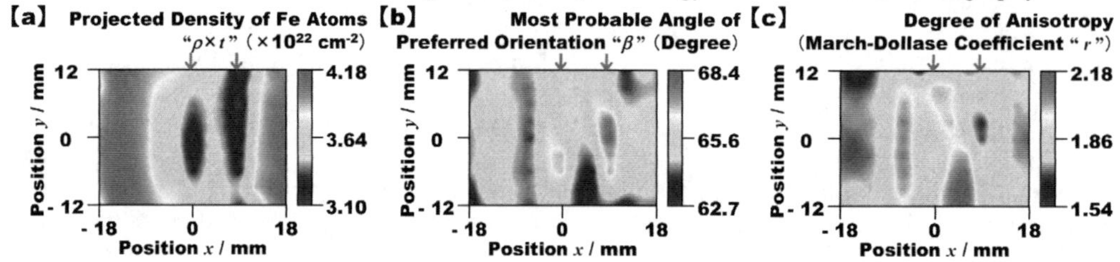

Figure 3. Experimental results of (a) image of the projection data of atomic number density of Fe, (b) image of the most probable angle of preferred orientation and (c) image of the degree of crystalline anisotropy around the stressed area. The allows on the tops of these images indicate the bent lines.

5. Conclusion

We succeeded in taking images of the projected density of atoms, the most probable angle of preferred orientation and the degree of crystalline anisotropy in a textured material simultaneously, using the pulsed neutron Bragg edge transmission method coupled with the Rietveld analysis of position-dependent TOF transmission spectra. These images are useful for crystallography, and this method will become a powerful tool for non-destructive material inspections.

Acknowledgements

The authors are greatly thankful to the NOBORU group [9] and Dr. T. Shinohara of Japan Atomic Energy Agency (JAEA) for experimental assistance at J-PARC. This work was partially supported by Grant-in-Aid for JSPS Fellows (No. 20002121) and Grant-in-Aid for Scientific Research (A) (No. 20246136) from Japan Society for the Promotion of Science.

References

[1] Kiyanagi Y, Kamiyama T, Iwasa H and Hiraga F 2004 *Key Engin. Mater.* **270** 1371
[2] Kockelmann W, Frei G, Lehmann E H, Vontobel P and Santisteban J R 2007 *Nucl. Instr. and Meth. A* **578** 421
[3] Iwase K, Sakuma K, Kamiyama T and Kiyanagi Y 2009 *Nucl. Instr. and Meth. A* **605** 1
[4] Sato H, Kamiyama T and Kiyanagi Y 2009 *Nucl. Instr. and Meth. A* **605** 36
[5] Santisteban J R, Edwards L, Fitzpatrick M E, Steuwer A, Withers P J, Daymond M R, Johnson M W, Rhodes N and Schooneveld E M 2002 *Nucl. Instr. and Meth. A* **481** 765
[6] Steuwer A, Withers P J, Santisteban J R and Edwards L 2005 *J. Appl. Phys.* **97** 074903
[7] Santisteban J R, Edwards L and Stelmukh V 2006 *Physica B* **385** 636
[8] Sato H, Takada O, Satoh S, Kamiyama T and Kiyanagi Y *Nucl. Instr. and Meth. A* (in press)
[9] Maekawa F, Oikawa K, Harada M, Kai T, Meigo S, Kasugai Y, Ooi M, Sakai K, Teshigawara M, Hasegawa S, Ikeda Y and Watanabe N 2009 *Nucl. Instr. and Meth. A* **600** 335
[10] Rietveld H M 1969 *J. Appl. Cryst.* **2** 65
[11] Kropff F and Granada J R 1977 *CAB-1977* Centro Atomico Bariloche, Institute Balseiro
[12] Vogel S 2000 *Ph.D. Thesis* Kiel University
[13] Iwase K, Nagata T, Sakuma K, Takada O, Kamiyama T and Kiyanagi Y 2007 *IEEE Nucl. Sci. Symp. Conf. Rec.* **2** 1716
[14] Dollase W A 1986 *J. Appl. Cryst.* **19** 267
[15] Marquardt D W 1963 *J. Soc. Indust. Appl. Math.* **11** 431

Contrast reduction due to small angle scattering in a neutron multilayer interferometer

Kaoru Taketani

High Energy Accelerator Research Organization,
1-1 Oho, Tsukuba, Ibaraki 305-0801, Japan

E-mail: kaoru.taketani@kek.jp

Abstract. We study the effect of small angle scatterings on the interference-fringe contrast in a Jamin-type multilayer interferometer by performing a model calculation. Our calculation can be the basis for not only choosing the etalon bases of Jamin-type multilayer interferometer for very cold neutron but also analyzing quantitatively neutron dark-field tomography.

1. Introduction

The phase shift in a neutron interferometer is proportional to the neutron wavelength. Although well-known interferometer using crystal cannot split long-wavelength neutron, a multilayer mirror can split long-wavelength neutron. Therefore, an interferometer using multilayer (multilayer interferometer) can become a sensitive tool to measure small interactions. A high contrast interference fringe was demonstrated using a multilayer interferometer with etalon plates for cold neutron [1]. Therefore, it is interesting to analyze the high contrast fringe can also be demonstrated using a multilayer interferometer for very cold neutron.

The broadening of the beam width due to small angle scattering becomes large when we used longer wavelength. Therefore, we may have to care that neutron beam width broadens due to small angle scatterings when they go through etalon plates, when we use the multilayer interferometer for very cold neutron. However, as far as we know, the analytical formula of contrast reduction due to small angle scatterings in a separated-path interferometer is still missing [2].

It is worth noting that the analysis is the basis for not only choosing the etalon bases of Jamin-type multilayer interferometer for very cold neutron but also analyzing quantitatively neutron dark-field tomography [3]. Because the dark-field tomography has a high potential for a great number of applications, a rigorous investigation of the property is important.

As a first step to solve the above mentioned problems, we analyze the contrast reduction due to small angle scatterings from a sample which is inserted in a Jamin type multilayer interferometer without considering the effects of small angle scatterings from the etalon plates.

2. Model calculation

Figure 1 shows the setup of our model calculation. We install a neutron spin interferometer (NSI) [4] between the source plane and the detector plane. A Jamin-type interferometer is installed in the NSI system and sample that has finite density-density correlation length was inserted in the Jamin-type interferometer. We let the z axis be parallel to the incident neutron beam, the y axis be parallel to the quantization axis of the spin, and the x axis be orthogonal to both the z and y axes as shown in figure 1.

The polarizer and the $\pi/2$ flipper both provide the superposition of the spin-up and the spin-down components. The magnetic mirror reflects only the spin-up component and the non-magnetic mirror reflects the spin-down component that goes through the magnetic mirror. The pair of mirrors separates the up and the down spin components spatially into the two parallel paths. The π flipper flips the spin direction. The waves penetrate the sample and are spatially superposed on each other on the second pair of mirrors. By scanning the relative phase of the oscillating magnetic field between the π flipper and the $\pi/2$ flippers, we observed interference fringes through the $\pi/2$ flipper and the analyzer.

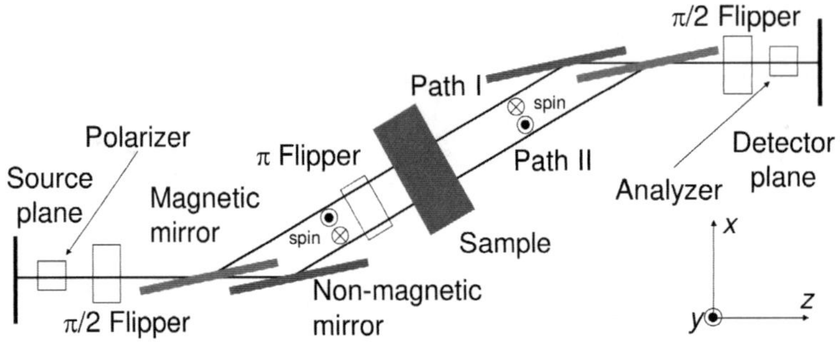

Figure 1. The setup for the model calculation

According to the partial coherence theory [5], the integrated intensity over the detector plane, I, is written as

$$I = \int \overline{G(\mathbf{r}_2, \mathbf{r}'_{0a})} G(\mathbf{r}_2, \mathbf{r}'_{0b}) \Gamma(\mathbf{r}'_{0a}, \mathbf{r}'_{0b}) d\mathbf{r}'_{0a} d\mathbf{r}'_{0b} d\mathbf{r}_2, \tag{1}$$

where

$$G(\mathbf{r}_2, \mathbf{r}'_{0l}) = G^{\mathrm{I}}(\mathbf{r}_2, \mathbf{r}'_{0l}) + G^{\mathrm{II}}(\mathbf{r}_2, \mathbf{r}'_{0l}). \tag{2}$$

Here $G^{\mathrm{I}}(\mathbf{r}_2, \mathbf{r}'_{0l})$ and $G^{\mathrm{II}}(\mathbf{r}_2, \mathbf{r}'_{0l})$ are the Green's functions for a neutron beam traveling along different interferometer paths, l can assume either a or b, $\mathbf{r}_2$ is the coordinates of the point on the detector plane and $\mathbf{r}'_{0a}$ and $\mathbf{r}'_{0b}$ are the coordinates of the points on the source plane, and $\Gamma(\mathbf{r}'_{0a}, \mathbf{r}'_{0b})$ is the cross-spectral density on the source plane.

Figure 2 shows the virtual source plane and the virtual samples of the setup. The distance from a point on the detector plane to a point on the source plane along the beam that propagates through multiple properly aligned mirrors and that from the point on the detector plane to the point on the corresponding virtual source plane are the same. To simplify the formula, we will discuss the beams that propagate from the virtual source plane in the following discussion. The position where neutrons

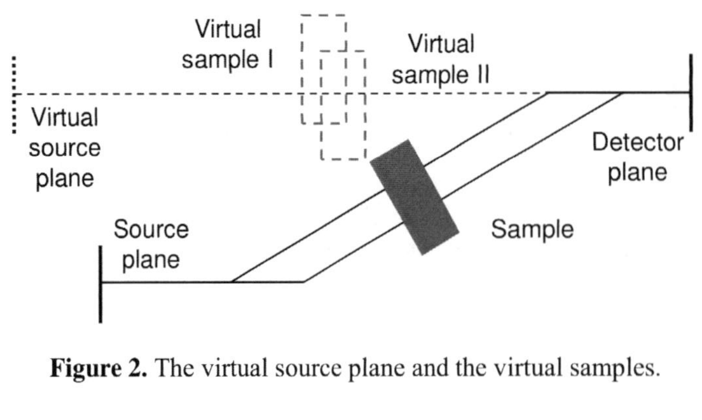

Figure 2. The virtual source plane and the virtual samples.

along path I go through the sample is different from the position where that along path II go through. Therefore, the positions of the virtual samples for each path are different as shown in figure 2.

Because Green's function for neutron through the NSI do not have strong $\mathbf{r}_2$, $\mathbf{r}_{0a}$, and $\mathbf{r}_{0b}$ dependence [6], we ignore it in the following analysis. In this case, the Green's function becomes

$$G_{\mathrm{J}}^{\mathrm{I}}(\mathbf{r}_2,\mathbf{r}_{0l})+G_{\mathrm{J}}^{\mathrm{II}}(\mathbf{r}_2,\mathbf{r}_{0l})e^{i\phi}, \tag{3}$$

where, $G_{\mathrm{J}}^{\mathrm{I}}(\mathbf{r}_2,\mathbf{r}_{0l})$ and $G_{\mathrm{J}}^{\mathrm{II}}(\mathbf{r}_2,\mathbf{r}_{0l})$ are Green's functions for neutrons through the Jamin type interferometer, l can assume either a or b, $\mathbf{r}_{0a}$ and $\mathbf{r}_{0b}$ are the coordinates on the virtual source plane, and ϕ is the relative phase of the oscillating magnetic field between the π flipper and the $\pi/2$ flippers.

When Born approximation is enough to describe the scattering from the sample, the Green's function is written as

$$G_{\mathrm{J}}{}^{p}(\mathbf{r}_2,\mathbf{r}_{0l})=G_{\mathrm{free}}^{p}(\mathbf{r}_2,\mathbf{r}_{0l})+G_{\mathrm{scat}}^{p}(\mathbf{r}_2,\mathbf{r}_{0l}) \tag{4}$$

where p can assume either I or II,

$$G_{\mathrm{free}}^{p}(\mathbf{r}_2,\mathbf{r}_{0l})=k\exp(ik|\mathbf{r}_2-\mathbf{r}_{0l}|)/|\mathbf{r}_2-\mathbf{r}_{0l}|, \tag{5}$$

and

$$G_{\mathrm{scat}}^{p}(\mathbf{r}_2,\mathbf{r}_{0l})=\int kb\rho_p(\mathbf{r}_1)\exp(ik(|\mathbf{r}_2-\mathbf{r}_1|)+|\mathbf{r}_1-\mathbf{r}_{0l}|)/(|\mathbf{r}_2-\mathbf{r}_1||\mathbf{r}_1-\mathbf{r}_{0l}|)d\mathbf{r}_1\,. \tag{6}$$

Here $\mathbf{r}_1$ is the coordinate on the virtual sample, $\rho_p(\mathbf{r}_1)$ is the number density of the virtual sample, b is the scattering length of the sample, and k is the wave-number of incident neutrons. In this case, the Green's functions in equation (1) can be transformed to

$$T_{\mathrm{f}}(\mathbf{r}_2,\mathbf{r}_{0\mathrm{a}},\mathbf{r}_{0\mathrm{b}})+T_{\mathrm{fs}}(\mathbf{r}_2,\mathbf{r}_{0\mathrm{a}},\mathbf{r}_{0\mathrm{b}})+T_{\mathrm{s}}(\mathbf{r}_2,\mathbf{r}_{0\mathrm{a}},\mathbf{r}_{0\mathrm{b}}). \tag{7}$$

Here

$$T_{\mathrm{f}}(\mathbf{r}_2,\mathbf{r}_{0\mathrm{a}},\mathbf{r}_{0\mathrm{b}})=\overline{G_{\mathrm{free}}(\mathbf{r}_2,\mathbf{r}_{0\mathrm{a}})}G_{\mathrm{free}}(\mathbf{r}_{\mathrm{d}},\mathbf{r}_{0\mathrm{b}}), \tag{8}$$

$$T_{\mathrm{sf}}(\mathbf{r}_2,\mathbf{r}_{0\mathrm{a}},\mathbf{r}_{0\mathrm{b}})=\overline{G_{\mathrm{free}}(\mathbf{r}_2,\mathbf{r}_{0\mathrm{a}})}G_{\mathrm{scat}}(\mathbf{r}_2,\mathbf{r}_{0\mathrm{b}})+\overline{G_{\mathrm{scat}}(\mathbf{r}_2,\mathbf{r}_{0\mathrm{a}})}G_{\mathrm{free}}(\mathbf{r}_2,\mathbf{r}_{0\mathrm{b}}), \tag{9}$$

$$T_{\mathrm{s}}(\mathbf{r}_2,\mathbf{r}_{0\mathrm{a}},\mathbf{r}_{0\mathrm{b}})=\overline{G_{\mathrm{scat}}(\mathbf{r}_2,\mathbf{r}_{0\mathrm{a}})}G_{\mathrm{scat}}(\mathbf{r}_2,\mathbf{r}_{0\mathrm{b}}), \tag{10}$$

$$G_{\mathrm{free}}(\mathbf{r}_2,\mathbf{r}_{0l})=G_{\mathrm{free}}^{\mathrm{I}}(\mathbf{r}_2,\mathbf{r}_{0l})+G_{\mathrm{free}}^{\mathrm{II}}(\mathbf{r}_2,\mathbf{r}_{0l}), \tag{11}$$

and

$$G_{\mathrm{scat}}(\mathbf{r}_2,\mathbf{r}_{0l})=G_{\mathrm{scat}}^{\mathrm{I}}(\mathbf{r}_2,\mathbf{r}_{0l})+G_{\mathrm{scat}}^{\mathrm{II}}(\mathbf{r}_2,\mathbf{r}_{0l}). \tag{12}$$

Because $T_{\mathrm{f}}(\mathbf{r}_2,\mathbf{r}_{0\mathrm{a}},\mathbf{r}_{0\mathrm{b}})$ is the transfer function for freely propagating neutrons, the contribution of $T_{\mathrm{f}}(\mathbf{r}_2,\mathbf{r}_{0\mathrm{a}},\mathbf{r}_{0\mathrm{b}})$ to I can be written as, using a constant J,

$$J(1+\cos\phi). \tag{13}$$

When the beam size on the sample is much smaller than the distance from the sample to the detector plane and much longer than the square root of the product of the neutron wavelength and the distance from the sample to the detector plane, the following equation is appropriate [7] :

$$G_{\mathrm{scat}}^{p}(\mathbf{r}_2,\mathbf{r}_{0l})=G_{\mathrm{free}}^{p}(\mathbf{r}_2,\mathbf{r}_{0l})bt\rho/ik\,, \tag{14}$$

where t and ρ are thickness and number density of the sample. Using equation (14) and assuming b is real, $T_{\mathrm{sf}}(\mathbf{r}_2,\mathbf{r}_{0\mathrm{a}},\mathbf{r}_{0\mathrm{b}})$ becomes zero. Therefore, the contribution of $T_{\mathrm{sf}}(\mathbf{r}_2,\mathbf{r}_{0\mathrm{a}},\mathbf{r}_{0\mathrm{b}})$ to I is zero.

$T_{\mathrm{s}}(\mathbf{r}_2,\mathbf{r}_{0a},\mathbf{r}_{0b})$ contains four terms. Each terms can be written as follows:

$$(kb/(z_2 - z_1)(z_1 - z_0))^2 \\ \times \int \rho_p(\mathbf{r}_{1a})\rho_q(\mathbf{r}_{1b})\exp(ik(|\mathbf{r}_2 - \mathbf{r}_{1a}| - |\mathbf{r}_2 - \mathbf{r}_{1b}| + |\mathbf{r}_{1a} - \mathbf{r}_{0a}| - |\mathbf{r}_{1b} - \mathbf{r}_{0b}|))d\mathbf{r}_{1a}d\mathbf{r}_{1b}, \tag{15}$$

where p and q can be assumed I and II. To analyze them, we introduce $\rho(\mathbf{r})$ that satisfy the following condition:

$$\rho_{\mathrm{I}}(\mathbf{r}) = \rho(\mathbf{r} + \mathbf{s}),\ \rho_{\mathrm{II}}(\mathbf{r}) = \rho(\mathbf{r} - \mathbf{s}) \tag{16}$$

where $\mathbf{s}$ is a constant vector and depends on the beam separation of the Jamin-type interferometer. Because $\rho_{\mathrm{I}}(\mathbf{r}_{1a})\rho_{\mathrm{I}}(\mathbf{r}_{1b}) = \rho_{\mathrm{II}}(\mathbf{r}_{1a})\rho_{\mathrm{II}}(\mathbf{r}_{1b}) = g(\mathbf{r}_{1a} - \mathbf{r}_{1b})$, the sum of (p,q) = (I,I) term and (p,q) = $(\mathrm{II},\mathrm{II})$ term can be written as, for a sufficiently fine collimation [8],

$$KS(\mathbf{Q}) \tag{17}$$

where K is a constant,

$$S(\mathbf{Q}) = \int g(\mathbf{r}_{1a} - \mathbf{r}_{1b})\exp(i\mathbf{Q}(\mathbf{r}_{1a} - \mathbf{r}_{1b}))d(\mathbf{r}_{1a} - \mathbf{r}_{1b}), \tag{18}$$

and

$$\mathbf{Q} = k((\mathbf{r}_2 - (\mathbf{r}_{1a} + \mathbf{r}_{1b})/2)/(z_2 - (z_{1a} + z_{1b})/2) - ((\mathbf{r}_{1a} + \mathbf{r}_{1b})/2 - (\mathbf{r}_{0a} + \mathbf{r}_{0b})/2)/((z_{1a} + z_{1b})/2 - z_0)). \tag{19}$$

On the other hand, because $\rho_{\mathrm{I}}(\mathbf{r}_{1a})\rho_{\mathrm{II}}(\mathbf{r}_{1b}) = \rho_{\mathrm{II}}(\mathbf{r}_{1a})\rho_{\mathrm{I}}(\mathbf{r}_{1b}) = g(\mathbf{r}_{1a} - \mathbf{r}_{1b} - 2\mathbf{s})$ the sum of (p,q) = (I,II) term and (p,q) = (II,I) term can be written as

$$KS(\mathbf{Q})\cos(\mathbf{Qs})\cos\phi. \tag{20}$$

Adding the above results, we obtain the interference intensity. Its contrast is written as

$$(J + KS(\mathbf{Q})\cos(\mathbf{Qs}))/(J + KS(\mathbf{Q})). \tag{21}$$

3. Discussion

In the above discussion, we have only discussed about the integrated intensity over the detector plane and assumed that beam collimation is sufficiently fine and Born approximation is sufficient. The effects of finite divergence and width of the neutron beam, the effects of multiple scatterings, detection position dependence of the contrast, and physical picture of our formula will be discussed in future papers.

4. Summary

We have performed a model calculation and found an analytical formula for contrast reduction due to small angle scattering in a Jamin type multilayer interferometer.

Acknowledgement

This work was supported by JSPS KAKENHI (19GS0120).

References

[1] Kitaguchi M, Funahashi H, Nakura T, Hino M and Shimizu H M 2003 *Phys. Rev.* A **67** 033609
[2] Rauch H and Werner S A 2000 *Neutron Interferometry* (Oxford: Oxford University Press) p 279
[3] Strobl M, Grünzweig C, Hilger A, Manke I, Kardjilov N, David C and Pfeiffer F 2008 *Phys. Rev. Lett.* **101** 123902
[4] Yamazaki D 2002 *Nucl.Instr. Meth.* A **488** 623
[5] Mandel L and Wolf E 1995 *Optical Coherence and Quantum Optics* (Cambridge: Cambridge University Press)
[6] Taketani K, Seki Y, Funahashi H, Hino M, Kitaguchi M, Otake Y and Shimizu H M 2007 *J. Phys. Soc. Jpn.* **76** 064008
[7] Newton R G 1982 *Scattering Theory of Waves and Particles* (New York: Springer-Verlag) p 25
[8] Gähler R, Felber J, Mezei F and Golub R 1998 *Phys. Rev.* A **58** 280

Ultra small angle scattering versus diffraction

O Ebrahimi[1,2], W Treimer[1,2,5], M Strobl[2,4], U Feye-Treimer[1], N Beul[2], E Jericha[3] and S O Seidel[1,2]

[1] Beuth Hochschule für Technik Berlin, University of Applied Sciences, Department Mathematics, Physics & Chemistry, D – 13353 Berlin, Germany
[2] Helmholtz Centre Berlin for Materials and Energy, SF1, D - 14109 Berlin, Germany
[3] Institute of Nuclear Physics, University of Technology, Stadionallee 2, A-1020 Wien, Austria
[4] Ruprecht Karls University Heidelberg, Department of Physical Chemistry, D-69120 Heidelberg, Germany

[5] Email: treimer@helmholtz-berlin.de , treimer@beuth-hochschule.de

Abstract. In the case of ultra small angle (neutron or x-ray) scattering (USANS, USAXS) it may happen that structures under investigations are not fully coherently illuminated by the incident wave. Despite this fact interference effects are observed similar to SAS data. In this case the measured scattering patterns must be different interpreted. We propose a procedure to calculate and adapt such scattering patterns to experimental data.

1. Introduction
The investigation of large particles by means of small angle scattering is the domain of ultra small angle neutron and x-ray scattering, USANS and USAXS. Small angle scattering assumes a coherent illumination of the particles (structures) under investigation and it covers a q–range from 10^{-2}nm^{-1} down to 10^{-5}nm^{-1} corresponding to app. 50nm and 50µm particle sizes. USANS is performed with double crystal diffractometers (DCD), but also new "classical" (no DCD) instruments are planned or under construction to cover a range down to 10^{-4}nm^{-1}. Therefore DCDs still keep the record of low momentum transfer measurements for USANS or USAXS and are used worldwide. The advantages of these instruments are the so-called decoupling of momentum transfer resolution from beam divergence and then their compact layout. Due to the high angular correlation of the beam between both crystals the transversal coherence length $\Delta_{c,transv}$ is much larger than of conventional instruments. For a DCD it is given by $\Delta_{c,transv} \sim (2\ \Delta_{Darwin} \cdot k_{transv})^{-1}$ [1], [2], i.e. it depends directly on the Darwin width of the reflecting crystals, which is of the order of sec of arc or in q-space 10^{-5}nm^{-1}. For classical small angle instruments one can estimate $\Delta_{c,transv} \sim 0.5\mu m$ up to $5\mu m$ depending on the beam collimation and geometry.

2. Theoretical considerations
If the size of structures to be investigated becomes of the order of µm small angle scattering investigations must be performed with so-called double crystal diffractometer (DCD) [3]-[5]. The characteristic entity of a DCD is its Darwin width or Darwin – plateau (~µrad), and therefore its extreme small FWHM of a rocking curve, usually of the order of some µrad or sec of arc. This strong

angular-wave length correlation exists only between monochromator and analyzer crystal, and it is used to detect tiny deviations of a beam scattered by the structure between these crystals. The beam itself can have a much larger divergence (e.g. 0.5°) which does not (nearly does not) change the width of the Darwin plateau and so the wavelength sharpness between both crystals. In the case of dynamical neutron diffraction the so-called Darwin width or Darwin plateau $\Delta\theta_B$ is given by [1]

$$\Delta\theta_B = \frac{N \cdot b_c \cdot \lambda \cdot |F_{hkl}|}{2 \cdot \sin(2\theta_B)} \cdot \frac{\lambda}{2\pi} \frac{1}{\sqrt{|b|}} \cdot DWF \qquad (1)$$

N = number of atoms/ unit volume, b_c = coherent scattering length, λ = wave length, F_{hkl} = structure factor, b = asymmetry factor of the reflection and DWF = Debye –Waller factor. If $\Delta k_y / k$ means the transversal k – spread of the beam one can write

$$\Delta\theta_B = \frac{\Delta k_y}{k} \cdot \frac{DWF}{\sqrt{|b|}} \qquad (2)$$

and from this [2]

$$\Delta k_y = \frac{N \cdot b_c \cdot \lambda \cdot |F_{hkl}|}{4 \cdot \sin(2\theta_B)} \qquad (3)$$

The inverse of $\Delta k_y = 1/\Delta k_y$ yields some μm, depending on the reflection and Darwin width, and it is a measure for the transversal coherence length $\Delta_{c,transv}$. Using classical considerations based on the Huygens-Fresnel principle one can calculate the coherent area of a beam given by the area of the first Fresnel zone A_{FZ} as

$$A_{FZ} = \frac{R \cdot r}{R + r} \cdot \pi \cdot \lambda \cdot \frac{1}{\sin(\theta)} \qquad (4)$$

In the case of a DCD, the parameters R = distance source – object, r = distance object – detector, and θ must be defined. θ can be assumed to be ~ 90°, R and r are of the order of meter or less. For x-rays these parameters are well defined [6], [7] in the case of neutrons a non perfect graphite crystal as pre-monochromator serves as virtual source and the calculation are a bit more complicated [8], however, again one yields a transversal coherence length of the order of 10μm. Contrary to these values we measured with the V12b DCD with two different independent methods a transversal coherence length of ~ 90(20)μm [9], [10]. To overcome this discrepancy between theoretical transversal coherence length $\Delta_{c,transv}$ and experimental measured values we started a new series of experiments using now the diffraction of neutrons by different thick Cu wires. The thicknesses (together with a coating) were larger than 90μm and depending on the interaction with the wire (fully coherent or partially coherent interaction) different diffraction patterns were expected. In the case of partially coherent interaction one can consider the resulting diffraction pattern to be built up by an incoherent sum of diffraction patterns stemming from different parts of the diffracting object. The particular diffraction patterns can be calculated on the basis of the Huygens-Fresnel principle taking into account - different to small angle scattering treatment - phase relationships between partial waves in the sample (details are given in [11]).

3. Experiments

The experiments were performed with the high resolution double crystal diffractometer V12b at the BER II reactor of the Helmholtz Centre for Materials and Energy in Berlin (HZB). The instrument was situated at the Nl3b neutron guide and had in the guide a graphite pre-monochromator with a mosaic spread of 0.5°. The mean wavelength λ of the reflected neutrons was 0.533(1)nm, the intensity in front of the Cu samples (between the crystals) 5 x10^3/cm^2.sec. The resolution function, i.e. the rocking curve, had a FWHM of 6.0(0.05)sec of arc (2.9x10^{-5}rad), in the case of the "Darwin reduction" [9] 1.6(0.05)sec of arc (7.76 x10^{-6} rad). These values correspond to a momentum transfer of $\Delta q = 3.4 \times 10^{-4} nm^{-1}$ and 9.1×10^{-5} nm^{-1}, respectively. The Cu wires (coated with an electric isolation) were mounted between the crystals as shown in figure 1.

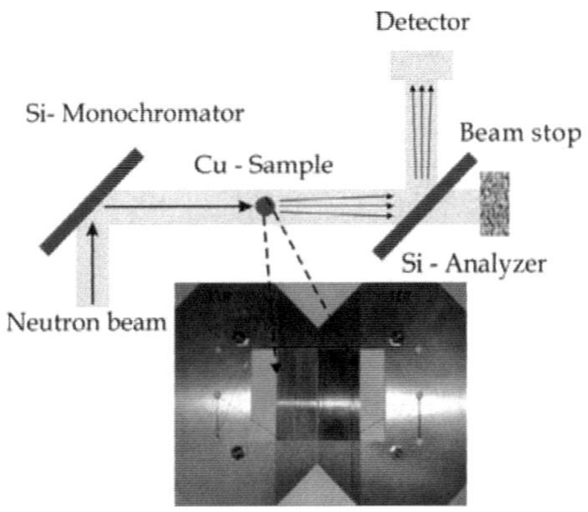

Figure1 Layout of the V12b DCD instrument and set of Cu wires (thickness = 130μm+ 9μm coating and 180 μm + 16μm coating). Si- monochromator and analyzer were seven-bounce channel cut crystal as in [9] (not shown here).

The distances of the 130μm thick wires from each other were 340μm – 410μm, in the case of the 180μm wires it was app 300μm, however for each individual Cu wire the distances to the adjacent ones were much larger than the expected transversal coherence length $\Delta_{c,transv}$. Because USANS scattering curves did not fit to the measured data the difference between different diffraction scattering curves basing on plane waves and Gaussian shaped waves having a finite lateral coherence length (FWHM) of 90μm (as was measured before [10]) were calculated assuming in both cases fully coherent illumination of the object. Figure 2 (a) and (b) shows the results, the agreement with experimental data was unsatisfactorily, it was worse assuming plane waves.

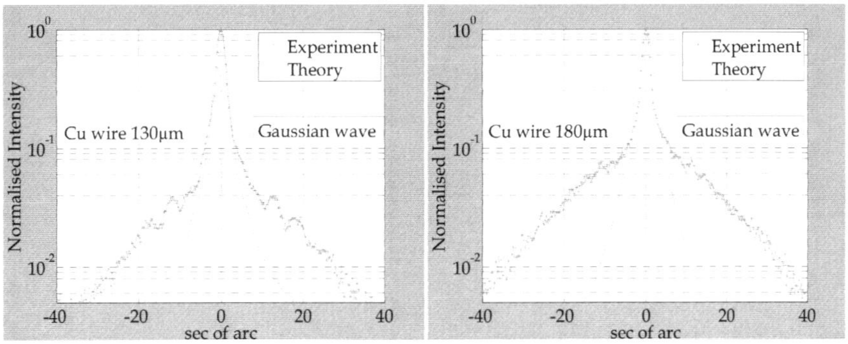

(a) (b)

Figure 2 Scattering curves by (a) 130μm and (b) 180μm thick Cu wires. Incident wave was assumed to have Gaussian shape (FWHM = 90μm), red = experimental data, blue = theory

The next step was to consider that neutron waves traverse the object at different parts. Based on this assumption one can calculate the incoherent sum of scattering curves due to such interactions. In Fig.3 the incident waves traverses the object at three different paths, for the 130μm thick Cu wire 65μm apart from each other and 90μm for the 180μm Cu wire. The different distances were scaled to the object width, i.e. the wider the object the larger the distance to the adjacent wire. These results are shown in figure 3. The now achieved agreement is much improved and it underlines that the proposed theoretical treatment describes the nature of interaction much better than standard theory of small angle scattering. If the distance of the wires from each becomes close to the FWHM of the Gaussian shaped wave the resulting scattering patterns converges to the plane wave treatment [11].

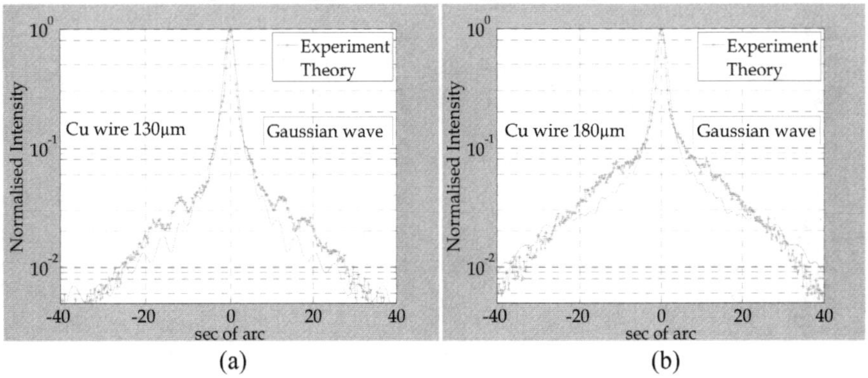

Figure 3 Diffraction pattern calculated by means of incoherent superposition of diffracted waves. The agreement of experimental data with theoretical curves (blue) is much better than in figure 2.

4. Summary

We presented a procedure to explain USANS scattering curves by means of incoherent superposition of diffracted waves emerging from different parts of the object (structure). In the case that structures are larger than the lateral coherence width of the incident (neutron) wave one has to apply a different procedure than classical (U)SANS theory to calculated the correct diffraction pattern. The incoherent superposition of diffracted partial waves apparently fits much better to scattering data than standard theory can yield.

This work was supported by the BMBF project no. 05KN7FK1

References

[1] Rauch H, Petraschek D 1978 *Topics Current Physics*, Springer Verlag **6**, 303–351,
[2] Jericha E, Baron M, Hainbuchner M , Loidl R , Villa M, Rauch H 2003 *J. Appl. Cryst.* **36** 778
[3] Schwahn D, Miksovsky A, Rauch H, Seidl E, Zugarek G,1985 *Nucl. Instr. & Meth. (A)* **239** 229.
[4] Treimer W, Berger G, Rückert B, Seidl E, 1987 *Z. Phys. B* **68** 527
[5] Aizawa K, Tomimitsu H, 1995 *Physica B* **213&214,** 884-886
[6] Kaganer V M, Jenichen B, Ploog K H, 2001 *J. Phys. D: Appl. Phys.* **34** 645
[7] Souvorov A, Drakopoulos M, Snigireva I, Snigirev A 1999 *J. Phys. D:Appl.Phys.***32** (1999) A184
[8] S. K. Sinha, M. Tolan, A. Gibaud, 1998 *Phys. Rev. B* **57**, 5 , 2740 - 2758
[9] Treimer W, Strobl M, Hilger A 2001 *Phys. Lett.* A **289** 151
[10] Treimer W, Hilger A, Strobl M 2006 *Physica B* **385–386** , 1388
[11] Treimer W, M. Strobl, Ebrahimi O, Seidel S O, Feye-Treimer U, (to be published)

First Observation of Neutron Deflections by Single Crystal Prisms

Sohrab Abbas[1], Apoorva G. Wagh, Markus Strobl* and Wolfgang Treimer*

Solid State Physics Division, Bhabha Atomic Research Centre, Mumbai - 400 085, India
*Helmholtz Zentrum Berlin, Glienicker Str. 100, D-14109 Berlin, Germany

[1]Email: abbas@barc.gov.in

Abstract. For a monochromatic neutron beam incident on a single crystal prism, we derived analytic expressions for variations of the deflection δ_{cr} and transmitted fraction I_O of forward diffracted neutrons, with the incidence angle θ. In the vicinity of a Bragg reflection, δ_{cr} deviates sharply from the deflection δ_{am} for an equivalent amorphous prism, reaching opposite extrema at either end of the total reflectivity domain and exhibits a 3 orders of magnitude greater sensitivity to θ than δ_{am}. We have experimentally observed the variation of δ_{cr} and I_O across a Bragg reflection for single crystal silicon prisms of various apex angles in several asymmetric Bragg configurations. The observed δ_{cr} deviates from δ_{am} by upto 27% with sensitivities $|d\delta_{cr}/d\theta|$ upto 0.43 arcsec/arcsec. The results are in harmony with theory. This observation has led to the design of a super-collimator Bragg prism producing neutron beams with sub-arcsecond angular widths.

1. Introduction

Amorphous prisms typically deflect thermal neutrons by upto a few arcsec [1, 2]. In a single crystal prism, for neutron incidence close to a Bragg reflection (Fig.1), the unreflected neutrons emerge from the side face of the prism into forward diffracted and diffracted beams [3-5]. As the Bragg reflection is approached from either side, the deflection δ_{cr} of the forward diffracted beam changes monotonically from its off-reflection (amorphous-prism) value δ_{am}, reaching an extremum at either end of the total reflectivity domain. We shall present here the first calculations and observations of deflections and intensity fractions of neutrons forward diffracted by single crystal prisms.

2. Theory

Let a neutron beam be incident on a single crystal prism surface at $\theta = \theta_B - \theta_S + LR/k_0$ outside the total reflectivity domain, viz. $|y| > 1$, in an asymmetric Bragg configuration (Fig.1) with a wave vector **RO** of magnitude k_O in vacuum. Here $y = (\theta - \theta_C) / w$ symbolises the reduced incidence angle, θ_C and w representing the centre and half-width respectively of the total reflectivity domain [4]. θ_B and θ_S denote the Bragg angle and the angle between the front face and the diffracting planes of the crystal respectively. Boundary conditions at the front face of the prism select a unique tie point T on the dispersion surface and at the exit face, dictate the emergent wave vector **SO** for the forward diffracted beam. The deflection, $-RS/k_O$, effected by the crystal prism is hence given by

$$\delta_{cr} = \delta_{am}\left\{1+\frac{F_H}{F_O}\sqrt{\frac{\sin(\theta_B-\theta_S)}{\sin(\theta_B+\theta_S)}}y\left(1-\sqrt{1-y^{-2}}\right)\right\}. \quad (1)$$

Here F_O and F_H signify the crystal structure factors for the **O** and **H** directions respectively and

$$\delta_{am} = -RS'/k_O = (n-1)\left\{\cot(\theta_B-\theta_S)-\cot(A+\theta_B-\theta_S)\right\}, \quad (2)$$

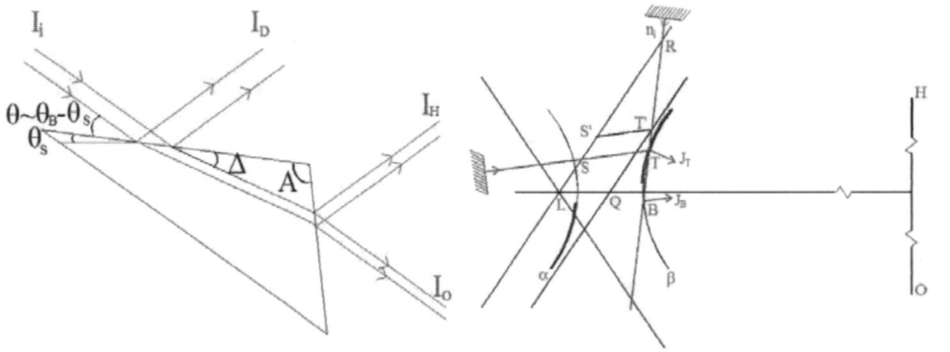

Fig.1 Neutron beam diffraction from a single crystal prism in real (left) and reciprocal (right) spaces.

is the amorphous prism deflection. δ_{cr} varies with θ in the arcsec/arcsec range just outside the total reflection domain ($|y| \gtrsim 1$), in contrast to the arcsec/deg variation of δ_{am}, implying a 3 orders of magnitude increase in sensitivity to the angle of incidence. As the Bragg reflection is approached from the low angle side, the magnitude of δ_{cr} decreases from $|\delta_{am}|$, reaching a minimum at $y = -1$. The deflection may even change sign (Fig.2b) for an appropriate choice of reflection and asymmetry. On approaching the Bragg reflection from the higher angle side, the magnitude of the deflection increases symmetrically by the same amount, attaining a corresponding maximum at the other extreme, viz. $y=1$, of total reflectivity (Figs. 2a and 2b) .

If the angle A between the front and side faces lies between $\theta_B+\theta_S$ and $\pi-(\theta_B-\theta_S)$, a fraction

$$I_O = \frac{\sin\Delta\sin(A+\theta_B-\theta_S)}{\sin(A+\Delta)\sin(\theta_B-\theta_S)} \quad (3)$$

of the incident intensity gets forward diffracted through the prism [5].

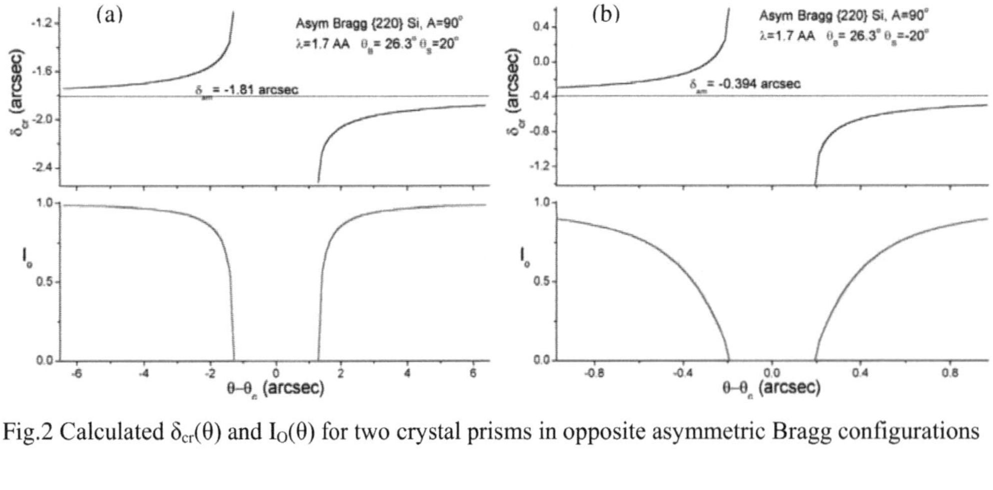

Fig.2 Calculated $\delta_{cr}(\theta)$ and $I_O(\theta)$ for two crystal prisms in opposite asymmetric Bragg configurations

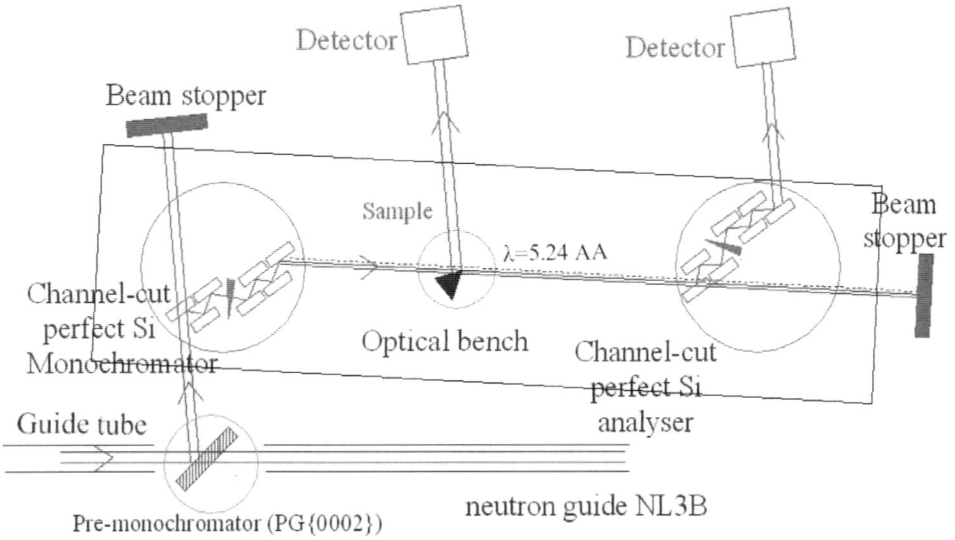

Fig.3 Schematic sketch of the experimental setup

3. Experiment

Shull [6] scanned a narrow slit across the diffracted beam from a crystal prism 36 years ago to deduce the $I_H(\theta)$ variation, yet could not measure its angular distribution due to nonavailability of arcsec wide neutron beams at that juncture. The advent of a 5.24 Å neutron beam with an angular FWHM ~ 2 arcsec [7] at the V12b double crystal spectrometer (Fig.3) of Hahn-Meitner-Institut in Berlin facilitated our experiment. Only a part of the neutron beam illuminated a silicon crystal prism, the remaining beam reaching the analyser directly to serve as a reference for the prism deflection measurement. One symmetric {111} prism of apex angle A equal to 56.7° and two asymmetric {111} prisms with asymmetry angles θ_S of 35° and 41° and apex angles of 94° and 98° respectively were investigated. Another detector was used to locate and continuously monitor the Bragg reflection from the prism. The analyser rocking curves were recorded for several angles of incidence θ at the prism covering a span of about 40 arcsec centred at its total reflectivity domain. At each θ, least-squares Voigt function fits were made to the direct and deflected neutron peaks in the rocking curve. The shift from the direct peak of and the area under the deflected peak (Fig.4) yielded δ_{cr} and I_O respectively.

4. Results and Discussion

Fig.5 depicts the experimental (points) and theoretical (curves) deflections δ_{cr} and transmitted neutron fractions I_O by the three silicon crystal prisms mentioned above. The sharp I_O curves (triangles) represent Eq. (3), incorporating the appropriate Debye-Waller factor. Each smooth I_O curve (squares) was obtained by convoluting the corresponding sharp curve with the incident 2 arcsec wide neutron angular profile, for comparison with the data. The neutron deflection and transmitted intensity fraction data agree well with predictions (cf. Section 2). The observed δ_{cr} deviate from δ_{am} by upto 27% with sensitivities $|d\delta_{cr}/d\theta|$ upto 0.43 arcsec/arcsec.

5. Conclusions

This paper has reported the first calculations and experimental observation of the deflection and intensity fraction of neutrons forward diffracted by a single crystal prism as a function of the angle of incidence θ. A crystal prism affords 3 orders of magnitude greater manoeuvrability over the deflection

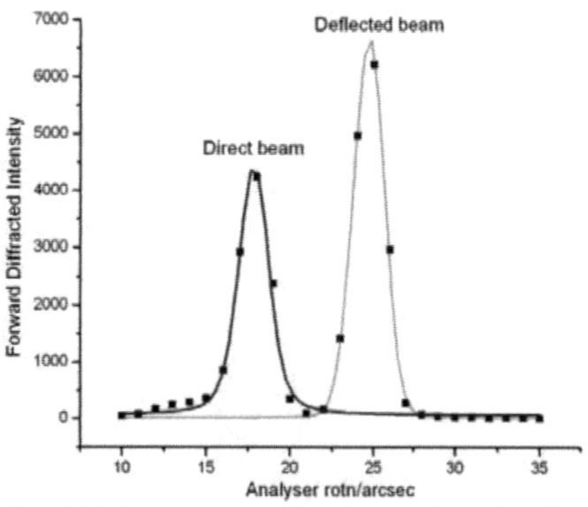

Fig.4 Voigt function fits (curves) to the direct and deflected beam scans (points).

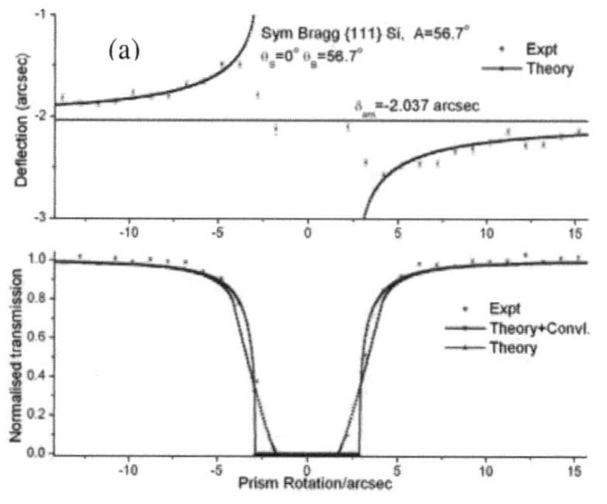

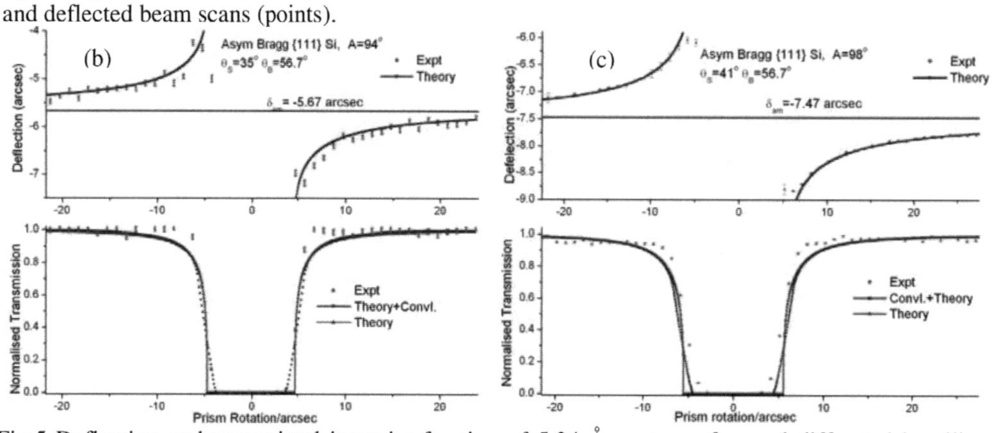

Fig.5 Deflection and transmitted intensity fraction of 5.24 Å neutrons forward diffracted by silicon crystal prisms: (a) symmetric {111} and asymmetric {111}: (b) $\theta_S = 35°$ and (c) $\theta_S = 41°$.

than an amorphous prism. Encouraged by this verification of our theoretical predictions, we have designed, fabricated and successfully operated a super-collimator Bragg prism monochromator [8] producing diffracted neutron beams (I_H in Fig.1) with sub-arcsec angular widths.

We acknowledge skilful assistance of M. Sharma of TP&PED, BARC and members of Chemie Lab, HMI (Berlin) in cutting a silicon prism each at the required apex angle and subsequently etching it.

References

[1] C S Schneider and C G Shull, *Phys. Rev.* **B3**, 830 (1971)
[2] A G Wagh and V C Rakhecha, *Pramana – J. Phys.* **63**, 381 (2004)
[3] B W Batterman and H. Cole, *Rev. Mod. Phys.* **36**, 681 (1964)
[4] A G Wagh, *Phys Lett* **A121**, 45 (1987); **A123**, 499 (1987)
[5] S Abbas and A G Wagh, *Sol. St. Phys. (India)* **50**, 317 (2005)
[6] C G Shull, *J. Appl. Cryst.* **6**, 257 (1973)
[7] A G Wagh, V C Rakhecha and W Treimer, *Phys. Rev. Lett.* **87**, 125504 (2001); W. Treimer, M. Strobl and A. Hilger, *Appl. Phys.* **A74** (Suppl.), S191 (2002)
[8] A G Wagh and S Abbas, *Sol. St. Phys. (India)* **51**, 353 (2006) and A G Wagh, S Abbas and W. Treimer, paper F2.5 in this volume (2009)

First Sub-arcsecond Collimation of Monochromatic Neutrons

Apoorva G. Wagh[†1], Sohrab Abbas[†] and Wolfgang Treimer*

[†]Solid State Physics Division, Bhabha Atomic Research Centre, Mumbai 400 085, India
*Helmholtz Zentrum Berlin, Glienicker Str. 100, D-14109 Berlin, Germany

[1]Email: nintsspd@barc.gov.in

Abstract. We have achieved the tightest collimation to date of a monochromatic neutron beam by diffracting neutrons from a Bragg prism, viz. a single crystal prism operating in the vicinity of Bragg incidence. An optimised silicon {111} Bragg prism has collimated 5.26Å neutrons down to 0.58 arcsecond. In conjunction with a similarly optimised Bragg prism analyser of opposite asymmetry, this ultra-parallel beam yielded a 0.62 arcsecond wide rocking curve. This beam has produced the first SUSANS spectrum in Q ~ 10^{-6} $Å^{-1}$ range with a hydroxyapatite casein protein sample and demonstrated the instrument capability of characterising agglomerates upto 150 μm in size. The super-collimation has also enabled recording of the first neutron diffraction pattern from a macroscopic grating of 200 μm period. An analysis of this pattern yielded the beam transverse coherence length of 175 μm (FWHM), the greatest achieved to date for Å wavelength neutrons.

1. Introduction

The sharpest [1] and narrowest [2] angular profiles for well separated up- and down-spin neutron[3] peaks were attained five years ago. For tightening the collimation further, we explored neutron diffraction from a single crystal prism. In the vicinity of a Bragg reflection, a fraction of neutrons incident on such a prism propagates through the crystal [4,5] and exits its side face [6-9] in forward diffracted and diffracted beams (Fig. 1). We have coined the term Bragg prism for this device.

2. Bragg prism

By judiciously optimising the Bragg reflection, asymmetric configuration, apex angle of the prism and incident beam position on the front face, the diffracted beam can be collimated down to a fraction of an arcsec [9]. An analyser prism can likewise be designed to accept an extremely narrow neutron angular profile. Fig. 2 depicts such an optimised Si {111} monochromator-analyser Bragg prism pair for 5.26 Å neutrons. A monochromator with the asymmetry angle $\theta_s = 50.1^o$ and apex angle $A = 172^o$ yields a beam collimated to 0.53 arcsec FWHM. An analyser with $\theta_s = -51^o$ and $A = 16^o$ is expected to accept a pair of 0.22 arcsec wide neutron peaks separated by 2.13 arcsec. Fig. 3 displays these predicted angular profiles, incorporating the appropriate Debye Waller factor and their convolution, comprising two 0.57 arcsec wide peaks separated by 2.35 arcsec. The single crystal prism combination thus produces and analyses a neutron beam with sub-arcsec collimation.

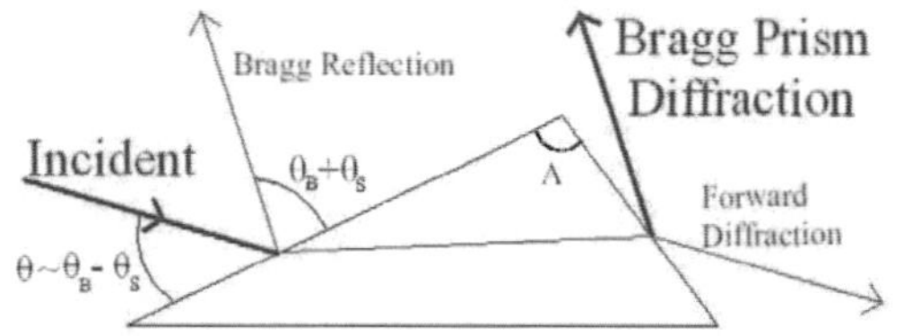

Fig. 1 Bragg prism diffraction (schematic). For incidence slightly outside the total reflectivity regime, unreflected neutrons emerge from the side face of the prism in forward diffracted and diffracted beams.

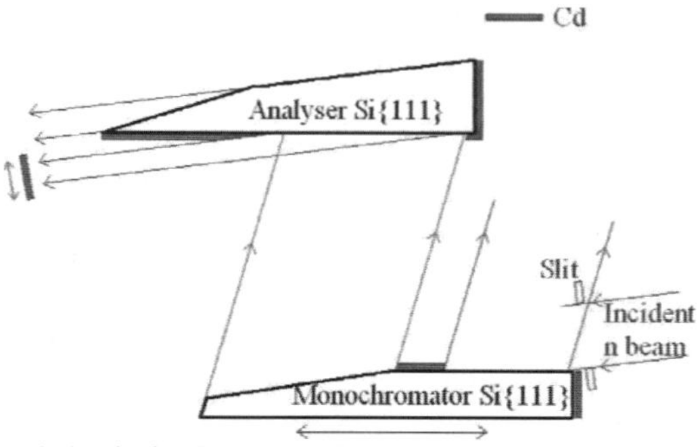

Fig. 2 Optimised Bragg prism monochromator-analyser pair for 5.26 Å neutrons.

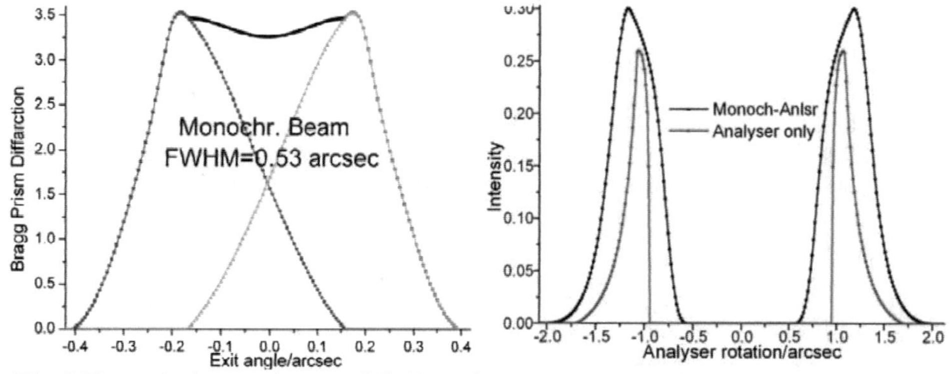

Fig. 3 Theoretical monochromated (left), analyser acceptance and combined neutron angular profiles.

3. Experimental

Several monochromators with θ_s close to 50° and one analyser with $\theta_s = -51°$ were fabricated with the specified apex angles at BARC. A <100> single crystal silicon ingot was aligned using {111} and {220} reflections of 1.2 Å neutrons at the triple axis spectrometer (TAS) in the Dhruva reactor. Cuts at the desired orientations and dimensions were made on this ingot on a Blohm precision grinding machine and subsequent surfacing achieved with a diamond polishing wheel at the Centre for Design and Manufacture (CDM). Final sub-micron polishing and a long (20 minute) and slow etching, to remove all residual strains, operations were performed at the Chemical Laboratory of Hahn-Meitner-Institut (HMI) in Berlin.

The experiment was carried out at the V12b Double Crystal Diffractometer set-up of BENSC (HMI) in Berlin. The analyser rotation could be adjusted in two stages; first in 1 arcsec steps with a geared step motor and then with a piezocrystal driven stage, with the smallest step size of 0.156 arcsec. Direct Bragg reflections, being much stronger and wider than prism diffractions, were first used, facilitating a quick and easy alignment of the analyser. For a monochromator prism ($\theta_s = 50.1°$), a 3.1 arcsec wide nearly triangular rocking curve was recorded (Fig. 4). The monochromator was then translated along the incident beam (see Fig. 2) to illuminate the analyser with its prism diffraction to obtain a 2.5 arcsec wide curve (Fig. 5). After optimising the analyser alignment, a cadmium sheet was introduced before the detector to stop Bragg reflected neutrons from the analyser. With these Bragg prism diffractions, the analyser tilt adjustment became even more critical and had to be made in 0.9 arcsec steps. The rocking curve (Fig. 6) comprising a pair of 0.62 arcsec wide peaks separated by 2.2 arcsec (squares), implies a 0.58 arcsec FWHM for each peak in the beam emanating from the monochromator and is in

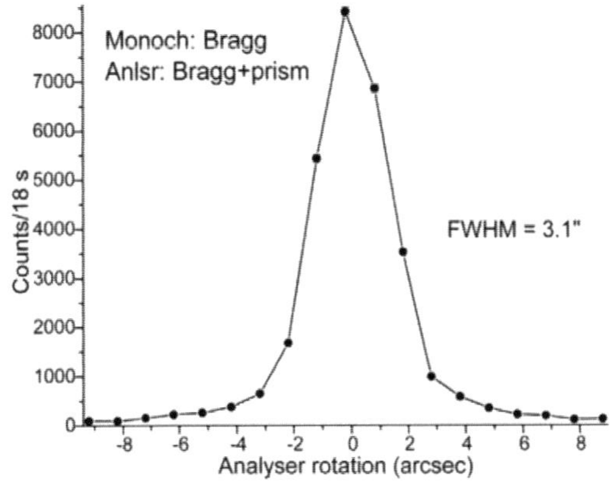

Fig.4 Monoch: Bragg, Analyser: Bragg+prism

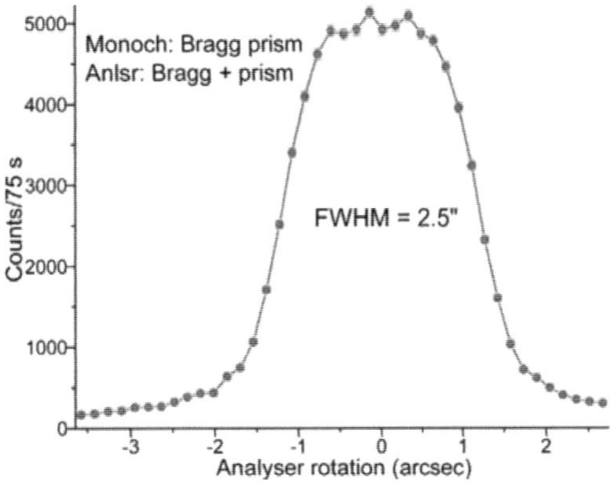

Fig.5 Monoch: Bragg prism, Analyser: Bragg+ prism

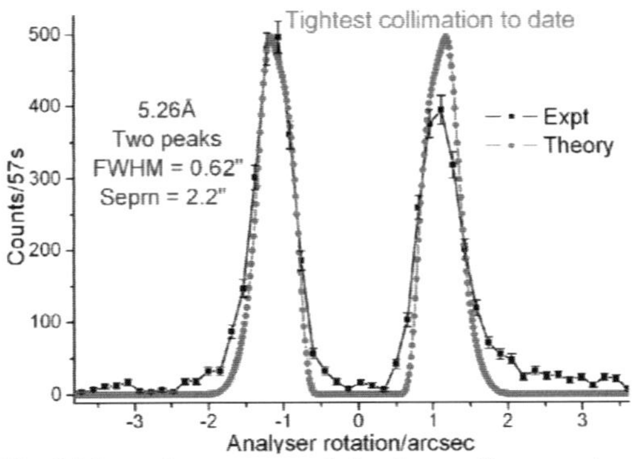

Fig.6 Monochromator and Analyser: Bragg prisms

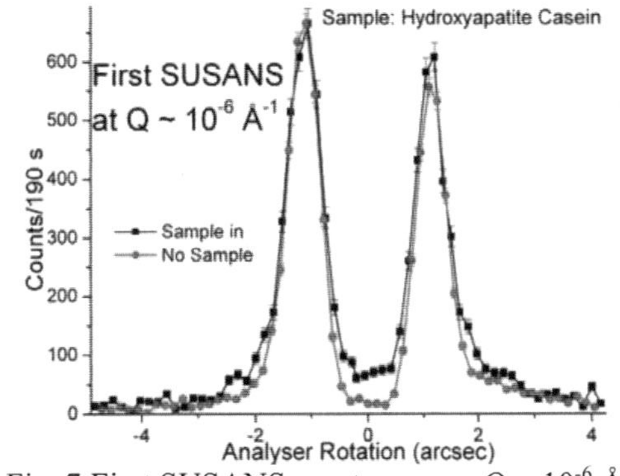

Fig. 7 First SUSANS spectrum over Q ~ 10^{-6} Å^{-1}

fair agreement with the theoretical prediction (circles). This constitutes the tightest collimation attained to date in the world for a monochromatic neutron beam.

This super collimated neutron beam can probe wave vector transfers Q ~ 10^{-6} Å^{-1}. The first SUSANS (Super Ultra-Small Angle Neutron Scattering) spectrum in this Q-range was recorded with a hydroxyapatite casein protein sample (Figs. 7-8) placed between the monochromator and analyser. The log-normal size distribution of spherical agglomerates in the sample inferred [10] from the SUSANS spectrum has a 53 μm median radius and 46 μm FWHM between half maxima of 27 and 73 μm. The greater half-maximum radius implies the instrument capability of characterising agglomerates upto about 150 μm in size.

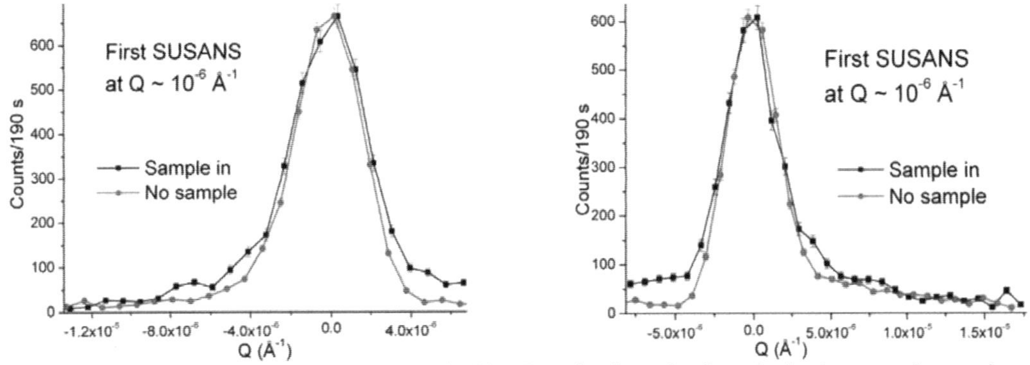

Fig. 8 Left (left) and right (right) peaks in Fig. 7 on the Q-scale. Sample: hydroxyapatite casein

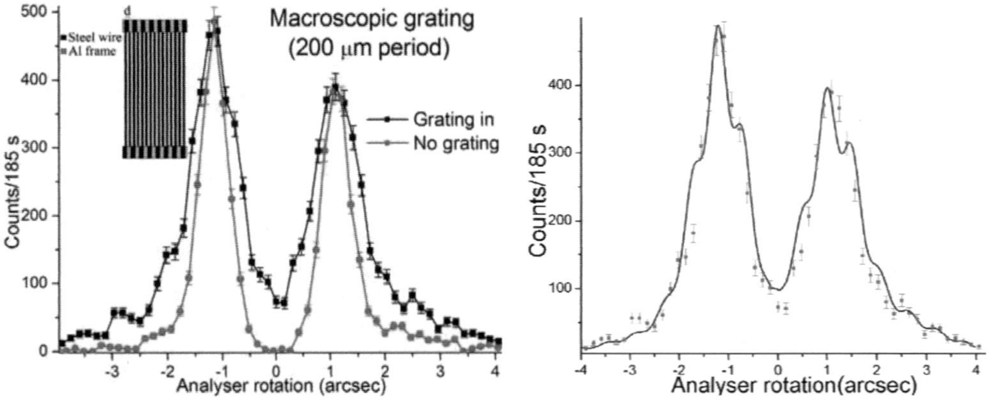

Fig. 9 First neutron diffraction pattern with a grating of 200 μm period (inset) and fitted curve (right)

Fig. 9 is a SUSANS pattern of a grating of ~ 200 μm period made by winding a steel wire of 100 μm diameter tightly on a 50x50 mm^2 aluminium frame (inset). The peaks, considerably broadened due to multiple scattering and refraction in the cylindrical wires, are modulated by clearly resolved diffraction oscillations. The transverse coherence length of 175 μm (FWHM) derived from the least squares fit (Fig.9 right) to the data far exceeds the previous best value of 80 μm [11] obtained for a 1.4 arcsec wide beam as well as the highest (5μm) observed in neutron interferometry [12].

6. Conclusion

We have presented the narrowest, viz. the first sub-arcsec, angular profile for a neutron beam having the highest transverse coherence length and affording SUSANS studies down to Q ~ 10^{-6} Å^{-1}.

Acknowledgements

We gratefully acknowledge help from R. Mittal (ingot alignment), S.S. Patil of CDM (prism fabrication), S.S. Jadhav of Spectroscopy Division (polishing one prism), D. Sahoo of TP&PED (etching facilities), members of Chemie lab, HMI (polishing and etching of prisms), Markus Strobl (supplying the protein sample) and O. Ebrahimi, S. Keil, P. Walter, R. Monka and other students of W. Treimer (skilful assistance) and HMI (hospitality during experimental runs).

References

[1] A G Wagh, V C Rakhecha and W Treimer, *Phys. Rev. Lett.* **87**, 125504 (2001); *Appl. Phys.* **A74 (Suppl.)**, S171 (2002)
[2] W Treimer, M Strobl and A Hilger, *Appl. Phys.* **A74 (Suppl.)**, S191 (2002)
[3] A G Wagh, V C Rakhecha, M Strobl and W Treimer, *Pramana – J. Phys.* **63**, 369 (2004); *J. Res. NIST (USA)* **110**, 231 (2005)
[4] B W Batterman and H. Cole, *Rev. Mod. Phys.* **36**, 681 (1964)
[5] A G Wagh, *Phys. Lett.* **A121**, 45 (1987); **A123**, 499 (1987)
[6] G Borrmann, G Hilderbrandt and H Wagner, *Z. Phys.* **142,** 406 (1955)
[7] A Authier, *J. Phys. Radium* **23**, 961 (1962)
[8] C G Shull, *J. Appl. Cryst.* **6**, 257 (1973)
[9] A G Wagh and S Abbas, *Solid State Phys. (India)* **51**, 353 (2006)
[10] S K Sinha, *Lecture notes on 'Interaction of X-rays and Neutrons with matter', www.dep.anl.gov/nx/lecturenotes.pdf*
[11] W Treimer, A Hilger and M Strobl, *Physica.* **B385-386**, 1388 (2006)
[12] H Rauch, H Wölwitsch, H Kaiser, R Clothier and S A Werner, *Phys. Rev.* **A53**, 902 (1996)

Neutron Beam Conditioning for Focusing SANS Spectrometers

János Füzi[1, 2] and László Rosta[1]

[1] Neutron Spectroscopy, Research Institute for Solid State Physics and Optics, Konkoly-Thege 29-33, H-1121 Budapest, Hungary
[2] Electrical Engineering, *Transilvania* University, Brasov, Romania

fuzi@szfki.hu

Abstract. Multibeam focusing offers an appealing compromise between high resolution and high incident flux configurations for SANS spectrometers. In fact so many "spectrometers" operate in parallel as the number of channels in the collimator. Each channel provides high resolution by small spot size on the detector and long sample-to-detector distance, involving significant limitation of the transmitted beam phase space volume, thus reducing the flux. The flux on the sample is increased by the large number of channels. In view of the multibeam collimation it is beneficial to increase the beam cross section and decrease the divergence at the same time. Two aspects related to the use of rotational velocity selectors are investigated. First the transmitted phase space is determined from the selector parameters. It is found that the beam azimuthal divergence with respect to the rotor axis has a significant effect on the selectivity. Neutrons flying along different paths are treated differently, leading eventually to energetic non-uniformity of the illumination of various collimator channels. Then the effect of the gap in the neutron guide at the selector location on the phase space uniformity at the collimator entrance is investigated and optimal selector location along the beam is proposed together with optimal neutron guide shape in the vicinity of the gap, which accommodates the selector.

1. Introduction

The angular resolution of a small angle scattering instrument with geometry defined in Figure 1 is given by [1, 2]:

$$(\sigma_\theta)^2 = \frac{1}{4}\left(\frac{r_1 f}{ld}\right)^2 + \frac{1}{4}\left(\frac{r_2(l+f)}{ld}\right)^2 + \frac{1}{12}\left(\frac{\Delta r}{d}\right)^2, \quad (1)$$

where Δr is the detector resolution. The smallest detectable scattering angle (Figure 2) is $\Delta\theta = r/d$. Small collimator aperture diameters and large collimation and focal distances ensure high angular resolution, but lead to significant limitation of the accepted phase space of the incoming beam.

The scattering vector resolution includes the selectivity of the velocity selector $\Delta\lambda/\lambda$:

$$(\sigma_q)^2 = k^2(\sigma_\theta)^2 + \frac{k^2}{12}\left(\frac{a}{d}\right)^2\left(\frac{\Delta\lambda}{\lambda}\right)^2, \quad (2)$$

where $k = 2\pi/\lambda$.

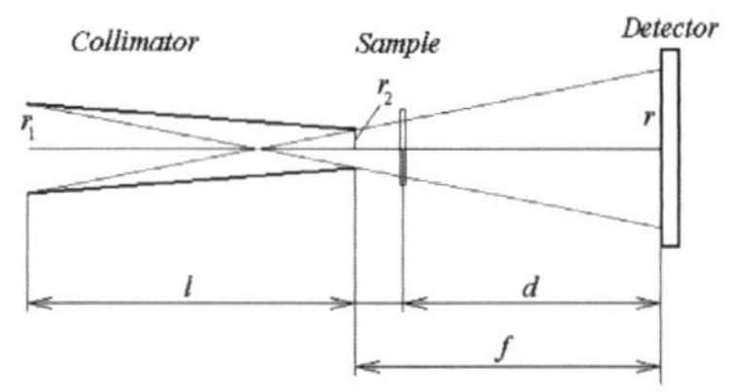

Figure 1. Dimensions

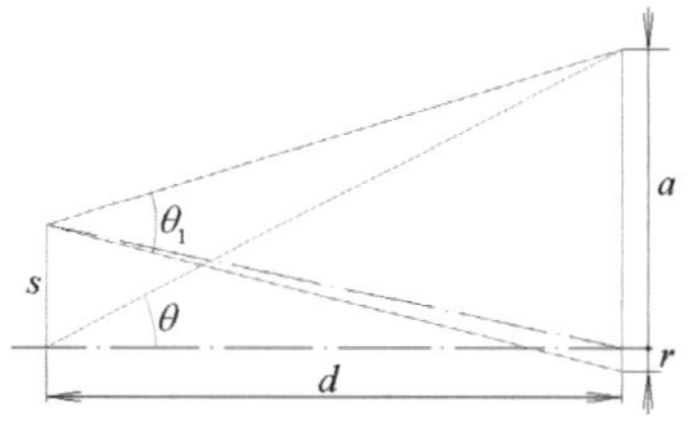

Figure 2. Scattering angles

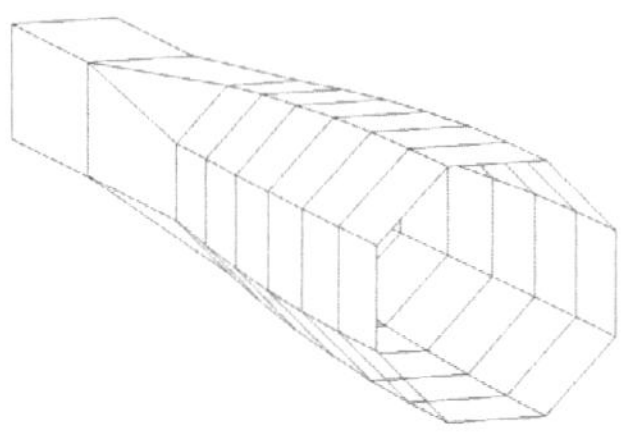

Figure 3. Beam expander

The use of multibeam collimation ensures the increase of the flux illuminating the sample by the number of channels n. The collimator transmission parameter becomes:

$$A\Omega = n\frac{\pi^2 r_1^2 r_2^2}{l^2} \tag{3}$$

2. Neutron beam phase space adjustment

In order to accommodate a larger number of channels, the beam cross section has to be increased at the entrance of the collimator, especially when a relatively small cross-section beam feeds the instrument. This can be done by means of a beam expander (Figure 3), on the expense of divergence decrease (Figure 4), to the extent of fully illuminating all channels. The condition is that the reflection angles seen from the sample upstream through the collimator do not exceed the critical angle of the incoming guide supermirror. The expander shown in Figure 3 is 5.6 m long and transforms the beam cross section from a 25 mm square to a 56 mm high octagon. The collimator is 4 m long and one channel diagonal is 2.8 mm at the inlet and 1.4 mm at the exit. The direct beam spot diagonal on the detector is 5.6 mm.

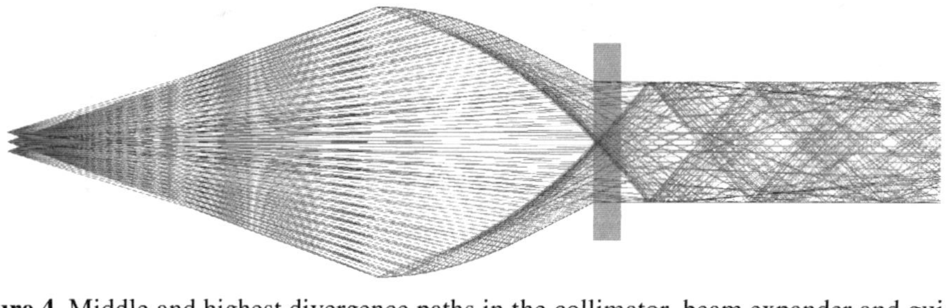

Figure 4. Middle and highest divergence paths in the collimator, beam expander and guide. The grey stripe is the optimal location of the velocity selector. Vertical : Horizontal scale: 100 : 1.

The highest reflection angle in the beam expander: 0.4°, corresponding to m = 1.33 at 4 Å; inside the incoming guide: 0.69°, corresponding to m = 2.29 at 4 Å.

3. Velocity selector location

Gaps in a neutron guide imply dark stripes in the beam image when viewed through a pinhole. The stripes become wider with increasing number of reflections (Figure 5). The gap accommodating the selector would cause partially or totally "blind" channels.

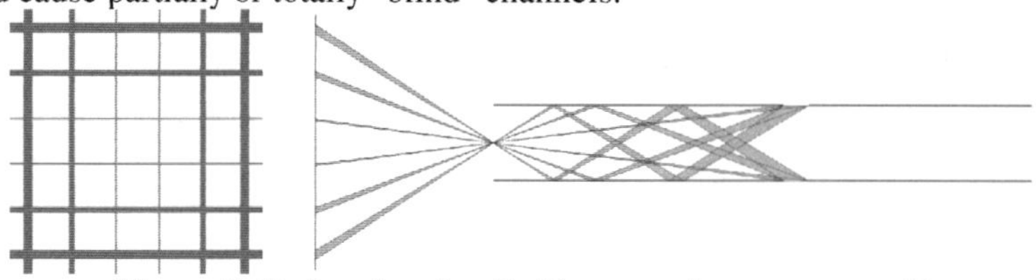

Figure 5. Dark stripes implied by a gap in a neutron guide

As shown on Figure 4, close to the exit end of the neutron guide there is a region with low reflection angles, below 0.06°. Thus here is the optimal location of the velocity selector. Increasing the guide cross section by 0.6 mm all around, there will be no line of sight of the guide end upstream of the gap (Figure 6).

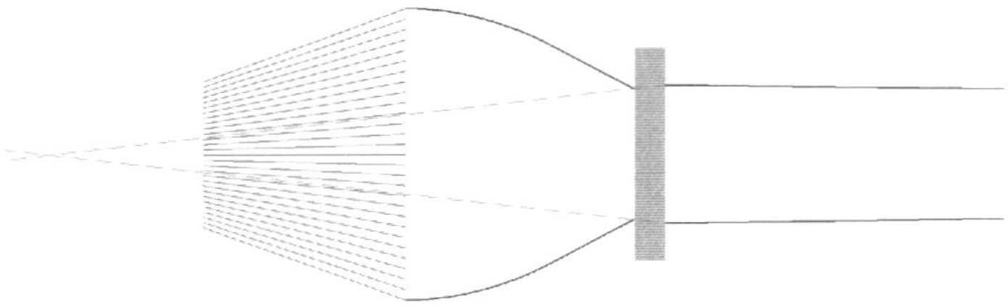

Figure 6. Guide shape at the selector location

4. Velocity selector characteristics

The characteristics of a multiblade velocity selector of constant 48000 rpm Å have been determined by means of time-of-flight pinhole imaging (Figure 7). The beam image taken at 9600 rpm in the wavelength range 4 – 6 Å can be seen on Figure 8, together with selected summation areas corresponding to +/- 1° (*a*), +/- 0.05° (*b*), - 0.64° ... - 0.54° (*c*) and 0.54° ... 0.64° (*d*) azimuthal (horizontal) respectively +/- 0.25° radial (vertical) divergence.

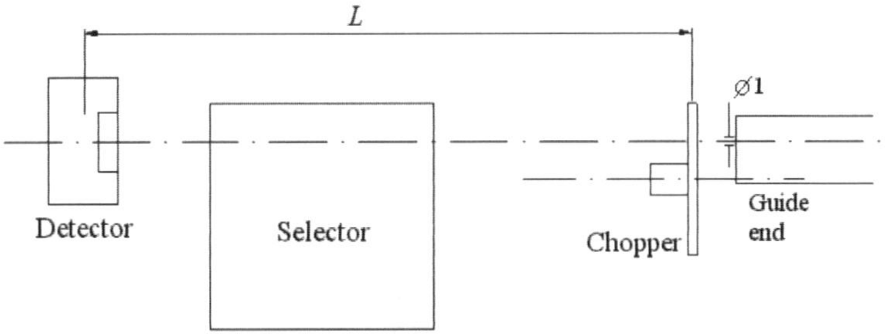

Figure 7. Measurement setup

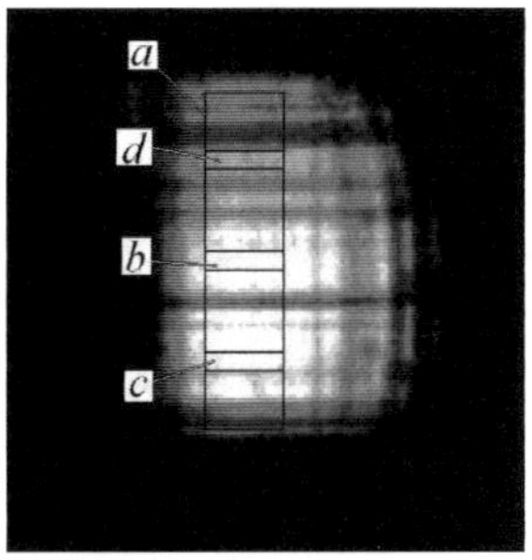

Figure 8. Beam image through the selector

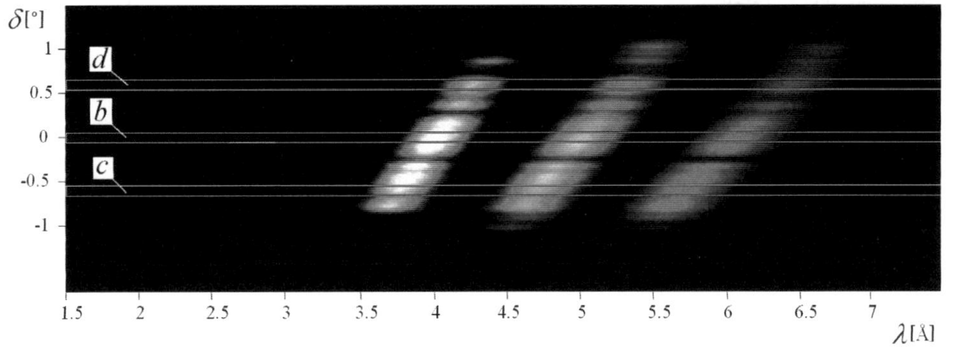

Figure 9. Transmitted intensity distribution with respect to wavelength and azimuthal divergence at 1200 rpm (leftmost stripe), 9600 rpm and 8000 rpm respectively, at zero rotor tilt angle

The transmitted intensity distribution with respect to wavelength, azimuthal divergence and rotor tilt angle is mapped on Figures 9 and 10. The results show, in accordance with theory [3], that azimuthal divergence has a significant effect on the selectivity (Figure 11). This means that different channels of the collimator will be illuminated by slightly different wavelength neutrons depending on the divergence angle it imposes with respect to the beam axis, in azimuthal direction to the selector rotor axis. Better and more uniform selectivity could be obtained placing the selector between the expander and the collimator (the divergence accepted by the collimator is 0.21°). This would require larger selector window – rotor diameter, and more space for radiation shielding. The selectivity curves as a function of rotor tilt angle and symmetrical azimuthal divergence range are plotted on Figure 12.

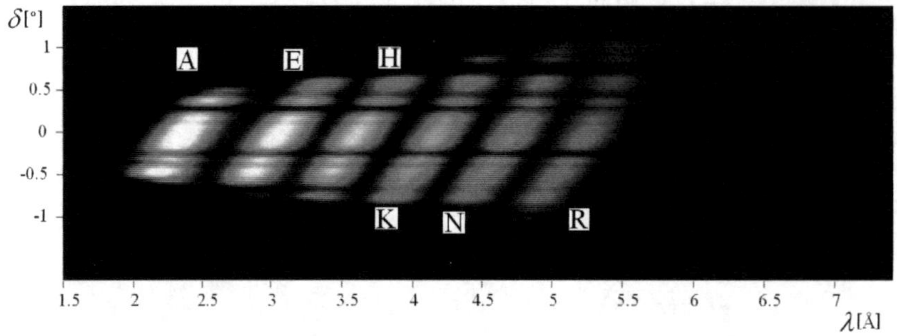

Figure 10. Transmitted intensity distributions at several tilt angles, with respect to wavelength and azimuthal divergence. Rotation speed: 12000 rpm

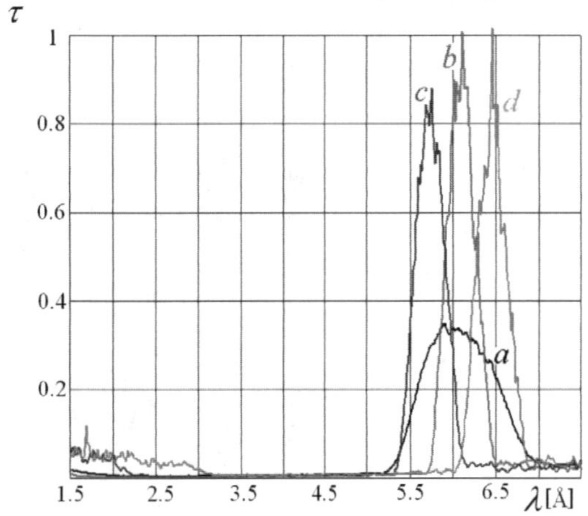

Figure 11. Transfer functions at 8000 rpm - the effect of azimuthal divergence

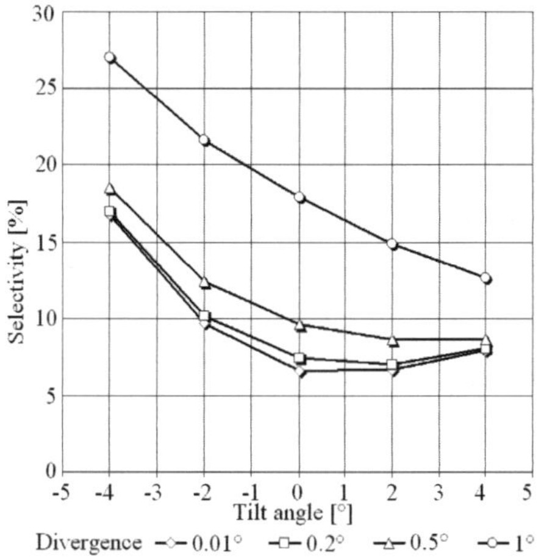

Figure 12. Selectivity as a function of rotor tilt angle

5. Conclusions

Aspects of neutron beam conditioning in view of the multibeam collimation for focusing SANS instruments have been investigated. Solutions for beam phase space adjustment are proposed together with őptimal selector location and neutron guide shape in the vicinity of the gap, which accommodates the selector. Thorough experimental verification of the selector transfer functions is presented.

Acknowledgment

The work has been supported by the *Jedlik Ányos* R&D project 00079/2008 granted by the Hungarian National Agency for Research and Technology.

References

[1] Mildner DFR, Carpenter JM: Optimization of the Experimental Resolution for Small-Angle Scattering, *J Appl Cryst* **17** (1984), 249-256

[2] Mildner DFR, Carpenter JM: Resolution of Small-Angle Scattering with Soller Collimation, *J Appl Cryst* **20** (1987), 419-424

[3] Rosta L, Füzi J, Hományi L: Benchmark testing of a multiblade neutron velocity selector; *Physica B* **385-386** (2006) 1283-1286

International Conference on Neutron Scattering 2009 IOP Publishing
Journal of Physics: Conference Series **251** (2010) 012076 doi:10.1088/1742-6596/251/1/012076

Neutron beam focusing using large-m supermirrors coated on precisely-figured aspheric surfaces

D. Yamazaki[1], **R. Maruyama**[1], **K. Soyama**[1], **H. Takai**[2], **M. Nagano**[2] **and K. Yamamura**[2]

[1] J-PARC Center, Japan Atomic Energy Agency, Tokai-mura, Ibaraki 319-1195, Japan
[2] Research Center for Ultra-precision Science and Technology, Osaka University, Osaka 565-0871, Japan

E-mail: yamazaki.dai@jaea.go.jp

Abstract. We have developed a 1-dimensional elliptic mirror combining a supermirror coated with ion-beam sputtering and precise elliptic surface figured with the numerically-controlled local wet etching process. In this study, NiC/Ti supermirror (m=4) was deposited on a precisely figured surface of synthesized quartz glass over 90 mm × 40 mm. Wideband neutrons of $\lambda > 3.64\text{Å}$ were focused with focal spot size down to 0.25 mm, peak intensity gain up to 6 without significant di use scattering. Time-of-flight measurements suggest that wideband neutrons are e ectively focused to the focal point.

1. Introduction

In recent years, there have been major developments of new neutron facilities including J-PARC, SNS and ISIS TS2. The new or upgraded sources will provide higher neutron intensity than ever. Nevertheless the intensities are still low compared to that of X-ray facilities. Neutron beam focusing devices are, therefore, indispensable in these facilities to accommodate micrometer-scale samples and second-resolution time-resolved measurements. As for focusing devices at spallation neutron sources, achromatic optics are particularly important because most experiments are performed with wideband beams. Re ective focusing is a prominent candidate for that purpose.

A focusing mirror needs to accept and re ect wideband beam with large divergence into small area with high e ciency and high signal-to-noise ratio. In order to meet these demands required are (1) large critical angle of total re ection, (2) high re ectivity, (3) low di use scattering, (4) large dimension and (5) precise surface gure of the focusing mirror. The low di use scattering (3) is particularly important to focus the beam on to a detector surface in order for small-angle scattering (SANS) or grazing-incidence SANS.

Ice and coworkers have developed focusing devices by bending a at supermirror into approximately elliptic shape [1, 2]. Their devices utilize $m = 3$ supermirror deposited on at substrates and have realized 2-dimensional focusing onto a 90 m spot. (m denotes the ratio of the total-re ection critical angle of the supermirror to that of natural nickel.)

We have been developing ultra-precise focusing mirrors combining high-performance neutron supermirror and very precise surface guring technique. We have already developed high-performance neutron supermirrors using ion-beam sputtering (IBS) process. Re ectivities 0.88 at $m = 3$, 0.82 at $m = 4$ and 0.40 at $m = 6$ have been achieved [3] with low di use scattering

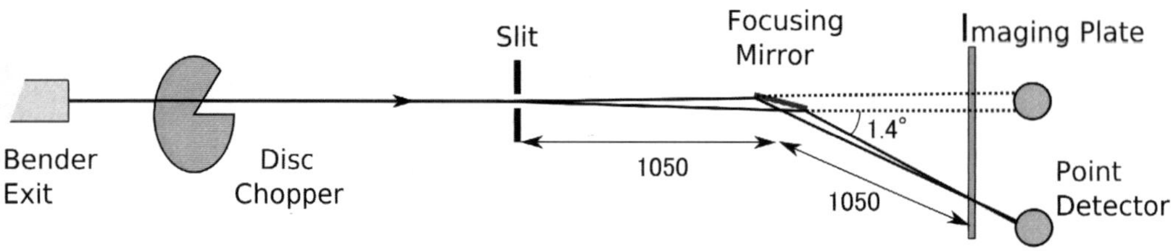

Figure 1. Experimental setup.

[4]. Critical angle up to $m = 6.7$ has also been demonstrated [5]. Mirrors can be deposited over a range of 500 mm in diameter. On the other hand, high-precision surfaces gurings have been realized over a large area using the numerically controlled local wet etching (NC-LWE) process [6, 7]. Peak-to-valley atness down to 69 nm has been achieved over 142 142 mm. In this letter, a fabrication of a 1-dimensional elliptic neutron supermirror and a performance test using wideband neutron beam are reported.

2. Mirror Fabrication

A mirror substrate was prepared by guring a surface of synthesized quartz glass into a plano-elliptical shape using the NC-LWE process. The ellipsoid is of $x^2/(1050.31)^2 + z^2/(25.66)^2 = 1$, where the units of x and z are mm. The focal length is 1050 mm. The aperture size is 90 mm (elliptic) 40 mm (straight) and it covers eccentric angle $87.55 < \ < 92.45$ deg of the ellipsoid. NiC/Ti supermirror ($m = 4$) was deposited using the IBS method on the elliptic quartz surface.

The elliptic mirror is supposed to accept a neutron beam with nominal glancing angle of 1.40 deg and re ect and focus wideband neutrons with wavelength $\lambda \geq 3.5$A. The acceptance angle, therefore, is 0.12 deg. Details of the fabrication are described elsewhere [8, 9] together with a test result of monochromatic beam focusing .

3. Experimental Results and Discussions

Wideband neutron beam focusing was demonstrated on the CHOP beam port of JRR-3. CHOP provides a neutron beam of $\lambda \geq 3$A which is bent by 20 degrees with a multi-channel cold neutron bender [10] at the end of the C-2 guide tube.

The beam-line is illustrated in Fig. 1. The focusing mirror was placed at an angle of 1.4 deg to the beam line. A horizontal beam slit was installed 1050 mm before the mirror. The slit width was 0.23 mm. Focused beam shape was observed with an imaging plate placed at 1050 mm after the mirror. Beam divergence was 0.09 deg, which results in the unfocused beam width 2.97 mm at the observation point.

Observed beam shapes and their horizontal pro les are shown in Fig. 2. Measurement time was 11160 seconds for each image. The intensities are expressed in "photo-stimulated luminescence (PSL)" per pixel of an imaging plate (50 50 m^2). Focused beam width was 0.25 mm in full width at half maximum, almost reproducing the initial beam size de ned by the slit. This result is adequate because the focusing mirror is based on the elliptic arc with eccentric angle around 90 degree. Focused peak intensity was 6 times the unfocused average intensity. Note that the tail of the focused beam is no more than that of the unfocused beam pro le, which suggests that the focusing mirror produces little di use scattering. Background PSL values randomly dispersed between 10 ^{4} and 10 ^{2} and were neglected in the analysis.

Figure 3(a) compares a time-of- ight (TOF) spectrum of the focused beam with that of the unfocused one, which were measured with a ^{3}He point detector. (Sharp dips are attributed to Bragg re ections by monochromators of upstream instruments in the C-2 beam line.) One

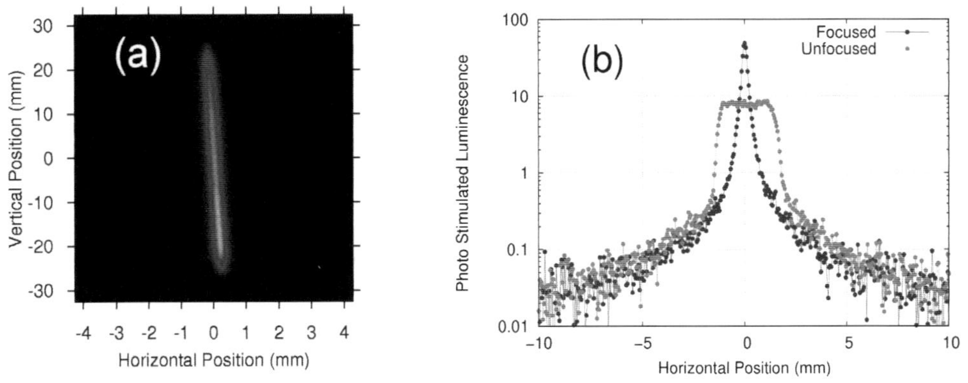

Figure 2. (a) 2-dimensional image of the focused beam. Note the scale di erence between the two axes. (b) Horizontal intensity pro les of focused and unfocused beam at the vertical center.

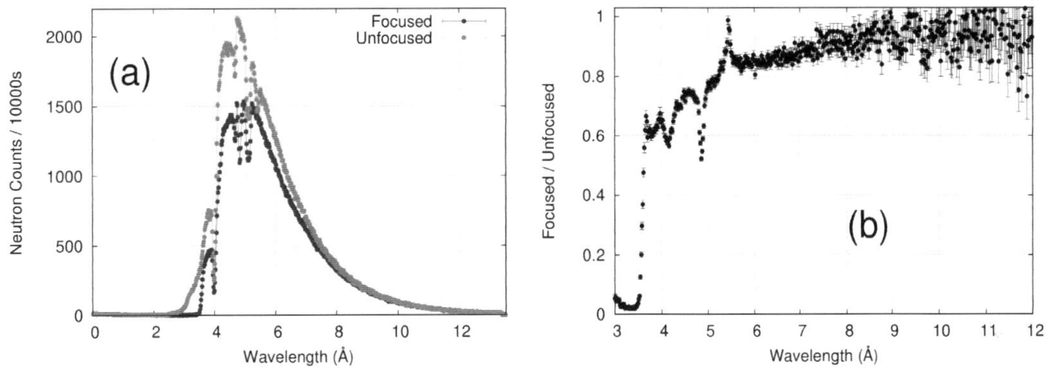

Figure 3. (a) TOF spectra of the focused and unfocused beam. (b) Ratio of the two spectra.

can see that neutrons are re ected and focused over a wide wavelength band. The rising of the focused beam pro le at wavelength $\simeq$ 3.64 A represents the critical wavelength of total re ection by the mirror. Assuming the glancing angle at the mirror is 1.40 degree, the critical wavelength corresponds to critical scattering vector $Q_z \simeq 0.084$ A ^{1} or $m \simeq 3.87$. This value is slightly smaller than the speci cation $m = 4$. This di erence could be ascribed to an error in the glancing angle because it was adjusted with manual stages and a wide detector-slit (5mm).

Figure 3(b) plots the ratio of the two TOF pro les.(The dips and peaks are attributed to shifts of dip-positions in the TOF spectra which result from changes in the monochromator angles.) We can see a typical re ectivity pro le of supermirror. The re ectivity $\simeq$ 0.6 at the critical wavelength also is a typical value. This result suggests that most re ected neutrons are transported to the focal spot without signi cant loss.

The same measurements were carried out with the slit width $\simeq$0.43 mm. Focused beam size was 0.42 mm and peak intensity was 4.6 times the unfocused intensity. Di use scattering once again was not signi cant. TOF and re ectivity measurement yielded similar pro les to the above case.

These results show that our focusing mirror meets the ve requirements mentioned above.

4. Future plans and summary

We are developing more e cient focusing mirrors with larger acceptance angles and smaller focal spot sizes. Acceptance angle can be improved by fabricating larger focusing mirrors and/or stacking them. The mirror size, which is now limited by the coating area of our IBS machine, is readily extendable to 400 mm or more in length. Smaller focused beam size can also be obtained by a mirror based on an elliptic arc with smaller eccentric angles. We are also developing two dimensional focusing by guring elliptic-cylinder surface or combining two elliptic mirrors (Kirkpatrick-Baez mirror).

In summary, we have developed a 1-dimensional elliptic mirror combining the supermirror deposition with ion-beam sputtering and precise elliptic surface gured with the numerically-controlled local wet etching process. Wideband neutrons of $\lambda > 3.64$A were focused with focal spot size down to 0.25 mm, peak intensity gain up to 6 and signal-to-noise ratio less than the non-focused beam. Time-of- ight measurements suggest that wideband neutrons are e ectively focused to the focal point.

Acknowledgments

Authors acknowledge K. Aizawa, S. Takata and I. Tamura for instructions of the CHOP instrument at JRR-3. This work is partially supported by the Industrial Technology Research Grant Program in 2005 from the New Energy and Industrial Technology Development Organization (NEDO) of Japan and a research grant from the Material Tool Engineering Foundation.

References

[1] Ice GE, Hubbard CR, Larson BC, Pang JWL, Budai JD, Spooner S and Vogel SC 2005, *Nucl. Instrum. Meth.* A **539** 312-20
[2] Ice GE, Hubbard CR, Larson BC, Pang JWL, Budai JD, Spooner S, Vogel SC, Rogge RD, Fox JH, Donaberger RL, *Mater. Sci. Eng.* A **437** 120-5
[3] Maruyama R, Yamazaki D, Ebisawa T, Soyama K 2009, *Nucl. Instrum. Meth* A **600** 68-70
[4] Maruyama R, Yamazaki D, Ebisawa T, Soyama K 2009, *J Appl. Phys.* **105** 083527
[5] Maruyama R, Yamazaki D, Ebisawa T, Hino M and Soyama K 2007, *Thin Solid Films* **515** 5704-6
[6] Yamamura K 2007, *Sci. Tech. Adv. Mater.* **8** 158-61
[7] Yamamura K 2007, *Annals of the CIRP* **56/1** 541-4
[8] Yamamura K, Nagano M, Takai H, Zettsu N, Yamazaki D, Maruyama R, Soyama K and Shimada S 2009, *Opt. Express.* **17** 6414-20
[9] Nagano M, Takai H, Yamazaki D, Maruyama R, Soyama K and Yamamura K, in this proceedings.
[10] Tamura I, Maruyama R, Yamamoto K, Nakamura K, Aizawa K and Soyama K 2008, *JAEA Review* 2008-004 p100

Journal of Physics: Conference Series **251** (2010) 012077 doi:10.1088/1742-6596/251/1/012077

Fabrication of plano-elliptical neutron focusing supermirror by numerically controlled local wet etching with ion beam sputter deposition

M Nagano[1], H Takai[1], D Yamazaki[2], R Maruyama[2], K Soyama[2] and K Yamamura[1]

[1]Research Center for Ultra-precision Science and Technology, Graduate School of Engineering, Osaka University, 2-1 Yamadaoka, Suita, Osaka, 565-0871, Japan
[2] J-PARC Center, Japan Atomic Energy Agency, Tokai, Ibaraki, 319-1195, Japan

E-mail: yamamura@upst.eng.osaka-u.ac.jp

Abstract. High-performance optical devices with a figure accuracy of sub-micrometer level and a surface roughness of atomic level are required to collect and/or to focus a neutron on a sample without scattering loss. To fabricate a high-performance neutron focusing supermirror, a two-stage numerically controlled local wet etching technique combined ion beam sputter deposition was developed. By applying this technique, a plano-elliptical substrate made of synthesized quartz glass with a figure accuracy of less than 0.5 μm was fabricated, and its surface was deposited with NiC/Ti multilayers by ion beam sputtering to produce an m=4 supermirror. The focusing performance of the supermirror was evaluated at the SUIREN of JRR-3M, and a focusing gain of 6 in peak intensity was achieved compared with a nonfocused direct beam.

1. Introduction

In recent years, very large proton accelerator facilities, such as J-PARC in Japan, ISIS in the UK and SNS in the USA, have been constructed. These facilities generate various quantum particles, including neutrons, by nucleus spallation. Neutron diffraction and scattering techniques are very useful for examining the physical, chemical, mechanical and magnetic properties of a material. However, the neutron intensity is very weak compared with synchrotron orbital radiation intensity, even in these facilities. Therefore, high-performance neutron collecting or focusing optical devices are required [1]. In these quantum optical devices, a figure accuracy of sub-micrometer level and a surface roughness of atomic level are essential for collecting and/or focusing neutrons onto a sample without scattering loss. However, in conventional machining processes, such as lapping or polishing, the accuracy of machining strongly depends on the stiffness of the machining equipment used and is also affected by external disturbances, such as the vibration or thermal deformation of the workpiece and the machine caused by the contact-removal mechanism. Therefore, it is very difficult to fabricate an ultraprecise optical device with a figure accuracy of sub-micrometer level with a high reproducibility.

To resolve these issues, we have developed a numerically controlled local wet etching (NC-LWE) process, which involves forced suction of an etchant with its volatile component using a vacuum pump, to fabricate an ultraprecise optical device [2]. The forced suction nozzle makes it possible to obtain a free configuration of the nozzle direction and prevents unwanted surface roughening by the adsorption of the volatile component of the etchant. Using this figuring system, although the cost of introduction is very low, highly stable and highly reproducible figuring properties are achieved more easily than when using other conventional machining systems, because the LWE process is insensitive to external disturbances, such as vibration or thermal deformation, owing to its noncontact chemical removal process.

2. Experimental procedure

2.1. Two-stage NC-LWE process

A plano-elliptical mirror substrate for focusing a neutron beam was fabricated by the NC-LWE figuring technique. The shape of the elliptical mirror is represented by equation (1).

$$\frac{x^2}{(1050.31)^2}+\frac{z^2}{(25.66)^2}=1 \quad \text{(unit: mm)} \tag{1}$$

The material, outer size, focal length and clear aperture size of the mirror substrate were synthesized quartz glass, 100 mm × 50 mm × 15 mm, 1050 mm, and 90 mm × 40 mm, respectively. The composition and temperature of the etchant were 37 wt% HF and 40°C, respectively. The curvature in the sagittal direction of this mirror was zero, and the neutron beam was focused one-dimensionally at the same magnification. The two-stage NC-LWE process was applied to fabricate a plano-elliptical mirror substrate. In this process, coarse figuring by one-dimensional numerically controlled scanning was performed using a large rectangular nozzle head with a size of 51 mm × 12 mm, then two-dimensional numerically controlled raster scanning was performed using a circular nozzle head with a diameter of 15 mm to correct the residual figure error [3]. Therefore, this novel figuring process makes it possible to fabricate an ultraprecise mirror substrate with a high efficiency.

2.2. Ion beam sputter deposition

The surface of the plano-elliptical substrate was deposited with NiC/Ti multilayers using ion beam sputtering (IBS) to fabricate a supermirror for increasing the critical angle of neutron reflection. The IBS system installed at the Japan Atomic Energy Agency (JAEA) had an effective deposition area of 0.2 m^2 (diameter of 500 mm) [4-6]. The NiC layer was obtained by Ni-C cosputtering [7].The carbon atoms added into the nickel layer suppressed interfacial roughness by reducing the crystal grain size of the nickel layer. A NiC/Ti supermirror with m=4, where m is the ratio of the effective critical angle of the supermirror to that of natural nickel, was fabricated. The thickness and total number of multilayers were 6 μm and 1201, respectively.

2.3. Evaluation of focusing performance

The focusing performance of the fabricated supermirror was evaluated at the Apparatus for Surface and Interface Investigations with Reflection of Neutrons (SUIREN) of Japan Research Reactor No. 3 Modified of Japan Atomic Energy Agency (JRR-3M).

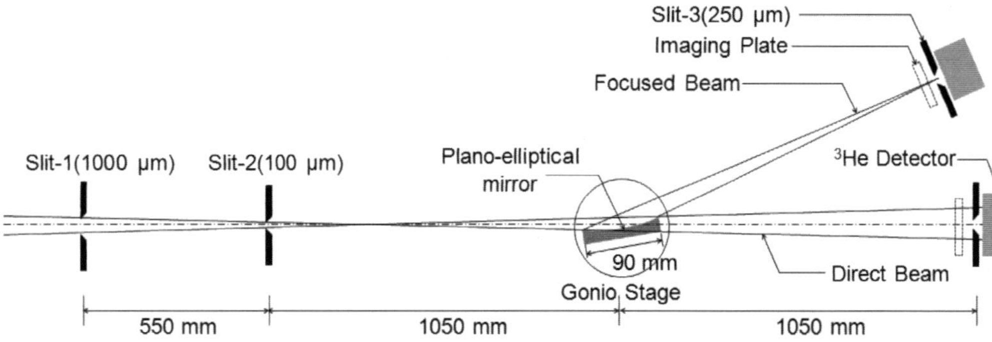

Figure 1. Schematic diagram of the measuring setup at the SUIREN. (Top view)

Figure 1 shows a schematic diagram of the setup for measuring the neutron beam profile at the SUIREN. The wavelength of the monochromatized neutron was 0.393 nm. The sagittal plane of the mirror was installed vertically on the goniostage to focus the neutron beam horizontally. The mean incident angle of the neutron beam was 1.40 degrees. Slit-2 was placed at the focal point of the ellipse, and the imaging plate or the ^{3}He detector was placed at the other focal point. The divergence of the neutron beam in the horizontal direction was set to 0.12 degrees by the combination of slit-1 and slit-2 to irradiate the neutron on the effective area of 90 mm length. The vertical divergence of the neutron beam was 0.39 degrees. The two-dimensional images of focused and nonfocused direct neutron beams were measured using an imaging plate (FUJIFILM BAS-ND; resolution: 50 μm), and detailed profiles of the beam were measured with a ^{3}He detector (Eurisys 43NH10/5X) and slit-3. The width and scanning pitch of slit-3 were 0.25 mm and 0.1 mm, respectively.

3. Results and discussion

Figure 2 shows the figure errors of the fabricated mirror in the meridional direction measured with a laser autofocus measuring machine (Mitaka Kohki NH-3SP; z resolution: 1nm). The two-stage NC-LWE process achieved a figure error of less than 0.5 μm with a total processing time of 318 min. However, the figure error degraded to about 1μm after the deposition of the NiC/Ti multilayers. It seems that the compressive stress of the deposited thin film deformed the substrate, which may be prevented by depositing films of the same thickness on both sides of the mirror substrate.

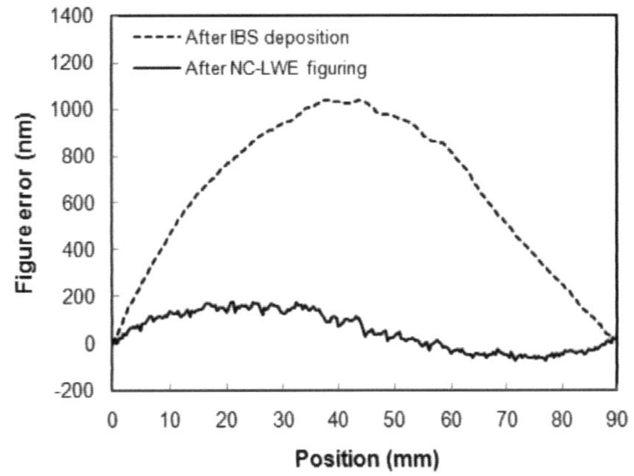

Figure 2. Figure errors of the fabricated plano-elliptical mirror.

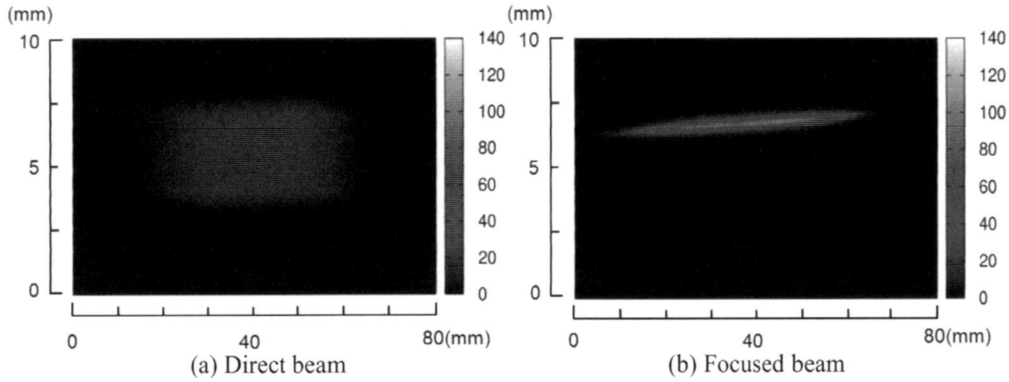

Figure 3. Neutron beam images detected by the imaging plate.

Figure 3 shows the two-dimensional images of the neutron beam detected using the imaging plate. Cold neutrons were uniformly focused at the effective mirror width of 40 mm. The neutrons detected beyond the mirror width were generated by the vertical divergence of the neutron beam. Figure 4 shows the detailed profiles of both focused and nonfocused direct neutron beams measured by the scanning of slit-3 placed in front of the ^{3}He detector. The full width at half maximum (FWHM) of the focused beam was 0.35 mm, and a focusing gain of 6 in peak intensity was achieved compared with the nonfocused direct beam.

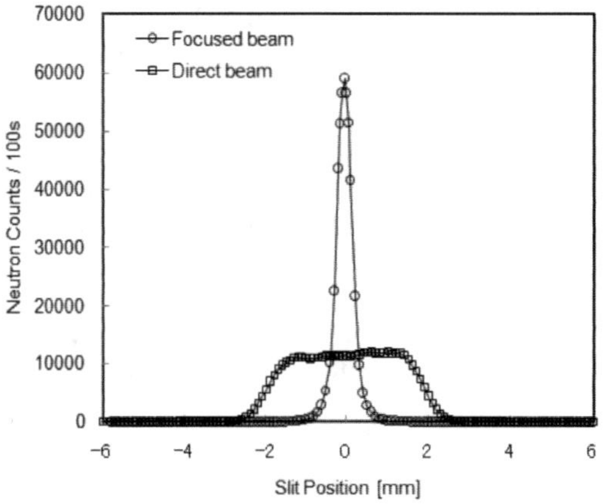

Figure 4. Neutron beam profiles measured by scanning of the slit-3.

4. Conclusions

A neutron focusing supermirror was fabricated by applying an NC-LWE technique combined ion beam sputter deposition. A plano-elliptical mirror substrate made of synthesized quartz glass with a clear aperture size of 90 mm × 40 mm was fabricated by applying a two-stage NC-LWE process. A large rectangular nozzle head and a small circular nozzle head were used in the first coarse figuring and second fine finishing stages, respectively, to reduce the fabrication time. The figure accuracy of the plano-elliptical substrate of less than 0.5 μm was achieved after NC-LWE figuring with a total fabrication time of 318 min. A supermirror with m=4 was produced by depositing of NiC/Ti multilayers on the substrate. The uniform one-dimensional focusing of cold neutrons and a focusing gain of 6 were achieved using the fabricated elliptical supermirror.

Acknowledgements

This work was partially supported by the Industrial Technology Research Grant Program in 2005 from the New Energy and Industrial Technology Development Organization (NEDO) of Japan, a Grant-in-Aid for the Global COE Program "Atomically Controlled Fabrication Technology" from the Ministry of Education, Culture, Sports, Science and Technology, Japan, and a research grant from the Machine Tool Engineering Foundation.

References

[1] Ice G E 2008 *Metall. Mater. Trans. A* **39** 3058
[2] Yamamura K 2007 *Annals of the CIRP* **56/1** 541
[3] Yamamura K, Nagano M, Takai H, Zettsu N, Yamazaki D, Maruyama R, Soyama K and Shimada S 2009 *Opt. Express* **17** 6414
[4] Maruyama R, Yamazaki D, Ebisawa T, Hino M and Soyama K 2006 *Physica B* **385-386** 1256
[5] Maruyama R, Yamazaki D, Ebisawa T, Hino M and Soyama K 2007 *Thin Solid Films* **515** 5704
[6] Maruyama R, Yamazaki D, Ebisawa T and Soyama K 2009 *Nucl. Instr. and Meth. A* **600** 68
[7] Wood J L 1992 *Proc. SPIE* **1738** 22

Development of a triplet magnetic lens system to focus a pulsed neutron beam

Takayuki Oku[1], Hiroshi Kira[1], Takenao Shinohara[1], Shin-ichi Takata[1], Masatoshi Arai[1], Jun-ichi Suzuki[1], Hirohiko M. Shimizu[2]

[1] J-PARC center, JAEA, Tokai, Ibaraki 319-1195, Japan

[2] KEK, Tsukuba Ibaraki 305-0801, Japan

oku.takayuki@jaea.go.jp

Abstract. A triplet magnetic lens system composed of three sextupole-magnets and two spin flippers was constructed to focus pulsed neutrons in a wide wavelength range with same focal lengths. In this study, we performed a pulsed neutron beam focusing experiment with the system. The design of the system and the experimental results are shown and discussed.

1. Introduction

A magnetic neutron lens based on a sextupole magnet functions as a focusing and defocusing lens for neutrons with positive and negative spin polarity, respectively [1-4]. By combining the focusing and defocusing functions of the magnetic lenses, we can control neutron beam shape and divergence with high degree of freedom (Fig.1) [5,6]. Adiabatic field connection and spin flipping device make it possible to realize such a multiplet magnetic lens system (Fig.1) [5,6]. As the magnetic lens has neutron wavelength-dependent focusing property, namely chromatic aberration, it has been proposed that the chromatic aberration can be compensated in the pulsed neutron beam focusing by changing the focusing and defocusing functions of the magnetic lenses of the multiplet magnetic lens system synchronously with time-of-flight of neutrons [7]. This idea was verified based on a numerical simulation [8]. In this study, we designed and constructed a triplet magnetic lens system composed of three sextupole permanent magnets and two spin flippers, and performed a pulsed neutron beam focusing experiment with the system. The design of the system and the experimental results are shown and discussed.

2. Experimental setup

In this study, we constructed a triplet magnetic lens system composed of three sextupole permanent magnets and two spin flippers, and performed a pulsed neutron focusing experiment at the beamline C3-1-2-1 (NOP) of JRR-3 at Japan atomic energy agency (JAEA). The sextupole magnet functions as neutron focusing and defocusing lens for neutrons with positive and negative spin polarity, respectively [1-4]. The three sextupole permanent magnets (M1, M2, M3) and two spin flippers (SF1, SF2) were located in series as shown in Fig. 1. The spin flippers were utilized to control spin polarity of neutrons incident into the sextupole magnet. Based on the numerical simulation, the system was designed so as to satisfy a focal condition for neutrons in the wavelength range as wide as possible for the beamline. The experimental setup is shown in Fig. 1. The neutron beam was pulsed by using a chopper, and then polarized by using a quadrupole magnet (Fig. 1). The neutrons which transmit through the quadrupole magnet is highly polarized with polarization degree P>0.99 [9, 10]. The pulsed polarized neutrons are delivered through the guide field and enter the triplet magnetic lens system. The spatial intensity distribution of the neutrons, which transmitted through the triplet magnetic lens system were measured by using a 2d-position sensitive detector [11].

The spin flippers are the Rf-gradient spin flippers which are designed to have flipping efficiency f > 0.99 for neutrons with λ > 4 Å [12]. The spin flippers are turned on and off by turning on and off the Rf-current. The Rf coils are connected to a LC circuits with the resonant frequency of about 80 kHz. Fig. 2 shows how fast the Rf-current rises up and down when the spin flipper is turned on and off. The rising and falling times are about 200 μsec which is fast enough for this experiment.

The spin flippers are controlled to be on and off synchronously with time-of-flight of the neutrons. A start signal is sent out from the chopper and sent to delay modules which delay the signals for some times. The delayed signals as triggers are sent to function generators to turn on the spin flippers (Fig. 3).

3. Results and discussion

At first, we performed the pulsed neutron focusing experiments of the triplet magnetic lens system without changing the spin flipper state synchronously with the time-of-flight of the neutrons. The spin flipper states were fixed during the single measurement. The measuring conditions are shown in Table 1. The measured full widths at the half maximum (FWHMs) of the neutron beam focused onto the detector position are plotted as a function of the neutron wavelength λ in Fig. 4. It was found that neutron focal wavelength changes depending on the spin flipper states. The focal condition of the magnetic lens based on the sexupole magnet is determined by the magnet length and λ as follows:

$$f = \frac{L_{mag}}{2} + \frac{h}{\sqrt{G\alpha}\, m_n \lambda} \cot\left(\frac{\sqrt{G\alpha}\, m_n \lambda}{h} \cdot \frac{L_{mag}}{2} \right) \tag{1}$$

Here, f is the focal length, L_{mag} the magnet length (Fig. 5), h Plank's constant, G the coefficient of the magnetic field gradient, $\alpha = | \mu_n/m_n | = 5.77 m^2 s^{-2} T^{-1}$, m_n the neutron mass and μ_n the neutron magnetic moment. By changing the states of the spin flippers at the wavelength λ_{CO} at which the FWHM curves of Fig. 4 cross over, we can satisfy the focal condition for the pulsed neutrons in a wide wavelength range. Then, we performed the pulsed neutron focusing experiment by turning on and off the spin flippers SF1 and SF2 synchronously with the time of flight of the neutrons at the cross over wavelengths $\lambda_{CO1\text{-}3}$ (Fig. 4). The obtained results are shown in Fig. 4. As the result, the neutron beam size at the detector position were realized to be the narrowest among the cases 1 to 4, and the FWHMs were kept less than 4 mm between λ = 7.4 Å and 10.8 Å (Fig. 4).

The magnetic lens system is applicable to a focusing geometry small-angle neutron scattering (FSANS) instrument with pulsed neutrons. In the application, pulsed neutrons in a wide wavelength range should be focused with a fixed focal length for efficient utilization of the neutrons. Moreover, the neutrons should be focused on the detector so sharp that the peak tail intensity goes down to the background level rapidly. Fig. 6 shows the radial averaged intensities of the neutrons focused onto the detector. It was found that the neutron intensities in the wavelength range from λ = 7.4 Å to 10.8 Å decrease sharply down to the background level with increasing distance from the peak center (Fig. 6) Therefore, the magnetic lens system is considered to be applicable to the FSANS instrument with pulsed neutrons.

4. Conclusion

We have constructed a triplet magnetic lens system composed of three sextupole magnets and two spin flippers to focus a pulsed neutron beam in a wide wavelength range with same focal lengths. We performed the pulsed neutron beam focusing experiment with the system. By choosing the focusing or defocusing functions of each sextupole magnet adequately and synchronously with the time-of-flight of the neutrons, we could focus the neutrons in a wavelength range of between λ = 7.4 Å and 10.8 Å with nearly same focal length. As the application of the magnetic lens system, it is considered to be applicable to a neutron focusing device for the FSANS instruments with pulsed neutrons, since we can focus the pulsed neutrons in a wide wavelength range so sharply onto the detector by using it.

References

[1] P.S. Farago, NIM. 30 (1964) 271.
[2] A. Steinhof. NIM-A 397 (1997) 371.
[3] H.M. Shimizu, et al., Physica B, 241-243 (1998) 172.
[4] H.M. Shimizu, et al., NIM-A 430 (1999) 423.
[5] T. Oku, et al., Appl. Phys. A 74[Suppl.] (2002) S261.
[6] T. Oku, et al., Proc. of SPIE 4509 (2001) 96.
[7] J. Suzuki et al., the presentation at the International symposium on Pulsed Neutron and Muon Science, IPS08.
[8] T. Oku, et al., NIM-A 600 (2009) 100.
[9] T. Oku, et al., Physica B 397 (2007) 188.
[10] T. Oku, et al., Meas. Sci. Technol. 19 (2008) 034011.
[11] K. Hirota, et al., Phys. Chem. Chem. Phys. 7 (2005) 1836.
[12] T. Oku, et al., Physica B 335 (2003) 226.

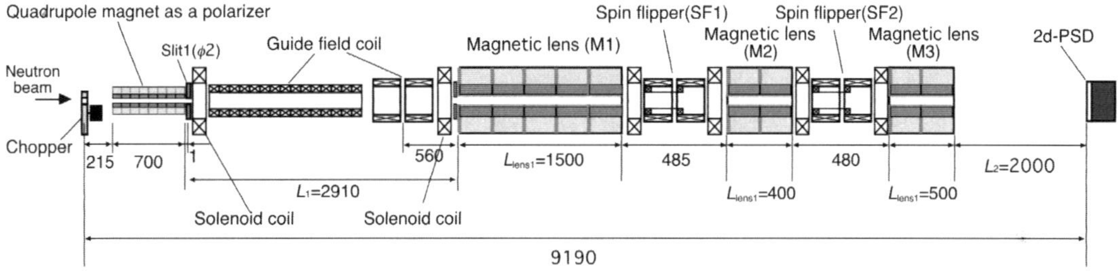

Fig. 1 Experimental setup.

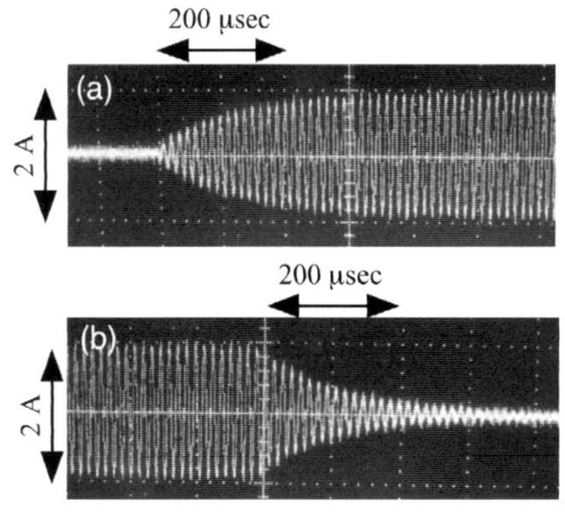

Fig. 2 The Rf-current of the spin flipper when the spin flipper is (a) on and (b) off.

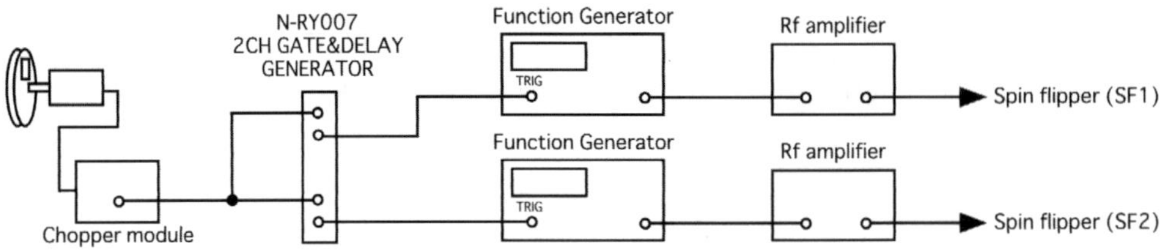

Fig. 3 A schematic signal processing circuit to control the spin flippers synchronously with the time-of-flight of neutrons.

Table 1 The states of the spin flippers and functions of the magnetic lenes

Case	States of the spin flippers		Functions of the magnetic lenses		
	SF1	SF2	M1	M2	M3
1	Off	Off	Focus	Focus	Focus
2	Off	On	Focus	Focus	Defocus
3	On	On	Focus	Defocus	Focus
4	On	Off	Focus	Defocus	Defocus

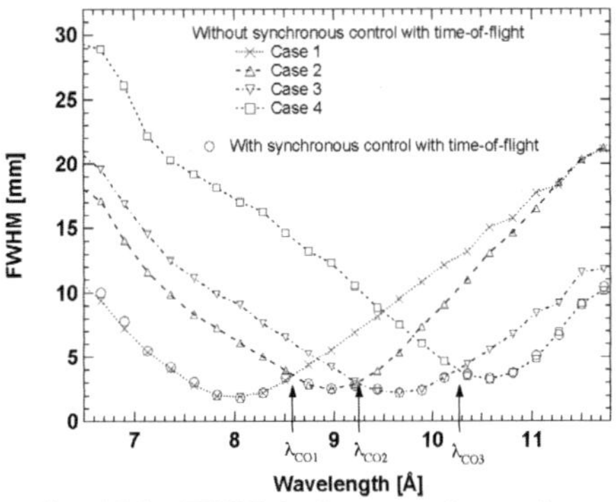

Fig. 4 The FWHM of neutron beam focused onto the detector position as a function of the time of flight of the neutrons.

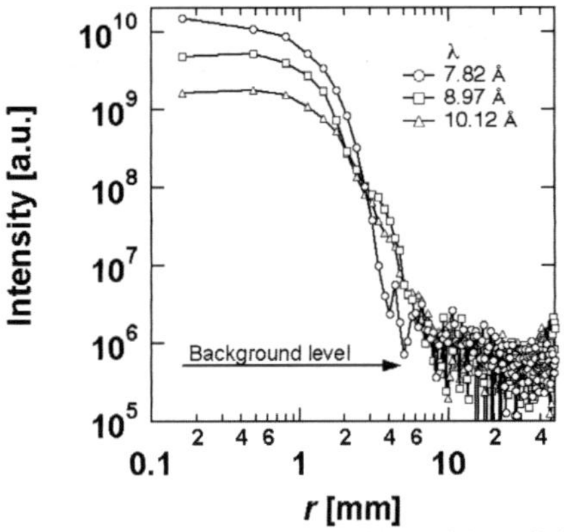

Fig. 6 Radial averaged intensities of the neutrons focused onto the detector.

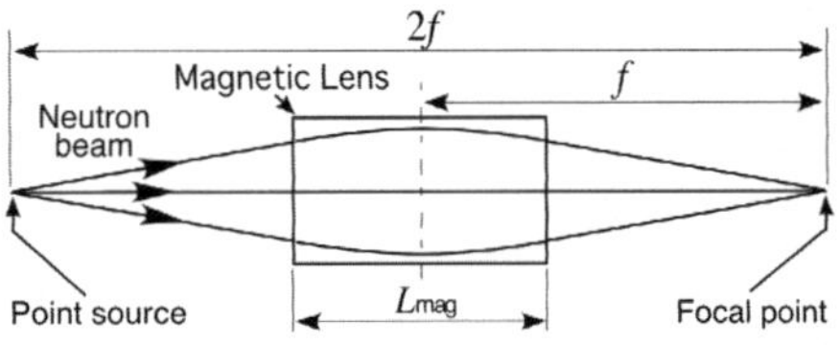

Fig. 5 A schematic layout of a magnetic lens which focuses the neutron beam with local length *f*.

International Conference on Neutron Scattering 2009 IOP Publishing
Journal of Physics: Conference Series **251** (2010) 012079 doi:10.1088/1742-6596/251/1/012079

Implementation of Repetition Rate Multiplication in Cold, Thermal and Hot Neutron Spectroscopy

Margarita Russina[1] **and Ferenc Mezei**[2,3]

[1]Helmholtz Zentrum Berlin, Hahn-Meitner-Platz1 100, 14109 Berlin, Germany
[2]Los Alamos National Laboratory, MS H805, Los Alamos, NM 87545, USA
[2]Budapest Neutron Center, SzFKI, Konkoly Thege u 29-33, 1121 Budapest, Hungary

E-mail: Margarita.russina@helmholtz-berlin.de

Conventionally, in time-of-flight (TOF) neutron spectroscopy neutron pulses of one single monochromatic wavelength are used. Repetition Rate Multiplication implies the use of a set of several monochromatic wavelengths coming from each source pulse, instead of a single one [1]. This makes it possible to overcome a major efficiency drawback of pulsed neutrons sources, i.e. the fact that the pulse repetition rate is too low for TOF spectroscopy, and allows us to freely choose the pulse repetition rate at the sample. The Repetition Rate Multiplication method was first developed in the framework of the IN500 project at LANSCE (USA) [2] and has in the meantime been implemented at several projects at ISIS (UK) [3] and J-PARC [4]. Here we describe the application of the Repetition Rate Multiplication method to hot, thermal and cold neutron scattering by using fixed and wavelength dependent pulse suppression. Further we present the first experimental realization of this novel technique using the flexible disc chopper system of the TOF spectrometer NEAT at BENSC.

1. Principle of Repetition Rate Multiplications

The typical features of "ideal" neutron chopper spectrometers are variable pulse frequency with rates ranging between 60 and 1000 Hz; variable pulse duration to optimize intensity to experimental needs and sharp pulse shape function to explore small signals next to the large ones. All these features are best realized at spectrometers at continuous sources. Most of the TOF instruments at pulsed sources suffer from inefficient data collection, which - due to the low source frequency varying from 10 to 50 Hz - can be as bad as only taking data at 10% of the time. The advent of the next generation of much more powerful spallation sources (SNS; J-PARC) is an additional motivation to make best use of the increased flux of these sources by more efficient data collection rates at TOF spectrometers.

To overcome the efficiency drawback due to the low repetition rate of pulsed spallation sources, the Repetition Rate Multiplication (RRM) method was proposed [1]. Contrary to the conventional method, where one single wavelength is used of each source pulse, RRM suggests to use pulses of several wavelengths emitted by each source pulse (Fig.1). Thus the instrument repetition rate increases and becomes fundamentally independent of the repetition rate of the source.

In cold neutron spectroscopy Repetition Rate Multiplication can efficiently be implemented by using a disc chopper system which was originally developed for the IN500 project at LANSCE, USA [2]. The IN500 disc chopper system (Fig.1) combines features of advanced continuous source instruments with conventional wavelength band choppers at spallation sources. First a wavelength band is defined by the slow choppers #1 and #2 rotating at the same frequency (20 Hz) as the source. This is later cut into a set of monochromatic pulses by the fast choppers #3 and #6, also defining the primary and secondary resolution of the instrument. While the slow choppers are designed to avoid overlap of neutrons coming from different source pulses, the fast chopper system removes the overlap of neutrons of different wavelengths coming from the same source pulse [2]. The frame overlap of scattered slow neutrons in the secondary spectrometer between the sample and the detectors can be further reduced, if necessary, in the similar way to continuous source TOF spectrometers by suppressing every second, third or fourth pulse by chopper #5 (not shown) of the basic 240 Hz pulse rate at the sample provided by chopper #6. Since the overlap of scattered spectra is defined by the time required for neutrons to cross the sample to detector distance the pulse repetition rate in an experiment is determined by the dimensions of the secondary spectrometer and not by the frequency of the source.

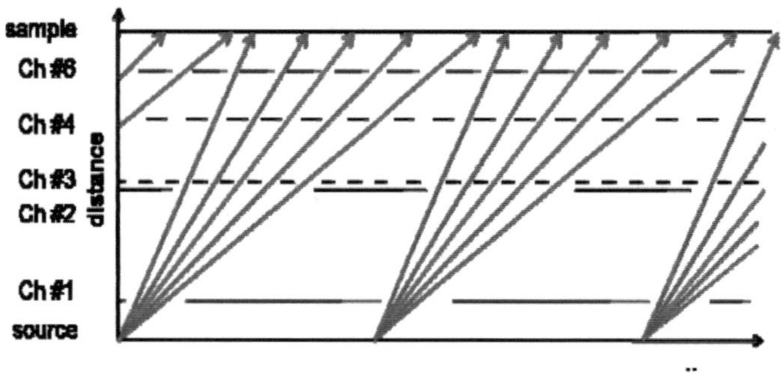

Fig.1. Principle of Repetition Rate Multiplication (RRM) as realized by a disc chopper system and compared to the conventional method of operation where only one single monochromatic wavelength (indicated by red arrows in the figure) is used. Basic layout parameters in the RRM development project IN500 at LANSCE: source pulse frequency 20 Hz, distance between source and choppers #1, 2, 3, 4, and 6, respectively, 7.3, 31.33, 31.35, 43.03 and 62.7 m.

For the design of the IN500 chopper system we applied the new approach of "wavelength filtering" [5], the main principle of which is to divide the chopper system into sub-systems with fewer choppers and to study the mathematical set of wavelengths transmitted with any finite probability by such a single sub-system. The transmission wavelength spectrum of the total system is limited by the common part of these sets and can be then controlled by the shifting in time (phasing) the sub-units relative to each other. Such an analysis of adequately selected sub-systems offers a very efficient way to assure that no unwanted, spurious neutrons leak through the chopper system.

The Repetition Rate Multiplication can also successfully be realized in thermal and hot neutron spectroscopy with the help of Fermi choppers with broad wavelength transmission band, such as the one developed for the powder diffractometer EXED in Berlin [6]. However, additional aspects have to be considered. For instance, the spectrometer MAPS at ISIS [7] operates at a source frequency of 50 Hz with the source to sample distance of 12 m and the sample to detector distance of 6 m. For these distances RRM pulses from all wavelengths up to 4 Å can be accommodated within the 20 ms time between source pulses. Taking into account the sample - detector distance, the time between pulses at the sample should typically be about 1 ms for 0.5 Å incoming neutron wavelength and 8 ms for 4 Å. Thus, for neutrons with short wavelengths using a pulse rate of some 1000 Hz (i.e. about 0.33 Å wavelength step between adjacent pulses at the sample) the frame overlap in the secondary spectrometer sets in for incoming neutron wavelengths above 1 Å. In contrast, with 200 Hz pulse rate this only occurs above 3 Å, but the wavelength step increases up to 1.7 Å between adjacent pulses. Thus, any fixed pulse repetition rate at the sample, as produced by choppers #5 and 6 , will typically allow for only 2-3 unperturbed RRM frames for an instrument like MAPS with relatively short primary and relatively large secondary flight paths. The pulse repetition in RRM mode on the sample can be improved further the with the help of a wavelength dependent pulse suppression disc chopper, which runs synchronously to the source pulses (10 cm/ms peripheral speed at 50 Hz) and has a number of unevenly placed slits. In the example of MAPS, at 1000 Hz basic pulse rate at the sample (i.e. up to

13 RRM pulses within 20 ms arriving at the detector), typically this chopper has to suppress the 4th, 6th, 7th, 9th, 10th, 11th and 13th RRM pulses coming from the same source pulse in order to avoid overlapping of spectra at the detector. As a consequence one can usefully extract 6 RRM pulses from each ISIS source pulse by this extended RRM scheme with wavelength dependent, selective pulse suppression. The use of this scheme, which is first described here, is particularly suited for spectrometers that have a conventionally short source to sample distance, which is advantageous for the delivery of incoming hot neutron wavelengths below 0.9 Å.

Using RRM for cold, thermal and hot neutrons opens up the crucial opportunity to optimize the design of pulsed neutron sources for all types of scattering experiments. Without RRM, the large data collection dead times in TOF spectroscopy at low pulse frequencies, which in the past used to appear as a requirement for source parameters contradictory to other key applications, such as high resolution diffraction or small angle scattering.

2. First experimental implementation of Repetition Rate Multiplication

For the first experimental implementation of Repetition Rate Multiplication we used the time-of-flight spectrometer NEAT at BENSC [8]. NEAT runs 7 chopper discs with two slits each rotating at 40-333 Hz. The distances between the first and the last choppers (Ch1-2 and Ch6-7, which are pairs of counter-rotating discs), the first chopper and the sample and the sample and the detectors are 12, 13.3 and 2.5 meters, respectively. A He^3 monitor is placed about 70 cm behind the last chopper.

In the two experimental runs studied the pulse of the source was emulated by the first chopper Ch1-2 operating at low speed and delivering the "source" pulses at 40 and 23.67 Hz, respectively. Choppers 3, 4 and 5 were stopped in open position so that the full cold neutron beam spectrum arrived to the last chopper. Sets of monochromatic neutron pulses were cut out from the pulsed white beam by the last chopper (Ch6-7) at frequencies ν that were equal to a multiple of the frequency of the first chopper pulses: $\nu_{Ch6\text{-}7} = m\nu_{Ch1\text{-}2}$, where m is integer. In addition to polychromatic RRM scans we took series of monochromatic measurements by using the complete NEAT chopper system for wavelength identification and for a frame-overlap correction.

We collected two data sets, one at a pulse repetition rate of 160 Hz at the sample and a source frequency of 40 Hz (data set A) and one at a pulse repetition rate of 165.67 Hz at the sample and a

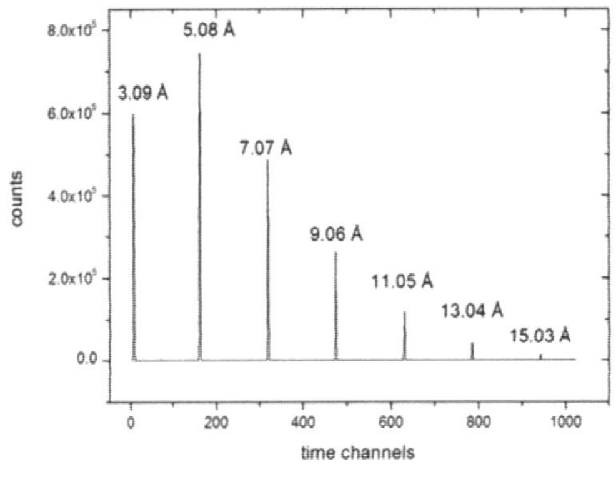

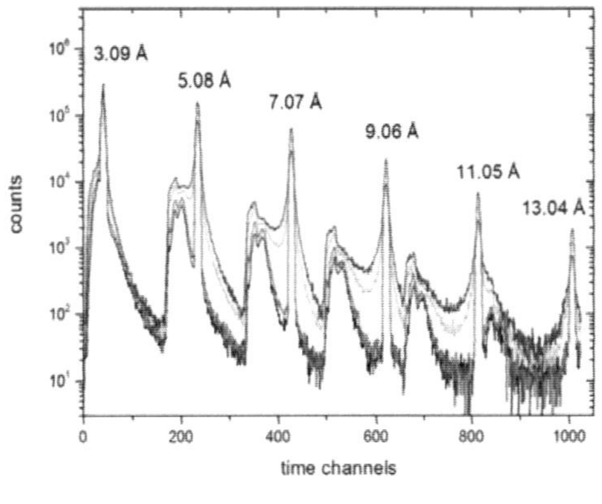

Fig. 2 Signal in RRM mode delivered to the monitor (left) and the detector (right) at 165.67 Hz pulse repetition rate at the sample and 23.67 Hz source frequency. Temperature dependent inelastic spectra from water sample confined into porous silica are shown as directly measured in the RRM mode. Temperatures of 160, 200, 260 and 300 K are indicated in black, red, green, and blue lines in figure, respectively.

source frequency of 23.67 Hz (data set B). In each case the length of the collected histograms was equal to the time between source pulses. The signal at the sample of the data set A consists of the wavelengths coming from the source pulse and providing the data collection trigger signal ("first frame" in pulsed source terminology) and neutrons with longer incoming wavelengths $\lambda_{inc} > 9.2$ Å emitted by the previous source pulse ("second frame"). To correct the data for this frame overlap we emulated the function of the frame overlap choppers by taking the spectra in the monochromatic mode at λ_{inc} = 11.33Å, 13.37Å and 15.34 Å and subtracting them from the polychromatic RRM spectra (longer wavelengths have negligible intensity). This approach enabled us to collect 3 useful RRM frames at the detector for each source pulse.

At lower source frequency the number of wavelengths which can be used from one source pulse increases. In the data set B taken at 165.67 Hz RRM pulse repetition rate and 23.67 Hz source frequency we were able to monitor 7 different wavelengths at the sample and to collect 6 wavelength frames at the detector. No frame overlap correction for the higher harmonics was required for the signal delivered to the sample, and the time T between pulses was long enough for neutrons up to 13.04 Å to arrive in the same data frame.

Both data sets demonstrate very well that there is enough time between pulses to record spectra for different wavelengths emanating from one source pulse. The shape of the scattered spectra reproduces those taken with monochromatic wavelength and no disturbing overlap is observed between the data sets for different wavelength RRM frames even at high temperatures and wavelengths except for very long ones, i.e. here 13.04 Å. Our RRM implementation study shows indeed that using low source pulse frequency enhances the efficiency of data collection also in TOF spectroscopy. The efficiency of the data collection can further be improved if the information collected is made more uniform, i.e. the difference between the individual RRM wavelengths extracted from one source pulse becomes smaller. This can be achieved by enhancing the distance between the source and the last choppers, which also improves the best available wavelength resolution.

2. Summary

Repetition Rate Multiplication can be successfully applied to hot, thermal and cold neutron scattering using fixed and wavelength dependent pulse suppression described here. The method offers a new way to operate TOF spectrometers at pulsed sources with efficiency similar to TOF spectrometers at continuous sources. The first experimental implementation of the Repetition Rate Multiplication technique was performed using the cold neutron TOF spectrometer NEAT at BENSC. The characteristics of RRM data collection in real life experiments were examined using a sample with a complex inelastic behaviour. Our results provide full experimental proof of the principle of the method. The study shows the technical capability to collect useful spectroscopic data with about an order of magnitude higher pulse rate and efficiency than the conventional single wavelength approach offers. Furthermore, the polychromatic RRM data collection provides higher information contents than the monochromatic one.

References

[1] Mezei F., J. Neutron Res. 6 (1997) 3, and references therein.
[2] Russina M., Meze F., Trouw F, Proc. of ICANS-XV, editors J.Suzuki and S.Itoh, (JAERI-Conf.2001-002), 2001, p. 349
[3] Mezei F., Russina M., LET chopper system optimization, Reports #1 and #2, HMI (2005); and http://ts-2.isis.rl.ac.uk/instruments/let/
[4] http://j-parc.jp/MatLife/en/instrumentation/bl14/bl14.html
[5] Mezei F., Russina M., in: I.S. Anderson, B. Guerard (Eds.) Advances in Neutron Scattering Instrumentation, Proc. of SPIE 4785 (2002) 24
[6] Peters J, Bleif H.-J., Kali G., Rosta L. and Mezei F., Physica B 385-386 (2006) p1019
[7] http://www.isis.rl.ac.uk/archive/excitations/maps
[8] Russina M., Mezei F., J. Nucl. Instr. and Meth., A604 (2009) pp. 624-631

Quantitative structure refinement from the ARCS chopper spectrometer

E S Božin,[1,2] P Juhás,[1] W Zhou,[1] M B Stone,[3] D L Abernathy,[3] A Huq,[3] and S J L Billinge[1,2]

[1]Department of Applied Physics and Applied Mathematics, Columbia University, New York, New York, 10027, USA
[2]Condensed Matter Physics and Materials Science Department, Brookhaven National Laboratory, Upton, New York, 11973, USA
[3]Neutron Scattering Sciences Division, Oak Ridge National Laboratory, Oak Ridge, Tennessee, 37831, USA

E-mail: sb2896@columbia.edu

Abstract. The new wide angular-range chopper spectrometer ARCS at the Spallation Neutron Source at Oak Ridge National Laboratory has been successfully used in white-beam mode, with no Fermi chopper, to obtain neutron powder diffraction based atomic pair distribution functions (PDFs). Obtained PDF patterns of Si, Ni, and Al_2O_3 were refined using the PDFfit method and the results compared to data collected at the NPDF diffractometer at Los Alamos National Laboratory. High quality resulting fits are presented, demonstrating that reliable powder diffraction data can be obtained from ARCS when operated in this configuration.

1. Introduction

Time-of-flight (TOF) instruments at spallation sources, such as the NPDF [1] at the Los Alamos Neutron Science Center (LANSCE) or the general materials diffractometer (GEM) [2] at ISIS, are routinely utilized to carry out neutron powder diffraction measurements. It is less common to exploit chopper spectrometers to obtain powder diffraction data suitable for structure refinements, since their primary role is inelastic neutron scattering measurements to obtain information about a system's dynamics. However, if operated without the monochromating Fermi chopper in the primary beam, a chopper spectrometer resembles the configuration of a TOF powder diffractometer. If there is no appropriate detector coverage at large scattering angles, the Q-resolution and Q-range, where Q is the magnitude of the scattering vector, may be insufficient for PDF analysis. However, the ARCS spectrometer at SNS has high detector coverage over a wide angular range, scattering angles between 2 and 135 degrees, resulting in an acceptable Q-resolution for the higher-angle detectors. The ARCS spectrometer is situated on a relatively short flight-path (13.6 m primary, and 3.0-3.4 m secondary) at the Spallation Neutron Source (SNS) at Oak Ridge National Laboratory (ORNL) in Tennessee, currently the brightest spallation neutron source in the world. The Fermi chopper mechanism at ARCS is mounted on a motorized translation table, allowing one to easily switch between two installed Fermi chopper slit packages and an open 'white-beam' position. This feature means that it becomes straightforward to analyze structure and dynamics from the same sample without dismounting it from the instrument. Here we present in greater detail the preliminary experiments, which demonstrate this capability on ARCS, that were reported in an earlier brief communication [3]. Quantitative structural information could be obtained from ARCS by obtaining reliable atomic pair distribution functions (PDFs) from the data [4] and refining models to them using the PDFgui software [5]. Here we describe in greater detail the procedure and the results.

Complete time-focussing of diffraction data from ARCS is not achievable at present due to lack of appropriate software routines. The data presented here were histogrammed to produce the 1D powder diffraction pattern using a generic TOF to d-space conversion protocol, without applying standard second order corrections. As a consequence, the data have a degraded resolution, and with poorly defined peak shapes. Attempts were made to refine the data by the Rietveld method using Fullprof [6, 7] and GSAS [8, 9] with poor results. Because the data were not fully time-focussed and inherently low resolution, the time was not spent to explore this issue any further and we concentrated on obtaining structure refinements from atomic pair distribution functions (PDFs) obtained from the data [4, 10] since the PDF is not very sensitive to details of the profile function. Successful refinements were performed as described below, demonstrating that this approach can be used even in the absence of software codes that do the full time-focussing, which is a significant challenge from a diffractometer of the complexity of ARCS. The data from Si and Ni samples are benchmarked against data from similar samples collected at the NPDF diffractometer at LANSCE with comparable statistics. NPDF is currently the highest resolution neutron powder diffractometer situated at a spallation source in the US [1] and is also a highly successful PDF instrument [11, 12, 13, 14, 15]. Finally, data for a corundum Al_2O_3 sample were collected and the PDF successfully refined, demonstrating the capability for carrying out research-quality diffraction experiments on nontrivial systems at the ARCS instrument.

2. Experimental

In this study we collected neutron TOF data at ARCS at SNS on three different commercially available standard samples: Ni, Si, and Al_2O_3. In addition, data from neutron TOF powder diffraction measurements on Ni and Si, carried out using the high-resolution NPDF diffractometer at LANSCE, are presented. All data were collected at room temperature.

For the ARCS experiment, loose powder samples (3.0 grams of Si, 8.3 grams of Ni, and 3.8 grams of Al_2O_3) were sealed in extruded vanadium tubes (1.11 cm in diameter, approximately 5.08 cm in height, and with wall thickness of 0.015 cm). Data for all the samples, as well as for the empty vanadium container (background), were collected for about 30 minutes each. A vanadium rod measurement was also performed to obtain incident spectrum information. The experiments were carried out using a neutron beam produced by 1.25 coulombs of charge impinging on a liquid mercury target surrounded by a decoupled water moderator [16]. In the case of standard NPDF measurements, to obtain the total scattering structure function, $S(Q)$, raw data are corrected for experimental effects, such as sample absorption and multiple scattering, and normalized by the incident spectrum [4], using the program PDFgetN [17]. The data from ARCS were converted from scattering events as a function of TOF and detector position to histogramed intensity as a function of d-spacing with bin sizes 0.0005 Å for $0.0625 \leq d \leq 0.25$ Å, 0.001 Å for $0.25 \leq d \leq 1.5$ Å, and 0.005 Å for $1.5 \leq d \leq 12.0$ Å using Distributed Data Analysis for Neutron Scattering Experiments (DANSE) software. Variable bin sizes were chosen because of the variation in resolution as a function of d-spacing. To get $S(Q)$, these data were minimally processed, using custom made programs to correct for the background parasitic scattering and to normalize by the incident spectrum only, without any other standard correction. This works because many of the corrections are the same for the sample and the vanadium standard and cancel during the normalization. We note that even with this coarse data treatment, it is possible to obtain useful structural information of reasonably high quality. A typical total scattering function of data collected at ARCS is shown in Figure 1(a), with the corresponding $F(Q)$ function from NPDF shown in Figure 1(b). The structure function from ARCS is accurate

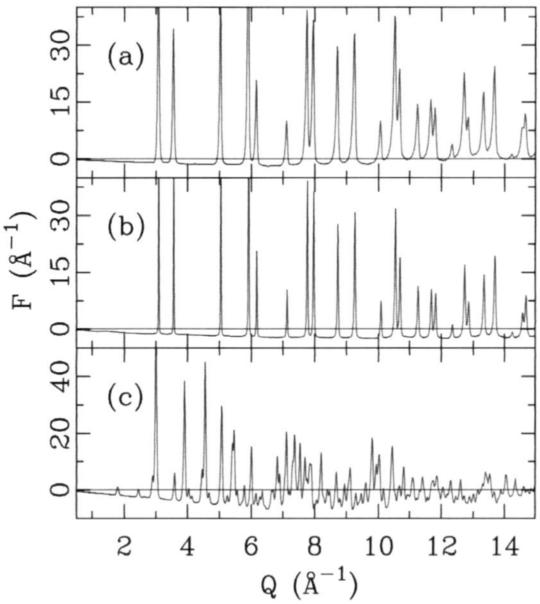

Figure 1. Total scattering function F(Q)=Q[S(Q)-1] of Ni from data collected at (a) ARCS and (b) NPDF, and Al_2O_3 obtained from ARCS data (c). All data were collected at room temperature for 30 minutes.

and measured with good statistics though clearly of much lower resolution than that from NPDF. The total scattering function of Al_2O_3 data collected at ARCS is shown in Figure 1 (c).

The PDF, $G(r)$, is obtained from the structure function, $F(Q)$, by a Fourier transformation according to $G(r) = \frac{2}{\pi}\int_{Q_{min}}^{\infty} Q[S(Q)-1]\sin Qr\, dQ$, where Q is the magnitude of the scattering vector and Q_{min} is a value of Q beyond the limit of the small angle scattering but below the limit of the lowest wide-angle scattering and $S(Q)$ is the properly corrected and normalized powder diffraction intensity [18]. The PDF gives the probability of finding an atom at a distance r away from another atom. The PDFs presented in this study were produced using various upper limits of integration in the Fourier transform, Q_{max}, as follows: Ni data Q_{max} = 25.0 Å^{-1}, Si data Q_{max} = 20.5 Å^{-1}, and Al_2O_3 data Q_{max} = 35.0 Å^{-1}. The upper Q-limits in these cases are dictated by factors such as signal to noise ratio, as well as by the imperfections at higher momentum transfers resulting from incomplete data processing procedures that were used.

Assessment of the structure information was carried out via refinements of the experimental PDFs, using the program PDFgui [5]. The details of the PDF method are provided elsewhere [4, 10].

3. Results and Discussion

Here we show the quality of PDFs that can be obtained from ARCS and compare them to PDFs from NPDF. Representative experimental PDFs of Si and Ni are shown in Figure 2 (open symbols), obtained using the same value of Q_{max} for both instruments. Simple visual comparison of the ARCS and the NPDF PDFs demonstrates that the ARCS data is of high quality. Furthermore, all ARCS datasets could be successfully refined. In Figure 2 fully converged structure models (solid lines) are plotted on top of the data. Standard structural models were used for Si ($Fd\overline{3}m$), and Ni ($Fm\overline{3}m$). The corresponding difference curves are shown offset for clarity.

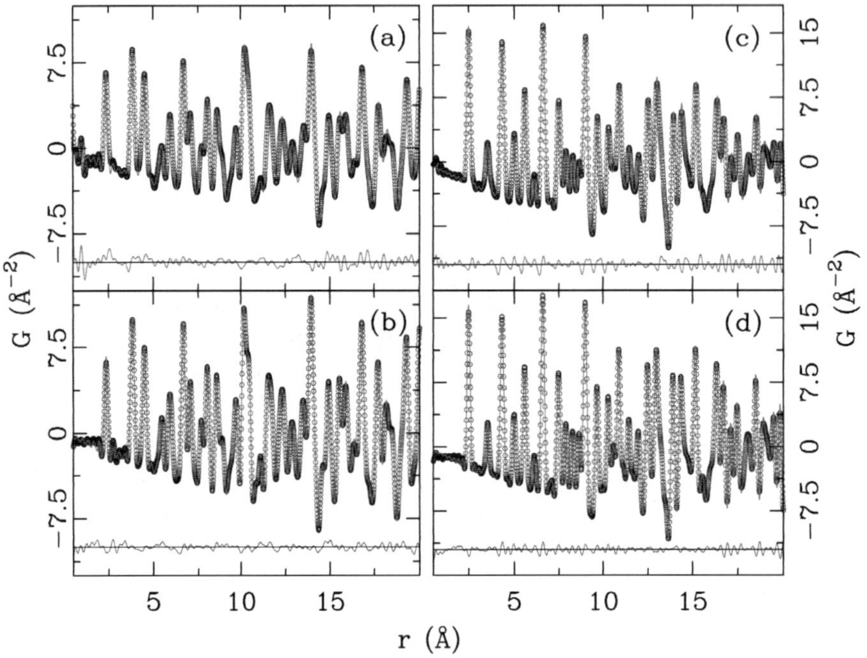

Figure 2. Experimental PDFs (open symbols), PDFgui fits of the structural models (solid lines) and corresponding difference curves (offset for clarity): (a) Si data collected at ARCS, (b) Si data collected at NPDF, (c) Ni data collected at ARCS, and (d) Ni data collected at NPDF. See text for details.

Refinements of Si, Figure 2(a) and (b), give lattice parameters of 5.4515(2) Å and 5.4361(2) Å, and U_{iso}^{Si} of 0.0086(6) Å^2 and 0.0060(7) Å^2 for ARCS and NPDF data respectively, with corresponding goodness of fit, r_W, of 0.120 and 0.086. The quantitative results for the other samples are reported elsewhere [3]. In PDF refinements the Q_{damp} parameter [5], which attenuates the PDF with increasing r, is the parameter that is most affected by the instrumental resolution. This parameter, as refined for the Ni standard, is markedly larger for ARCS (0.046(11) Å^{-1}) than it is for NPDF data (0.017(4) Å^{-1}) [3]. This is illustrated in Figure 3 where two datasets for Si, one from ARCS and the other from NPDF are plotted on top of each other over a wide range of r. The PDF from the ARCS data clearly damps out more quickly with increasing-r due to the lower resolution of these ARCS data. To emphasize this, a dashed line is shown that depicts the Gaussian envelope function that is multiplied with the model-PDF to simulate this effect in the calculated PDFs. However, a significant signal persists up to 60 Å, which is beyond the range over which most PDF refinements are carried out. This suggests that powder diffraction data from ARCS are quite appropriate for PDF analysis.

Interestingly, the PDF peak width in the ARCS data also appears to be larger than that for the NPDF data for the same Q_{max} used. This effect is known to appear when data have a Q-dependent peak broadening [19, 20] but appears to be particularly marked in these ARCS data. If it is not accounted for in the model, this results in an overestimate for atomic displacement parameters. It is also apparent that there exists a small but systematic offset of about +0.01 Å in the lattice parameters obtained from the refinement of ARCS data of both Si and Ni [3], compared to those obtained for NPDF data. This is presumably an artifact that comes about due to the inadequate TOF conversion used in our procedure. In a real experiment this is readily

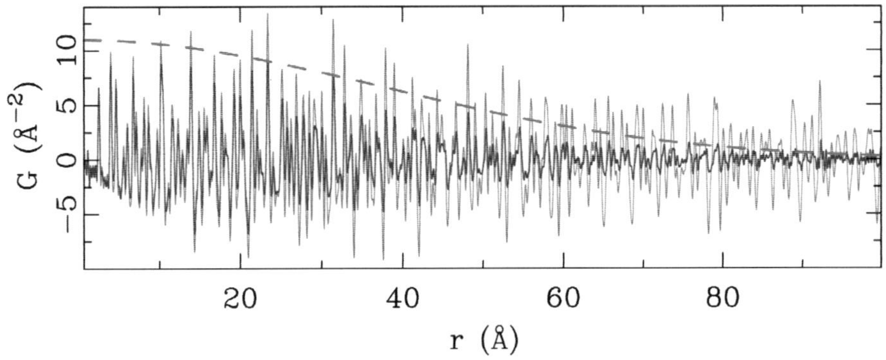

Figure 3. Comparison of experimental PDF data for Si 640c standard obtained at ARCS (red solid line) and NPDF (gray solid line) at room temperature, processed using the same value of Q_{max}. Thick dashed green line depicts the effect of diminishing PDF profile of ARCS data due to limited reciprocal space resolution.

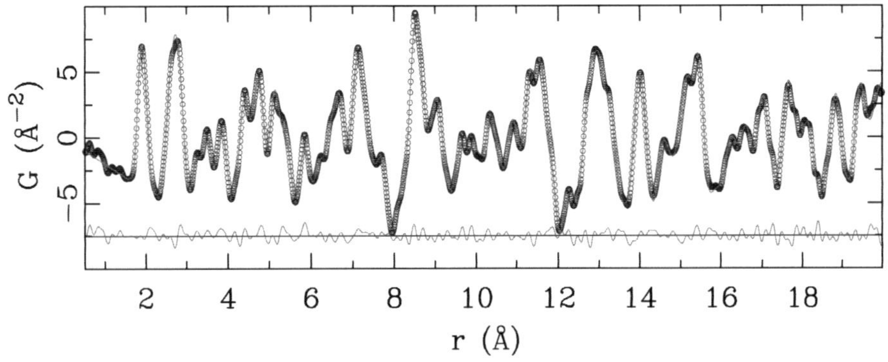

Figure 4. PDFgui fit of the Al_2O_3 structural model to ARCS data collected at room temperature. The data are shown as open symbols and the calculated curve as a solid line. Offset below is the difference curve.

corrected for using a calibration measurement such as carried out here.

In Figure 4 we show an example of a structural refinement to data from corundum Al_2O_3, which has more structural parameters than the simple Ni and Si structures. The results [3] are in reasonable quantitative agreement with the literature values [21].

4. Summary

We have refined the first powder diffraction data from the ARCS chopper spectrometer instrument at the Spallation Neutron Source, operating in white-beam mode. The PDF data have been obtained and also refined using PDFgui. The refinements were compared to similar refinements from an established neutron powder diffractometer, NPDF at LANSCE. They yielded high quality fits and quantitatively reliable structural data, although some systematic offsets have been identified. Thus, we verify that high quality powder diffraction data, suitable also for nano-scale structure determination using the PDF, can be obtained from ARCS when used in white-beam mode without the Fermi chopper operating. This may be useful, for example, when it is desired to measure both structure and dynamics of a sample *in situ* without removing it from the instrument. The ARCS Fermi chopper is able to be remotely removed from the beam to switch between white-beam diffraction and monochromatic dynamics measurements quickly.

Acknowledgements
We thank Thomas Proffen for providing standard data from NPDF that is used in this study. Work in the Billinge-group was supported by by the Office of Science, U.S. Department of Energy, under contract no. DE-AC02-98CH10886. DANSE software was developed under NSF award DMR-0520547. This work has benefited from the use of NPDF at the Lujan Center at Los Alamos Neutron Science Center, funded by DOE Office of Basic Energy Sciences. Los Alamos National Laboratory is operated by Los Alamos National Security LLC under DOE Contract DE-AC52-06NA25396. The research at Oak Ridge National Laboratory's Spallation Neutron Source was sponsored by the Scientific User Facilities Division, Office of Basic Energy Sciences, U. S. Department of Energy.

References

[1] Proffen T, Egami T, Billinge S J L, Cheetham A K, Louca D and Parise J B 2002 *Appl. Phys. A* **74** s163–s165
[2] Williams W G, Ibberson R M, Day P and Enderby J E 1998 *Physica B* **241-243** 234–236
[3] Božin E S, Juhás P, Zhou W, Stone M B, Abernathy D L, Huq A and Billinge S J L 2009 *J. Appl. Crystallogr.* In press. Preprint at arXiv:0903.1268v2
[4] Egami T and Billinge S J L 2003 *Underneath the Bragg peaks: structural analysis of complex materials* (Oxford, England: Pergamon Press, Elsevier)
[5] Farrow C L, Juhás P, Liu J W, Bryndin D, Božin E S, Bloch J, Proffen T and Billinge S J L 2007 *J. Phys: Condens. Mat.* **19** 335219
[6] Rodriguez-Carvajal J 1990 *Abstracts of the Satellite Meeting on Powder Diffraction of the XV Congress of the IUCr, Toulouse, France*
[7] Rodriguez-Carvajal J 1993 *Physica B* **192** 55–69
[8] Larson A C and Von Dreele R B 2000 General structure analysis system report No. LAUR-86-748, Los Alamos National Laboratory, Los Alamos, NM 87545
[9] Toby B H 2001 *J. Appl. Crystallogr.* **34** 201–213
[10] Billinge S J L 2008 *Powder diffraction: theory and practice* ed Dinnebier R E and Billinge S J L (London, England: Royal Society of Chemistry) pp 464 – 493
[11] E S Božin, Schmidt M, DeConinck A J, Paglia G, Mitchell J F, Chatterji T, Radaelli P G, Proffen T and Billinge S J L 2007 *Phys. Rev. Lett.* **98** 137203
[12] Sundaram N, Jiang Y, Anderson I E, Belanger D P, Booth C H, Bridges F, Mitchell J F, Proffen T and Zheng H 2009 *Phys. Rev. Lett.* **102** 026401
[13] Greedan J E, Gout D, Lozano A D, Derahkshan S, Th Proffen, H-J Kim, Božin E S and Billinge S J L 2009 *Phys. Rev. B* **79** 014427
[14] Page K, Kolodiazhnyi T, Proffen T, Cheetham A K and Seshadri R 2008 *Phys. Rev. Lett.* **101** 205502
[15] Kim Y, Cadars S, Shayib R, Proffen T, Feigerle C S, Chmelka B F and Seshadri R 2008 *Phys. Rev. B* **78** 195205
[16] Mason T E, Abernathy D, Anderson I, Ankner J, Egami T, Ehlers G, Ekkebus A, Granroth G, Hagen M, Herwig K, Hodges J, Hoffmann C, Horak C, Horton L, Klose F, Larese J, Mesecar A, Myles D, Neuefeind J, Ohl M, Tulk C, Wang X and Zhao J 2006 *Physica B* **385-395** 955–960
[17] Peterson P F, Gutmann M, Proffen T and Billinge S J L 2000 *J. Appl. Crystallogr.* **33** 1192–1192
[18] Farrow C L and Billinge S J L 2009 *Acta Crystallogr. A* **65** 232–239
[19] Toby B H and Egami T 1992 *Acta Crystallogr. A* **48** 336–46
[20] Qiu X, Božin E S, Juhás P, Proffen T and Billinge S J L 2004 *J. Appl. Crystallogr.* **37** 110–116
[21] Lewis J, Schwarzenbach D and Flack H D 1982 *Acta Crystallogr.* **38** 733

New polarizing guide for neutron wavelengths above 2.5 Å

Th Krist[1], C Pappas[1], A Teichert[1], C Fehr[1], D Clemens[1], E Steichele[2] and F Mezei[1]

[1]Helmholtz Zentrum Berlin for Materials and Energy, Glienicker Str. 100, 14109 Berlin, Germany
[2]NTK GmbH, Brauerstr. 12, 86825 Bad Wörishofen, Germany

E-mail: krist@helmholtz-berlin.de

Abstract. We present a new polarizing system built for the relocated wide angle Neutron Spin Echo instrument SPAN. The new instruments at the second Guide Hall of BENSC and the relocation of SPAN to this hall of BENSC required a new beam extraction system and a new polarizer for SPAN, which replaced the old beam splitter produced in 1994 with FeCo-Si supermirrors with m=2. The new polarizer uses Fe-Si supermirrors, which do not run the risk to become activated as the old FeCo-Si supermirrors and was designed to deliver a polarized beam for wavelengths above 2.5 Å. The final polarizing cavity has a length of 9 m with a cross section of 60 mm x 100 mm. Si wafers coated on both sides with m=2.5 Fe-Si polarizing supermirrors are glued into the guide at an angle of 0.38° to the walls.
The guide was installed during the second half year of 2006 and the first tests in early 2007 revealed excellent polarization efficiency over the whole wavelength range of the spectrometer of 2.5 Å to 9 Å, amounting to above 95% at 4.5 Å.

1. Introduction

In 2005 the second neutron guide hall at our institute was completed and the relocation of the Spin-Echo instrument SPAN into the new hall required a new polarizer system. At the old position (1992-2004) in the first guide hall the neutron beam was polarized by means of a beam splitter made from FeCo-Si supermirrors with m=2 [1], where m is the supermirror critical angle in units of the nickel critical angle. The old mirrors could not be used again due to their high activation and their size did not fit. Therefore it was decided to build for the new polarizer a transmission cavity [2] with Fe-Si supermirrors. According to the requirements of SPAN the cross section was 60mm x 100mm and the lower wavelength limit was set to 2.5 Å. The old analyser only polarized down to 4 Å.

Fe-Si polarizing supermirrors reliably exhibit in transmission average flipping ratios above 40 or a polarization above 95% for coatings with m=2.5. Since there was enough space in the guide hall it was decided to use a single line cavity instead of a V-shaped cavity to avoid the loss at the tip of the V, which must be shielded, and to simplify the construction. The cavity was designed on the basis of such coatings with VITESS simulations [3]. To check them as well as the cavity construction, first a test cavity was built and measured. After successful simulation of the experimental results the final cavity was built.

2. Simulations

The simulations were performed with the supermirrors assembly tool of VITESS.

The simulations showed that for these supermirrors a polarization for wavelengths of 2.5 Å and above can be achieved with an inclination angle of the wafers to the guide walls of 0.38°, cf. Figure 1.

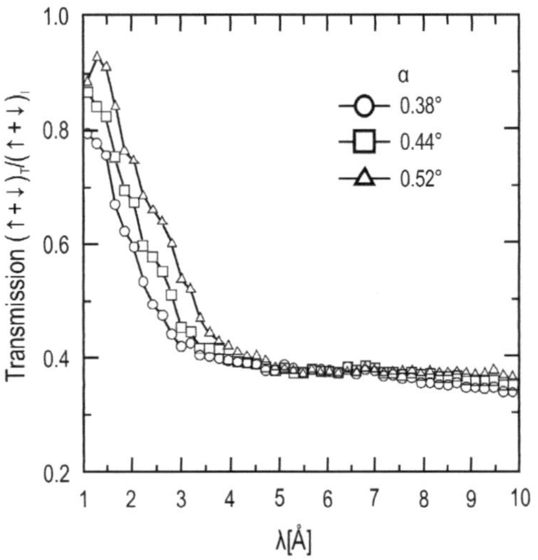

Figure 1 VITESS simulations for the ratio of the transmitted to the incoming neutrons through a cavity. The curves refer to different angles of the Si wafer in the cavity which are coated with polarizing Fe-Si supermirrors with m=2.5.

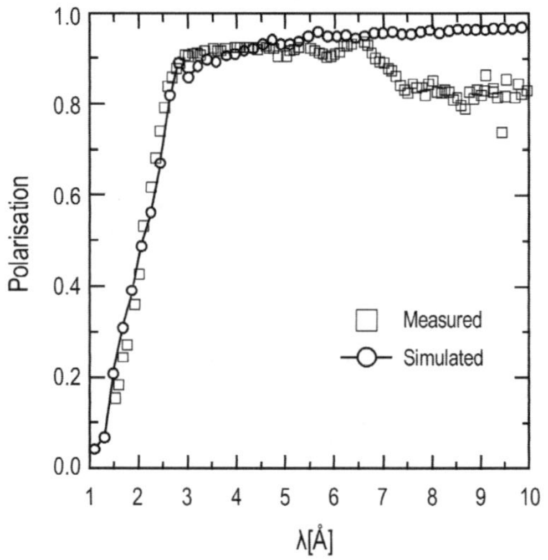

Figure 2 VITESS simulations for the polarization of neutrons transmitted through the test cavity together with experimental data.

3. Test cavity

To verify the simulations and to test the mechanical feasibility of the construction a test cavity with the wafers provided by our institute was built by the company NTK to the following specifications: Length: 1m, width: 7.1 mm, height: 100 mm, wafer angle: 0.38°, wafer coating: both sides with m=2.5 Fe-Si supermirrors, wall coating: ^{58}Ni on the sides, natural Ni on top and bottom.

The test cavity was equipped with a magnetic field with a minimum strength of 300 G at each point of the cavity. The ensemble was measured at the reflectometer AMOR at PSI.

Figure 2 shows the measured polarization together with the simulated one for the test cavity in the wavelength range of 2 Å – 9 Å.

The results show that the cavity polarizes for wavelengths above 2.5 Å as simulated. They also verify that this can be achieved for a wafer height of 100mm. The deviation of the experimental data from the simulated curve for wavelengths above 5.5 Å is due to a reduced efficiency of the polarizer used at AMOR for this experiment: The small size of the detector used on AMOR misses some of the neutrons at higher divergence angles. Thus the measured polarization data differs from the simulated values.

4. Final cavity

Finally a 9 m long polarizing cavity with a cross section of 60mm x 100mm was built with Si wafers coated on both sides with Fe-Si polarizing supermirrors, the walls were coated with ^{58}Ni on the sides and natural Ni on top and bottom. The total supermirror area was 1.8 m^2. The angle of the wafers to the walls was 0.38°. Figure 3 shows a sketch.

The requirements for the wafers were:

- a critical angle of the supermirrors of above m=2.5,
- a flipping ratio in transmission above 20 at m=2.5,
- a bending radius of the coated wafer larger than 7 m.

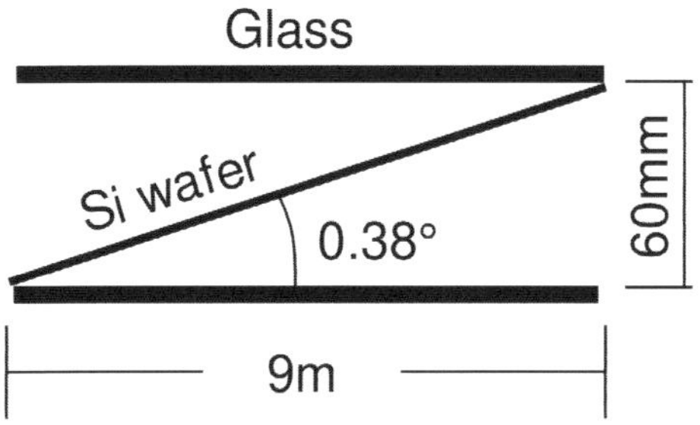

Figure 3. Sketch of the final cavity, not to scale. The glass walls are coated with ^{58}Ni, the silicon wafers are coated with polarizing Fe-Si supermirrors with m=2.5.

The value of the minimum acceptable bending radius was determined from VITESS simulations where the effect of the waviness of surfaces can be determined. Since the wafers are spherically deformed the value refers to vertical as well as horizontal bending. Neutron measurements to test the wafers were performed at the BENSC reflectometer V14 with a wavelength of 4.8 Å.

Figure 4 shows the curves for the transmitted intensity of the two spin states of neutrons through a typical wafer, which has a radius of 11 m, together with the polarization and - for the better visibility - also the flipping ratio. The polarization is given by the ratio of the difference of the transmission values for the two spin components over their sum while the flipping ratio is the ratio of the transmission values for the two components. The average values are 96% for the polarization and 50 for the flipping ratio

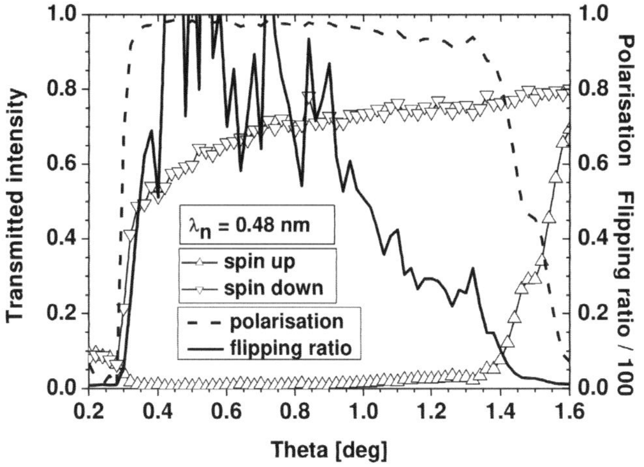

Figure 4: Transmitted neutron intensity through a silicon wafer coated on both sides with a Fe-Si supermirror for both spin components together with flipping ratio and polarization. Theta is the angle of the sample to the incoming neutron beam.

The bending measurements were performed on a profilometer DEKTAK 3030. Figure 5 shows a profilometer scan for a wafer with a concave surface having a radius of 325±4m which corresponds to a tensile stress of 3 MPa.

All the wafers for SPAN had radii between 8 m (350 MPa) and 3260 m (<1 MPa). They all met or exceeded the standards given above.

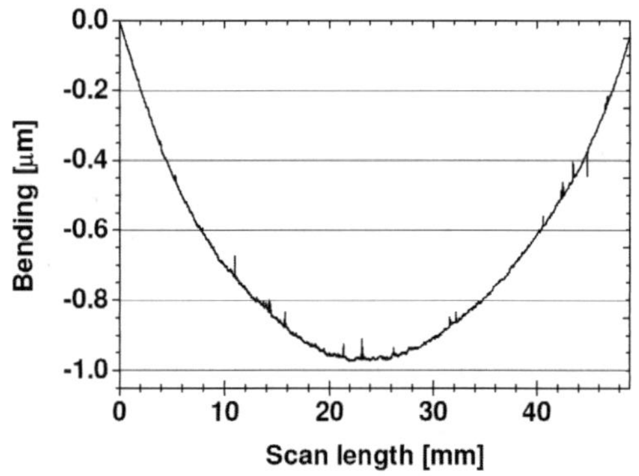

Figure 5: Profilometer measurement of a Fe-Si supermirror on a silicon wafer substrate showing a radius of 325±4m which corresponds to a tensile stress of 3 MPa. It indicates the precision of the measurement as well as the uniformity of the bending.

The coated wafers were again provided by our institute and built into the cavity by the company NTK.

The guide was installed during the second half year of 2006 and the result of a test in 2007 is shown in Figure 6. The cavity revealed excellent polarization efficiency over the whole wavelength range of the spectrometer of 2.5 Å to 9 Å, amounting to above 95% at 4.5 Å. These values are very well suited to do spin echo experiments. The decay of the polarization for larger wavelengths is an inherent feature of cavities and can be avoided only in a cavity only by a double polarization.

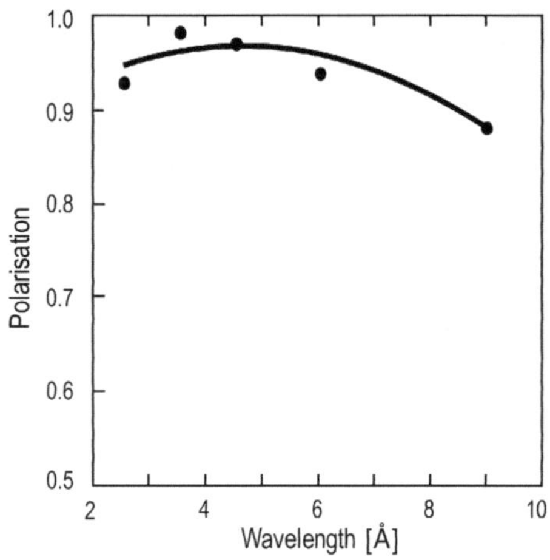

Figure 6. Polarisation of the final cavity for SPAN measured in March 2007. The curve is to guide the eye.

5. References

[1] Krist Th, Pappas C, Keller Th and Mezei F, 1995 Physica **B 213 & 214** 939-941
[2] Mezei F, 1989 Proc. SPIE **983** 10
[3] Lieutenant K, Zsigmond G, Manoshin S, Fromme M, Bordallo H, Champion D, Peters J and Mezei F, 2004 Proc. SPIE **5536** 134

High Performance Supermirrors on Metallic Substrates

C. Schanzer[1], P. Böni[1], and M. Schneider[1,2]

[1] SwissNeutronics, Bruehlstrasse 28, CH-5313 Klingnau, Switzerland
[2] Laboratory for Neutron Scattering ETH & PSI, CH-5232 Villigen PSI, Switzerland

E-mail: christian.schanzer@swissneutronics.ch

Abstract. Recently, we have developed an optimized process for polishing Al substrates leading to an extraordinary low surface roughness comparable to float glass of high quality. Indeed, supermirror coatings with R = 91% at the critical angle of reflection of m = 2 were produced demonstrating the excellent quality of the Al surface. Supermirror coated Al substrates open new options for advanced neutron optical devices. As one example neutron guides can start very close from the moderator. We investigated this option by Monte-Carlo simulations for i) a conventional, curved guide and ii) an elliptic guide. In case of i) an increase of flux at long wavelengths is obtained because of complete illumination of the phase space accepted by the guide. The elliptic guide starting close to the moderator (ii) has significantly enhanced neutron transport properties since only useful neutrons are extracted and transported to the sample. As a result such a guide concept is superior in terms of flux when compared with a conventional guide system and in terms of signal to background and well defined beam focusing compared with an elliptic guide starting at a larger distance from the moderator. In particular, experiments on novel materials, which are often only available in small quantities or samples under extreme conditions will profit from neutron beams of such high quality.

1. Introduction

The recent years have demonstrated that almost all modern neutron scattering instruments rely on the efficient transport of neutrons by means of neutron guides. Present neutron guides are usually assembled from flat glass substrates. The natural surface of floated glass provides an extremely small roughness enabling for high reflectivity even for very large angles of reflection, i.e. up to m = 6 times the critical angle of Ni. For example, reflectivities R > 80% for m = 4 and R > 70% for m = 5 are regularly obtained [1].

Unfortunately, glass has limited stability against irradiation, temperature, and mechanical stress, effects which are particularly severe close to the moderator. Therefore neutron guides can only be used for distances larger than approximately 1.5 m from the moderator causing under illumination of the phase space of the guide with increasing wavelength. In order to improve the illumination some guide systems have reflectors at their front end, which are assembled from polished metal plates. In the case of steel mirrors [2] the phase space of the reflector is smaller than the neutron guides accept, thus the illumination of the guides is still incomplete. Supermirror coated, polished aluminum assemblies [3] still suffer from low reflectivity of typically R = 60% at m = 2 due to a poor surface roughness. Alternative replica techniques for neutron mirrors on metallic substrates are mostly applied to simple Ni mirrors [4] because the supermirrors lack in terms of reflectivity [5]. Moreover, advanced optics, e.g. elliptic guides, require high m supermirrors at their entrance to fully exploit their performance.

Recent developments in neutron supermirrors [1, 6] enable for excellent reflectivities, but require atomically smooth substrates. Hence metal substrates of a similarly high quality are desired for demanding applications, which are discussed here.

2. Supermirror coatings on metallic substrates

For the application of metallic substrates it is desirable to use similar formats as it is common for glass based guides. Therefore the format of 250 mm × 130 mm × 10 mm was selected and aluminium as the material to avoid severe activation by neutrons in the long term. On such substrates the processes to refine the surface were steadily developed and the results were monitored by atomic force microscopy (AFM) on a scan area of 10 µm x 10 µm, similar to the in-plane coherence length of the neutrons. Floated glass served as the reference since, either non-borated or borated, it has been proven to be the best substrate for supermirror coatings when prepared properly. Fig. 1a shows a typical image of the topography of a float glass. The analysis results in a RMS roughness of $\sigma_{RMS} \approx 1.8$ Å. Fig. 1b depicts the AFM image of the aluminium surface as it is obtained from the present state of the art of the refinement process. A roughness of $\sigma_{RMS} \approx 2.0$ Å is obtained, which is rather close to the typical value for floated glass. Measurements on various locations confirm the homogeneity of the quality indicating that the refinement process is constant across the complete substrate area.

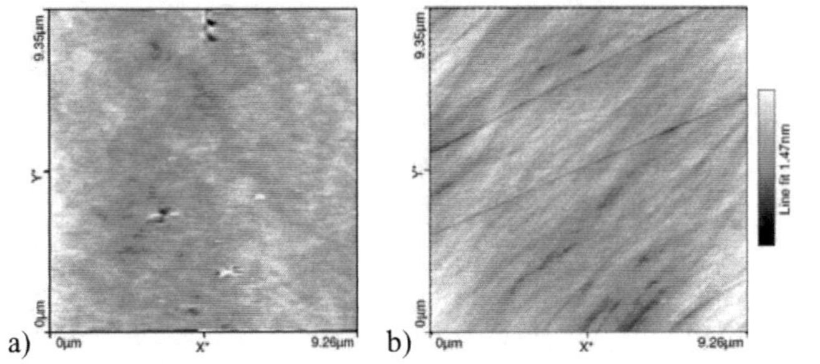

Fig. 1: AFM images of the surface of a) float glass and b) aluminum substrate. The surface of the aluminum exhibits fine scratches from the polishing. The color code is identical for both images. The measurements are performed with the AFM easyScans 2 from Nanosurf.

Finally, the substrate was coated with a Ni/Ti supermirror $m = 2$. The measured reflectivity profile is presented in Fig. 2 showing $R = 91$ % at $m = 2.1$. For comparison the reflectivity of a similar supermirror on float glass is included in Fig. 2 reaching $R = 94$ % at the edge of the supermirror. The excellent neutron reflectivity of the supermirror on the aluminium substrate confirms the excellent smooth surface. However, it becomes also clear that the float glass is still slightly superior to the refined aluminium substrate, which is in good agreement with the trend seen from the AFM data. The absolute roughness values σ_{RMS} may not be real because of the finite lateral size of the AFM tip, which results in a smoothed measured profile.

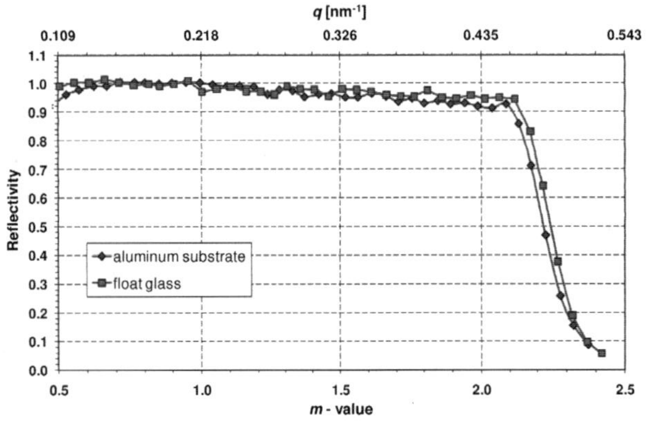

Fig. 2: Neutron reflectivity of a $m_{nominal} = 2$ supermirror on a refined aluminum substrate (diamonds) and on float glass (squares). The m-value is slightly larger than the targeted value due to a small deviation of the thickness calibration of the deposition process.

3. Applications – a Monte Carlo study

3.1. Conventional, parallel guide

The conventional neutron guide NL1 of the cold guide system at the FRM II is considered for a comparison of various designs for the transport of cold neutrons. NL1 has a constant aperture of 60 mm × 120 mm, a length of 36 m and starts about 2 m from the moderator. Geometrical details, flux measurements and simulations of the actual guide are presented in [3]. For the present simulations the initial section of NL1 is extended towards the moderator starting at a distance of 0.1 m.

Fig. 3 plots the gain which is obtained by the extended guide starting closer at the moderator compared to the presently installed guide. The plot shows two distinct ranges with a cross over at a wavelength of about 5 Å. Below that the extension of the guide does not improve the transmission since the size of the moderator is large enough to provide complete illumination of the phase space accepted by the actual guide. The incomplete illumination for larger wavelengths is eliminated when the guide starts close to the moderator resulting in an increasing gain with increasing wavelength.

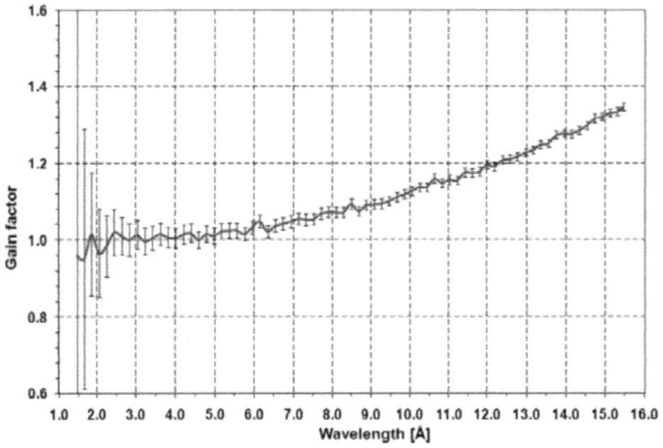

Fig. 3: Gain obtained from the extension of the neutron guide NL1 (FRM II) towards the moderator. The extended guide starts at a distance of 0.1 m while the original guide starts at 2 m.

3.2. Elliptic guide

In [7] an elliptic guide is compared with other guide concepts, in particular with the existing conventional guide that delivers the neutrons of the Swiss Spallation Neutron Source (SINQ) to the instrument TASP. As a result, the elliptic guide was shown to be *the* superior guide concept. For the present study, we used the same parameters for the elliptic guide as in [7], however, we extended it at the entrance towards the moderator, while maintaining the long and short axis of the ellipses. Hence the entrance has a reduced cross-section of 15.85 mm × 54.38 mm starting at a distance d = 0.3 m to the moderator, compared to the original aperture of 35 mm × 120 mm at d = 1.5 m.

Fig. 4 compares the gain factors of the original and extended elliptic guides with respect to the conventional guide, identically computed like in [7]. The original elliptic guide shows the largest gain in terms of flux of about 5 in the range of $3 < \lambda < 6$ Å. For an identical coating of $m = 3$ the extension of the guide leads to a reduced gain factor of about 2.4 in the respective range. This is understood since the smaller entrance reduces the number of neutrons which are collected from the finite size of the moderator and can propagate via multiple reflections to the sample (10 mm x 40 mm).

The reduced gain of the extended elliptic guide can be partially recovered by utilizing supermirrors with $m = 4$. Those allow larger reflection angles, especially at the ends of the elliptic guide, thus increasing the phase space accepted at the entrance and focused at the end of the guide.

In addition to the transmission properties, issues of the radiation background shall be discussed. A part of the background depends on the size of the opening in the biological shielding of the source, which is essentially determined by the cross-section at the entrance of the guide. As an indicator for signal/background the ratio of transmitted flux normalized to the entrance cross-section is computed.

Fig. 4 shows that the extended elliptic guides have a gain in signal/area of guide entrance more than one order of magnitude, thus making them superior in terms of signal/background.

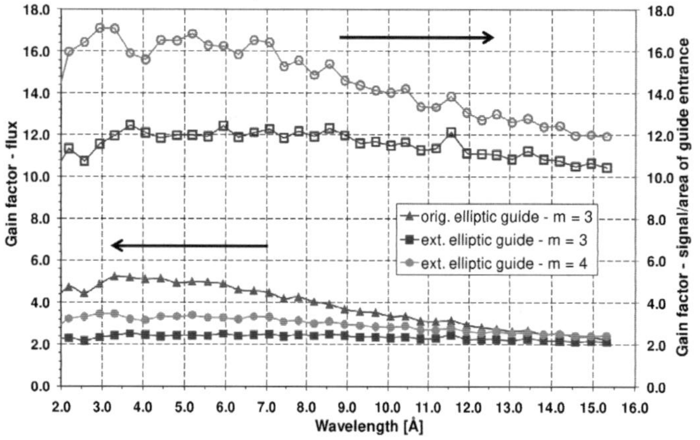

Fig. 4: Comparison of the gain factor of various elliptic guides with respect to a conventional guide. Details of the computation are given in [7]. The gain in terms of flux is assigned to the left ordinate (filled symbols). The comparison in terms of signal/area of guide entrance, normalized to the conventional guide is assigned to the right ordinate (open symbols with corresponding type).

4. Conclusion

Metallic substrates are envisaged for neutrons optics in order to overcome limitations implied by glass. Benefits are: i) increased durability against large changes in temperature and intense irradiation, ii) enhanced mechanical deformation possible to impose more complicated geometrical designs of guides, for example for focusing, and iii) robustness against mechanical damages and implosions.

In this paper we present a breakthrough in the refinement of the surface of aluminum substrates enabling for a supermirror reflectivity $R = 91$ % at $m = 2.1$. The low roughness is confirmed by AFM measurements and it is almost identical to float glass. Hence, even supermirrors with excellent reflectivity at larger m-values appear feasible and are the goal of future developments.

Dedicated applications of neutron guides with metallic substrates are initial guides very close to the moderator. Here we demonstrate by simulations that an enormous gain is obtained for a conventional guide that starts close to the moderator, e.g. 30 % at a wavelength of 15 Å. For the concept of an elliptic guide, the extension towards the moderator provides an improved transport of neutrons. While the transmission is moderately reduced, significant improvements in the signal-to-background ratio can be obtained. Moreover the point-to-point transformation can be more closely approximated thus reducing chromatic aberration and providing a "clean" beam with well defined, compact phase space.

Acknowledgement: Work was supported by the Swiss National Science Foundation through the National Centre of Competence in Research MaNEP.

References

[1] www.swissneutronics.ch
[2] S.J. Cho, private communication
[3] K. Zeitelhack, C. Schanzer, A. Kastenmüller, A. Röhrmoser, C. Daniel, J. Franke, E. Gutsmiedl, V. Kudryashov, D. Maier, D. Päthe, W. Petry, T. Schöffel, K. Schreckenbach, A. Urban and U. Wildgruber, Nuclear Instruments and Methods in Physics Research A **560** (2006) 444
[4] F. Lang, *Metallic guide using replica technics*, ILL guide workshop (2006)
[5] Yuji Kawabata, Masatoshi Suzuki, Seiji Tasaki, Kazumasa Somemiya, Nuclear Instruments and Methods in Physics Research A **420** (1999) 213
[6] M. Hino et al., Nuclear Instruments and Methods in Physics Research A **529** (2004) 54
[7] C. Schanzer, P. Böni, U. Filges, T. Hils, Nuclear Instruments and Methods in Physics Research A **529** (2004) 63.

International Conference on Neutron Scattering 2009 IOP Publishing
Journal of Physics: Conference Series **251** (2010) 012083 doi:10.1088/1742-6596/251/1/012083

Automatic sample changer for IBARAKI materials design diffractometer (iMATERIA)

Akinori Hoshikawa[1], Toru Ishigaki[1], Masao Yonemura[1], Kenji Iwase[1], Hidetoshi Oguro[1], Dyah Sulistyanintyas[1], Takashi Kamiyama[2], Makoto Hayashi[3]

[1]. Ibaraki University, Tokai-mura, Ibaraki, Japan.
[2]. Neutron Science Laboratory, KEK, Tsukuba, Ibaraki, Japan.
[3]. Ibaraki Prefecture, Mito, Ibaraki, Japan

E-mail: akinori.hoshikawa@j-parc.jp

Abstract. IBARAKI Materials Design Diffractometer (iMATERIA) is a high-throughput powder diffractometer. iMATERIA has a vacuum chamber, which is enclosed in a shield. Each sample is measured in 10 minutes, and must be placed in the vacuum chamber before being measured. If the vacuum in the chamber cannot be maintained while exchanging samples, the process of re-establishing the vacuum would become a bottleneck when samples are exchanged. To reduce exchange time, we developed and manufactured an automatic sample changer that can handle a large number of samples (up to 672) through both the vacuum chamber and shielding blocks.

1. Introduction

The Japan Proton Accelerator Research Complex (J-PARC) is a high-intensity proton accelerator facility. Neutron beams are produced at the Materials and Life Science Experimental Facility (MLF) from megawatt-class high-power pulsed proton beams. IBARAKI Materials Design Diffractometer (iMATERIA) [1] is built at the 20th beam line (BL20) at MLF as a versatile powder diffractometer. This diffractometer is aimed at facilitating advanced material science and is promoted by Ibaraki Prefecture—the local government. This diffractometer has a high throughput so as to handle a large number of samples at a resolution of $\Delta d/d \sim 0.16\%$; each sample is measured in just 10 minutes. Thus, neglecting the time taken to exchange samples, it should be possible to measure about 150 samples a day. iMATERIA has a vacuum chamber in which samples are placed, and this chamber is enclosed by a shield. If a vacuum cannot be maintained in the chamber while exchanging samples, the process of re-establishing the vacuum would become a bottleneck.

Small automatic sample changers have been used in the past, e.g., the 10-sample changer used for Vega [2] and Sirius [3] at the High Energy Accelerator Research Organization (KEK). The largest, which is used for High-Pressure-Preferred Orientation (HIPPO) [4] at Los Alamos Neutron Science Center (LANSCE), can handle 100 samples simultaneously. However, these sample changers cannot add new samples or exchange stocked samples without breaking the vacuum. At minimum, we need an automatic sample changer that is capable of changing 600 samples (over 4 days), but without breaking the vacuum in the main chamber. Accordingly, we have developed and manufactured an automatic sample changer that can handle a large number of samples through both the vacuum chamber and shielding blocks.

Fig. 1. Overview of automatic sample changer.

2. Automatic sample changer

Figure 1 shows an overview of the automatic sample changer, which is organized into five parts: sample storage room, articulated robot system, elevator, pre-vacuum chamber system, and sorting system. Both the sample storage room and the articulated robot are installed on an instrument shield. The elevator, the pre-vacuum chamber system, and the sorting system are attached to the flange (port diameter 800 mm) for the main vacuum chamber, and these parts can be removed easily as a unit. The pre-vacuum chambers connect to the main vacuum chamber through gate valves, and the sorting system is set into a small chamber connected directly to the main chamber.

There are two lines from the sample storage to the sorting mechanism (two elevators and two pre-vacuum chambers). The pressure in each line is independently controlled between atmospheric pressure and vacuum. Thus, samples can be transported to the main chamber while other samples are being measured. Moreover, new samples can be set in place in the sample storage area while one set of measurements is being made. Thus, the sample-exchange time is substantially reduced.

We also developed a new sample holder

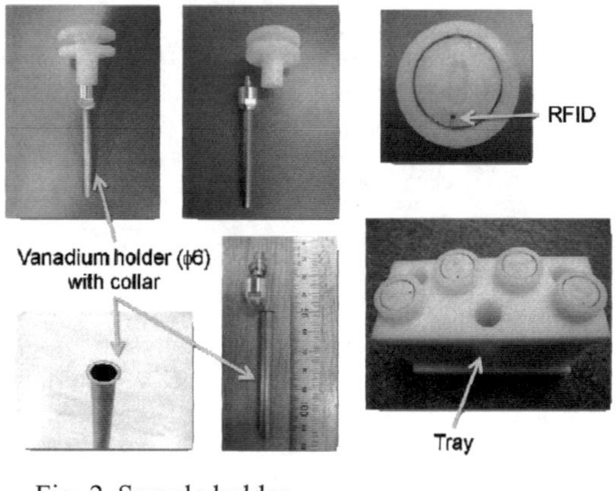

Fig. 2. Sample holder.

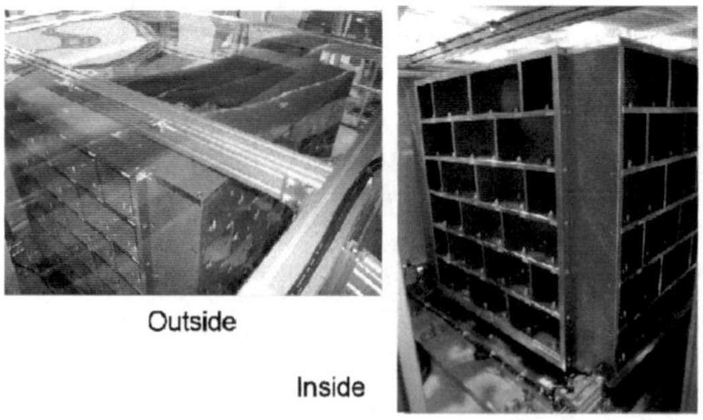

Fig. 3. Sample storage room.

for this automatic sample changer. This sample holder is made of vanadium (0. 1 mm thick) and has a collar. The samples are packed into this sample holder, which has an Al cap and an In seal as shown in Fig. 2. A radio frequency identification (RFID) tag is placed on the top of the holder cap (made of poly oxy methylene) with two collars for handling. Each sample is controlled using RFID.

Fig. 4. Articulated robot.

To allow the addition of new samples, or the replacement of stocked samples, during measurement, the sample storage room is located outside the instrument shield. In the sample storage room, there is a shelf for samples (see Fig. 3). The shelf rotates, and the front side of the shelf presented is automatically changed according to need. The sample storage system can position 168 trays, each storing 4 samples. Thus, a maximum of 672 samples can be stored in this room.

A sample access point is located to the side of the sample storage room on the instrument shield as shown in Fig. 1. We can position only one tray (4 samples) in this access point at a time. Information on samples (name, weight, height, ID number, position in the tray, etc.) is held in a sample database, and RFID is used to identify samples. After checking the RFID information, the tray is put in the sample storage room by the articulated robot. Sample handling uses the tray (4 samples) as a unit. When we want to retrieve a sample after measurement, the articulated robot picks up the tray containing the required sample.

According to a user-defined schedule, the articulated robot picks up a tray from the sample storage room and places it on the sample stage for transport to the elevator as shown in Fig. 4. The sample stage moves

Fig. 5. Elevator and pre-vacuum chamber.

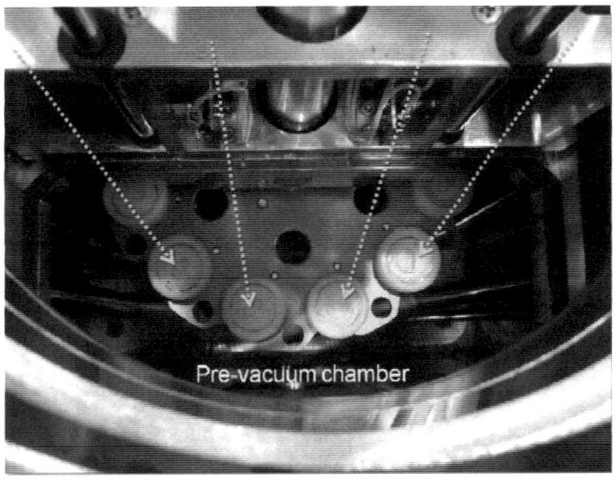

Fig. 6. Interior of pre-vacuum chamber.

horizontally to the elevator. The elevator picks up sample holders from the tray, and carries these sample holders vertically through through-holes in the shielding blocks to the pre-vacuum chamber, as shown in Fig. 5. The empty tray then returns to the sample storage room.

The pre-vacuum chamber has a volume of approximately 50 L. A half-round opening provides a pathway for the samples, with a gear-like stock holder at the bottom, as shown in Fig. 6. This stock holder can store 12 samples. Accordingly, the articulated robot and the elevator transports 3 trays (or a total of 12 samples) to the pre-vacuum chamber.

The vacuum in the main vacuum chamber is maintained. However, the pressure in the pre-vacuum chamber is controlled between atmospheric pressure and vacuum. In addition, the top side of the pre-vacuum chamber is at atmospheric pressure, while the bottom side is at a vacuum. Two gate valves are positioned above and below the pre-vacuum chamber. In addition, there are two pre-vacuum chambers in this system, and there is a stock holder for 12 samples in each pre-vacuum chamber. We can control the pressure for up to 12 samples during the measurement process. Thus, no time need be spent re-establishing a vacuum, except when first evacuating the chamber.

After evacuating the chamber, the underside gate valve opens, and the gear-like stock holder moves from the pre-vacuum chamber into the main vacuum chamber as shown in Fig. 7.

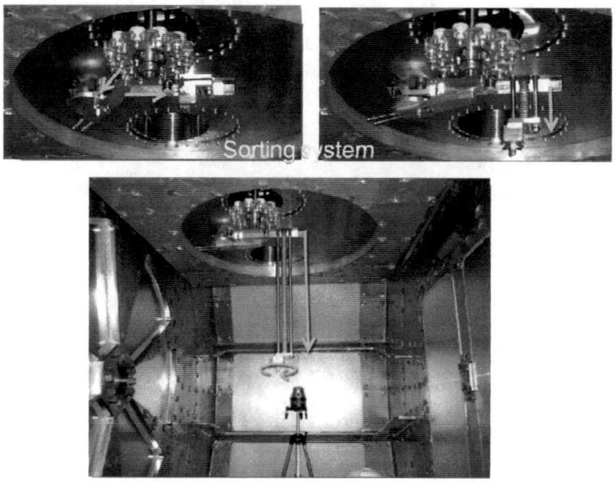

Fig. 7. Sorting system in the vacuum chamber.

The sorting system is situated within the main vacuum chamber. A sample is picked from the gear-like stock holder (containing 12 samples) and is moved to the neutron beam center as shown in Fig. 7. Two collars are needed on the sample holder at this point: one for holding, and the other for handling. The sample height can be changed for each sample. Optionally, the sample can be rotated during measurement.

We manufactured an automatic sample changer with the capacity to handle more samples than any previous design. We have already installed this automatic sample changer, and it has passed an operational check. Currently, we are developing control software in order to work with a data-acquisition system. The sample exchange time is substantially reduced using this automatic sample changer.

References

[1] T. Ishigaki et.al., Nucl. Instr. Meth. Phys. Res. A, 600 (2009) 189-191.
[2] T. Kamiyama et.al., Physica, B213&214 (1995) 875-877.
[3] T. Kamiyama et.al., Mat. Sci. Forum, 321-324 (2000) 302-307.
[4] H. –R. Wenk, L. Lutterotti, S. Vogel, Nucl. Instr. & Meth. Phys. Res., A515 (2003) 575-588.

Initial testing of a Compact Crystal Positioning System for the TOPAZ Single-Crystal Diffractometer at the Spallation Neutron Source

Matthew Frost[1], Christina Hoffmann[1], Jack Thomison[1], Mark Overbay[1], Michael Austin[2], Peter Carman[2], Robert Viola[2], Echo Miller[2], and Lisa Mosier[2]

[1]Oak Ridge National Laboratory, Oak Ridge, TN, USA
[2]Square One Systems Design, Jackson, WY, USA

E-mail: frostmj@ornl.gov

Abstract. A precise, versatile, and automated method of orienting a sub-millimeter crystal in a focused neutron beam is required for efficient operation of the TOPAZ Single Crystal Diffractometer at the Spallation Neutron Source at Oak Ridge National Laboratory. To fulfill this need, a Compact Crystal Positioning System (CCPS) has been developed in collaboration with Square One Systems Design in Jackson, Wyoming. The system incorporates a tripod design with six vacuum-compatible piezoelectric linear motors capable of $< 1\mu m$ resolution. National Instruments LabVIEW provides a means of system automation while at the same time accommodating the modular nature of the SNS sample environment control software for straightforward system integration. Results from an ambient test at the Advanced Photon Source at Argonne National Laboratory will be presented.

1. Introduction

The TOPAZ single-crystal neutron diffractometer is scheduled to start commissioning early in 2010. It is designed to use crystal sizes that have been used traditionally by x-ray diffraction in order to maximize scientific impact. High solid-angle detector coverage on TOPAZ will allow the instrument to maximize the capture of neutrons scattered from a crystal sample, thus minimizing the number of crystal positions required to gather enough data to refine a crystal structure. As a result of this innovative detector orientation concept however, one will require a precise method of crystal orientation that will not interfere with detector coverage, operate in a vacuum, and still be versatile enough to cover as much of the crystal reciprocal space as possible.

A three-point, six-axis design was conceived and prototyped using high-resolution, piezo-electric actuators. The design was chosen due to its compact nature and the additional capability of translational motion. The design implements the six axes in the form of three vertical axes, and three horizontal axes, one of which is orthogonal to the other two. This motor concept, along with the unique three-point plate interface allows a flexibility that cannot be achieved with a traditional goniometer alone. In addition to this flexibility, the bulk of the positioner is offset from the sample position itself, minimizing the shadowing of positioner components in the detectors. A solid model of the design concept can be seen in Figure 1, and the prototype setup inverted on a table for characterization purposes can be seen in Figure 2. This prototype

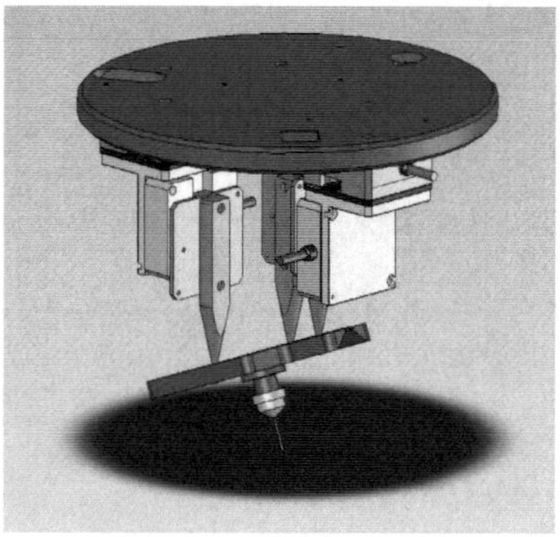

Figure 1. A CAD generated solid model of the ambient design concept.

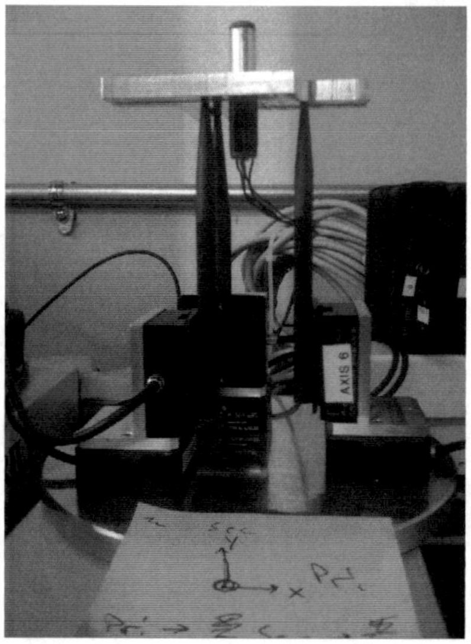

Figure 2. A photograph of the prototype. The prototype was inverted for testing purposes.

is capable of tilting up to a 15° incline in any direction, while still maintaining the sample in its original position. By adding a small rotational motor to this setup, this positioning concept compliments the TOPAZ instrument detector coverage.
Complementary to the versatility of the hardware, control software was developed that can take full advantage of the positioner functionality. A GUI was designed using LabVIEW that is flexible, will integrate with the SNS control system, and provides a level of control abstraction that would be familiar to the crystallography community.

2. Control Method and Implementation

Basic motor control of the system is relatively trivial, however emulating the desired goniometric parameters is not. This challenge was solved by taking advantage of the tripod nature of the hardware and developing code that takes a vector along the sample pin as the normal of an abstract plane. This plane can then be used to determine the appropriate location of each of the tripod's three points. This setup allows the use of traditional goniometer orientation angles, while at the same time preserving the flexibility of three axes of translational motion enabled by the hardware design. The aforementioned level of control abstraction is employed in the logical flow chart as seen in Figure 3.

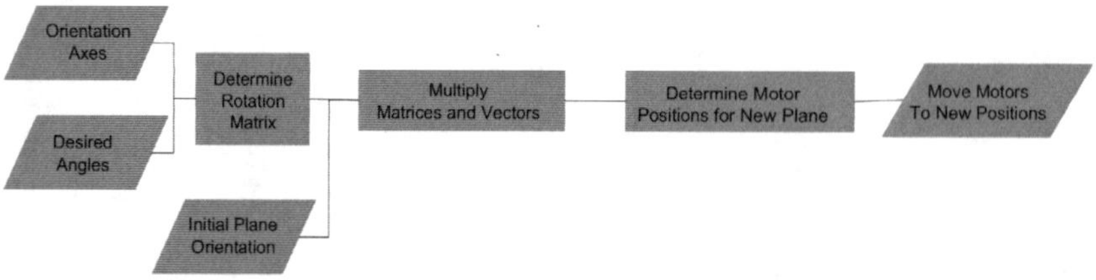

Figure 3. A flow chart describing the motor positioning algorithm.

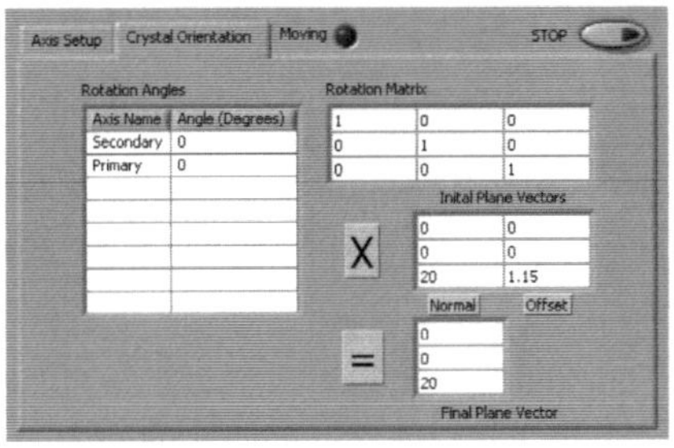

Figure 4. A view of the positioner control software.

A set of desired angles in a pre-programmed orientation space is determined by the user, then the software will determine a rotation matrix that represents the orientation space to be multiplied by the initial plane vectors. Once the new plane vectors are determined, the software then determines the new tripod positions to be set by the motors.

Other features implemented into the software include the ability to save and recall orientation spaces, three axis translation for fine-tuning crystal location and the ability to have as many as ten unique rotation axes in one orientation space.

3. Ambient System Testing

Tests were performed at the Advanced Photon Source in Chicago, IL, USA to determine the functionality of the ambient system. The system was adapted to be installed at Sector 15 of the APS, in place of a traditional goniometer on an x-ray single-crystal diffractometer. The repeatability of the system was tested by mounting a test crystal on the positioner, centering the crystal sample in the x-ray beam, and then repeatedly tilting the crystal in between 0° and 5° around the beam axis.

As seen in Figure 5, Reflections were observed from the test sample. For every repetition back to zero, the same spot was located and a fit performed in order to determine the location of the reflection on the detector. The repeatability of the system is shown to be in the range of $30\mu m$, although a general upward trend was observed. This trend was ultimately attributed to an insufficient holding force and low encoder resolution.

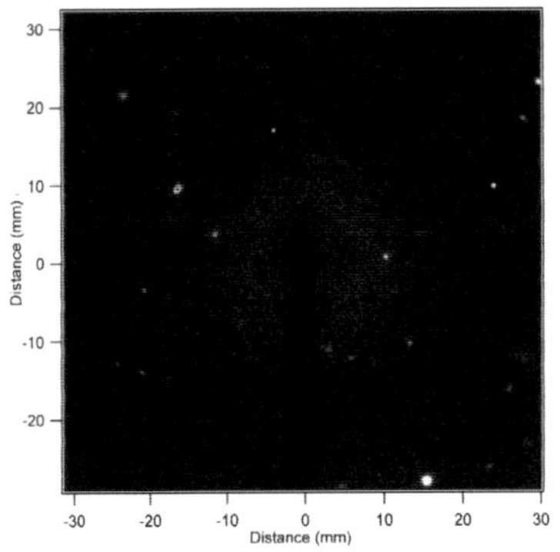

Figure 5. X-ray diffraction observed at 0°. Special thanks to X.P. Wang for test sample.

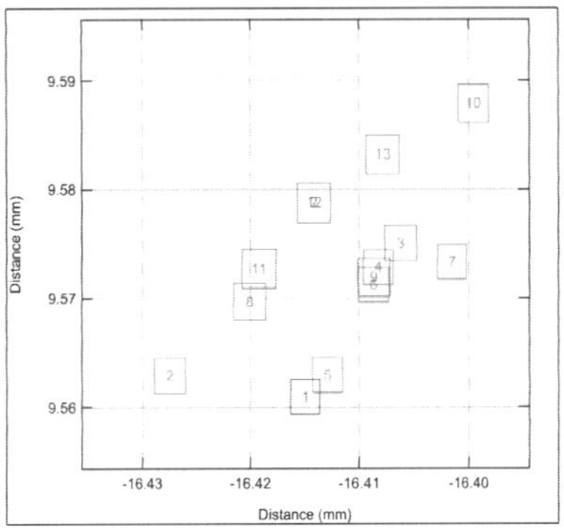

Figure 6. Reflection fit locations for repeated tilting from 0° to 5° and back to 0°.

4. Cryogenically Compatible CCPS

Testing has begun on a cryogenically compatible version of the initial design concept. The design tested at APS was adapted to use vacuum-compatible attocube Linear Nano-positioners with resistive encoders. These encoders allow better then $100nm$ precision, depending on the setup of the encoder. In addition to the change in motors, a means of cooling the oriented sample needed to be addressed. Initially, a braid will be tested as seen Figure 8, but exchange gas methods could be implemented as well in the future.

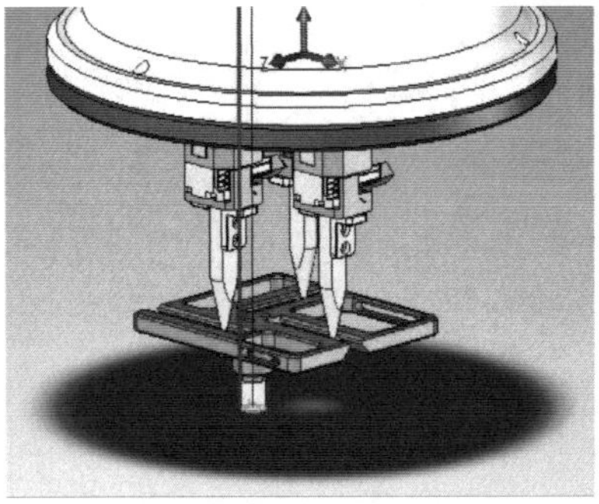

Figure 7. A CAD view of the Cryogenic CCPS. The different motors used provide the same functionality with a slight change in design.

Figure 8. A photograph of CCPS during vacuum testing.

5. Acknowledgments

This research is supported by DOE Small Business Innovation Research grant number DE-FG02-05ER84257.
ORNL is managed by UT-Battelle, LLC under Contract No. DE-AC05-00OR22725 for the U.S. Department of Energy, Office of Science.
Use of the Advanced Photon Source was supported by the U.S. Department of Energy, Office of Science, Office of Basic Energy Sciences, under Contract No. DE-AC02-06CH11357.
LabVIEW is instrument virtualization software created by National Instruments Corporation, Austin, TX, USA.
atto**cube** systems AG manufactures and priovides ultra-high precision spatial positioning systems and is located in Munich, Germany.

A low-temperature sample orienting device for single crystal spectroscopy at the SNS

T E Sherline[1], L Solomon[1], C K Roberts II[1], D Bruce[1], B Gaulin[2] and G E Granroth[1]

[1] Neutron Sciences Directorate, Oak Ridge National Laboratories, Oak Ridge, TN, USA
[2] Department of Physics and Astronomy, McMaster University, Hamilton, ON, Canada

E-mail: sherlinete@ornl.gov

Abstract. A low temperature sample orientation device providing three axes of rotation has been successfully built and is in testing for use on several spectrometers at the spallation neutron source (SNS). Sample rotation about the vertical (ω) axis of nearly 360° and out of plane tilts (ϕ and ν) of from -3.4° to 4.4° and from -2.8° to 3.5°, respectively, are possible. An off-the-shelf closed cycle refrigerator (CCR) is mounted on a room temperature sealed rotary flange providing ω rotations of the sample. Out-of-plane tilts are made possible by piezoelectric actuated angular positioning devices mounted on the low temperature head of the CCR. Novel encoding devices based on magnetoresistive sensors have been developed to measure the tilt stage angles. This combination facilitates single crystal investigations from room temperature to 3.1 K. Commissioning experiments of the rotating CCR for both powder and single crystal samples have been performed on the ARCS spectrometer at the SNS. For the powder sample this device was used to continuously rotate the sample and thus average out any partial orientation of the powder. The powder rings observed in S(**Q**) are presented. For the single crystal sample, the rotation was used to probe different regions of momentum transfer (**Q**-space). Laue patterns obtained from a single crystal sample at two rotation angles are presented.

1. Introduction

The need for a device capable of orienting and cooling a single crystal sample to temperatures significantly below 10 K on a time-of-flight instrument is well known in the neutron scattering community. A device capable of both of these tasks, which has been built and is in testing, will be described.

When performing single crystal experiments, the ability to align the crystallographic axes with respect to both the incident neutron beam and the detectors is necessary in order to fully probe **Q**-space and to take advantage of the symmetry of the detector bank. Though it is highly desirable to maximize the **Q**-space available for measurement, technical difficulties limit the ability to tilt the crystal out of the horizontal plane of the instrument. Typically, the orientation of a single crystal used in an inelastic neutron scattering experiment is known *a priori*, so only small out-of-plane tilts are needed to bring the crystal into proper alignment.

When considering measurements on a powder sample, concern arises that the individual grains may preferentially align along one crystallographic axis, creating a texture in the resultant scattering pattern. Rotating or oscillating the powder sample during the neutron measurement tends to average out any texture which may be present. Since this device has the ability to rotate from 0° to 360° and back, texture effects may be reduced or eliminated by oscillating between these two values.

2. Aparatus

The low temperature sample orienting device consists of a CCR [1] mounted vertically on a room temperature sealed rotary flange [2] providing ω rotations. Two piezoelectric actuated tilt stages [3] for ϕ and ν rotations with in-house developed encoders are mounted within an exchange gas can which is thermally anchored to the bottom of the CCR. The resolution of the encoders is sufficient to determine the angle of each tilt stage to better than 0.1° over much of the angular range. Further developments will improve the performance of the encoders. An aluminum cold shield mounted to the nominally 50 K intermediate stage of the CCR surrounds the low-temperature cold head and exchange gas can. The entire assembly, as shown in Fig. 1, rests in a stainless steel thimble whose flange forms the upper seal of the sample vacuum vessel of the spectrometer, e.g. that described in Ref. [4]. The neutron beam, shown as a dashed horizontal blue line in Fig. 1, is 749.3 mm below the thimble flange. For operation on the tilt stage, mount to sample center distance is 52.3 mm while CCR mount to sample distance is 151.9 mm. The tilt stage is capable of carrying a total of 60 g.

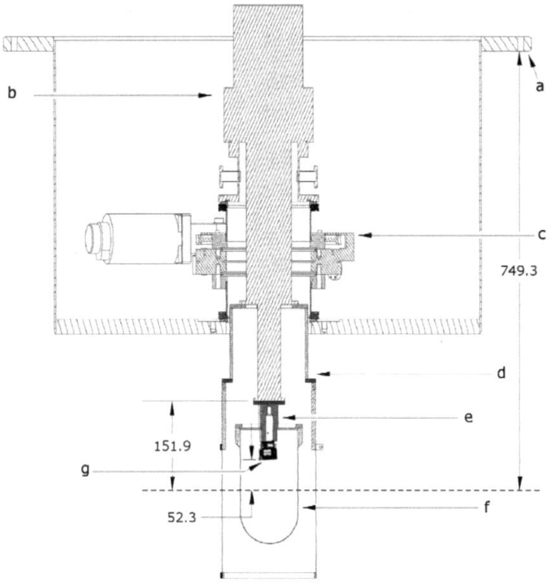

Figure 1. Schematic representation of the low temperature sample orientation device showing a) the thimble (arrow indicates the vacuum sealing surface), b) the CCR, c) the sealed rotary flange, d) the Al cold shield, e) the Cu thermal link and upper flange of the exchange gas can, f) the bottom portion of the exchange gas can, and g) the tilt stages. Note that everything below the sealed rotary flange is within the sample vacuum space. Dimensions are represented in mm. The dashed horizontal blue line represents the neutron beam.

The thermal performance of the CCR has been tested both with and without a load and with two different heat shields. Results shown in Fig. 2 indicate that the addition of a thermal load increases the total cooling time from less than 40 minutes to well over two hours. Additionally, the base temperature increases from 3.1 K to 6.5 K and creeps up over time. Based on experience with other heat shield configurations on identical CCRs, this likely originates from a degrading thermal link between the upper and lower portions of the heat shield. Nevertheless, this device has been effective for measuring several samples [5].

3. Neutron Scattering Results

The rotating cryostat has been used on the ARCS [6] and SEQUOIA spectrometers at the SNS. The intensity vs. detector bank position measured using the rotating cryostat on two example systems are shown.

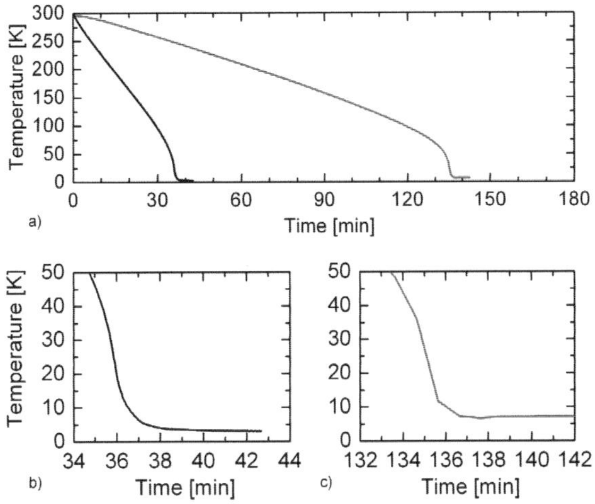

Figure 2. Two cooldown curves showing the thermal performance of the CCR. a) The black curve shows the cooldown of the CCR without any thermal load and using a solid thick walled cold shield. The red curve shows the cooldown of the CCR in the configuration shown in Fig. 1. b) and c) Expanded views showing the behavior of the final temperatures achieved for the two configurations.

Diffraction rings from a power sample of $NaNiO_2$ contained within a cylindrical Al can mounted via a Cu thermal link to the bottom of the CCR can clearly be seen in Fig. 3. The measurement was made using an incident energy of 60 meV while the sample, maintained at 5 K, was rotated from 0° to 360° and back in order to negate possible texture effects.

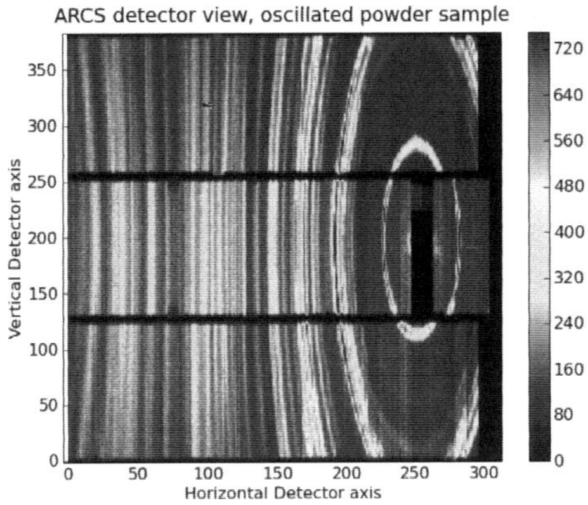

Figure 3. Intensity vs. detector bank position taken an the ARCS spectrometer for the powder sample $NaNiO_2$.

In a single crystal measurement on $La_{2-x}Ba_xCuO_4$ [X≈0.09], the pre-aligned crystal was mounted to the bottom of the CCR via a Cu thermal cold finger extension simlar to that used in the powder experient. While being maintained at 6 K, the rotating cryostat was used to scan through one plane of reciprocal space in order to obtain Laue patterns at two distinct values of ω as shown in Fig. 4 and Fig. 5.

4. Conclusion

The low-temperature rotating cryostat has been successfully used on an operating beamline at the SNS. Observations of both diffraction rings from a powder sample and Laue patterns from multiple **Q** values in a single crystal sample have been made possible using this device. Further development of the tilt stages will enable more extensive single crystal investigations in the near future.

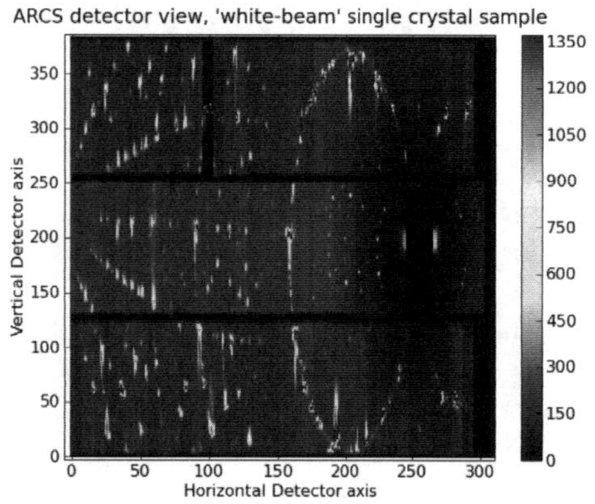

Figure 4. Laue image for $La_{2-x}Ba_xCuO_4$ [X≈0.09] in one rotational orientation.

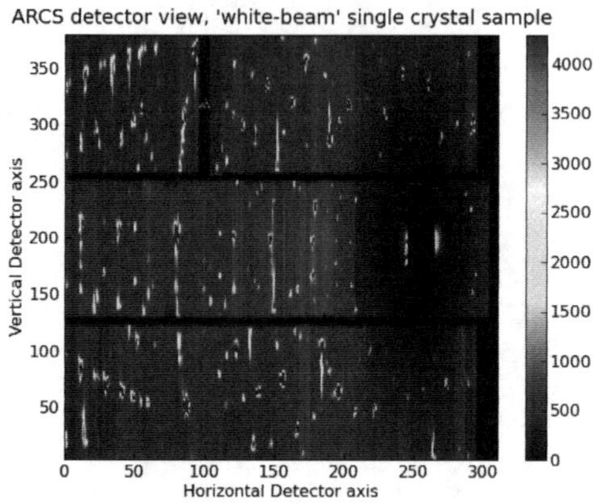

Figure 5. Laue image for $La_{2-x}Ba_xCuO_4$ [X≈0.09] in a second orientation rotated 40° from the pattern shown in Fig. 4.

Acknowledgments
The authors would like to acknowledge Doug Abernathy, Matt Stone, and Mark Loguillo of the ARCS instrument team for providing the data and plots for both the powder and intensity plots. This submission was sponsored by a contractor of the United States Government under contract DE-AC05-00OR22725 with the United States Department of Energy.

References
[1] ARS Model DE-210S. Advanced Research Systems, Inc., 7476 Industrial Park Way, Macungie, PA 18062 USA. (610)-967-2120, `http://www.arscryo.com`.
[2] Thermionics Model RNN-600MS. Thermionics Vacuum Products, 231-B Otto Street, Port Townsend, WA 98368 USA. (800)-962-2310, `http://www.thermionics.com`.
[3] Attocube Models ANGp100 and ANGt100. attocube systems AG, Koenigstrasse 11aRGB, 80539 Muenchen, Germany. 49(0)89-2877800-0, `www.attocube.com`.
[4] G. E. Granroth, D. H. Vandergriff, and S. E. Nagler, Physica B **385-386**, 1104 (2006).
[5] M. D. Lumsden *et al.*, Phys. Rev. Lett. **102**, 107005 (2009).
[6] D. L. Abernathy, Notiziario Neutroni e Luce di Sincrotrone **13**, 4 (2008).

In-situ Polarized ^{3}He-Based Neutron Polarization Analyzer for SNS Magnetism Reflectometer

W-T Lee[1], X Tong[1], J Pierce[1], M Fleenor[1], A Ismaili[1], JL Robertson[1], WC Chen[2], TR Gentile[2], A Hailemariam[1], R Goyette[1], A Parizzi[1], V Lauter[1], F Klose[3], H Kaiser[4], C Lavelle[4], D V Baxter[4], GL Jones[5], J Wexler[6], and L McCollum[1].

[1] Oak Ridge National Laboratory, Oak Ridge, TN 37831, USA

[2] National Institute of Standards and Technology, Gaithersburg, MD 20899, USA

[3] Australian Nuclear Science and Technology Organisation, PMB 1 Menai NSW 2234 Australia

[4] Indiana University Cyclotron Facility, Bloomington, IN 47408, USA

[5] Hamilton College, Clinton, NY 13323, USA

[6] Univesity of Massachusetts, Amherst, MA 01003, USA

E-mail: leewt@ornl.gov

Abstract. We report here the construction and neutron transmission test results of an *in-situ* polarized ^{3}He-based neutron polarization analyzer system for the Magnetism Reflectometer at the Spallation Neutron Source, Oak Ridge National Laboratory. The analyzer uses the Spin-Exchange Optical Pumping method to polarize the ^{3}He nuclei of a cell of ^{3}He gas. Polarized neutrons scattered from the sample are intercepted by the polarized ^{3}He gas which strongly absorbs neutrons in one spin-state while allowing most neutrons in the other spin-state to pass through. To maintain a stable analyzing efficiency during an experiment, the ^{3}He gas is continuously polarized *in-situ* on the instrument. Neutron transmission measurements showed that 73% ^{3}He polarization was reached in this setup.

1. Introduction

Polarized ^{3}He neutron spin filters are based on the spin-dependence absorption of the neutrons by ^{3}He. If the ^{3}He nuclear spin and the neutron spin are anti-parallel, the absorption is very strong. If the spins are parallel, there is considerably less absorption. The transmission of spin+ (spin-parallel) neutrons and spin- (spin-anti-parallel) neutron beam through a cylindrical cell of polarized ^{3}He gas are

$$T_{\pm} = T_e \exp\left(-n\sigma_0 l\lambda \pm n\sigma_0 l\lambda P_{He}\right), \quad (1)$$

where T_e is the transmission of the empty cell, n is the number density of the ^{3}He gas, the absorption cross-section σ_0 = 2966 x 10^{-24} cm^2 for λ = 1 Å [1], l is the path length through the gas, and P_{He} is the

^{3}He polarization. The empty cell transmission is weakly dependent on the neutron wavelength and is taken to be a constant in thermal and cold neutron scattering applications. With an incident polarized neutron beam, placing the polarized ^{3}He after a sample analyzes the polarization of the scattered neutrons. The neutron polarization analyzing efficiency P of a cell of polarized ^{3}He gas is

$$P = \frac{T_+ - T_-}{T_+ + T_-} = \tanh\left(n\sigma_0 l\lambda P_{He}\right). \tag{2}$$

The wavelength dependence of the ^{3}He neutron spin filter analyzing efficiency is gradual. It can be used as a broadband neutron spin-filter. In addition, the spin filter can accommodate a large beam cross-section. For typical beam divergence used in neutron scattering, its analyzing efficiency is virtually independent of the angular divergence of the neutron beam. Because of these characteristics, polarized ^{3}He has increasingly been used in neutron scattering works.

2. *In-situ* analyzer

We have been working with the polarized ^{3}He research community to develop the use of polarized ^{3}He based neutron spin filters [2-4]. Drawing on the experiences of the previous development, we have developed an *in-situ* polarization analyzer for the SNS Magnetism Reflectometer [5].

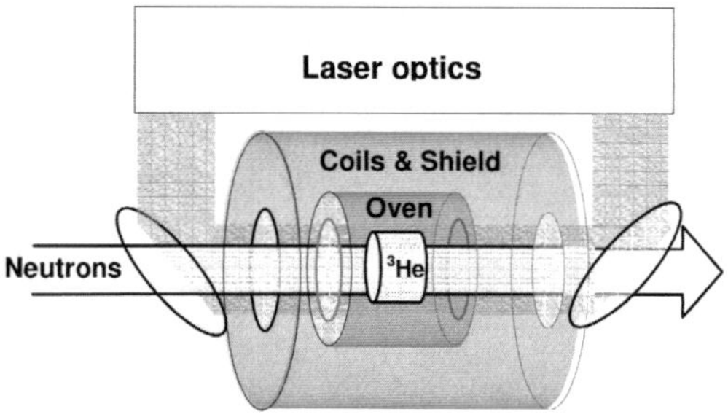

Figure 1. Schematic diagram of the analyzer. The details are explained in the text.

Figure 2. Analyzer setup at the reflectometer. Upper: Laser optics. Lower: Analyzer.

Figure 1 illustrates the setup of the analyzer system. The key component is a cell of polarized ^{3}He gas in the neutron flight path. The cell, called "Barbera", was made by the National Institute of Standards and Technology team [6]. It is made of aluminosilicate glass and is 11.8 cm in inner diameter and 7.5 cm in length. The ^{3}He gas pressure in the cell is 1.52 bar. As we used the Spin-Exchange Optical Pumping (SEOP) method to polarize the ^{3}He gas [7], the working substances included alkali metals and 120 mbar of nitrogen gas. Barbera was a "hybrid cell" which used both rubidium and potassium [8]. Figure 2 shows a picture of the setup.

The cell was placed in an oven that heats it to 200ºC. In a departure from the convention, non-inductive electric heater pads driven by direct current instead of hot air were used [9]. The cylindrical shape oven was made of Teflon. Double c-axis sapphire windows on both ends of the oven allowed passages of neutrons and light. The temperature controlled the vapor pressure of the alkali, which determined the polarizing speed of the ^{3}He. Additional thermal insulation was used outside the oven. A compensated solenoid provided the uniform magnetic field to maintain the ^{3}He polarization. The solenoid was enveloped by a cylindrical μ-metal enclosure with openings for passages of neutrons and light. The enclosure improved the field uniformity at the cell and shielded the ^{3}He from external magnetic interference. Compressed air flow kept the enclosure at near room temperature.

In the SEOP method for polarizing ^{3}He, circularly polarized light tuned to the $5s_{1/2} - 5p_{1/2}$ D1 transition of the rubidium at 794.7 nm is shined on the cell along the field direction. The light optically pumps the rubidium atoms from one magnetic sublevel of the $5s_{1/2}$ ground state to the $5P_{1/2}$ state. The excited rubidium atoms return to either sublevel with equal probability through a collisional de-excitation from the nitrogen gas molecules. Optically pumping on one of the sublevels depletes its population and polarizes the rubidium vapor in milliseconds. The use of a hybrid cell reduced the demand of light power to compensate the loss of rubidium polarization. In our setup, we used a 150 W 3-bar stack solid state laser. A Volume Bragg Grating in front of each bar centered the emissions at 794.7 nm and narrowed the bandwidth to 0.6 nm full-width-at-half-maximum. An electronically controlled liquid crystal retarder was used to control the chirality of the light. We used two similar laser optics setups to shine light with opposite chirality from both sides of the ^{3}He cell. The laser optics was placed above the μ-metal enclosure. 350 μm-thick (002) oriented silicon wafers with polarization-preserving dielectric coating reflected the light towards the ^{3}He cell while allowing neutron passage with little attenuation.

When a polarized rubidium or potassium atom collides with a ^{3}He atom, there is a finite probability that the ^{3}He nuclei and the alkali atomic electron undergo spin-exchange hyperfine interaction. The result is polarized ^{3}He nuclei. It takes tens of hours for the ^{3}He gas to reach its equilibrium polarization, typically in the range of 70-80%.

A sine-coil was placed in the space between the oven and the solenoid to provide a transverse radio-frequency field for a nuclear magnetic resonance (NMR) method called "adiabatic fast passage" (AFP). By sweeping the r.f. frequency through the ^{3}He Larmor frequency at a selected sweep rate, the ^{3}He polarization was flipped between parallel and anti-parallel to the solenoid field. This enables the measurement of spin+ and spin- neutron intensities. We tested one flip every 2 minutes.

The analyzer system was covered with laser shielding panels and mounted on an elevator at the detector table of the reflectometer. It was lowered into the neutron flight path when used. The ^{3}He was continuously polarized during an experiment to maintain a stable analyzing efficiency. Free Induction Decay (FID) of the ^{3}He NMR signal was used to monitor the ^{3}He polarization.

3. Neutron Transmission Measurement

Neutron transmission through the analyzer was determined by measuring the transmitted intensities with the analyzer in the neutron beam and with the analyzer removed. To avoid the uncertainty due to the efficiency of neutron spin-transport, we used iron plates to depolarize the incident neutrons. Each transmission was measured using spin+ and spin- incident neutron beam. The results with either spin state of the incident beam were virtually identical. This confirmed that the neutron beam was depolarized. The transmission of unpolarized neutrons through the analyzer system is

$$T_N = (T_+ + T_-)/2 = T_e \exp(-n\sigma_0 l\lambda)\cosh(n\sigma_0 l\lambda P_{He}), \tag{3}$$

where T_e includes the transmission through the cell walls, the silicon mirrors, and the sapphire windows. To separately determine the "cell thickness" $n\sigma_a l$, we also depolarized the cell and measured the unpolarized neutron transmission:

$$T_0 = T_N(P_{He} = 0) = T_e \exp(-n\sigma_0 l\lambda). \tag{4}$$

Measuring the intensities of cell-polarized and cell-unpolarized also allow a cross-check that is independent of T_e. The analyzing efficiency is related to the two transmissions by

$$P = \left(1 - \frac{T_0^2}{T_N^2}\right)^{1/2}. \tag{5}$$

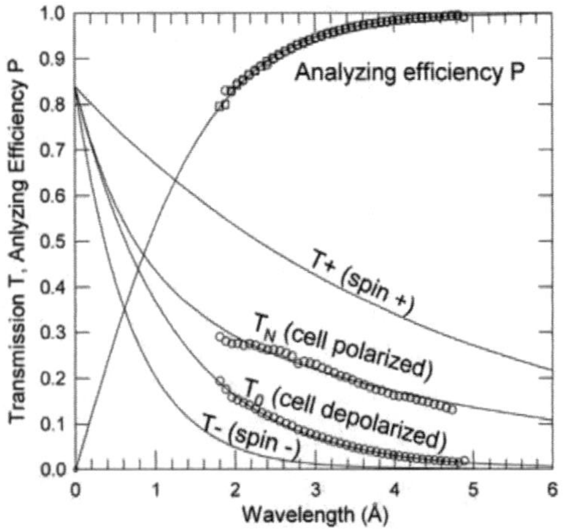

Figure 3. Measurements of unpolarized neutron transmission through the ^{3}He cell.

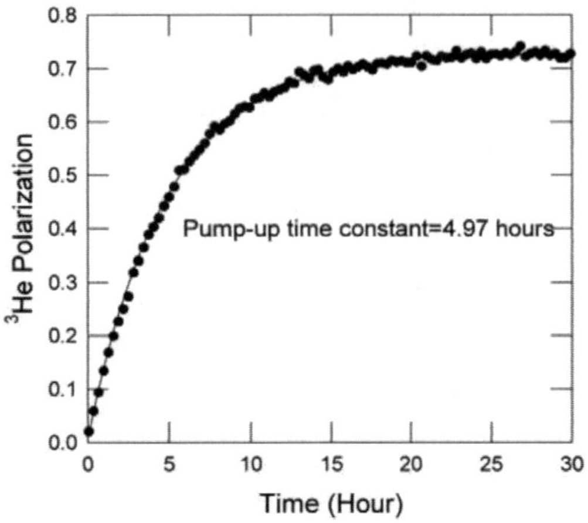

Figure 4. ^{3}He polarization as a function of time during the polarizing process.

Figure 3 shows the measured T_N and T_0 after the ^{3}He polarization reached equilibrium. Each dataset was measured for 1 hour counting time. The 1 hour counting time was chosen to reduce the amount of data during more than 2 days of continuous measurements. The data was normalized to the proton count of the accelerator to account for accelerator fluctuations. The analyzing efficiency was obtained by processing the data using equation (5). The data were fitted to equation (3) and (4) from which we determined the ^{3}He polarization to be 73%. Given the high accumulated neutron counts, the statistical error was 0.1%. We believe systematic uncertainties were the primary source of uncertainties in these measurements. Given the limited time available, however, we were not able to carry out addition measurements for determining the uncertainties. The cell thickness $n\sigma_a l$ = 0.825 at λ = 1 Å agreed with a previous NIST result ($n\sigma l$ = 4.14 at λ = 4.96 Å) and we obtained T_e = 0.84. The curve-fitting to the analyzing efficiency confirmed the parameters. We include also the expected T_+ and T_- for this system in the figure. The ^{3}He polarizing process was monitored by FID measurement. The data was scaled to the ^{3}He polarization determined from the neutron measurements. Fitting the data to a saturating exponential curve gives a pump-up time-constant of 4.97 hours. This is considerably shorter than the 10-15 hour time-constant we expected at 200°C. We believe laser heating of the cell as a result of the SEOP process resulted in the cell temperature higher than temperature measured by a resistance temperature detector the oven.

In conclusion, we have constructed a polarized ^{3}He based neutron polarization analyzer for the SNS Magnetism Reflectometer. Neutron transmission measurement showed a saturation ^{3}He polarization of 73%. The polarizing process in this setup had a pump-up time-constant of 4.97 hours.

4. References

[1] Mughabghab SF, *et al.* 1981 Neutron resonance parameters and thermal neutron cross sections, *Neutron Cross Sections* Vol **1** Academic Press, New York
[2] Jones GL, *et al.* 2005 *Physica* B **356** 86
[3] Jones GL, *et al.* 2006 *Proc. Int. Collaboration on Adv. Neutron Source (ICANS-XVII)* **III** 838
[4] Jones GL, *et al.* 2006 *Physica* B **385-386** 1131
[5] Lauter V, *et al.* 2009 *Physica* B *accepted for publication.*
[6] Chen WC, *et al.* 2007 *Phys. Rev.* A **75** 013416.
[7] Gentile TR, *et al.* 2005 *J. Res. Natl. Inst. Stand. Technol.* **110** 215 and references therein.
[8] Babcock E, *et al.* 2003 *Phys. Rev. Lett.* **91** 123003
[9] Tong X, et. al. 2009 *this proceedings.*

Electrical heating for SEOP-based polarized ^{3}He system

X Tong[1], J Pierce[1], W T Lee[1], M Fleenor[1], W C Chen[2], G L Jones[3], and J L Robertson[1]

[1] Oak Ridge National Laboratory, Oak Ridge, TN 37831, USA
[2] National Institute of Standards and Technology, Gaithersburg, MD 20899, USA
[3] Department of Physics, Hamilton College, Clinton, NY 13323, USA

E-mail: tongx@ornl.gov

Abstract. Development of neutron spin filters based on polarized ^{3}He is underway at Spallation Neutron Source (SNS). We report the progress of electrical heating tests in polarized ^{3}He based on Spin-Exchange Optical Pumping (SEOP) method. We first test the system performance based on electrical heating via non-inductance heating pads. We observe a contribution of 955 hours to the relaxation time T_1 from the heating pads. We then test the electrical heating SEOP pumping system at the SNS beamline Magnetic Reflectometer. We currently obtain 73% ^{3}He polarization in a cell with 820 cm^3 in volume.

1. Introduction

Polarized ^{3}He neutron spin filters have been used both as neutron polarizers and polarization analyzers in various polarized neutron scattering experiments [1, 2]. We have been developing the use of ^{3}He based on the Spin-Exchange Optical Pumping (SEOP) method [3]. During the SEOP pumping process, the gaseous ^{3}He cell is heated between 160^oC to 210^oC in order to create enough alkali vapor pressure for the spin-exchange process. The conventional way to heat the oven which contains the ^{3}He cell is to blow hot air in through silicone, teflon or copper tubes. This is simple to implement in an offline SEOP setup. However, constraints of space, heating, cooling and noise at a neutron scattering instrument make it cumbersome to use a hot air oven for *in situ* polarized ^{3}He based polarizers and analyzers [4]. To solve the problem we propose a new approach of heating source – none-inductive electrical heater pads. The heating pads are resistive heaters made of a printed circuit where the current path is laid out to enclose only a very small area, which might interfere with all the coils that are wound around the cell.

Our tests focused on comparing the system performance of using non-inductive electrical heating pads to hot air in the same oven in our laboratory based ^{3}He polarizing system. We find that both methods can reach the same range of temperature in the oven. Using the free induction decay (FID) nuclear magnetic resonance (NMR) method to monitor the ^{3}He polarization, we found a small but acceptable reduction of the ^{3}He polarization relaxation time T_1 with the heating pads energized. There exists little change of the final polarization that the cell can reach under both situations, which is insignificant. We also measure the impact of the heater pads toward using adiabatic fast passage (AFP) method to flip the ^{3}He polarization and find little interference. Another test is also carried out in our *in situ* ^{3}He polarizing system to find out its polarizing ability. The system with heating pads installed successfully polarize ^{3}He to 73%. We will report the details of these tests.

2. Laboratory based test

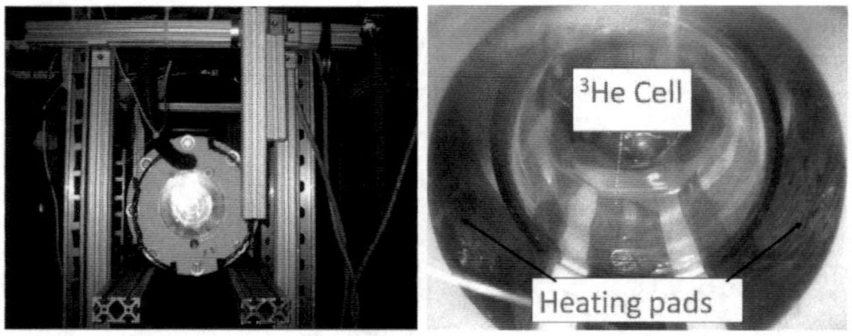

(a) The frame (b) The oven

Figure 1. (a), The lab based SEOP system, laser optics excluded; (b), ^{3}He cell sits on teflon stands which are placed in the oven. Two heating pads wrap around the oven wall to provide uniform heating.

The laboratory based ^{3}He polarizing system with the oven is shown in Fig. 1. The system sits on an 80-20 frame. A pair of Helmholtz coils of 16" apart provide a field of 13.5 G at the center of the coils. A teflon based oven is placed at the center of the coils. The oven wall is 2" thick of teflon, two double-sided sapphire window are placed on each side for better heat insulation. The ^{3}He cell, Bordeaux, is a hybrid cell containing gaseous ^3He, rubidium and potassium, is made at National Institute of Standards and Technology (NIST). Two pieces of Mica heating pads [5] are curved and placed inside the oven 5 cm far from the cell to provide a uniform heating to the oven. Both heating pads have the same dimension of 4" x 6" and each of them have a resistance of 12.6 Ohms. An Agilent power supply sends in about 200 W power to the heating pads which heats up the cell to 190°C. Optical pumping is then performed by a 150 W diode laser shining from one side of the oven. One should be aware that the heating from Laser is comparable to the heating pads. In our setup the laser can heat the cell up to 90°C. A 150-turn 18 AWG coil is placed at the top of the ^{3}He cell for FID polarization measurement. Because this test is carried out in the lab, the FID signal, which is proportional to ^{3}He polarization, is our only monitor for ^{3}He polarization.

The cell is first polarized till its polarization saturates. The measured pump up time constant is about 33 hours. Then we turn off the laser and monitor the FID signal continuously to extract the ^{3}He polarization relaxation time T_1. We first get the data for the normal condition, by turning off the heating pads. We haven't started to take data until the oven reaches room temperature due to the fact that the relaxation contribution from the rubidium vapor is too high and high temperature. Also, the FID signals change by temperature, which may bring systematic errors to our result. After we gather all the relaxation FID signals, we fit the data using Eq. (1).

$$P(t) = P_0 exp(-t/T_1) \tag{1}$$

where P(t) is the ^{3}He polarization at time t from the start time, P_0 is the initial ^{3}He polarization from the start time, and T_1 is the relaxation time.

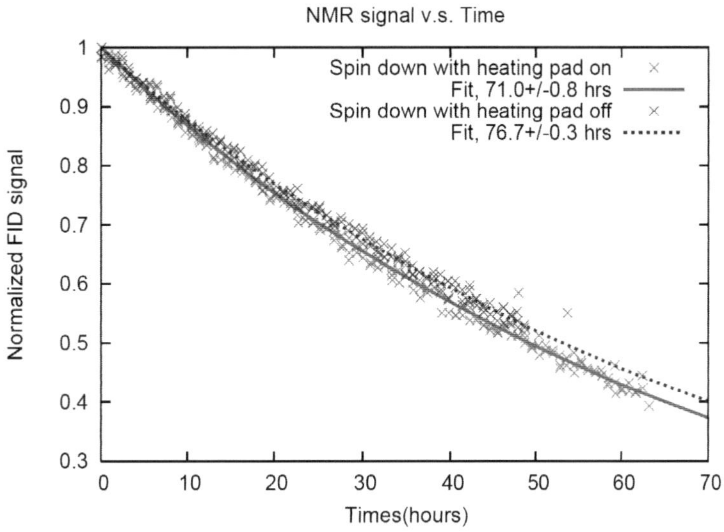

Figure 2. Relaxation time T_1 fits, with heating pad on and heating pad off.

After the T_1 is determined, we re-pump the cell till its polarization saturates. Then we measure the cell relaxation data while the heating pads are on. In order to provide the same temperature condition as when we turn off the heating pads, we blow 80 PSI compressed air into the oven to cool it down to the room temperature. We then fit the data to get a second T_1. Both fits are plotted in Fig. 2. Although the cell orientation is not changed for the re-pump process, the measured relaxation time might be slightly different from the first measured relaxation time [6]. Unfortunately we don't have the time to finish this measurement at that point.

As we can see from Fig. 2, when the heating pad is turned off, the fitted T_1 is 76.7 ± 0.3hours; when the heating pad is on, the fitted T_1 is 71.0 ± 0.8hours. We find less than 10% of relaxation time loss caused by the electrical heating pads. Yet his loss does little harm to the final saturation polarization.

We also test the impact that the heating pads bring to the AFP process. A 600 mV V_{pp} radiofrequency (RF) signal sweeping from 30 kHz to 120 kHz in 1.5 seconds is sent into a RF amplifier which amplifies the signal by 100. The output signal is then sent into the AFP coils wound outside of the oven. Each AFP flip takes less than 2 seconds to complete. Because for each AFP flip the ^{3}He polarization loss is very low ($< 0.1\%$), the FID signal loss is too little to be measured. Thus we take one FID signal measurement after 20 continuous AFP flips. 15 FID signal data points are taken. We then fit the data and find that with heating pad on, the fitted AFP loss per flip is $0.061 \pm 0.002\%$; with heating pad turned off, the fitted AFP loss per flip is $0.055 \pm 0.002\%$. The impact is obviously negligible.

3. *In situ* test

After we test the heating pads in the lab, we move the oven to our *in situ* SEOP pumping system and install the system at the SNS beamline 4A - magnetic refractometer. Our setup consists of two levels of equipments sitting on an 80-20 frame. The upper level is the laser and optics level. Two 150 W diode lasers are circularly polarized by liquid crystals. Laser lights are then reflected from two dielectric mirrors into the cell to optical pump the ^{3}He. The lower level contains the magnetic shielding and the oven. The cell Barbera was made at NIST and it contains 1.52 bar ^{3}He with a cell length of 8 cm. The heating pads heat the oven to around 200°C.

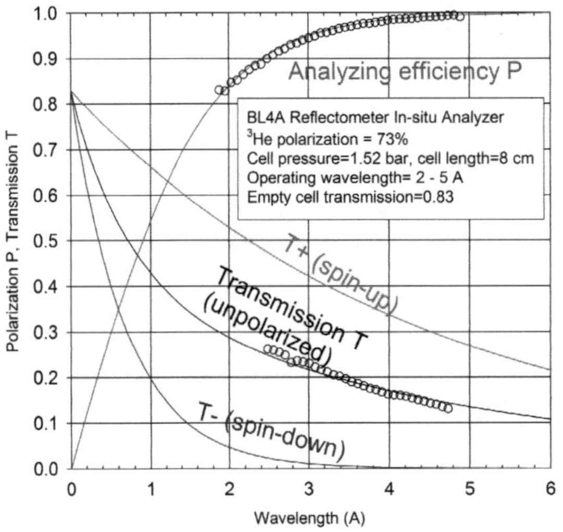

Figure 3. Magnetic reflectometer polarized ^{3}He test, neutron polarization versus neutron wavelength fit.

We start the test by extracting the cell thickness from two neutron transmission runs [7]. The transmissions are related in Eq. (2).

$$T_0(\lambda) = T_E exp(-n\sigma_0\lambda l) \tag{2}$$

where T_0 is the neutron transmission through unpolarized ^{3}He cell, and T_E is the transmission contributed from the sapphire windows of the oven and the dielectric mirrors. The cell thickness is then defined by $n\sigma_0 l$, where n is the number density of the ^{3}He, σ_0 = 29,662 b/nm, λ is the neutron wavelength in nanometers, and l is the cell length [8]. The fitted thickness is 8.35 nm^{-1}, which agrees with the value from NIST.

For an unpolarized neutron beam transmitting through a ^{3}He cell, the transmitted neutron polarization of $P_n(\lambda)$ is given by Eq. (3).

$$P_n(\lambda) = tanh(P_{He} n\sigma_0\lambda l) \tag{3}$$

Thus by fitting the outgoing neutron polarization versus the neutron wavelength, we are able to extract the ^{3}He polarization to be 73% ± 0.1%. A plot of above fitting is shown in Fig. 3.

4. Conclusion

We have tested an alternative heating method, electrical heating, for the SEOP pumping method for polarized ^{3}He. The non-inductive heating pads installed in the oven provide enough heating for the spin exchange pumping process. At our lab based SEOP pumping setup, we have tested the relaxation time based on two situations: with the heating pads energized and with the pads off. Compressed air is blown into the oven to maintain the same temperature condition. The T_1 are 71.0±0.8 hours and 76.7±0.3 hours for heating pads on and off, accordingly. We have also tested the heating pads' impact on the AFP flip loss. The AFP loss per flip is 0.061% with heating pads on comparing to 0.055% with heating pads off. From these lab based tests, we conclude that the negative impact that the electrical heating pads bring to the SEOP process is negligible. We have also tested the heating pads' performance on our *in situ* pumping system. The final ^{3}He polarization is 73%, indicating that heating pads are a good alternative to the hot air heating.

Acknowledgements
We would like to acknowledge Thad Walker for his earlier related work on high frequency heaters for use with spin-exchange optical pumping.

References
[1] Gentile T R, et al. 2006 *Physica B* **356** 96
[2] Jones G L, et al. 2006 *Physica B* **385-86** 1131
[3] Lee W T, et al. 2009 *submitted*
[4] Jones G L, et al. 2005 *Physica B* **356** 86
[5] Minco Corporation, www.minco.com
[6] Jacob R E, et al. 2004 *Phys. Rev. A* **69** 021401(R)
[7] Rich D R, et al. 2002 *Nucl. Instrum. Methods A* **481** 431
[8] Keith C D, et al. 2004 *Phys. Rev. C* **69** 034005

Dynamically polarized sample at the Spallation Neutron Source

J Pierce[1], D G Crabb[2], S Tomanicek[1], N Demarse[1], J Maxwell[2], J Mulholland[2] and J K Zhao[1]

[1] Neutron Scattering Sciences Division, Oak Ridge National Laboratory, Oak Ridge, TN
[2] Physics Department, University of Virginia, Charlottesville, VA

E-mail: piercejj@ornl.gov

Abstract. We report on the progress of the Dynamically Polarized Sample at the Spallation Neutron Source. The SNS DPS is a collaborative project with the Univ. of Virginia constructed for the purpose of polarized neutron scattering and diffraction. The project aims at significantly enhancing the signal-to-noise ratio by utilizing the strong spin dependent scattering cross section of hydrogen. One of the areas that are expected to benefit from this project is neutron protein crystallography, where a 10 fold gain in diffraction intensity and 10 fold reduction in incoherent background are within reach. The first step of the SNS DPS project uses a simple one Kelvin refrigerator and a 5 Tesla magnet. Our final goal is to design and build optimized setups for instruments such as the neutron protein diffractometer at the SNS (MaNDi). A brief overview of the DPS project will be given.

1. Introduction

The spin polarization of the nuclei in samples has the potential to greatly improve current experimental techniques. This is due to polarization dependence of neutron scattering. This polarization dependence is expressed by the spin dependent scattering amplitude, shown in Equation 1, where **I** is the spin vector for the nucleus, and **s** is the spin vector for the neutron.

$$b = b_0 + b_1(\mathbf{I} \cdot \mathbf{s}) \tag{1}$$

The coherent (Equation 2) and incoherent (Equation 3) scattering cross sections vary as a function of sample polarization P_I, neutron polarization P_n and nuclear spin I, as well as the quantites b_0 and b_1 shown in Equation 1. The effect is especially strong for the case of the hydrogen nucleus, which can be highly polarized in a variety of materials, including biological ones [1].

$$\sigma_{coh} = 4\pi(b_0^2 + P_nP_I \cdot Ib_0b_1 + \frac{1}{4}P_I^2I^2b_1^2) \tag{2}$$

$$\sigma_{inc} = \pi(I(I+1)b_1^2 - P_nP_I \cdot Ib_1^2 - P_I^2I^2b_1^2) \tag{3}$$

For the case of polarized neutrons, as the polarization of the hydrogen in the sample increases (where positive polarization is defined as polarization parallel to that of the neutron), the coherent scattering increases dramatically, while the incoherent scattering cross section drops quickly, until the case of 100% sample hydrogen polarization, for which the incoherent scattering from hydrogen vanishes. For material which contains large quantities of hydrogen, notably biological materials such as proteins, the incoherent scattering from hydrogen is often the dominant source of background. This means that a possible ten fold increase in signal associated with the increase in the coherent scattering can be coupled with a ten fold decrease in the incoherent scattering, resulting in an overall factor of 100 gained in the signal to noise ratio. The cross section for a polarized neutron scattering off a proton as a function of proton polarization are shown in Figure 1.

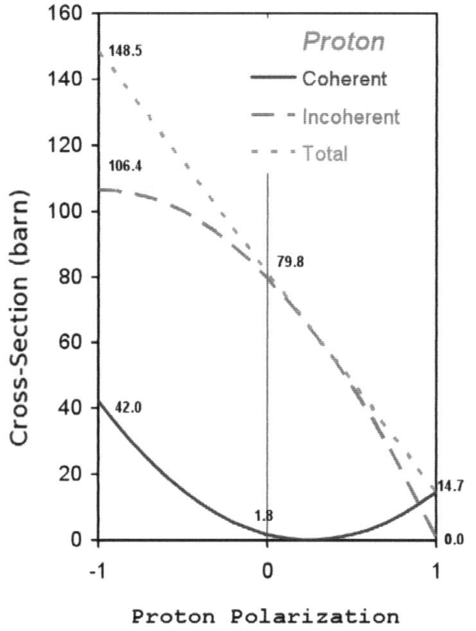

Figure 1. Cross section of polarized neutrons scattering off of a proton, as a function of proton polarization, where positive proton polarization is defined as the proton be polarized along the direction of the neutron polarization [2]. Hydrogen (proton) polarizations of 97% have been achieved in optimized materials, with polarizations in the 75% to 85% range being achievable in a variety of materials, including biological ones [1].

2. Dynamically Polarized Sample

The Dynamically Polarized Sample (DPS) program at the SNS hopes to develop a process and apparatus for polarizing sample materials in support of the beamlines at the Spallation Neutron Source (SNS). The polarization of samples will be done though a process known as Dynamic Nuclear Polarization (DNP) [3]. A full description of DNP is beyond the scope of this report, but briefly, it uses high magnetic fields (in this case, 5 Tesla), low temperatures (1 Kelvin or lower) and microwave radiation to induce high nuclear polarizations in specially prepared sample materials.

Materials to be polarized (with some exceptions) must be constantly in the high magnetic field and low temperature region, as well as being constantly subjected to microwave radiation in order to create or maintain polarization. This requires the design and construction of a magnet and cryostat that meet the necessary conditions, but still can accommodate the requirements of a neutron scattering instrument, such as overall size, construction materials, stray magnetic fields, and acceptance for scattered neutrons.

There is an additional requirement that the sample material be specially prepared before it can be polarized. This preparation amounts to the addition of a small number of paramagnetic centers (unpaired electrons that are free to align in the magnetic field). These centers will be added primarily by introducing commercial chemical dopants to the material. Polarization is generated in the immediate vicinity of the paramagnetic centers. The polarization then diffuses throughout the bulk of the sample. The optimal concentration of these centers will vary from material to material. It is therefore desirable to have a test setup, not on a beamline, which would allow for the polarization of samples in parallel with any running polarized sample experiment, in order to carry out tests of dopant levels. The test setup also has the advantage of being much easier to build and operate, since it does not share any of the design requirements of the in-beam apparatus. A test setup has been assembled at the SNS, which consists of a 1 Kelvin helium-4 cryostat, a 5 Tesla magnet, a 140 GHz microwave EIO tube, and a 213 MHz NMR system, which is used to measure the polarization.

3. Science Projects

3.1. Polarized Protein Crystals

Neutron scattering with protein crystals has been used widely to explore protein structure. Typical techiques involve growing crystals as large as possible, and possibly deuterating the sample in order to increase the signal size and lower the backgrounds. However, some proteins do not lend themselves well to these techniques. They may not form very large crystals, or the crystal structure may be altered or destroyed by deuteration. These protein systems could benefit from the DPS program. The improvement in signal to noise ratio associated with nuclear polarization is very evident in proteins, which generally have a very high hydrogen content. By polarizing the protein crystals, it would be possible to get more data, with lower backgrounds in the same amount of time, even with a smaller crystal and without the need to deuterate the sample.

Polarization tests have recently been carried out at the University of Virginia, where we polarized samples composed of lysozyme crystals dialyzed in a solution with TEMPO (a commercially available paramagnetic dopant). Tests with lysozyme are done as a proof of principle, since lysozyme is inexpensive, and crystalizes relatively easily. Initial tests were done with just one concentration of TEMPO. While enhanced polarization was observed, the polarization reached (approximately 20%) shows that it is necessary to produce samples with a wide variety of TEMPO concentrations to determine what the optimum dopant level is. This will be carried out both at the University of Virginia, and at the Spallation Neutron Source, using the test setup, which is currently being commissioned. The process of finding the optimum concentration for the paramagnetic centers may have to be performed for many different samples, requiring that the test setup be run frequently.

3.2. Site Specific Polarization

It may be advantageous to polarize only the nuclei in a specific part of the sample. For instance, one may want to polarize only the nuclei near a specific part of a protein, thereby highlighting that site. The paramagnetic centers in the material cause the nuclei nearby them to experience a different local magnetic field. This means that there will be a shift in thier NMR frequency. The bulk of the sample can be depolarized with continuous, high power RF radiation at the NMR frequency associated with the DNP magnetic field. While this is done, the polarization near the paramagnetic centers will remain high, so long as the DPS system remains operating.

A test of this technique was carried out at the University of Virginia, using samples prepared at Oak Ridge National Laboratory. The samples were a solution containing the GAPDH protein, along with a spin label derived from TEMPO. This spin label has an affinity for the active site of the GAPDH, which will cause it to bind there. The sample was polarized to a reasonable level to verify the usefulness of the spin label (for this technique high bulk polarization is not necessary). Figure 2 shows the NMR signal when the sample was polarized to -40% (DNP is capable of producing negative polarization as easily as positive, in this test negative was used), and Figure 3 shows the polarization of the sample as a function of time. After the sample was polarized, it was subjected to a de-polarizing NMR pulse, which succeeded in destroying the bulk polarization, leaving the polarization near the spin-label undisturbed. This was verified by watching the polarization near the spin label diffuse through the bulk. In a neutron scattering experiment, the depolarizing NMR pulse and DNP would be used continously, resulting in a steady and high polarization near the spin-label, and zero polarization in the bulk of the sample.

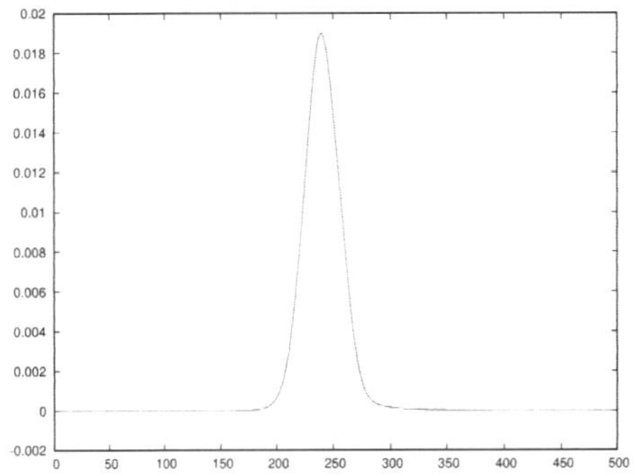

Figure 2. Hydrogen NMR signal for polarized, spin labeled GAPDH. The NMR signal for enhanced polarization is a large and distinct peak, with a broad, parabolic background which is easily subtracted.

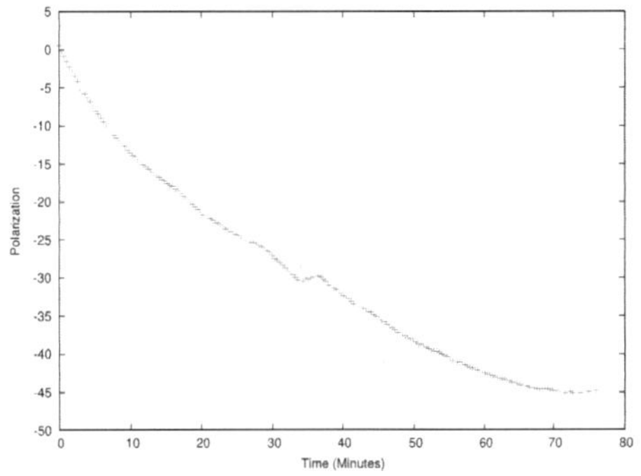

Figure 3. Hydrogen polarization of spin labeled GAPDH as a function of time.

4. Development

Work is underway to design and build a Dynamically Polarized Sample, which can be used in a a variety of the neutron scattering instruments at Oak Ridge. The goal is to design a system which could be used on instruments such as MaNDi, the SNS neutron protein crystallography diffractometer. Such a system would have to be very compact, yet have large enough scattering angles available to acquire useful diffration data. This may lead to the design of a 2.5 Tesla, dilution based system, or a frozen spin system [3]. This system would be used along side the test system that has already been assembled.

Acknowledgments

SNS is managed by UT-Battelle, LLC, under contract DE-AC05-00OR22725 for the U.S. Department of Energy

References

[1] Stuhrmann H B, et al 1986 *Eur.Biophys. J* **14** 1-6
[2] Zhao J K 2005 *Physica B* **10** 1237
[3] Crabb D G and Meyer W 1997 *Annu. Rev. Nucl. Part. Sci.* **47** 67-109

Journal of Physics: Conference Series **251** (2010) 012089 doi:10.1088/1742-6596/251/1/012089

Development of customised environmental chambers for time-resolved *in situ* diffraction studies

M J Styles[1] **and D P Riley**[1,2]

[1] The University of Melbourne, Melbourne, VIC 3010, Australia

[2] Australian Institute of Nuclear Science and Engineering (AINSE), PMB 1 Menai, NSW 2234, Australia

E-mail: m.styles2@pgrad.unimelb.edu.au, driley@unimelb.edu.au

Abstract. In an effort to mitigate the expense and broaden the applicability of customised environment chambers, researchers at the University of Melbourne and the Australian Nuclear Science and Technology Organisation (ANSTO) have designed and are currently commissioning a modular reaction chamber, capable of separating the necessities of diffraction methodologies from those of the desired sample environment. The *In Situ Reaction Chamber* (ISRC) abstracts many of the details intrinsic to the diffractometer, allowing users to design inexpensive environmental inserts that may be readily customised to their individual needs. The first insert to be developed for use with the ISRC is a high temperature furnace capable of providing an oxidising sample environment up to 1600°C.

1. Introduction

Progress in many technological fields including energy production, aerospace and bio-medicine is increasingly dependent upon novel materials with exceptional properties. Often these materials are produced via complex and energy intensive processes, making them costly both economically and environmentally. However, there is increasing experimental evidence to suggest modern time-resolved diffraction techniques, given sufficient time-resolution, can substantially reduce the time and cost involved in process optimisation and validation by providing vital insight into process kinetics [1]. This can be achieved through an examination of the fundamental mechanisms by which a material is synthesised, on a time scale commensurate with the rate limiting kinetic mechanisms [2].

Time-resolved *in situ* diffraction is a particularly powerful technique in materials research, capable of observing the evolution of phases as they change in quantity and composition, and correlating these changes with processing parameters such as temperature, pressure and time [3]. The time resolution of an *in situ* diffraction experiment is determined by the flux of radiation at the sample position and the speed at which detectors acquire the data. Modern position sensitive detector (PSD) technology has enabled diffraction pattern acquisition and readout times to be reduced on several high flux instruments (e.g. D20, GEM and WOMBAT), enabling diffraction studies of rapid material reactions and transformations with the same or better time resolutions as conventional methodologies [4]. Consequently time-resolved experiments can now generate extremely large data sets, with 10,000 diffraction patterns or more becoming common place. While quantitative analysis of data sets of this size (for information such as phase composition and reaction kinetics) has traditionally been laborious and time consuming, new software packages with the ability to either sequentially refine patterns or simultaneously fit 2-dimensional data sets (i.e. Rietveld surface fits) have begun to make quantitative analysis of these data sets feasible.

Overall, the development of reliable sample environment instrumentation remains a limiting factor in time-resolved *in situ* studies. If the results of a time-resolved *in situ* experiment are to be applied to the interpretation of a real process, it is critical that the sample environment be representative of the system under consideration. Often the environment chambers available at major research facilities are optimised for particular instruments and are restricted in their capabilities, or alternately are developed at great expense by user communities for a single, specialised experiment. These approaches often result in the development of chambers customised to the needs of a specific instrument or user and hence restrict the wider application of these techniques. Unfortunately, the difficulty and expense of producing a customised chamber has frequently resulted in poorly performing experiments or has deterred further experimental development.

These issues can be addressed by identifying and separating the experimental aspects associated with a particular instrument or technique, from those associated with the sample environment or process simulation. The result is a modular system in which the relatively expensive, instrument-specific technology can be developed once by a research institution, while the highly customised environment chambers are developed as needed by individual users to accurately simulate their system of interest and improve the quality of data obtained.

2. The *In Situ* Reaction Chamber

Researchers at the University of Melbourne in collaboration with ANSTO have designed, constructed and are currently commissioning a modular reaction chamber capable of achieving this separation of technologies. The ISRC (Figure 1) has been designed specifically for use on the high intensity powder diffractometer WOMBAT, located at the OPAL Research Reactor (Lucas Heights, Sydney, Australia), and is representative of the instrument specific chamber required by the modular technique.

The ISRC has been designed to address the following issues associated with high-flux neutron powder diffraction:

- Air scatter has been identified as a major contributor to instrumental background and loss of incident neutron flux and as such, the main chamber of the ISRC can be either evacuated or filled with helium. Additionally, the beam inlet window will allow for the fitting of a neutron guide. In general, the present design aims to maximise chamber diameter, as limited by the collimator and PSD, providing the largest sample volume, while minimising air-scatter.
- Flux attenuation is critical for ultra fast acquisitions, where time resolution is defined by the ultimate flux at the sample position. Aluminium has been used for incident and diffracted beam windows to reduce flux attenuation, which have been reduced to a constant minimum thickness over the full 160° range.
- Sample alignment is achieved on the ISRC by the use of standard, self-locating vacuum flanges, which are set at a specific height from the centre of the incident beam.
- Concurrent experimentation can be used to gather complimentary information or verify data while the diffraction measurement is being taken. The ISRC features vacuum ports set at various angles and heights to facilitate these experiments.
- Thermal loadings have been addressed by incorporating active water cooling within the double wall of the main chamber, permitting the use of very high temperature sample environments.

Figure 1. The assembled *In Situ* Reaction Chamber at the Australian Nuclear Science and Technology Organisation.

3. Customised environmental insert: The high temperature furnace

The first custom insert device to be developed for the ISRC is a high temperature furnace for the assessment of oxide ceramic materials (Figure 2). It has been designed to operate at temperatures up to 1600°C with partial oxidising or reducing atmospheres and is intended for use in the investigation of catalytic reactions, ceramic processing (sintering) and structural integrity of oxide ceramics as a function of time.

- Radiative band heaters mounted external to the sample environment (in the vacuum of the ISRC), above and below the beam height. These heaters allow very high temperatures to be reached accurately and repeatedly without contaminating (or being contaminated by) the sample environment or affecting the diffraction pattern.
- High thermal mass of the environment tube assists in thermal stability. This is the only material other than the sample (and sample vessel) to be placed in beam.
- Gas atmospheres are possible with ports for gas flow in and out of chamber. Controlled circulation is achieved via a second internal tube located out of the beam path. Vacuum ports also allow a vacuum to be maintained within the sample environment if required.
- Active cooling of the sample port allows for the use of standard vacuum fittings, facilitating accurate sample alignment.

4. Discussion

Traditional methods of performing *in situ* diffraction have been limited by the availability of customised reaction chambers. Cost has been a significant factor in limiting this development due to the unique design constraints of each reaction vessel. The present modular design of the ISRC deliberately separates the unique experimental aspects from the generic diffraction requirements. This allows customised inserts can be developed to meet the requirements of each reaction, and then placed into a diffraction specific chamber via a standard interface. Designed carefully, a single insert may be used on a variety of instruments. An additional benefit of the modular technique is the ability to easily characterise each insert prior to use, minimising setup time and maximising the likelihood of a successful experiment. For time-resolved studies in particular it is essential to minimise the effect of the environment chamber on the passage of the

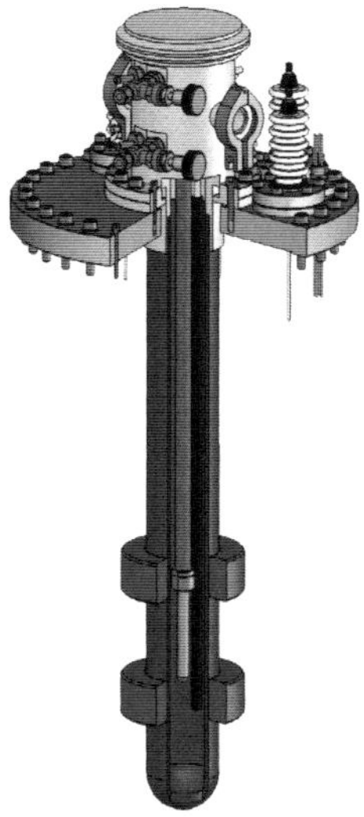

Figure 2. Section view of the high temperature furnace insert. It has been designed for an operating temperature of 1600°C under partial oxidising or reducing atmospheres. Note the sample cannister located between radiative band heaters.

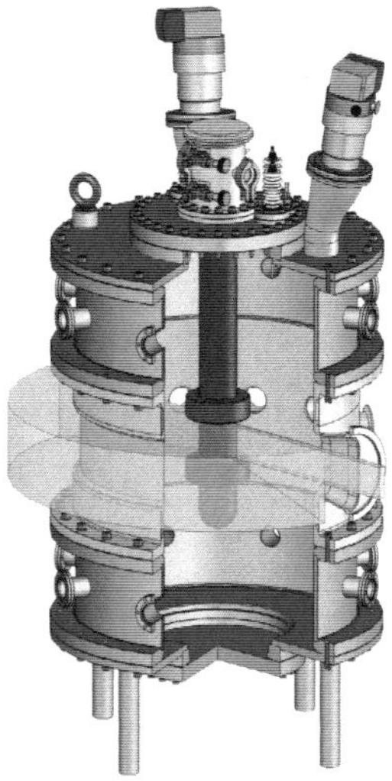

Figure 3. Section view of the assembled ISRC and high temperature insert, illustrating passage of incident and diffracted beams.

diffracting radiation. Customised inserts allow users to do this via methods most appropriate for their circumstances, using only the equipment necessary for their experiment. Since process optimisation typically involves experimentation using multiple techniques, the ability to use a single, well characterised environment chamber multiple times can significantly reduce errors.

5. Concluding remarks

High time resolution *in situ* diffraction is capable of providing a vital insight into the kinetics of materials processing. The information gathered by this technique can be applied to the optimisation of materials synthesis and the enhancement of operational performance. Successful material development is dependent upon the quality of the sample environment, which must accurately simulate the conditions of the system under consideration. The feasibility of developing quality sample environments is increased by separating the technology associated with particular diffraction techniques from the technology necessary for environment simulation.

Acknowledgments

The authors would like to acknowledge the advice and assistance of Dr Thomas Hansen, Dr Paul Henry and Prof. Erich Kisi and acknowledge funding from the Australian Institute of Nuclear Science and Engineering (AINSE) and the Australian Nuclear Science and Technology Organisation (ANSTO), including the assistance of Mr Scott Olsen and Mr Jason Christoforidis.

References

[1] Riley D P, Kisi E H 2008 and Hansen T C *J. Am. Ceram. Soc.* **91** 3207
[2] Riley D P and Kisi E H 2007 *J. Am. Ceram. Soc.* **90** 2231
[3] Kisi E H and Riley D P 2002 *J. Appl. Christallogr.* **35** 664
[4] Riley D P, Kisi E H, Hansen T C and Hewat A W *J. Am. Ceram. Soc* **85** 2417

New high temperature gas flow cell developed at ISIS

R Haynes[1], S T Norberg[1,2], S G Eriksson[2], M A H Chowdhury[1], C M Goodway[1], G D Howells[1], O Kirichek[1], and S Hull[1]

[1] ISIS Facility, STFC Rutherford Appleton Laboratory, Chilton, Didcot, Oxfordshire, OX11 0QX, UK
[2] Department of Chemical and Biological Engineering, Chalmers University of Technology, SE-412 96 Gothenburg, Sweden

E-mail: oleg.kirichek@stfc.ac.uk

Abstract. A flow-through quartz gas cell, together with a gas flow control and monitoring system, has been designed and constructed at ISIS. This equipment allows neutron powder diffraction data to be collected on samples at temperatures up to around 1300 K when exposed to user chosen mixtures of O_2, Ar, CO_2, and CO. By exploiting the sensitivity of neutrons to the presence of light atoms such as oxygen, it is possible to probe the crystal structure of oxide materials as a function of oxygen partial pressures down to $\log_{10}p(O_2)$ of about -20. The resultant structural information can then be correlated with the bulk properties of the materials, whose research and technological interests lie in fields such as energy production, storage materials, catalysis, and earth science.

1. Introduction

Neutron powder diffraction is a well established tool for investigations of the structural properties of crystalline solids, with the widespread availability of sample environment devices which allow studies to be performed as a function of temperature, pressure and magnetic field (see, for example, [1]). Facilities which permit diffraction studies to be performed under controlled atmospheres have been developed at a number of neutron sources (see, for example, [2-3]) but tend to be rather less routinely used. Nevertheless, they offer an extremely powerful approach to probe the relationship between crystal structure and bulk properties of materials, particularly if the latter are simultaneously monitored [4]. This paper describes the design and construction of a flow-through quartz gas cell for use at ISIS, together with its complementary flow control and monitoring system which allows the oxygen partial pressure within a flowing gas stream around the sample to be varied. Its use is illustrated using data collected on nonstoichiometric ceria, $CeO_{2-\delta}$, using the Polaris medium resolution powder diffractometer [5].

2. Technical aspects

Figure 1 illustrates the gas flow quartz cell and its location within the 'Leicester' furnace used at ISIS. With reference to figure 1, the furnace uses a cylindrical vanadium foil heater element (4), surrounded by three vanadium heat shields (5) which minimise radiation losses to the outer jacket. Modifications to this furnace were necessary to allow for the location of the gas flow quartz cell (6). These modifications include the addition of a bottom flange which houses two 'O'-rings to seal the quartz cell; the thermocouple feed-through (9) to allow the thermocouple (type K, used to control the temperature of the furnace) to be placed inside the quartz cell; the gas inlet (2); and an increased vanadium heater element diameter to accommodate the quartz cell. The top flange locates onto the NW40 flange on the top plate of the furnace and has three functions: to hold and seal the quartz cell by means of a single 'O'-ring, to allow the gas to flow out (3) and to hold the oxygen and temperature sensor (type R) (1). The oxygen sensor (type DS zirconia sensor supplied by Australian Oxytrol Systems) is located just above the sample pellets (7), which sit on a porous quartz frit (pore size 16-40 μm). The oxygen sensor monitors the content of O_2 within the atmosphere flowing past the sample pellets, with the oxygen partial pressure, $p(O_2)$ (in units of atm.), calculated using the voltage, V (mV), and the temperature, T (K), of the sensor using the Nernst equation, $p(O_2) = p_{ref}(O_2)e^{-46.421V/T}$, where $p_{ref}(O_2)$ is the oxygen content within the reference air, ($p_{ref}(O_2) = 0.209$).

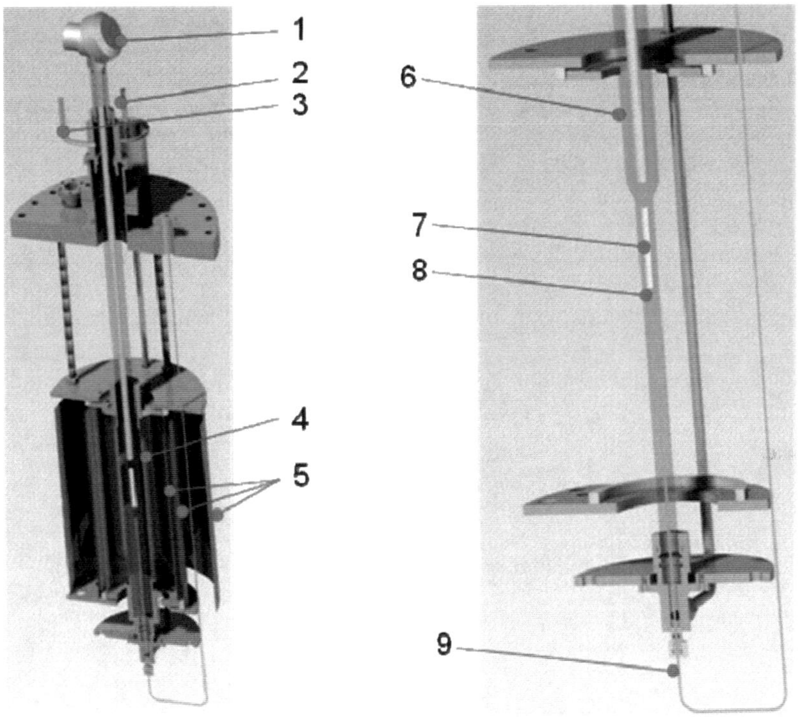

Figure 1. The schematic diagram of the flow-through gas cell (with the right hand portion highlighting the central portion of the left hand figure), showing (1) oxygen and temperature sensor; (2) gas inlet; (3) gas outlet; (4) heater element; (5) heat shielding; (6) quartz sample holder; (7) sample; (8) quartz frit; (9) thermocouple.

The composition and flow of gas through the quartz cell are controlled by means of a gas panel constructed at ISIS which consists of four mass flow controllers (Hastings 300 series). For the initial testing and commissioning of the apparatus, the four gases used were O_2, Ar, CO_2, and CO. Each mass flow controller is controlled by a single channel digital display and can be set manually or *via* a Labview® user interface running on the powder diffractometer's control computer. The latter allows changes in the gas flow and composition to be synchronised with the neutron data collection, so that powder diffraction spectra can be collected as the atmosphere surrounding the sample (and, hence, the oxygen partial pressure) are varied. A pressure transducer and Baratron® monitor the pressure in the system and the panel and cell gas circuit are fitted with relief valves to protect the system from the over pressure.

The assembly of the cell for a neutron experiment proceeds as follows. Firstly, the sample pellets are loaded in the quartz cell, which is then placed inside the furnace. Once located in the bottom flange and the 'O'-ring nut tightened, the top flange can be fitted. The thermocouple is then pushed through the feed-through until it touches the underside of the quartz frit and the 'O'-ring nut is tightened. The oxygen sensor is inserted into the quartz cell and is sealed down by means of an 'O'-ring and brass nut. At this point the sample circuit is checked for leaks and the furnace placed into its outer jacket and evacuated. Once loaded onto the neutron diffractometer the gas panel with the desired gases can be connected via a capillary line. The capillary joint is checked for leaks and the system is ready to introduce gas. The furnace can subsequently be heated to the required temperature, at a rate dependent upon the furnace vacuum. Cooling of the furnace is achieved by reducing the maximum power output and at approximately 400 K helium can be introduced to the furnace vacuum to speed up the cooling process.

3. Preliminary Results

Initial tests of the quartz cell and gas panel were performed on the Polaris powder diffractometer at ISIS [5] using a stack of 20 pellets of CeO_2, each of 6 mm diameter and 2-3 mm thickness. Diffraction data were collected at a temperature of 1273(2) K as a function of oxygen partial pressure, with a gas flow of 80 cubic centimetres per minute and a succession of gas mixtures in the sequence $O_2 \rightarrow Ar \rightarrow CO_2 \rightarrow CO$ used to provide increasingly reducing atmospheres (see figure 2). Data collection times of 15 minutes were used, with data collection started 5 minutes after changing the gas composition, to allow equilibration.

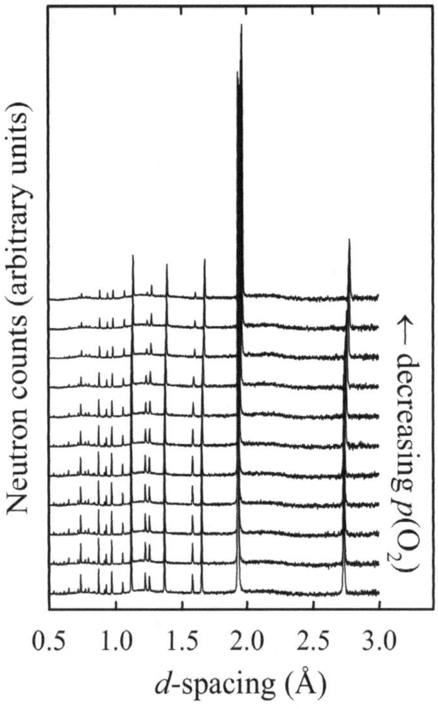

Figure 2. The neutron powder diffraction pattern of $CeO_{2-\delta}$ measured at 1273(2) K on decreasing oxygen partial pressure, $p(O_2)$. From bottom to top, the spectra were obtained at $\log_{10}p(O_2)$ values of -0.1, -3.3, -8.3, -10.2, -12.1, -14.1, -14.8, -15.5, -16.4, -17.0, and -17.6.

$CeO_{2-\delta}$ and its chemical derivatives are of current interest as possible electrolyte materials for Solid Oxide Fuel Cells (SOFCs) [6]. Within an SOFC, the electrolyte operates at temperatures in the range from 800 K to 1200 K and transmits O^{2-} ions from the cathode (air side) to the anode (fuel side). Thus, the material must be stable under both oxidizing and reducing atmospheres. The data shown in figure 2 illustrate a potential drawback associated with the use of CeO_2 because, at oxygen partial pressures less than around 10^{-14} atm there is a gradual increase in lattice parameter (associated with the reduction of some Ce^{4+} to the larger Ce^{3+} species), a loss of oxygen (which causes the changes in the relative intensities of the Bragg peaks, such as the pair at ~1.25 Å in figure 2) and the onset of significant disorder within the crystalline lattice (seen as a reduction in the intensities of the Bragg peaks at low *d*-spacings). The latter also cause diffuse scattering, observed as weak, broad undulating features in the intensity between the Bragg peaks (and not visible in figure 2). A detailed analysis of the diffuse scattering, to probe the nature of the ionic disorder within reduced $CeO_{2-\delta}$, has recently been published [7].

4. Conclusions

Many common cation species, such as Fe, Cu, Co, and Mn, can adopt different valence states. As a consequence, oxides containing these species often exhibit significant anion non-stoichiometry, which can have a profound influence on the bulk properties of the material. Thus, in addition to the field of SOFC materials mentioned above, the apparatus described in this paper is likely to find widespread applications within research areas such as high temperature superconductors, multiferroic compounds, and mineralogy. A number of development projects are also underway to expand the range of science to be addressed, including the use of controlled humid atmospheres for the study of proton conducting ceramics for fuel cell applications and the combined use of the gas panel with an *in-situ* resistance measurement cell [4] to simultaneously probe the structural and conducting properties of materials as a function of oxygen partial pressure and temperature.

5. Acknowledgements

The gas flow cell and control panel were constructed with financial support from the Swedish Research Council, Vetenskapsrådet.

References

[1] Bailey I F 2003 *Z. Kristallogr.* **218** 84
[2] Turner J F C, Done R, Dreyer J, David W I F and Catlow C R A 1999 *Rev. Sci. Instrum.* **70**, 2325
[3] Tonus F, Bahout M, Henry P F, Dutton S E, Roisnel T and Battle P D 2009 *Chem. Comm.* 2556
[4] Engin T E, Powell A V, Haynes R, Chowdhury M A H, Goodway C M, Done R, Kirichek O and Hull S 2008 *Rev. Sci. Instrum.* **79**, 095104
[5] Hull S, Smith R I, David W I F, Hannon A C, Mayers J and Cywinski R 1992 *Physica B* **180-181**, 1000
[6] Ishihara T, Sammes N M and Yamamoto O 2003 '*Electrolytes*' in '*High Temperature Solid Oxide Fuel Cells: Fundamentals, Designs and Applications*', ed. Singhal S C and Kendall K, Elsevier, Oxford.
[7] Hull S, Norberg S T, Ahmed I, Eriksson S G, Marrocchelli D and Madden P A 2009 *J. Solid State Chem.* **182**, 2815.

The development of a sapphire cell for neutron scattering experiments at high temperature and high pressure

L. Walker, C. Luttrell, L. Solomon, and A. Church

Oak Ridge National Laboratory, Oak Ridge, TN 37831 USA

E-mail: walkerl@ornl.gov

Abstract. A high pressure, high temperature sapphire cell has been designed to provide users at the Spallation Neutron Source with the ability to perform in-situ supercritical bulk water studies. The design criteria for the cell required the achievement of a maximum pressure of 150 MPa at the maximum temperature of 500°C. Other conceptual constraints, such as a 360° viewing angle and top loading integration further complicated and limited available cell models. The sapphire cell was tested and modified due to pre-mature failure. Due to convenient geometry and ease of operation, interest in the cell for more moderate operating requirements has increased.

1. Introduction

High temperature and high strength characteristics make sapphire a perfect candidate for neutron scattering experiments where the material to be observed exhibits interesting behavior outside of typical experiment parameters (4 to 300 K, 0.1 to 5 MPa). One of these moderately extreme environment experiments was proposed for a high energy resolution neutron backscattering spectrometer at which Quasi-Elastic Neutron Scattering (QENS) can be performed. QENS probes diffusion and relaxation processes, and a typical requirement is an extended range of available temperatures. One long standing application of QENS has been the study of bulk and confined water [1].

Supercritical bulk water has generated interest for applications such as cleaning and dissolving pollutants and other non-polar substances like oils, as well as, nuclear reactor applications [1]. Both high pressure and high temperature are required to reach the supercritical phase. The thermodynamics of confined liquids, liquid confined in nano sized pores, are also of interest in this regime.

The original operational requirements needed to support these studies were a maximum temperature of 500°C (773 K) and pressures to 400 MPa. Practical consideration required modification of these goals to 500°C and 153 MPa. This article describes the design, fabrication, and test process of a high pressure and high temperature sapphire sample cell.

2. Cell Material Selection

The sample cell was manufactured from sapphire grown by Crystal Systems Incorporated [2], and machined by Insaco Incorporated [3]. Sapphire is often used in high pressure applications because of its strength, transparency to visible light, and low incoherent neutron scattering and absorption cross sections [4]. The relevant properties of sapphire, which where extrapolated from indexed single crystal data, are given in Table I. To further minimize reflections in the diffraction plane, the sapphire cell was machined in such a way that its vertical axis is oriented 2.5° off the c-axis of the parent material [4].

Material Properties	Value	Reference
Average Tensile Strength @500°C	324 MPa	[5]
Modulus of Elasticity	344 GPa	[2]
Density	3.98 g/cm^3	[2]
Maximum Working Temperature	2040°C	[2]

Table I. Material properties for single crystal sapphire.

As designed, the sample cell has a total length of 93.75 mm, a minimum wall thickness of 12.5 mm, and a bore that is 5mm wide by 81.25 mm tall as shown in figure 1. To minimize pressure build up within the cell the sample bore was modified from a design with sharp corners to a design with 1.5 degree radius corners. To optimize the amount of sample in the beam, sapphire inserts were designed. These inserts provide annular gap sizes of 0.1 mm and 1 mm. Gap sizes, or sample thicknesses, will vary based on the strength of the neutron scattering from the sample and are required in order to minimize multiple scattering from the sample.

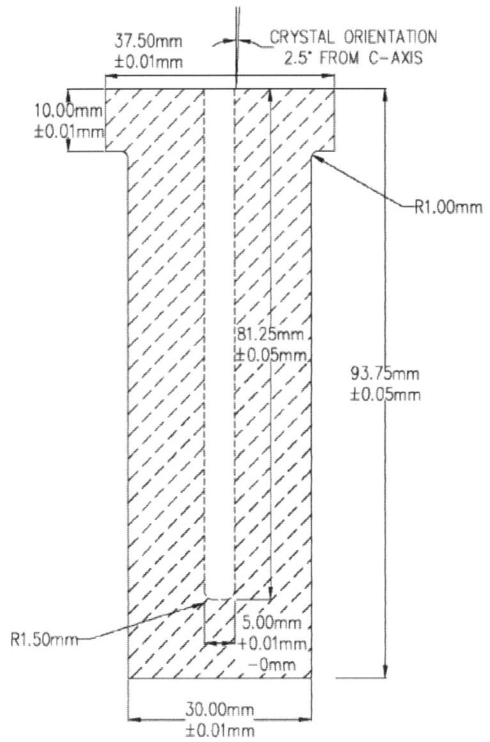

Figure 1. Sapphire cell dimensions

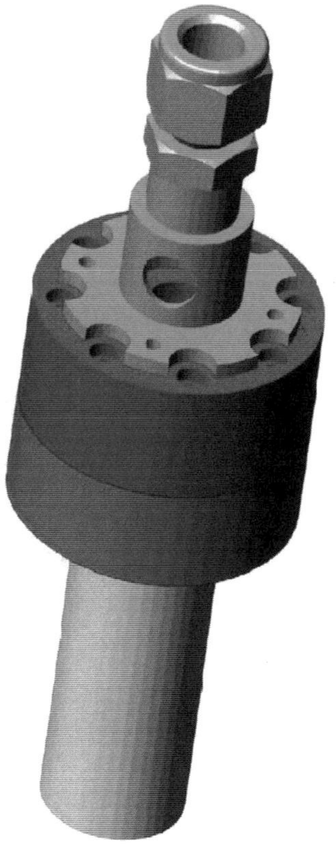

Figure 2. 3-D Assembly of sapphire cell, stainless steel fittings, and Swagelok [6] sample stick adapter

3. Pressure Requirements for Cell

The design of the cell body is principally a flanged tube of sapphire ($\alpha - Al_2O_3$). The stainless steel fittings are designed in two parts. One part is a cradle for the cell flange and mates with the second part which acts as a cap for the sapphire cell. The assembly, as shown in figure 2, uses a metal c-ring as a sealing surface, and a HiP [7] high pressure fitting to connect the assembly to high pressure tubing and apparatus. Fully assembled the sapphire cell is designed to be operated at a pressure of 153 MPa at 500°C. Safety considerations at the SNS facility require a testing protocol at 1.5 times the maximum allowable working pressure (MAWP) at room temperature. The cell was thus designed for 270 MPa.

The design of the sapphire cell proceeded in two stages. First, a finite element analysis was performed to verify the theoretical calculations and pinpoint any weaknesses in the cell structure when under the maximum anticipated pressure. The results of the finite element analysis show (Figure 3) that without a radius of curvature on the bottom inside corner of the cell bore, the maximum stresses in the cell loaded with an internal pressure of 162.5 MPa are at or above the yield strength for sapphire at 500°C. By including a radius of 1 mm at this location the maximum stress is reduced to 218 MPa and by including a radius of 1.5 mm the maximum stress is reduced to 161 MPa. A radius of 1.5 mm reduces the maximum stress in the corner to approximately the same value as the stress through the wall.

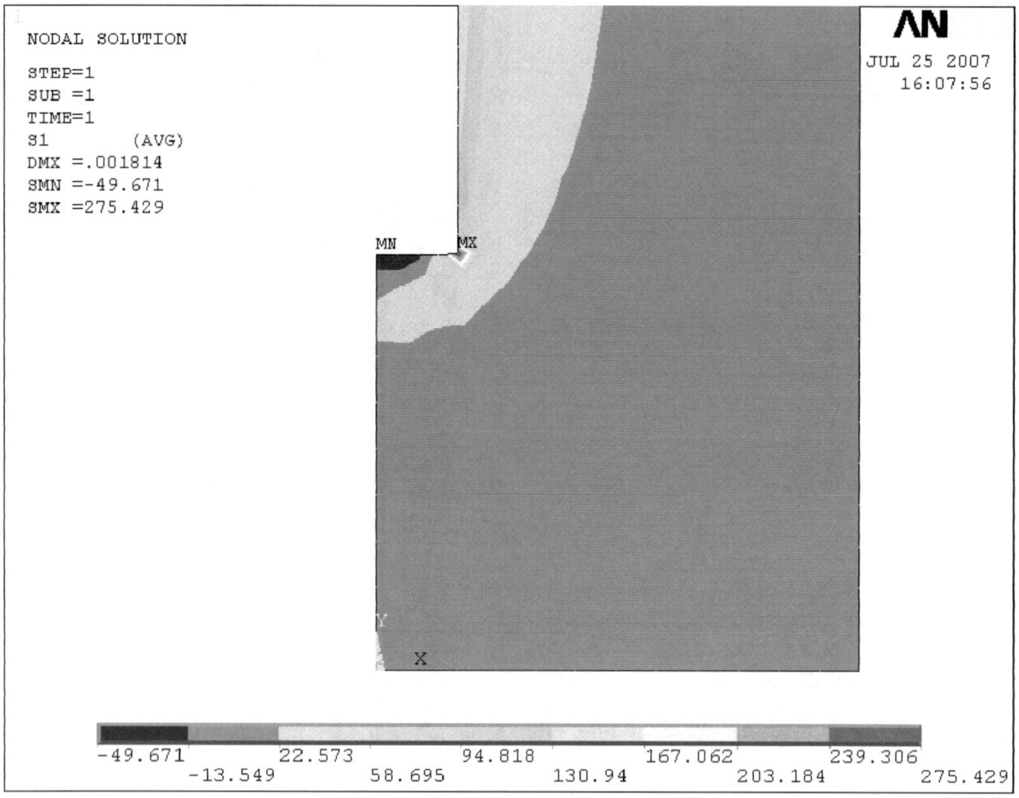

Figure 3. Finite Element Analysis shows Maximum Principal Stress distribution in the cell

Secondly, an FEA modified Barlow's Equation (1) was used to calculate the bursting pressure of the sapphire cell. The working pressure is typically defined as ½ the bursting pressure in systems where a safety factor of 2 is required [8].

$$P_b = S\frac{\left(d_o^{\,2} - d_i^{\,2}\right)}{\left(d_o^{\,2} + d_i\right)} \quad (1)$$

where P_b is defined as the bursting pressure, S is the tensile strength of the sapphire, and d_o and d_i are the outer and inner cell diameter, respectively. With S = 324 MPa, d_o = 0.030 mm, and d_i = 0.005 mm, the MAWP for the sapphire cell is 153 MPa.

4. Cell Verification and Testing

Testing parameters were established to ensure safe work practices. As mentioned previously, test pressure will be 1.5 times the design pressure. In each instance pressure tests will be organized by the system owner. Also, pressure tests must be repeated after any repair, modification, or failure. Most importantly, during testing and operation, pressure equipment must be protected by at least one pressure limiting device.

The sapphire cell was inspected and tested upon delivery. The original cell failed unexpectedly after 10 MPa of pressure was applied. A design investigation was conducted and possible cell modifications were suggested. As a result, the cell body was returned to Insaco, Inc. [3] for exterior polishing (original cell had rough finish). During that time, interest in the use of the cell at lower pressures (1 to 5 MPa) increased. After polishing, the cell was tested and verified for operation at pressures up to 10 MPa and temperatures up to 80°C. The cell successfully performed at these conditions. After experiments at these conditions, the cell will be tested to allow operation at its maximum working pressure. Plans are also in place to test one or more cell bodies to failure. These tests will provide necessary feedback for future gas pressure cell designs.

5. Conclusions

A sapphire cell that maintains pressure up to 10 MPa at 80°C was fabricated at SNS to perform in-situ supercritical bulk water studies. A similar sapphire cell was fabricated at HFIR for operation at 35MPa and 12K [4]. Having both cells at ORNL provides a full temperature range to users wishing to perform QENS experiments. Two recent user experiments at the SNS facility have provided insight into background issues associated with the thick walled sapphire cell design. Feedback of this nature assists cell designers in making modifications that improve the cell's strength, durability, and neutron transparency.

Acknowledgments

We acknowledge Eugene Mamontov for providing primary cell design specifications and Bryan Chakoumakos and Claudia Rawn for providing inspiration for cell design and material. Research sponsored in part by the Laboratory Directed Research and Development Program of Oak Ridge National Laboratory (ORNL). Work at the Spallation Neutron Sources was sponsored by the Scientific User Facilities Division, Office of Basic Energy Sciences, U.S. Department of Energy.

References

[1] Mamontov E, Cole D R 2006 *Phys. Chem. Chem. Phys* **8** 4908-14
[2] http://www.crystalsystems.com/sapphys.html
[3] http://www.insaco.com/
[4] Rondinone A J, Jones C Y, Marshall S L, Chakoumakos B C, Rawn C J, Lara-Curzio E 2003 *Can. J. Phys.* **81** 381-85
[5] Schmid F, Harris D C 1998 *J. Am. Ceram. Soc.* **81** (4) 885-93
[6] http://www.swagelok.com/
[7] http://www.highpressure.com/
[8] Kohan A L 1997 *Boiler Operator's Guide* ed 4 (New York: McGraw-Hill Professional) chapter 8 pp 287-8

Cryogen free high magnetic field sample environment for neutron scattering.

R B E Down[1], G Kouzmenko[2], O Kirichek[1], R Wotherspoon[2], J Brown[2], Z A Bowden[1]

[1]ISIS Facility, STFC, Rutherford Appleton Laboratory, Harwell, Didcot, UK

[2]Oxford Instruments plc, Tubney Woods, Abingdon, Oxon OX13 5QX, UK

E-mail: o.kirichek@rl.ac.uk

Abstract. Cryogenic equipment can be found in the majority of neutron scattering experiments. Recent increases in liquid helium cost caused by global helium supply problems lead to significant concern about affordability of conventional cryogenic equipment. However the latest progress in cryo-cooler technology offers a new generation of cryogenic systems in which the cryogen consumption can be significantly reduced and in some cases completely eliminated. These systems also offer the advantage of operational simplicity, require less space than conventional cryogen-cooled systems and can significantly improve user safety. At the ISIS facility it is possible to substitute conventional cryostats with cryogen free systems. Such systems are based on the pulse tube refrigerator (PTR) which possesses no cold moving parts. Oxford Instruments in collaboration with ISIS have developed new high magnetic field sample environment equipment based on re-condensing technology. This project includes 9T wide angle chopper magnet for spectrometry and 14T magnet for diffraction. The main advantage of these systems is that all magnet operating procedures, for example cooling, running up to the field and quenching remain the same as for a standard magnet in a bath cryostat. This approach also provides a homogeneous temperature distribution, which is crucial for optimum magnet performance.

1. Introduction

It is probably no exaggeration to say that cryogenic equipment is used in the majority of neutron scattering experiments. Most neutron facilities have a fleet of various cryostats providing low and ultra-low temperature sample environment and superconducting magnets for high magnetic fields [1]. Such large scale usage of cryogenic equipment requires significant resources and creates a number of logistical problems including health and safety issues and the considerable cost of the required cryogens.

The latest progress in cryo-cooler technology offers a new generation of cryogenic systems in which the cryogen consumption can be significantly reduced and in some cases completely eliminated [2, 3]. These systems also offer the advantage of operational simplicity, require less space than conventional cryogen-cooled systems and significantly improve user safety.

At the ISIS facility it is possible to substitute conventional cryogenic systems with cryogen free systems. A preference has been given to systems based on the pulse tube refrigerator (PTR) which possesses no cold moving parts. This fact reduces generated noise and vibration [4], increases reliability and decreases service efforts.

In this presentation we discuss the design and test results of two superconducting split pair magnet systems supplied recently to the ISIS by Oxford Instruments [5]. These systems were successfully tested at Oxford Instruments and currently are being prepared for commissioning at the ISIS facility.

Figure 1. Wide angle chopper instrument 9T superconducting magnet in the cryostat with helium re-condensed by PTR.

2. PTR actively-cooled superconducting magnets

The design of re-condensing magnet cryostats is usually based on the design of similar bath cryostats [2]. As shown in Fig.1 the superconducting magnet is immersed in the liquid helium. The radiation shield is cooled by the cooler's 1st stage and the 2nd stage re-condenses helium directly in the helium vessel. Thus the re-condensing system does not consume any liquid helium in normal operation.

The main advantage of this system is that all magnet operating procedures, for example cooling, running up to the field and quenching remain the same as for a standard magnet in a bath cryostat. This approach also provides a homogeneous temperature distribution, which is crucial for optimum magnet performance. All standard system accessories like current leads, magnet power supply, helium transfer siphon, etc. can be used with this system. One more important advantage is the ability of the magnet to stay at field for more than forty eight hours before quenching in the case of a cryo-cooler power failure. This feature also makes it possible to move the cold magnet between different neutron scattering instruments and/or the cryogenic laboratory.

3. Factory test results at Oxford Instruments

Both the wide angle chopper magnet for spectrometry and the 14T magnet for diffraction measurements share the similar top parts of the re-condensing cryostat with Variable Temperature Inserts (VTI) but have different bottom parts accommodating high field split pair magnets and sample space. Some comparative magnet characteristics are given in the table below.

In order to achieve maximum re-condensation and provide spare cooling power for possible future VTI flow re-condensation the cryostat is equipped with one of commercially available 1W @ 4.2K PTR manufactured by Cryomech Inc. This PTR has substantial spare cooling power for the size of the system and therefore in normal steady state operation this spare cooling power needs to be counteracted by built-in electric heaters in order to stabilize system pressure, temperature and liquid level. Alternatively the heater can be switched off and the system will reach equilibrium state at sub atmospheric pressure and lower temperature when PTR cooling power (which is dropping with lowering temperature) will be balanced by the heat load. Experiment shows that equilibrium temperature in this case is of order of 3.5K however this operational mode is not desirable due to inconvenience of topping the system up and danger of atmospheric air ingress into helium volume.

Table 1. The main features of split pair superconducting magnets.

	Chopper Magnet	14T Magnet
Central Field Strength at 4.2K	±9T	±14T
Central field homogeneity (over 10mm x 10mm cyl.):	<0.4%	<0.4%
Neutron aperture in horizontal plane	2 x 90° & 2 x Ø40mm	340°
Neutron aperture in vertical plane	±15°	+10° and -5°
Current decay in persistent mode	30 ppm/hr	20 ppm/hr
Sample space diameter	50mm	25mm
Sample temperature range	1.5-300K	1.5-300K

The system is equipped with an automatic pressure control system keeping the He bath pressure at a set level (usually 15 mbar above atmospheric pressure) by means of supplying required power to the built-in heater. It was demonstrated during the system test that temperature, pressure and liquid He level can be kept constant in steady-state operation with approximately 600-650mW supplied to the heater. The plot of the above mentioned parameters in steady state operation is shown on Fig.2.

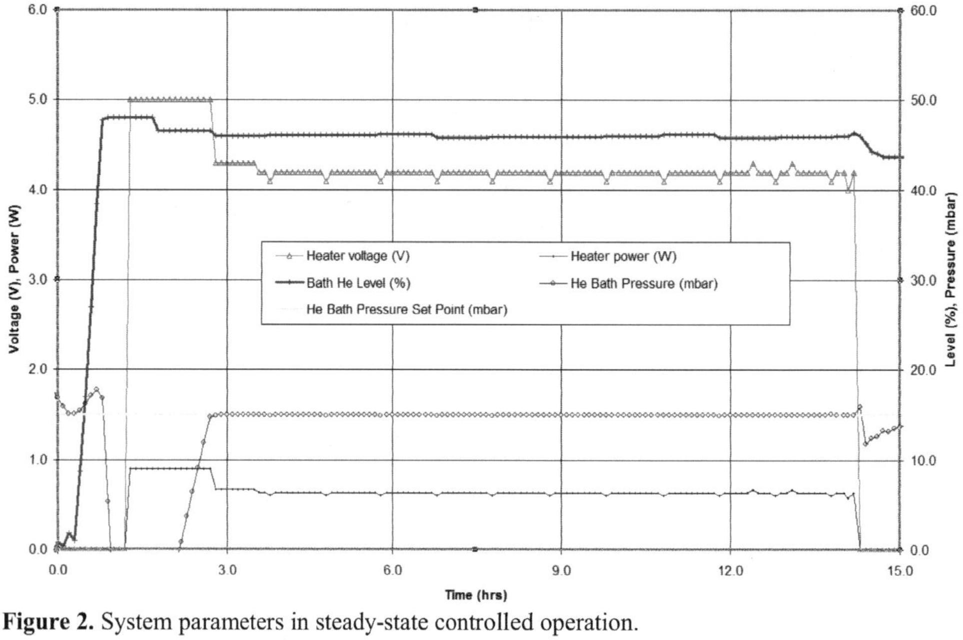

Figure 2. System parameters in steady-state controlled operation.

One of the very useful operational modes of the system available due to substantial spare cooling power and provision of special gas return path is external gas re-condensing. Gaseous He at room temperature and a constant flow rate of approximately 2.5 gL/min was supplied to the system via a set of filters and a cold trap. The system was capable of re-condensing its own boil-off plus additional external gas, still controlling bath pressure at a set level using the heater (though apparently with less power). The He level rising was observed confirming the fact of room temperature gas re-condensation. The plot of the system parameters in the described operational mode is shown on Fig.3. The VTI performance exceeding specification was demonstrated in the course of system test as well. Because the dry scroll 35 m^3/hr pump supplied by Edwards-BOC is used for gas circulation in the VTI it was possible to demonstrate a few modes with He gas flow re-introduction back to the system from the pump exhaust. This mode is similar to the external gas re-condensing described above but He level remains stable – i.e. the system works in zero-boil-off regime with the VTI running at a set temperature. Some further work in assessing necessity of gas purification for long term operation is still to be done but the first results are very promising and indicate that the system is capable of working in fully re-condensing mode under different operational conditions.

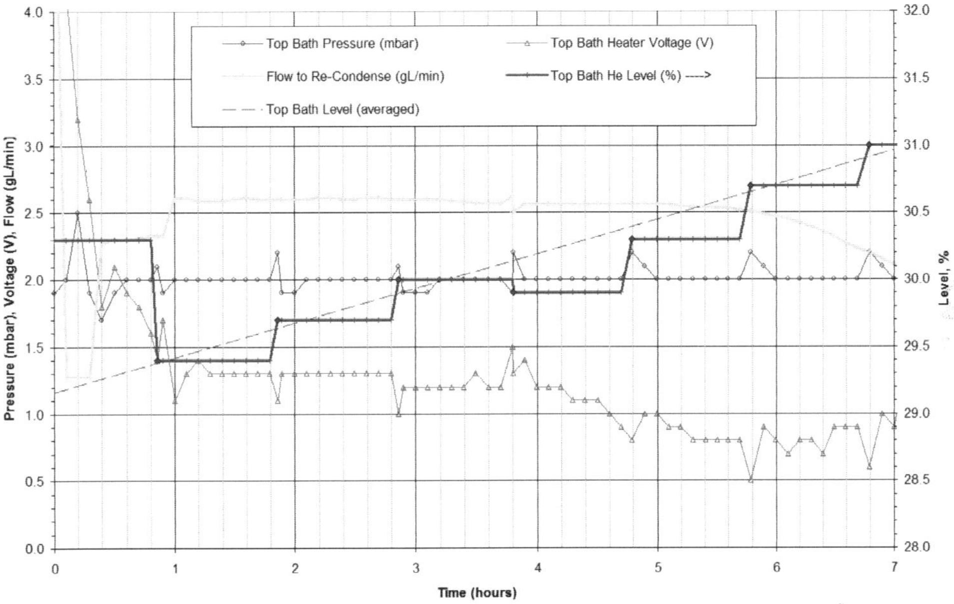

Figure 3. System parameters in external gas re-condensation mode.

Currently ISIS uses a number of dilution and ^{3}He refrigerator inserts used with standard VTI. All these systems can easily be made cryogen-free if used in a re-condensing cryostat with VTI [4].

4. Conclusions

Driven by global helium supply problems and by the desire to optimize technical resources, the ISIS facility is carrying out an internal development program to substitute gradually all conventional cryogenic systems with cryogen-free systems. As explained above the helium consumption of these systems can be dramatically reduced and in some cases completely eliminated. These systems are also more convenient to operate and require less maintenance. First two large split pair superconducting magnet systems are received at the ISIS facility after successful factory tests at Oxford Instruments. Full gas re-condensation in steady state and external gas supply was demonstrated during the tests as well as the possibility to fully re-utilise the VTI flow.

References

[1] Bailey I F 2003 *Z. Kristallogr.* **218** 84
[2] Kirichek O, Carr P, Johnson C and Atrey M 2005 *Rev. Sci. Instrum.* **76** 055104
[3] Evans B E, Down R B E, Keeping J, Kirichek O and Bowden Z A 2008 *Meas. Sci. Technol.* **19** 034018
[4] Kirichek O, Evans B E, Down R B E and Bowden Z A 2009 *Journal of Physics: Conference Series* **150** 012022
[5] Oxford Instruments plc: http://www.oxford-instruments.com

Aspects of superconducting magnet design for neutron scattering sample environments

F J Brown

Oxford Instruments NanoScience, Tubney Woods, Abingdon, Oxon, OX13 5QX, UK

E-mail: joe.brown@oxinst.com

Abstract. The majority of superconducting magnets for neutron scattering experiments take the form of split pairs with magnetic field vector in the vertical plane. Sample environment access is along the vertical magnetic axis whilst neutron access is in the horizontal plane. Split pair superconducting magnets present significantly more challenges in terms of design than the simpler solenoid type arrangement and the addition of requirements for neutron access further complicates the situation. Many of the requirements of split pair magnets for neutron scattering are conflicting and often compromises have to be made. Presented here are some of the more important design criteria and the ways in which these are met in practical magnet designs. Topics covered range from the choice of superconducting material through to the control of magnetic flux density profiles and mechanical aspects of the magnet former providing the neutron access between the coils. Most of the information presented is based on recent or current production magnets manufactured by Oxford Instruments for a range of neutron related applications.

1. Introduction

Although simple solenoid geometry magnets are employed in neutron scattering experiments the majority take the form of split pair solenoids. The greater proportions of these are orientated with field vector and sample environment access in the vertical direction whilst neutron access is in the horizontal plane.

The 'split pair solenoid' configuration takes its name from the fact that the windings are circularly symmetric but split into two regions in the axial direction arranged either side of a central gap. Conventionally the solenoid part of the name is dropped and such magnet arrangements are simply known as 'split pairs'.

Many of the requirements of split pair magnets for neutron scattering are conflicting and often compromises have to be made. Presented here are some of aspects of split pair superconducting magnet design and engineering from basic operating parameters to mechanical design and control of flux density profiles.

Data is taken from magnets designed by Oxford Instruments to meet a wide range of individual requirements for laboratories around the world. A typical magnet system for neutron scattering is shown in figure 1 whilst figure 2 shows a section through the magnet and lower portion of the cryostat

Figure 1. A typical superconducting split pair magnet system for neutron scattering

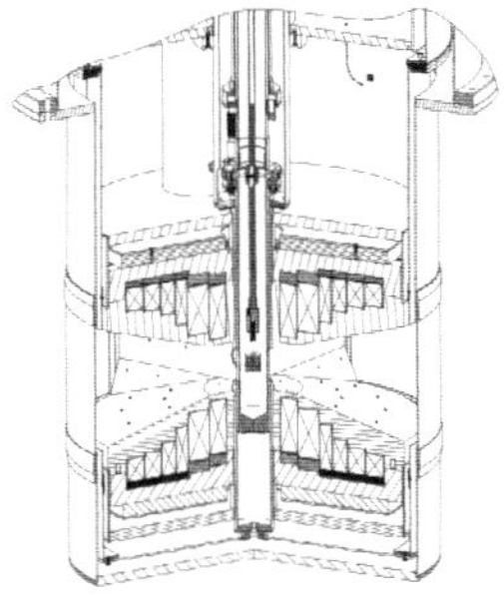

Figure 2. Sectional view through the magnet and lower cryostat of the system shown in figure 1

2. Superconducting materials

The main commercial superconductors are Niobium Titanium (NbTi) and Niobium Tin (Nb3Sn). The first is a ductile alloy and conductors including this material are easily handled and wound into coil form. The latter is a brittle intermetallic compound which is normally formed from constituent components by high temperature reaction in the as-wound form in a process known as "wind and react". Both conductor types contain high conductivity normal metal to act as stabilisation and to provide a normal current path in the case where the superconductor passes into the normal regime in the event known as a "quench". The normal stabilisation material is copper.

Engineering current density is defined as the critical current divided by the total area of the conductor. In high field magnets the two materials are employed in various sizes of conductor at positions within the magnet to suit the local conditions.

In general the flux density within the magnet windings falls with increasing radius and the conductor sizes are graded by size to achieve similar critical current operating fractions throughout the magnet

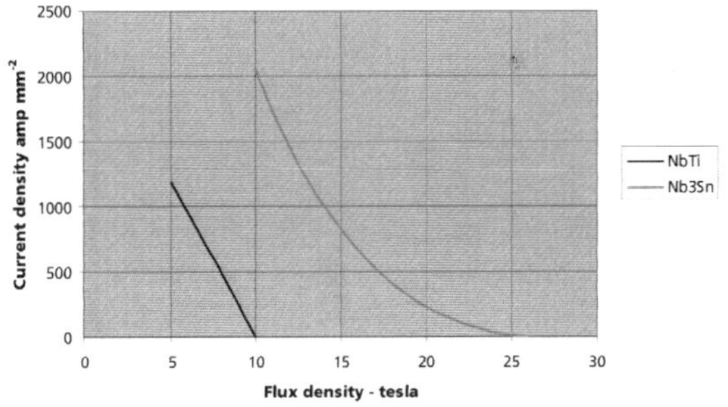

Figure 3. Engineering current density as a function of flux density for NbTi and Nb_3Sn

Critical current is measured by testing short lengths of conductor at various flux densities. For a standard conductor the curve generated is known as the short sample curve and a magnet is designed to operate at a certain fraction of this.

3. Principal Specifications and Magnet Operating Parameters

3.1. Flux Density Homogeneity

In a solenoid magnet it is relatively simple to produce flux density uniformity or homogeneity over the required sample volume down to the PPM level. In split pairs the geometry constraints applied by the neutron aperture make the creation of very uniform flux density much more difficult. There are configurations of split pair windings that allow for higher uniformities but in general, over typical neutron scattering sample sizes, the homogeneity is normally limited to the range 0.1 to 5%.

3.2. Operating Flux Density

The available current density for a given conductor decreases with increasing flux density (as shown in the figure above). The flux density seen by the superconductor at the inside of the magnet windings is greater than the operating central value.

This is particularly so in split pair magnets where the ratio can be large. In the example shown in figure 4 the ratio is almost 1.6 giving a maximum flux density of 14.3 tesla for a central value of 9 tesla

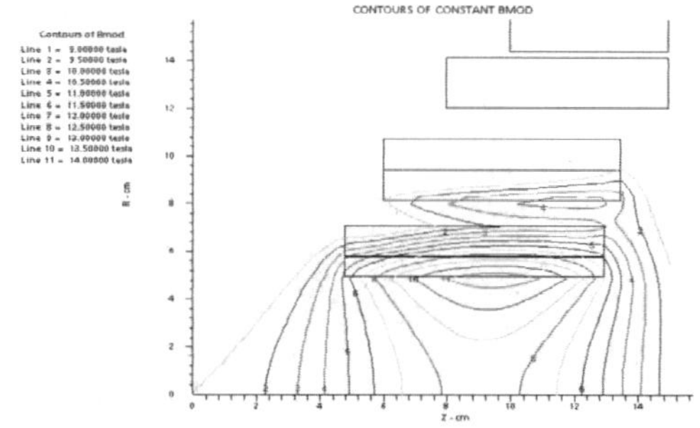

Figure 4. Flux density map showing the peak value in a 9 tesla split pair magnet

3.3. Forces and Stresses

For circular symmetric windings it is normal to describe the Lorentz forces in two directions; axial and radial.

Radial components of flux density react with the current flowing in the conductor to produce axially directed forces. The radial flux density generally has a maximum value at the ends of the windings resulting in axial forces directed towards the mid-plane. For simple solenoids this is not generally a serious issue as the associated axial stress is well within the compressive strength of the winding. In a split pair however these forces have to be reacted by the components that form the gap between the windings. There are also issues related to the interfaces between winding blocks and plates that form the sides of the gap.

Axial components of flux density react with the current flowing in the conductor to produce radially directed forces. These forces, over the majority of the windings, are outwardly directed resulting in the conductor being placed in tension and producing winding stress in the hoop direction. After the available current density limited by the peak flux density, this parameter is generally the next most important limiting factor. Gradients in the hoop stress distribution may also present problems as these result in radially directed stresses which, if positive in direction, tend to pull the winding apart.

4. Split pair magnets for neutron studies

The main purpose of the "split" in split pair magnets is to give access normal to the magnet axis. For many applications involving the use of split pair magnets access through the central region may be satisfied by simple holes through the spacer. In the case of magnets for neutron studies the access aperture requirements are generally greater and it is normal to use either rings of aluminium alloy between the two halves of the magnet to provide a "window" transparent to neutrons or wedge shaped openings between the coils. The axial forces outlined in the previous section clearly have to be reacted between the two sets of windings by the components that make up the space between them to provide the required support to maintain mechanical integrity with minimal distortion.

In the aluminium alloy ring type arrangement the available opening angle can be close to 360° with only a single dark angle of 10 to 20° being required to allow cryogenic and electrical connections between the two halves.

The thickness of the rings depends on the magnetic forces and the mechanical properties of the aluminium alloy employed. The magnetic forces are not only dependant on the central flux density but also to some extent on the geometry of the coils.

The mechanical properties of aluminium alloys are composition dependent with only a restricted number of types being suitable for neutron transmission and having suitable mechanical properties. A typical example of such a magnet is shown in figure 5.

Figure 5. A typical aluminium alloy spaced split pair magnet shown with one side plate removed

In the wedge shaped opening approach it is clear that the rigidity of the former plates that form the sides of the openings has to be significantly greater than in the window ring case outlined above.

As with the ring case approximate calculations may be made using appropriate formulae to determine stresses and local distortions but only finite element analysis can provide the detail required to allow optimised design. Figure 6 shows a typical wedge aperture type magnet assembly.

Figure 6. Typical wedge shaped aperture magnet

5. Summary

The number of variables involved with a split pair is greater than for simple solenoids due to the added access through the mid-plane. As described this leads to increased complexity of superconducting split pair magnets for neutron scattering.

The number of potential specification variables is large and magnets of this type tend to be specifically designed for each application. For instance the size of the neutron aperture (height, vertical divergence angles and horizontal angle) has a very large impact on the other magnet parameters such as available flux density, homogeneity and magnet size / cost.

The mechanical aspects of split pair magnets for neutron scattering studies are important if the overall magnet is to be stable and reliable in operation.

The exploitation of the latest developments in superconductor design, together with robust magnet manufacturing techniques, makes possible high flux density magnets for neutron scattering studies of up to 16 tesla. The number of competing parameters makes defining the range of apertures and divergences difficult however in general higher flux densities require smaller neutron apertures and divergences to keep the magnet within reasonable size and cost.

Motofit – integrating neutron reflectometry acquisition, reduction and analysis into one, easy to use, package.

Andrew Nelson[1]
The Bragg Institute, Australian Nuclear Science and Technology Organisation,
New Illawarra Rd, PMB1, Menai, NSW2234, Australia

E-mail: andrew.nelson@ansto.gov.au

Abstract. The efficient use of complex neutron scattering instruments is often hindered by the complex nature of their operating software. This complexity exists at each experimental step: data acquisition, reduction and analysis, with each step being as important as the previous. For example, whilst command line interfaces are powerful at automated acquisition they often reduce accessibility by novice users and sometimes reduce the efficiency for advanced users. One solution to this is the development of a graphical user interface which allows the user to operate the instrument by a simple and intuitive "push button" approach. This approach was taken by the *Motofit* software package for analysis of multiple contrast reflectometry data. Here we describe the extension of this package to cover the data acquisition and reduction steps for the Platypus time-of-flight neutron reflectometer. Consequently, the complete operation of an instrument is integrated into a single, easy to use, program, leading to efficient instrument usage.

1. Introduction

Neutron scattering instruments at modern facilities are now able to collect large volumes of high quality data faster than ever before. A large amount of effort is spent on constructing the complex instrumental hardware, with additional resources being used to develop software to acquire (operate), reduce and analyse data from the instrumentation. One example is the approach used in the DANSE project [1] which, in part, was initiated to develop software infrastructure for the reduction and analysis of data from the high intensity Spallation Neutron Source facility.

The software created for this purpose has to cope with complex data analysis problems that mirror the sophisticated nature of a neutron scattering experiment. Existing analysis tools have been more than adequate but often show deficiencies, such as their inability to cover new technique aspects, or the use of a command line interface to control the program. This latter point is often invisible to expert users, but can be a huge stumbling block for novices. At best this leads to inefficient use of the instrument, at worst the user obtains bad data and may not return.

The *Motofit* reflectometry analysis package [2] was originally designed to solve such a problem, offering sophisticated analysis of multiple contrast data, but in an easy to use manner.

Recently we faced the ease-of-use problem during the construction of the new *Platypus* time-of-flight (TOF) neutron reflectometer [3] at OPAL. To address this *Motofit* was extended and is now an integrated approach for the acquisition, reduction and analysis of reflectometry data. This integration means that the user is only ever exposed to a single program throughout the experiment lifecycle,

[1] To whom any correspondence should be addressed.

leading to simple operation, a better understanding of the data and efficiencies in preparing data for publication. *Motofit* was designed to operate in the *Igor Pro* (Wavemetrics Inc) data analysis program [4], offering rapid development using the instrinsic data analysis functionality already present.
Here we describe each of the modules of *Motofit* to give an overview of how Platypus operates.

2. Acquisition module – FIZZY

The Platypus reflectometer [3], like most instruments at the OPAL facility, is operated using a server-client paradigm. The instrumental hardware is controlled by a command line SICS control server [5], which is responsible for moving motors, speaking to the detector acquisition electronics and software, as well as saving the data in NeXUS format. The FIZZY module was designed to provide a user interface (GUI) for a generic SICS server and have the following functionality: 1) easy visualisation and control of instrument parameters (motor positions, etc), 2) single button press data acquisition and alignment (figure 1a), 3) realtime visualisation of multidimensional detector images (figure 1b) and 4) batch mode for measurement of multiple samples.
Figure 1 demonstrates the simplicity of the user interface. There are no other menus to FIZZY, with all instrument operation taking place in a single multi-tabbed window. Even the potentially difficult batch mode operation is simplified via popup suggestions for the different kinds of instrument operations available.
Behind the scenes FIZZY communicates with SICS using the *Sockit* and *EasyHttp* plugins for IGOR [6], which enable TCP/IP and http communication. FIZZY simply abstracts the SICS command line functionality into GUI controls, such as buttons.

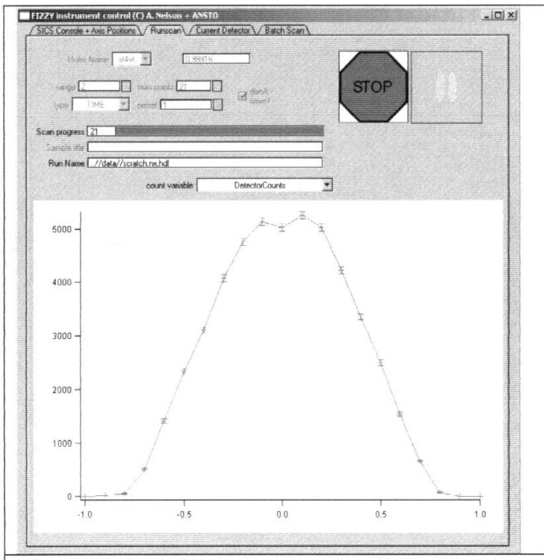

Figure 1a) Alignment scans and acquisitions require a single button press

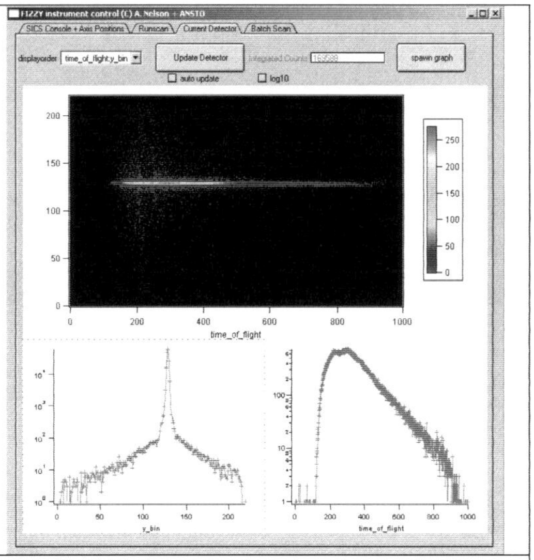

Figure 1b) Real time detector view showing 2D image (ypixel vs TOF) with projections along each of the axes.

3. Reduction module – SLIM

Whilst most of the data analysis is done offline, real-time assessment of the data streaming off the instrument is vital, as this often determines the course followed in an experiment. The SLIM module offers this access in a simple manner. All the user has to do is select the runs they need to visualise, with reflectivity curves being generated automatically. The user has many different choices for the plot system, both 1D (R vs Q, R vs lambda, R vs TOF, Intensity vs TOF, Intensity vs lambda, etc) and 2D detector images.

The SLIM module accepts NeXUS files as input and outputs both specular and offspecular reflectivity. The output files are text based and have two forms, a columnar dat file and an XML file that is related to the canSAS XML format for SAS data. All data files contain experimental uncertainties - standard deviation for reflectivity and full width half maximum for Q resolution.

The exact reduction process will be detailed in a later paper, although we give a short overview here. Both direct and reflected 2D (detector y pixel vs TOF) spectra are corrected for spatial detector efficiency. A gravity correction is applied to both spectra, as Platypus is a vertical scattering plane instrument and the angle of incidence is wavelength dependent. Since data is oversampled in time-of-flight the 2D spectra are rebinned using dT/T bins that reflect the instrumental resolution. The data are then converted to a wavelength axis. The actual angle of reflection is calculated using the positions of the specular reflection and direct beam on the detector. An instrumental background is subtracted by linear fit between regions either side of the spectral ridge, for a given wavelength bin (although constant Q subtraction is being developed).

The direct beam wavelength spectrum is then obtained by summing detector y pixels over the incident beam width, for a given wavelength.

In contrast each pixel in the reflected beam image has a Q_z, Q_y value calculated for it, with the pixel intensity being divided by the direct beam intensity for that wavelength. Rebinning of pixels with equal Q_z in the specular beam yields a reflectivity curve. Experimental uncertainties are propagated throughout the reduction process, with the Q resolution calculation according to van Well [7].

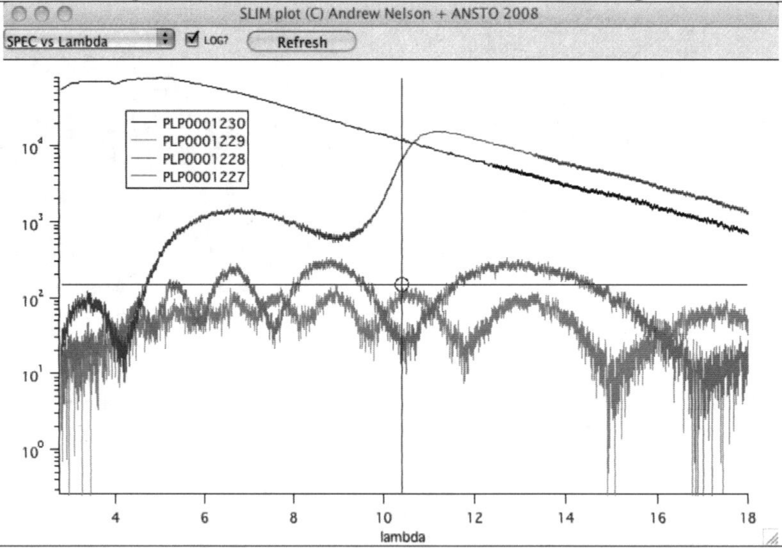

Figure 2) Realtime data visualisation of direct and reflected beams. The uppermost curve is a direct beam run, the other curves are reflection runs at different incident angles.

4. Analysis module - *Motofit*

The reflectivity analysis module was the first to be developed and is described in more detail by Nelson [2]. It's overriding design criteria were the simultaneous least squares analysis of multiple contrast neutron and X-ray reflectometry data, with an interface that was easy to use by novices and experts alike. Since the original description further functionality has been added to extend the analysis capabilities including: genetic optimisation [8], batch fitting of multiple datasets [9], monte carlo error analysis [10], exploration of χ^2 for different parameter combinations, analytical SLD profiles.

The main advantage of integrating all the aspects of the experimental software is that the data files from SLIM are immediately available for analysis in *Motofit*. However, data analysis in *Motofit* is not restricted to data obtained on Platypus, data from any neutron and X-ray reflectometry data can be also be used.

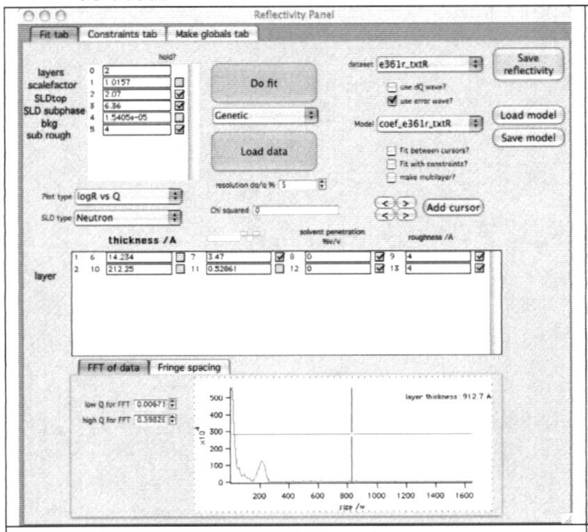

Figure 3a) *Motofit* analysis panel

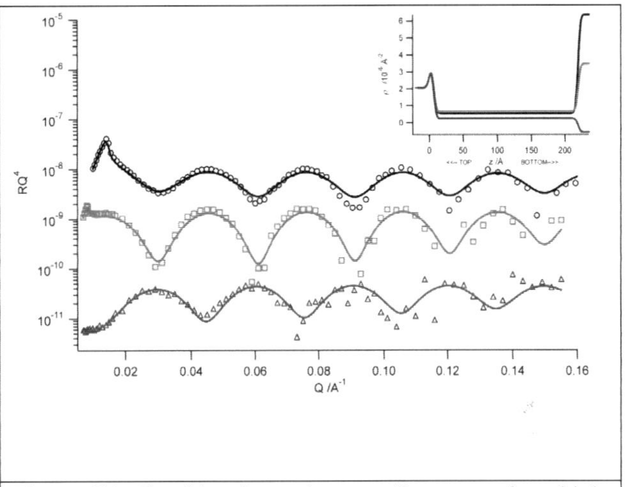

Figure 3b) Graphics from the co-refinement of multiple contrast data are publication ready.

5. Conclusions

We have streamlined the user operation of the Platypus neutron reflectometer by integrating data acquisition, reduction and analysis into a single program. This leads to efficient use of Platypus by novice and expert users alike.

The *Motofit* program is available from http://motofit.sourceforge.net or from http://dav1-platypus.nbi.ansto.gov.au.

6. References

[1] http://danse.us, The DANSE project

[2] Nelson A 2006 Co-refinement of multiple contrast neutron / X-ray reflectivity data using MOTOFIT *J. Appl. Crystallogr.* **39** 273-276

[3] James M; Nelson A, Brule A and Schulz J 2006 Platypus: A time-of-flight neutron reflectometer at Australia's new research reactor *Journal of Neutron Research* **14** 91-108

[4] http://www.wavemetrics.com IGOR Pro Wavemetrics Inc.

[5] Heer H, Konnecke M and Maden D 1997 The SINQ instrument control software system *Physica B* **241** 124-126

[6] available from http://www.igorexchange.com

[7] van Well A A and Fredrikze H 2005 On the resolution and intensity of a time-of-flight neutron reflectometer *Physica B* **357** 204-207

[8] Wormington M, Panaccione C, Matney K and Bowen D 1999 Characterization of structures from X-ray scattering data using genetic algorithms *Philosophical Transactions of the Royal Society of London*, **1999**, *357*, 2827-2848

[9] Mueller-Buschbaum P, Bauer E; Maurer E, Nelson A and Cubitt R. 2007 Competitive swelling and de-swelling of this polystyrene films by toluene *Physica Status Solidii - RRL* **1** 68-70

[10] Heinrich F, Ng T, Vanderah D, Prabhanshu S, Mihailescu M and Losche M 2009 A New Lipid Anchor for Sparsely Tethered Bilayer Lipid Membranes *Langmuir* **25** 4219-4229

International Conference on Neutron Scattering 2009 IOP Publishing
Journal of Physics: Conference Series **251** (2010) 012095 doi:10.1088/1742-6596/251/1/012095

Doing Your Science While You're in Orbit

Mark L. Green

Tech-X Corporation

Stephen D. Miller, Sudharshan S. Vazhkudai, and James R. Trater

Oak Ridge National Laboratory

mlgreen@txcorp.com

Abstract. Large-scale neutron facilities such as the Spallation Neutron Source (SNS) located at Oak Ridge National Laboratory need easy-to-use access to Department of Energy Leadership Computing Facilities and experiment repository data. The Orbiter thick- and thin-client and its supporting Service Oriented Architecture (SOA) based services (available at https://orbiter.sns.gov) consist of standards-based components that are reusable and extensible for accessing high performance computing, data and computational grid infrastructure, and cluster-based resources easily from a user configurable interface. The primary Orbiter system goals consist of (1) developing infrastructure for the creation and automation of virtual instrumentation experiment optimization, (2) developing user interfaces for thin- and thick-client access, (3) provide a prototype incorporating major instrument simulation packages, and (4) facilitate neutron science community access and collaboration. The secure Orbiter SOA authentication and authorization is achieved through the developed Virtual File System (VFS) services, which use Role-Based Access Control (RBAC) for data repository file access, thin- and thick-client functionality and application access, and computational job workflow management. The VFS Relational Database Management System (RDMS) consists of approximately 45 database tables describing 498 user accounts with 495 groups over 432,000 directories with 904,077 repository files. Over 59 million NeXus file metadata records are associated to the 12,800 unique NeXus file field/class names generated from the 52,824 repository NeXus files. Services that enable (a) summary dashboards of data repository status with Quality of Service (QoS) metrics, (b) data repository NeXus file field/class name full text search capabilities within a Google like interface, (c) fully functional RBAC browser for the read-only data repository and shared areas, (d) user/group defined and shared metadata for data repository files, (e) user, group, repository, and web 2.0 based global positioning with additional service capabilities are currently available. The SNS based Orbiter SOA integration progress with the Distributed Data Analysis for Neutron Scattering Experiments (DANSE) software development project is summarized with an emphasis on DANSE Central Services and the Virtual Neutron Facility (VNF). Additionally, the DANSE utilization of the Orbiter SOA authentication, authorization, and data transfer services best practice implementations are presented.

1. Background and Development

The Spallation Neutron Source (SNS) [1] and the High Energy Isotope Reactor (HFIR) [2] support worldwide users on neutron instruments for scientific research utilizing neutron scattering science. The Neutron Science Teragrid Gateway (NSTG) [3] has initiated a work force consisting of High Performance Computing (HPC) software experts and SNS instrument scientists to make TeraGrid resources available to facility researchers. A limited software facility infrastructure has been designed, developed, and deployed on the prototype NSTG. It is being tested on select resources across TeraGrid and available to SNS/HFIR as well as other neutron scientists. The SNS Portal project is based on a Java Applet application and is fully functional serving the SNS/HFIR user communities [4].

While SNS is operating at full power, it will offer unprecedented performance for neutron-scattering research, with more than an order of magnitude higher flux than any existing facility. To realize the potential this offers for research in many fields, a world-class suite of instruments is being developed that makes optimal use of the SNS beams and that is suited to the needs of users across a broad range of disciplines. Instruments will be available to researchers with varying degrees of experience, from new graduate students and first-time neutron users to experienced users with an interest in instrument design. This broad range of users, novice to sophisticated, provides an ideal user community for integration and development of the Multitier Portal Architecture (MPA). Additionally, the user community requirements are also quite disparate based on their science domains. The Tech-X Corporation [5] has been working with the Scientific Computing Group at SNS in developing a MPA for this purpose.

It is extremely important that this system be as flexible as possible to support the wide range of use cases presented by the user community. The SNS software services are available from remote web clients (thin-clients) and desktop clients and user applications (thick-clients). Instrument scientists responsible for the operation of the instruments during experiments require the use of thin- and thick-clients for performing their duties and subsequent calibration and data analysis.

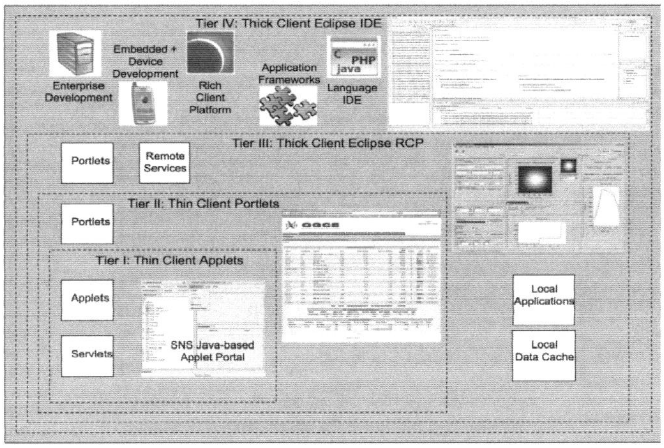

Figure 1: Multitiered Portal Architecture (MPA) layered presentation tier.

The basic portal architecture is modular with well-defined interfaces facilitating a distributed development environment with the ability to incorporate new technologies seamlessly. Today many portals are required to perform in an extremely dynamic environment where the quantity, sophistication, and requirements of portal users change rapidly. It is highly desirable to scale the portal capabilities with the requirements and sophistication of the user. In order to effectively and efficiently provide a scalable, robust, and flexible presentation tier we present the following multitiered approach providing thin-client servlets, applets, portlets, and web services in addition to thick-client Rich Client Platform (RCP) and Integrated Development Environment (IDE) support. The MPA system capabilities are shown in Figure 1. The Multitier Portal Architecture has the potential to scale to the expertise and requirements level of the user while maximizing the reuse of the developed infrastructure and services in successive tiers. With a carefully planned and defined service architecture each subsequent tier can significantly reuse and leverage existing services provided by the less sophisticated tiers. The simplest MPA Tier 1 can be accessed from a commodity browser whereas the feature rich MPA Tier 4 requires the local installation of an Integrated Development Environment (IDE) that has the capability of modifying/enhancing/customizing the entire MPA architecture.

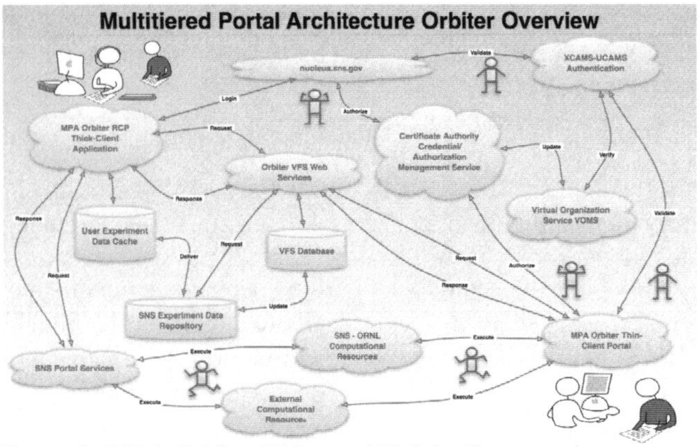

Figure 2: MPA Orbiter Thin- and Thick-client service infrastructure.

Tech-X has obtained a Small business Innovative Research (SBIR) grant from the Department of Energy for the development of a prototype MPA at the SNS (award #: DE-FG02-08ER85000). As illustrated in the system architecture diagram, Figure 2, the Multitiered Portal Architecture (MPA) Orbiter Overview the "MPA Orbiter RCP Thick-Client Application" and the "MPA Orbiter Thin-Client Portal" share several system components within the Orbiter Service Oriented Architecture (SOA) implemented in this prototype development effort. The cyber security measures implemented at the SNS are quite extensive and have required constant attention during the Orbiter development process. The XCAM-UCAMS [6] authentication system is an in-house authentication/authorization authority that serves as the core cyber security identity management service at ORNL (we have adopted this system as our identity management system for the Orbiter project). The SNS Portal Services provide valuable system components for integrating with the SNS-ORNL computational Resources and External Computational Resources (i.e. Application Manager) [7] and is a project funded by the SNS Scientific Computing Group. We have worked in close collaboration with this group during our entire Orbiter development project.

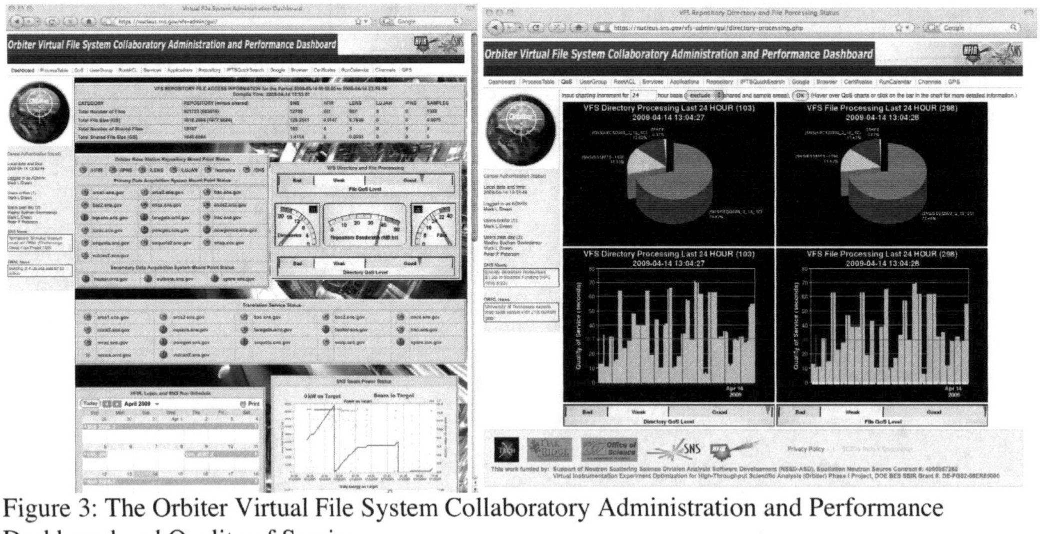

Figure 3: The Orbiter Virtual File System Collaboratory Administration and Performance Dashboard and Quality of Service.

One of the most valuable core Orbiter components is the Orbiter Virtual File System (VFS) SOA services. The VFS Relational Database Management System (RDMS) consists of approximately 45 database tables describing 498 user accounts with 495 groups over 532,000 directories with 904,077 repository files. Over 59 million NeXus [8] file metadata records are associated to the 12,800 unique NeXus file field/class names generated from the 52,824 repository NeXus files. Services that enable (a) summary dashboards of data repository status with Quality of Service (QoS) metrics, (b) data repository NeXus file field/class name full text search capabilities within a Google like interface, (c) fully functional RBAC browser for the read-only data repository and shared areas, (d) user/group defined and shared metadata for data repository files, (e) user, group, repository, and web 2.0 based global positioning with additional service capabilities are currently available. The baseline system in place utilizes the free open-source Structured Query Language (SQL) full-text search engine Sphinx [9] for queries. The Orbiter VFS is a key component, enabling both a robust system development and feature-rich user interface and system metrics, Figure 3. The Amazon Web Service (AWS) [10] model is used for securing RESTful web service requests within the Orbiter SOA.

2. Conclusion and Future Work

The current prototyping project conducted in collaboration with the SNS was a tremendous success. Several of the Orbiter SOA services are performing within the SNS production environment providing value to the scientific researchers. In particular the Orbiter find NeXus file service has received nearly 6,000 service requests over a 24-hour period and the Orbiter search NeXus file service provides full index searches, typically taking on the order of tenths of seconds, on over 39 million distinct field/class values. Using the Sphinx indexer that is approximately 5,000 times faster than a comparable SQL based search in this case enhances the Orbiter search capabilities. Our collaboration with the Scientific Computing Group at the SNS has been extremely productive and we continue to leverage the Orbiter infrastructure for other related projects.

The Tech-X Corporation SBIR Principal Investigator, Dr. Green, has completed the DOE SBIR Phase I feasibility and prototype that is currently served by the SNS. The subsequent SBIR Phase II proposal has been funded and proposes several Orbiter system enhancements based on the prototype collaboration investigation. The key features that will be integrated into the Orbiter system will be a) enhanced support for virtual instrumentation, b) development of the thick-client user interface design, c) integrate ISAW [11] and DANSE [12] project applications as Software as a Service (SaaS), and d) build support for the Orbiter VFS SOA distributed base station deployment. This proposed work as well as additional enhanced capabilities will be developed in the DOE SBIR Phase II effort over a two-year project term.

References

[1] SNS: http://neutrons.ornl.gov/facilities/facilities_sns.shtml
[2] HFIR: http://neutrons.ornl.gov/hfir/hfir_facilities.shtml
[3] NSTG: http://www.teragrid.org/gateways/projects.php?id=56
[4] SNS Portal: http://neutronsr.us
[5] Tech-X Corporation: http://www.txcorp.com/
[6] ORNL XCAMS: https://xcams.ornl.gov/xcams/
[7] OIC Clusters: https://oic.ornl.gov/
[8] NeXus: http://www.nexusformat.org/Main_Page
[9] Sphinx: http://www.sphinxsearch.com/
[10] Amazon Web Services: http://aws.amazon.com/
[11] ISAW: http://www.pns.anl.gov/computing/isaw/index.shtml
[12] DANSE: http://danse.us/

Journal of Physics: Conference Series **251** (2010) 012096 doi:10.1088/1742-6596/251/1/012096

The SNS/HFIR Web Portal System – How Can it Help Me?

Stephen D. Miller, Al Geist, Kenneth W. Herwig, Peter F. Peterson, Michael A. Reuter, Shelly Ren, Jean-Christophe Bilheux, Stuart I. Campbell, James A. Kohl, Sudharshan S. Vazhkudai, John W. Cobb, Vickie E. Lynch, Meili Chen, and James R. Trater

Oak Ridge National Laboratory

Bradford C. Smith, Tom (William) Swain, and Jian Huang

University of Tennessee, Knoxville

Ruth Mikkelson, Dennis Mikkelson

University of Wisconsin, Stout

Mark L. Green

Tech-X Corporation

millersd@ornl.gov

Abstract. In a busy world, continuing with the status-quo, to do things the way we are already familiar, often seems to be the most efficient way to conduct our work. We look for the value-add to decide if investing in a new method is worth the effort. How shall we evaluate if we have reached this tipping point for change? For contemporary researchers, understanding the properties of the data is a good starting point. The new generation of neutron scattering instruments being built are higher resolution and produce one or more orders of magnitude larger data than the previous generation of instruments. For instance, we have grown out of being able to perform some important tasks with our laptops – the data are too big and the computations would simply take too long. These large datasets can be problematic as facility users now begin to grapple with many of the same issues faced by more established computing communities. These issues include data access, management, and movement, data format standards, distributed computing, and collaboration among others. The Neutron Science Portal has been architected, designed, and implemented to provide users with an easy-to-use interface for managing and processing data, while also keeping an eye on meeting modern cybersecurity requirements imposed on institutions. The cost of entry for users has been lowered by utilizing a web interface providing access to backend portal resources. Users can browse or search for data which they are allowed to see, data reduction applications can be run without having to load the software, sample activation calculations can be performed for SNS and HFIR beamlines, McStas simulations can be run on TeraGrid and ORNL computers, and advanced analysis applications such as those being produced by the DANSE project can be run. Behind the scenes is a "live cataloging" system which automatically catalogs and archives experiment data via the data management system, and provides proposal team members access to their experiment data. The complexity of data movement and utilizing distributed computing resources has been taken care on behalf of users. Collaboration is facilitated by providing users a read/writeable common area, shared across all experiment team members. To date, these shared areas are the fastest growing data spaces. The portal currently has over 370 registered users, almost 5TB of experiment and user data, approximately 660K files cataloged, and had almost 10,000 unique visits last year. Future directions for enhancing portal robustness include examining how to mirror data and portal services, better facilitation of collaborations via virtual organizations, enhancing disconnected service via "thick client" applications, and better inter-facility connectivity to support cross-cutting research. The portal has established itself in the SNS/HFIR user community, and the development team strives to continue to improve the quality of features and services provided in order to better serve the community.

1. Background

With the advent of the Spallation Neutron Source (SNS) at Oak Ridge National Laboratory, an opportunity occurred to revisit how to provide software and services via the facility. Under the guidance of Ian Anderson, then Director of the Experimental Facilities Division (XFD), a number of Neutron Science Software Initiative (NeSSI) workshops were held in 2003 and 2004 to solicit feedback from the community on its requirements and wishes for next generation software [1]. The primary outcome was the desire to have a portal system which would help manage and process data for users. For easy access, it was decided that the Neutron Science Portal (portal) would have a web interface via a web browser not requiring users to install software. Portal design work commenced in 2004 and 2005 with an operational portal deployed by April 28, 2006 when first neutrons were detected at SNS. These first data were incorporated into the portal and are still available today.

The core portal development team has stayed in-tact ever since the design phase, and is now an active part of the Neutron Scattering Science Division (NSSD) operations headed by Deputy Division Director, Ken Herwig. Portal development has been a good collaborative effort both within Oak Ridge National Laboratory, and in including external contributors from academic and commercial settings. Initial design and development effort was undertaken by the ORNL Computer Science and Mathematics Division (CSMD) under the guidance of Al Geist with early work done by Sudharshan Vazhkudai and Jim Kohl. Upon establishment of the SNS Analysis Software Team in 2004 when Stephen Miller became the team leader, additional people were added to the SNS team, Pete Peterson, Michael Reuter, Shelly Ren, and Jean Bilheux, and began to work with the CSMD personnel. Tom Swain from The University of Tennessee Software Quality Research Laboratory (SQRL) [2] joined the team early on to help establish portal requirements and specifications. Since then, Bradford Smith of SQRL has joined to lead test and integration efforts. These requirements and specifications documents were important for establishing a direction and are relevant even today. As the development effort progressed, it became apparent that the ISAW visualization tools could play an important role in providing visualization via the portal, and Ruth and Dennis Mikkelson joined the team, initially funded via an NSF grant. Leading the effort to architect and establish the computing infrastructure has been Jim Trater of the ORNL IT group.

Leading the way in High Performance Computing (HPC) has been John Cobb and his software development team comprised of Vickie Lynch and Meili Chen, who have been funded via the NSF Neutron Science TeraGrid Gateway (NSTG) project to make TeraGrid HPC resources available to SNS researchers [3]. This team has been instrumental in helping to navigate between NSF and DOE computing worlds by working to establish a community account usable by the SNS portal users on TeraGrid resources thus eliminating the need for portal computing users to have to obtain TeraGrid accounts unless they so desired. The NSTG team has deployed McStas instrument simulation software [4], data fitting software, and will soon deploy data reduction software on TeraGrid computing resources.

Since the early days, the portal has grown considerably. It holds data from SNS [5] and High Flux Isotope Reactor (HFIR) [6] as well as the Low Energy Neutron Source (LENS) at the University of Indiana Bloomington [7], some example data from the NPDF instrument at the Lujan Center [8], and the entire IPNS [9] data repository. The portal has approximately 370 registered users, contains almost 4TB of data (much of it is compressed) in almost a million files. On an average day, the portal has about 20 individual visitors, receives about 4500 web site "hits", and performs about 260 compute jobs. Thus for SNS and HFIR users, the portal has established itself as a regular fixture as part of performing experiments and data analysis.

More recently Mark Green of Tech-X corporation [10], funded via a DOE SBIR project, joined the team to assist with data management. He worked with development team members, IT and cybersecurity personnel to develop programmatic interfaces to identify and access data contained within the repository. Also, Jian Huang of The University of Tennessee SEELAB [11] joined the team recently working on rebinning software to transform data on an irregular grid to a regular grid as part of work supporting single crystal inelastic data reduction and analysis. Stuart Campbell recently joined the SNS team from the Diamond X-ray facility in the U.K [12] primarily to develop software for the time of flight inelastic science group and to help with NeXus data format software development.

2. What is the portal?

A user accesses the portal via a web browser. In order to do this and to utilize portal computing resources, the user must first obtain an XCAMS computer user account [13], which must be requested and granted. Once approved, users can login to the portal whereupon they will have access to their "workspace", consisting of pointers to experiment data that they are allowed to view. Data are stored by facility, instrument, proposal IDs in order to support providing a multi-facility data repository.

In addition to providing data access, the portal also provides access to applications and computing resources. These same resources are available both to on-site users at SNS and HFIR, and to users external to the facility. The portal provides a suite of applications which the user is able to launch remotely, but the applications appear locally. This capability is made available via NX remote desktop technology [14] accessed via the browser. The NX remote desktop feature enables the portal to present an application to the user as though it runs locally, while also taking advantage of remote computing resources. This approach has allowed the portal team to leverage significant resources for performing computation. The application backend can run on a portal server, create and submit jobs to a job queue, and return results which the GUI and the user can access at their sites.

Segmenting computation from the user interface layer enables utilization of a wide array of computing resources. GUI applications run by users at SNS on analysis computers can run locally. Applications launched by remote users can be run on separate portal servers and computation servers at SNS. Other resources include the Oak Ridge Institutional (OIC) cluster [15] and the NSTG cluster [16]. Both of these resources are not managed by SNS staff and thus have specific rules for accessing these resources which could be difficult for most users to traverse. To facilitate using these resources, community accounts are utilized where a job for a portal user is run as the community user on these resources. The Application Manager is the intelligent subsystem within the portal which handles transparent execution of user jobs on a variety of aforementioned target environments. In addition, it handles seamless data staging in and out of these resources, job monitoring, logging, failover and recovery.

Thus the portal includes data browsing and search, but also provides GUI tools that are connected to powerful computing resources. Applications include experiment data histogramming tools, data reduction tools, data visualization tools, and McStas Monte Carlo simulation. Due to the ability to run applications remotely, the portal also supports hosting of community produced applications such as ISAW [17], DAVE [18], DANSE [19] applications, EXPGUI [20], FullProf [21], GSAS [22]. Database tools such as the Cambridge Structural Database [23] and the Inorganic Crystal Structure Database [24] are also available. Experiment support tools such as a Sample Activation Calculator are also available which enables users to compute the estimated radioactive dose rate that specified samples will produce for a given instrument.

The portal development effort has produced a significant code base. Currently the composite portal development effort has produced over 650,000 lines of code since the onset of portal development.

Data reduction applications account for approximately half of this code base. Software development follows a rigorous process for defining the content of each release. A centralized software build process is utilized to coordinate building correct versions for test and deployment. Testing is done to ensure that features meet requirements, and upon successful completion, the new version is deployed. Software are maintained in a subversion (SVN) [25] repository with a web based issue tracking system called Trac [26] placed atop SVN.

3. Why is the portal usage growing?

A quick and naive answer for why the portal usage is growing is that the quantity of data is growing. However, it is not merely the quantity and scale of data that is attracting portal usage, but a fundamentally convenient way to work with that data. The facility creates the data and thus has significant influence over how convenient it is for users to acquire, access, and process data. In the past, users collected raw data at their instruments, but this is no longer the primary mode of operation at SNS, which has invested considerable effort into these major capabilities that significantly improve convenience:

1. Experiment metadata are automatically captured and associated with the experiment data. These data are translated into a self-describing file format called NeXus [27] which utilizes the underlying HDF-5 scientific data format [28].
2. The live cataloging process takes experiment data as it is streamed from the instruments, catalogs pertinent metadata, and stores the data for access. The entire process is quick and places data in the centralized repository in a matter of seconds to minutes.
3. There is a high level of integration between user experiment support systems and the data management system. For instance, the data management system integrates with the proposal system (IPTS) [29] in order to capture and associate user and experiment information with experiment data.
4. The facility curates the data on behalf of the user, thus addressing worries users may have for where they store their data, and the integrity of this data.

These functions are all conducted automatically on behalf of the user, in a transparent manner and their data is available for access via the portal using a basic web browser. Thus the complexities of preparing, moving, cataloging, and discovering data are all handled for the user. The portal offers both data browsing and data search for discovering data. A path structure had to be determined in order to facilitate locating data contained within a multi-facility data repository. The generalized path stores data by facility, instrument, proposal ID, collection, and run number. In this way, users can uniquely locate their data across facilities and instruments. The search interface offers users a powerful means for quickly locating data. Search is particularly useful for those users who have performed a number of experiments and are challenged to remember where to browse to locate data. The search interface enables users to locate data by a number of metadata tags such as proposal ID, run number, Principle Investigator (PI) name, sample name, and date. But the user can only locate data which they have permission to access according to the proposals with which they are affiliated. Thus, in addition to the XCAMS authentication system, there is an authorization layer that grants fine-grained access. The facility is capable of providing this data privacy as it has knowledge of the experiment team members via the proposal system. Additional users can be added to the proposal as the PI desires using the proposal check-in utility of the portal.

Once data are located there are a number of things which a user can do such as inspect the data, examine the metadata, or visualize the data. The portal facilitates examining a number of standard file formats such as ASCII text, XML, NeXus, along with a number of imaging formats such as jpg. In particular, one can inspect the NeXus metadata of interest by digging into the hierarchy of attribute groups. The data within the NeXus files can also be visualized. The ISAW visualization tools have been adapted to work via the portal, but needed to be adapted to handle the case where the data could be too large to send over the internet. Thus the visualization tool operates in a client-server mode where only the necessary data are sent over the network link. To enhance data exploration, data can be viewed in tabular form, 1D or 2D plot. The 2D plot can present the data in either linear or log color scales.

The portal also provides users the ability to both upload and download data, and users can freely move data to and from the portal. This begs the question for why users do not simply take all of their data home with them to reduce and analyze. They can certainly do so. however, users do not usually take the data with them out of convenience, but out of necessity. One feature of the NeXus files is that they utilize internal compression for the data much like a ".zip" file. The difference being that the metadata within the file are still readable. The HDF-5 utilities perform compression and decompression on behalf of the user. It is common to achieve 90-95% compression with the NeXus files. Thus copying a NeXus file to one's laptop and reading the data can produce an out-of-memory error. For instance, a typical 40MB to 50MB ARCS NeXus histogram file expands to approximately 4GB. A Sequoia file is approximately 50% bigger at 6GB, and a TOPAZ [30] data file will expand to almost 15GB. Thus the data are simply too large to process efficiently via typical user laptops, especially when one considers that data reduction typically requires processing a number of files concurrently – perhaps as many as six.

Realizing that the data are large, the facility also provides another convenience – facility produced and maintained data reduction applications are also made available via the portal. Integrated within the portal is a mechanism which enables GUI applications to be displayed locally on one's own computer, however the application is actually running back on facility computers[1]. Utilizing this web based "thin client" approach has a number of significant advantages:

1. The user always has access to the most recent version of the application without having to perform installation.
2. Data are kept proximal to computing thus the amount of time to move data is minimized.
3. The facility can provide access to substantial computing resources which users may not have available to themselves otherwise.

The portal is useful for providing individual researchers convenient access to computing and data, however it is already providing more than that by facilitating collaboration. In the past, data sharing required the interested parties to figure out where to place the data in a mutually accessible, and safe location. Depending upon the size of the data, finding such a location is not always easy else we would simply email it to our collaborators. Within the portal data hierarchy, each experiment proposal subdirectory has within it a 'shared' subdirectory to facilitate collaboration. These are the only read-write subdirectories within the repository as all others are read-only. In addition, only the experiment team that can access the experiment data can access these shared areas. Thus experiment team members can place data in an area common among themselves in order to share data. The shared areas are the fastest growing regions in the data repository. All of the above features ensure that the portal attracts a steady influx of users.

4. What's next?

There are several notable growth areas for the portal ranging from data, computing and networking, and new application development. Integration with X-ray sources to make X-ray data available for co-analyses with neutron scattering data has good scientific impact potential. Working with the X-ray facilities is one step towards developing a user facility network (UFNet) which could be used to facilitate cross-cutting scientific research. Other portals exist in the X-ray and neutron scattering communities, and new portals are likely on the horizon. Inter-operating with these portals will be

[1] Note that paging techniques do exist which would make it possible to process such data on a laptop, however there is a question of the feasibility for processing very large data sets this way.

important but will bring significant challenges in federating their respective user authentication systems.

The future also looks to hold a new way of working by enabling users to leverage the ever-increasing portable computing power available to them. To illustrate, the web browser thin client approach does not require the user to install applications and relies upon the server-side for most of the work. But a "thick client" would enable a user to perform work on their local computer while also providing a remote interface to portal infrastructure enabling data movement and remote computation at the discretion of the user. Thus computing via a thick client would shift work away from a remote server onto a user's own computer based upon what makes sense to be done locally. Work has begun in producing a thick client via an SBIR project lead by Mark Green of Tech-X Corporation. Similarly, the ISAW package is also being adapted to leverage remote data and computing resources.

Successful next generation applications will enable domain scientists to more fully focus on their research and less on the nuances of computing, yet will provide access world class cyberinfrastructure and scientific computing applications. Users will have more freedom of choice and flexibility for how and where they perform data analysis and computing. The future of computing for facility users looks bright.

References

[1] NeSSI: http://neutrons.ornl.gov/workshops/nessi/NeSSI_agenda_linked.pdf
[2] SQRL: http://sqrl.eecs.utk.edu/
[3] NSTG: http://www.teragrid.org/gateways/projects.php?id=56
[4] McStas: http://www.mcstas.org/
[5] SNS: http://neutrons.ornl.gov/facilities/facilities_sns.shtml
[6] HFIR: http://neutrons.ornl.gov/hfir/hfir_facilities.shtml
[7] LENS: http://www.iucf.indiana.edu/lens/
[8] Lujan: http://lansce.lanl.gov/lujan/
[9] IPNS: http://www.pns.anl.gov/
[10] Tech-X Corporation: http://www.txcorp.com/
[11] UTK SEELAB: http://www.cs.utk.edu/~seelab/
[12] Diamond Synchrotron Facility: http://www.diamond.ac.uk/
[13] ORNL XCAMS: https://xcams.ornl.gov/xcams/
[14] Nomachine: http://www.nomachine.com/
[15] OIC Clusters: https://oic.ornl.gov/
[16] NSTG Cluster: http://www.ornl.teragrid.org/guides/userguide.html
[17] ISAW: http://www.pns.anl.gov/computing/isaw/index.shtml
[18] DAVE: http://www.ncnr.nist.gov/dave/
[19] DANSE: http://danse.us/
[20] EXPGUI: http://www.ncnr.nist.gov/programs/crystallography/software/expgui/expgui.html
[21] FullProf: http://www.ill.eu/sites/fullprof/php/programs.html
[22] GSAS: http://www.ccp14.ac.uk/solution/gsas/
[23] CSD: http://www.ccdc.cam.ac.uk/products/csd/
[24] ICSD: http://icsdweb.fiz-karlsruhe.de/
[25] SVN: http://subversion.tigris.org/
[26] Trac: http://trac.edgewall.org/
[27] NeXus: http://www.nexusformat.org/Main_Page
[28] HDF: http://www.hdfgroup.org/HDF5/
[29] IPTS: http://www.ornl.gov/sci/iums/ipts/
[30] TOPAZ : http://neutrons.ornl.gov/instrument_systems/beamline_12_topaz/index.shtml

Neutron Science TeraGrid Gateway

Vickie Lynch, Meili Chen, John Cobb, Jim Kohl, Steve Miller, David Speirs, and Sudharshan Vazhkudai

Oak Ridge National Laboratory, Oak Ridge, TN

lynchve@ornl.gov

Abstract. The unique contributions of the Neutron Science TeraGrid Gateway (NSTG) are the connection of national user facility instrument data sources to the integrated cyberinfrastructure of the National Science FoundationTeraGrid and the development of a neutron science gateway that allows neutron scientists to use TeraGrid resources to analyze their data, including comparison of experiment with simulation. The NSTG is working in close collaboration with the Spallation Neutron Source (SNS) at Oak Ridge as their principal facility partner. The SNS is a next-generation neutron source. It has completed construction at a cost of $1.4 billion and is ramping up operations. The SNS will provide an order of magnitude greater flux than any previous facility in the world and will be available to all of the nation's scientists, independent of funding source, on a peer-reviewed merit basis. With this new capability, the neutron science community is facing orders of magnitude larger data sets and is at a critical point for data analysis and simulation. There is a recognized need for new ways to manage and analyze data to optimize both beam time and scientific output. The TeraGrid is providing new capabilities in the gateway for simulations using McStas and a fitting service on distributed TeraGrid resources to improved turnaround. NSTG staff are also exploring replicating experimental data in archival storage. As part of the SNS partnership, the NSTG provides access to gateway support, cyberinfrastructure outreach, community development, and user support for the neutron science community. This community includes not only SNS staff and users but extends to all the major worldwide neutron scattering centers.

1. Introduction

A gateway can bring advanced cyberinfrastructure capabilities to an entire science community. The gateway can provide an intuitive, effective, efficient organized architecture to provide an information interface to facility users while at the same time making available a large data archive (metadata and disk), high-end computing resources and, powerful analysis capabilities. The services and resources are coordinated, general purpose, and reliable. A science gateway enables scientific communities of users with a common scientific goal to use high performance computing, have a common interface, and leverage community investments. Some gateways are web portals with a web-based user interface in front and a service oriented architecture in the back. The gateway is not just a webpage. Rather it is a unified architecture for the planning and deployment of applications, services, and tools of interest to the community.

2. Gateway Support

The TeraGrid [1] is a network of high performance computers and resources at eleven different supercomputer centers supported by the US National Science Foundation. The partner facilities are connected via a dedicated wide area 10 Gigabit per second with over 1.5 petaflops of peak computing performance across 136,740 CPU-cores. In addition non-traditional resources are also available on the TeraGrid including three specialized high-performance visualization clusters, storage services including 2.9 PetaBytes (PB) of spinning disk and 60 PB of tape archive as well as a Condor pool for capacity computing with an additional 99,000 available cpu-cores. TeraGrid resources continue to grow with additional computing, visualization, storage, capacity computing, and experimental architectures expected in the near future. Today there are over 4,000 TeraGrid user accounts for over twenty resources at the eleven facilities. Traditional high performance computing access requires accounts be created on every computing resource for every SNS proposal user. This could imply creating thousands of accounts in a few years of SNS operation.

A solution to this problem is a science gateway. Instead of multiple accounts, a community user account with a community certificate is used for access to high performance computing. A "community user" is a dedicated account identity that stands for the often routine resources usage requests that would come from a specific scientific community. Examples could be web-services providing genomic searches, a general chemical modeling and simulation capability, or a general, repeatable workflow required as part of a facilities support of its experimental users. With a gateway, only one community account is needed on each distributed resource. The users do not have a login shell for the community account but rather use the community account as a role based access mechanism. The gateway developers take care of account management, accounting, certificate management using a community certificate, application code deployment on parallel processors, and user support.

Neutron Science TeraGrid Gateway [3] access is available via the Neutron Scattering Portal [2]. It is one of over 31 existing TeraGrid science gateways. It is charting new territory for wide cyberinfrastructure services such as federated user management, callback end-user identification with automatic service auditing across federated identity space gateway community accounts, grid interoperability, and participation in attribute-based authorization. The gateway user has privileges to run a limited set of neutron science application codes via the community account on the TeraGrid computers. The use of the Jimmy Neutron community account is a success of NSTG because if enables researchers to do their work by lowering the cost of entry to do high performance computing. These jobs are launched using Globus [4] commands from the portal. Input from the portal to the TeraGrid and output from the TeraGrid to the portal is done automatically using the GridFTP [5] servers deployed over a dedicated 10 Gigabit per second TeraGrid network. NeXus [6] which is based on hdf5 [7] is the standard format used in the portal. There is also a visualization capability in the portal including many features designed by the ISAW team [8].

3. Simulations

NSTG is coupling neutron science and cyberinfrastructure with McStas [9] simulations (Figure 1) that use the TeraGrid's parallel computing resources. These simulations have the ability to simulate a neutron scattering experiment including the entire instrument and the sample material inside the instrument. They require accurate neutron scattering simulations of realistic experimental arrangements including all items in the beam scattering path (containers, cryostats, neutron optics ...) as well as modeling the chosen sample under study. It is necessary to accurately simulate neutron scattering off of all these materials. Theory driven simulation allows comparison of the number of neutrons that hit the detector with the experiment with few or no free parameters to empirically fit.

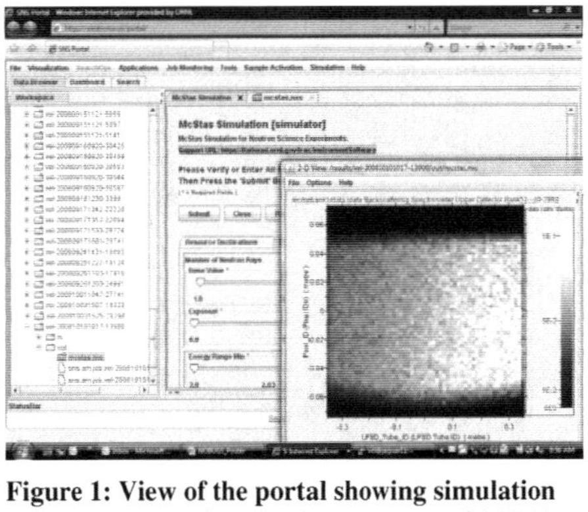

Figure 1: View of the portal showing simulation submission and visualization based on ISAW [8] of resulting NeXus [6] file.

This approach is new for the neutron community and opens up new vistas for experiment planning and experimental analysis. Experiments can be computationally designed and evaluated to test feasibility and optimization at the conceptual stage. It will also allow experimenters to "practice" in order to be better prepared to respond flexibly when the actual experiment is underway during the valuable allocated beam time, such as exploring observational optimizations beforehand. Simulations also allow more detailed analysis of experimental results by testing hypotheses about sample structure and dynamics to develop explanations of experimental results, for example by using a non-physical partial model to understand the significance of different sources of signal to the measured scattered neutron counts.

4. Fitting Service

Fast analysis of the data obtained from SNS instruments is critical for quick data comparison to resolution-convolved models and for making automated on-the-fly experimental adjustment/control possible. Collaborating with SNS instrument scientists, the Neutron Science TeraGrid Gateway team at ORNL has been designing and developing software tools, which use parallel computing techniques to perform fast data fitting (Figure 2) on the TeraGrid resources. The fitting service does a fit of theoretical models to the data files from the experiments. The user decides which model to use, which fitting algorithm to use, which parameters to vary, and gives initial guesses to the fitting parameters. An adaptive nonlinear least squares algorithm is implemented in parallel on the TeraGrid and shows linear scaling at least up to 32 processors. It minimizes a nonlinear sum of squares using an analytic or numerical Jacobian matrix. The fitting service uses the NL2SOL [10] or Dakota [11] software packages.

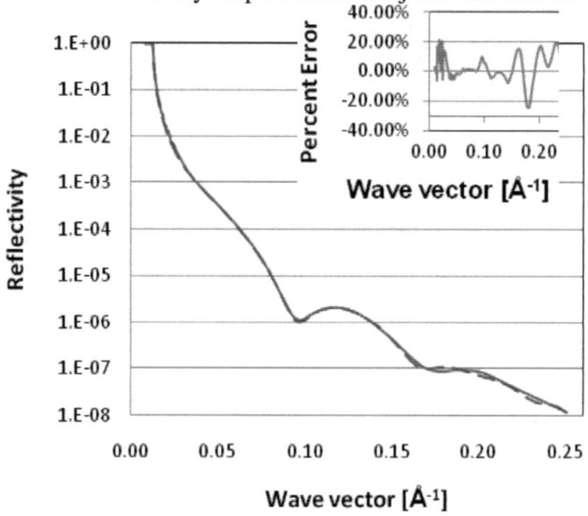

Figure 2: Fit of reflectivity data with the fitting service.

5. Data Replication

One need of the SNS is to create replicated copies of its data archive both for resiliency of availability and advanced data placement to support close co-ordination with other computational tasks such as simulations and remote visualization. We have created a re-usable workflow to transfer data from SNS record copies to other, different locations. The first goal is to replicate the data to the HPSS storage system at the Leadership Computing Facility. This workflow includes moving the data, maintaining and updating archive status while observing proper data access controls. Follow-on tasks include replicating to other storage targets and technologies such as the OIC at ORNL, HPSS at Indiana, global file systems and logistical networking. Replication plans must address data policy and data ownership. The same person must own the data as it crosses into different user authentication zones such as the TeraGrid. Also privacy of the data must be maintained using data encryption.

6. Conclusions

A discipline specific Science Gateway can help facilities scale to a large number of users, give facilities access to high performance computing such as the TeraGrid, and enable a scientific community to use community software through a common interface. Thus researchers are more productive with a gateway because they use the same tools, use a common data format, and share data easily.

Closer integration of the SNS and the TeraGrid is the goal of this work. In the future, we plan for the NSTG to flow streaming data from SNS for data reduction workflows on the TeraGrid and move data from the SNS data repository for further processing and analysis on the TeraGrid. Also universities with SNS proposals will use the high bandwidth interconnect of the TeraGrid to support data movement between SNS and their institution. Real-time simulations of experiments can use TeraGrid resources to predict experimental results as they are running or between experiments.

7. Acknowledgements:

We are thankful to the McStas developers, Emmanuel Farhi, Kim Lefmann, and Peter Willendrup, for their on-going collaboration with us and Mark Hagen from SNS for his help. We also would like to acknowledge our SNS-SCG collaborators including Shelly Ren, Jens Schwidder, and Jim Trater. Finally we need to acknowledge NSTG staff including Greg Pike and Dave Giles.

This material is based upon work supported by the National Science Foundation's Office of Cyberinfrastructure TeraGrid project (OCI-0525789 and OCI-0504074) as well as TeraGrid resource allocations under the following NSF programs: Partnerships for Advanced Computational Infrastructure, Distributed Terascale Facility (DTF) and Terascale Extensions: Enhancements to the Extensible Terascale Facility under current TeraGrid project allocation, TG-DMR060000T.

Prepared by Oak Ridge National Laboratory, P.O. Box 2008, Oak Ridge, Tennessee 37831-6285, managed by UT-Battelle, LLC for the U.S. Department of Energy under contract number DE-AC05-00OR22725.

References

[1] Catlett C , et al 2007 "Teragrid: Analysis of organization, system architecture, and middleware enabling new types of applications" *Advances in Parallel Computing* (Amsterdam: IOS Press)

[2] Website for Spallation Neutron Source information http://neutronsr.us

[3] Cobb J W, Geist A, Kohl J A, Miller S D, Peterson P F, Pike G G, Reuter M A, Swain T, Vazhkudai S S, and Vijayakumar N N 2007 *J. of Concurrency and Computation: Practice and Experience* **19** 809-826

[4] Foster I, "Globus Toolkit Version 4: Software for Service-Oriented" 2006 *IFIP International Conference on Network and Parallel Computing*, (New York: Springer-Verlag LNCS) 3779, 2-13

[5] Allcock W, Bresnahan J, Kettimuthu R, Link M, Dumitrescu C, Raicu I, and Foster I "The Globus Striped GridFTP Framework and Server" 2005 *Proc. SC05*

[6] Website for NeXus information http://www.nexusformat.org

[7] Website for hdf5 information http://www.hdfgroup.org/

[8] Mikkelson D J, Mikkelson R L, Worlton T G, Chatterjee A, Hammonds J P, Peterson P F, and Schultz A J, "Coordinated, Interactive Data Visualization for Neutron Scattering Data" 2002 *Proc. NOBUGS*

[9] Willendrup P, Farhi E, and Lefmann K 2004 *Physica B* **350** 735

[10] Dennis J E, Gay D M, and Welsch R E, 1981 *ACM Transactions on Mathematical Software* **7** 369-383

[11] Eldred, M S, Adams B M, Haskell K, Bohnhoff W J, Eddy J P, Gay D M, Griffin J D, Hart W E, Hough P D, Kolda T G, Martinez-Canales M L, Swiler L P, Watson J P, and Williams P J, 2007 "DAKOTA: A Multilevel Parallel Object-Oriented Framework for Design Optimization, Parameter Estimation, Uncertainty Quantification, and Sensitivity Analysis. Version 4.1 Users Manual" Sandia Technical Report SAND 6337

AUTHOR INDEX

AUTHOR INDEX

AUTHOR INDEX

AUTHOR INDEX

AUTHOR INDEX